Artificial or Constructed Wetlands
A Suitable Technology for Sustainable Water Management

Editors

María del Carmen Durán-Domínguez-de-Bazúa
Universidad Nacional Autónoma de México
Facultad de Química, Departamento de Ingeniería Química
Laboratorios de Ingeniería Química Ambiental y de Química Ambiental
Ciudad Universitaria, Ciudad de México
México

Amado Enrique Navarro-Frómeta
Universidad Tecnológica de Izúcar de Matamoros
Barrio de Santiago Mihuacán
Izúcar de Matamoros, Puebla
México

Josep M. Bayona
Department of Environmental Chemistry
IDAEA-CSIC, c/Jordi Girona, 18
E08034-Barcelona
Spain

CRC Press
Taylor & Francis Group
Boca Raton London New York

CRC Press is an imprint of the
Taylor & Francis Group, an **informa** business

A SCIENCE PUBLISHERS BOOK

Cover credit: Three of the cover illustrations belong to Xochimilco "natural" wetland reproduced by kind courtesy of Cand. Dr. Víctor Jesús García-Luna, the first author of Chapter 1. The fourth one in the hexagon was taken by Prof. Dr. Amado Enrique Navarro Frómeta, the second editor, during one of his academic visits to colleagues in South America.

CRC Press
Taylor & Francis Group
6000 Broken Sound Parkway NW, Suite 300
Boca Raton, FL 33487-2742

First issued in paperback 2020

© 2018 by Taylor & Francis Group, LLC
CRC Press is an imprint of Taylor & Francis Group, an Informa business

No claim to original U.S. Government works

ISBN-13: 978-1-138-73918-5 (hbk)
ISBN-13: 978-0-367-78114-9 (hbk)

Library of Congress Cataloging-in-Publication Data

Names: Durân Domâinguez de Bazâua, Marâia del Carmen, editor.
Title: Artificial or constructed wetlands : a suitable technology for
    sustainable water management / editors, Marâia del Carmen
    Durâan-Domâinguez-de-Bazâua, Universidad Nacional Autânoma de Mâexico,
    Facultad de Quâimica, Departamento de Ingenierâia Quâimica, Laboratorios
    de Ingenierâia Quâimica Ambiental y de Quâimica Ambiental, Ciudad
    Universitaria, Ciudad de Mâexico, Mâexico, Amado Enrique Navarro-Frâometa,
    Universidad Tecnolâogica de Izâucar de Matamoros, Barrio de Santiago
    Mihuacâan, Izâucar de Matamoros, Puebla, Mâexico, Josep M. Bayona,
    Department of Environmental Chemistry, IDAEA-CSIC, c/JordiGirona,
    Barcelona, Spain.
Description: Boca Raton, FL : CRC Press, Taylor & Francis Group, [2018] |
    Includes bibliographical references and index.
Identifiers: LCCN 2018011966 | ISBN 9781138739185 (hardback : acid-free paper)
Subjects: LCSH: Constructed wetlands.
Classification: LCC TD756.5 .A78 2018 | DDC 628.2/5--dc23
LC record available at https://lccn.loc.gov/2018011966

Visit the Taylor & Francis Web site at
http://www.taylorandfrancis.com

and the CRC Press Web site at
http://www.crcpress.com

Preface

Over the last years, there has been a boom concerning the publication of new insights on the use of artificial wetlands or constructed wetlands since they constitute a highly promising technology for the removal of a wide spectrum of pollutants present in wastewaters.

This book is directed to a broad audience interested in the water cycle management, such as policy makers, local and regional authorities and universities, teaching and learning institutions. It also will be of interest to any individual interested in rural, suburban, and urban areas who are constructing artificial or constructed wetlands systems at the household and/or municipal levels. The general public that wishes to learn about this ecofriendly technology may find it useful.

This book stands out among those already available due to its holistic point of view and treatment of the different approaches used to understand the principles of ecological treatment systems.

The first chapter deals with the ever-increasing problem of the transformation of natural wetlands into artificial ones due to the omnipresence of human beings that are more concerned with the survival aspects of their lives than for the protection of the natural environment surrounding them. The example of a UNESCO site such as *Xochimilco* in the heart of Mexico City is certainly important in this context.

The second chapter deals with the prevalence of the black-box approach to wetland design leading to a description of a new methodology for CW experimental research. The motivation for this proposed methodology is two-fold. Firstly, it is a way of understanding the processes occurring within the CW black-box and, secondly, it is a useful tool for the development of design guidelines which show lower reliance on local wetland conditions and would be more widely, or generically, applicable.

The third chapter deals with research on the transformation of atrazine, an herbicide that is still widely used, based on microbial consortiums isolated from wetland water samples in *Xochimilco*. Analytical methods were established *ad hoc* to identify and measure atrazine and its metabolic products. Also, the study of the use of indigenous bacteria and its potential to clean wastewaters and to produce electricity considering the combination of artificial wetlands (AW) with microbial fuel cells (MFC) is included.

The next chapter (Chapter 4) deals with the analysis of the latest documents of the United Nations World Water Assessment Program, which leads to the conclusion that, in the face of water scarcity, the purification and reuse of wastewater is essential. In addition, the analysis of available technologies and their costs show that constructed wetlands (CWs) represent a viable and functional option for the tertiary polishing of secondarily treated wastewaters that can alleviate problems of scarcity through the use of municipal effluents.

Chapter 5 approaches the role of the support media on the removal of pollutants (i.e., micropollutants and heavy metals) since for many years, support medium was considered an inert material that was only used for the bacterial consortia establishment or for vegetation support.

Chapters 6 to 9 report the application of this technology to case studies in three continents, namely, Africa, Asia, and America, particularly Cuba and Mexico. Overall, successful results were obtained which illustrates that well designed constructed wetlands can function in different climates and under varied socioeconomic constrains in order to achieve the goal to meet regulations on treated wastewater discharge.

Chapters 10 to 13 deal with its use for livestock effluents in a Colombian case study funded by LIFE Programme, the EU's financial instrument that supports projects about environment, nature conservations and climate actions throughout the EU as well as aquaculture effluents in the Pacific Ocean near Mexico. In these three chapters the fate of emerging pollutants is given a consideration.

Chapter 14 explores the application of artificial wetlands to generate bioelectricity from microbial processes in an electrochemically assisted constructed wetland aiming to reduce greenhouse emissions while wastewater is decontaminated.

Chapter 15 presents evaluations of empirical and mechanistic models developed to predict the removal of organic micro-contaminants and bacteria and genes that are resistant to antibiotics through plants from the perspective of the agricultural reutilization of treated water. Several models are presented, from simple linear regressions based on physical, chemical, and/or physical-chemical properties of contaminants to unsteady state mechanistic models or transient multi-compartmental ones that are commonly used in human risk evaluations.

Chapters 16 and 20 are focused on the role of macrophytes in the uptake and detoxification of organic micropollutants occurring in wastewaters.

Chapters 17 and 18 overview the effect of design parameters on the organic load and nutrient removal in constructed wetlands of different configuration as well as the use of anaerobic reactions as a pretreatment prior to CW treatment.

Finally, Chapter 19 concerns a particularly important and relevant subject—it deals with the strategies of the operation of these ecosystems under the perspective of the global climate change scenario.

We hope that this book will become a major reference work for the widespread dissemination and further uptake of this successful ecotechnology.

María del Carmen Durán-Domínguez-de-Bazúa
Amado Enrique Navarro-Frómeta
Josep M. Bayona

Contents

Introduction

The future of research and applications of artificial or constructed wetlands has specific needs. Further progress requires a better understanding of the biogeochemical processes in these systems while abandoning the black box approach. Only in this way can treatment systems be used and appropriated by local end users in a significant and progressive fashion. In this sense, we can highlight the following aspects:

- The future of technology will be intimately linked to the local innovation, especially in the aspects related to flow management, the configuration of the systems, coupling with other treatment methods, substrate media, the used plant species, taking advantage of the local flora and the use of local waterproofing materials. The use of enhanced removal with specific enzymatic reactions and/or bioaugmentation should be considered. Altogether, it will contribute to the reduction of costs and make this ecotechnology more versatile, resilient and adaptable to local conditions.

- The growing salinization of global freshwater resources and saline intrusion into the different coastal systems is due to several factors such as the increase in sea level (affected by global warming), mining processes that generate significant volumes of water with high salt content, and the disposal of ultramembrane filtration waste and energy resources. This requires research and exploration of specific plants and micro-organisms that can remove the afore-mentioned contamination. In this sense, research concerning the proper management of natural wetlands provides other opportunities to develop technological solutions for conservation and sustainable approaches for using their capacity for wastewater depuration.

- The use of bioengineered systems for obtaining an optimal balance of aerobic and anaerobic processes to reduce or mitigate pollution levels and specific contaminants, as well as adequate nutrient levels in the effluent, according to the use that it will be given. Special attention deserves the improvement in the microbiological quality of the treated water. In this sense, we can mention the use of design features and devices that increase the oxygen supply to the different parts of the wetlands, the use of photocatalytic processes and of functionalized nanoparticles hoisted for the capture of specific contaminants.

- The management of the artificial wetlands (AWs) or constructed wetlands (CWs): Innovation in the use and/or safe disposal of plant material that has accumulated certain contaminants and development of methods for the online monitoring of wetlands functioning. It is also necessary to consider the technological solutions to address changes in the quality of the water that is treated, to increase the AW or

CW lifetime by decreasing the substrate clogging phenomena as well as to restore these systems once they reach the end of their useful life, which will contribute to their resilience and acceptance.

- The very nature of wastewater as a carrier of energy and of many valuable substances. The production of bioenergy through the coupling of wetland and microbial fuel cells is a promising field that develops rapidly at the present time.

Finally, the first editor dedicates this book to the memory of Dr. Peter Kuschk, a great pioneer in research about artificial or constructed wetlands. Thanks to his bonhomie and open mind young people from all over the world were warmly received in his laboratories. Thus, the dissemination of the knowledge of this ancient technology, used by the Aztecs to maintain pristine and clean the lakes of the Mexico basin, has been possible. This eco-technology was not only rediscovered in Germany in the middle of the 20th century but thanks to Dr. Peter Kuschk it has acquired a scientific approach. He will remain in the heart of all those interested in the use of artificial or constructed wetlands.

María del Carmen Durán-Domínguez-de-Bazúa
Amado Enrique Navarro-Frómeta
Josep M. Bayona

1

Xochimilco, Mexico, a Natural Wetland or an Artificial One?

Víctor Jesús García-Luna,[1,] Marisela Bernal-González,[1]
Federico Alfredo García-Jiménez[2] and María del Carmen
Durán-Domínguez-de-Bazúa[1]*

INTRODUCTION

In Mexico, most of the climates of the planet are represented, largely due to its complex topography and geographical position as an area of confluence and boundary of neartica and neotropical bioregions. This, together with the geological and soil variability, makes it possible to represent, in just under 2,000,000 km^2, practically all the great ecosystems of the world. Due to these characteristics, Mexico is part of a select group of five megadiverse countries hosting between 60 and 70 percent of species diversity of plants and animals on the planet (CONANP 2015, 2016a). In fact, for some authors, the group incorporates 12 countries: Mexico, Colombia, Ecuador, Peru, Brazil, Zaire, Madagascar, China, India, Malaysia, Indonesia and Australia. Others increase the list to more than 17, adding Papua New Guinea, South Africa, US, Congo, Philippines, and Venezuela (CONABIO 2017).

---

[1] Universidad Nacional Autónoma de México, Facultad de Química, Departamento de Ingeniería Química, Laboratorios de Ingeniería Química Ambiental y de Química Ambiental (UNAM-FQ-DIQ-LIQAyQA), Ciudad Universitaria, Ciudad de México, México.

[2] Universidad Nacional Autónoma de México, Instituto de Química, Ciudad Universitaria, Ciudad de México, México.

* Corresponding author: ing.vgarcial@yahoo.com.mx

Wetlands

Wetlands are transitional environments between terrestrial and aquatic ecosystems and are considered one of the most biologically diverse ecosystems (CONAGUA 2008). A simple description of a wetland is that of soils that are permanently or periodically flooded or saturated, in environments with fresh water or with some degree of salinity (López et al. 2010). According to the dictionary of the Royal Spanish Academy (*RAE*, in Spanish, for *Real Academia Española*), a wetland is defined as "surface water and groundwater shallow" (RAE 2017). The Mexican National Water Law (Act) (*LAN*, in Spanish, for *Ley de Aguas Nacionales*), in Fraction XXX of the third article, defines wetlands as "the transition zones between the aquatic and terrestrial systems which constitute areas of temporary or permanent flooding, whether or not subject to the influence of tides, such as swamps, bogs and marshes, whose limits are the type of hydrophyte vegetation of permanent or seasonal presence, the areas where the soil is predominantly water and, the lacustrine or soil areas permanently moist by the natural discharge of aquifers" (LAN 2014). Finally, the Convention on Wetlands held in Ramsar, Iran, in 1971, makes use of a broader definition, since in addition to considering swamps, marshes, lakes, rivers, peatlands, oases, estuaries and deltas, it also considers artificial sites such as reservoirs, salt marshes (*salinas* in Spanish), and marine areas near the coasts, whose depth at low tide does not exceed 6 m, which may include mangroves and coral reefs (CONANP 2016b; Ramsar Convention Secretariat 2016).

Wetland Types

Berlanga-Robles et al. (2007) reviewed 18 classification systems around the world and considering the advantages and disadvantages of such classification systems, in order to contribute to the Mexico's National Inventory of Wetlands (*INH*, in Spanish, for *Inventario Nacional de Humedales*). They proposed a general classification scheme based primarily on the proposal of Cowardin et al. (1979). This applied some of the geomorphological (landforms) and hydrological (water regime) criteria of Semeniuk and Semeniuk (1995). It also added artificial wetlands proposed by the Ramsar Convention. Thus, in general terms, the rest of the classifications analyzed are modifications of any of these three. Table 1 shows this proposed classification of wetlands in Mexico, which, in general, is presented in a hierarchical way with three areas, five systems, six subsystems, and 26 classes. The use of keys at different levels is proposed, in order to easily identify any type of wetland within the classification system, regardless of the common name received.

Importance of Wetlands

Worldwide, wetlands are vital to human survival (Hu et al. 2017). They are one of the most productive environments in the world, cradles of a great biological diversity, sources of water, and primary productivity. Innumerable species of plants and animal depend on wetlands for survival. Wetlands are home to a large number of birds, mammals, reptiles, amphibians, fish, and invertebrates and are also important reservoirs of plant genetic material (Ramsar Convention Secretariat 2016).

Table 1. General classification scheme wetlands in Mexico (Berlanga-Robles et al. 2007).

Ambient	System	Subsystem	Class
Marine-coastal	Marine system Estuarine system	Subtidal subsystem Intertidal subsystem	M/a/m1. Coastal water strip M/a/m2. Littoral M/b/m1. *Estero*, coastal lagoon M/c/m2. Marshes
Continental	River system	Permanently flooded subsystem	C/d/h1. River, cavern C/d/h2. Creek, rivulets C/d/h3. Canal C/b/h1. Lakes, *ciénega* or *ciénaga* (a wetland system unique to the American Southwest), *cenote* (a deep natural well or sinkhole, especially in Mexico and Central America)
	Lacustrine system	Seasonal flooded subsystem	C/b/h2. Pond C/b/h3. Flooded basin
	Palustre (marsh, swamp) system	Intermittently flooded subsystem	C/d/h4. Ditch C/b/h4. Wet depression
		Seasonally saturated subsystem	C/c/h2. Swamp, peatbog, *popal* (wetland in Nahuatl) C/c/h3. Flooded plain C/c/h4. Damp plain C/e/h4. Wet slope C/f/h4. Wet mountain
Artificial			A1. Aquaculture ponds A2. Artificial ponds A3. Irrigated land A4. Agricultural land A5. Areas of salt exploitation A6. Areas of storage water A7. Excavations, quarries, pools A8. Water treatment areas A9. Transportation and drainage Canals

Ecotopes:

M/a/m1. Rocky and unconsolidated bottom/seabed/reef

M/a/m2. Water bed/reef/rocky and unconsolidated littoral

M/b/m1. Rocky and unconsolidated bottom/water bed

M/c/m2. Rocky and unconsolidated bottom/emerging wetland/shrub-forest wetland

C/d/h1. Rocky bottom/bottom unconsolidated/water bed/rocky littoral/littoral unconsolidated/ emerging wetland/shrub-forest wetland

C/d/h2. Bed of current

C/d/h3. Bed of current

C/d/h4. Rocky and unconsolidated bottom/water bed/littoral unconsolidated/wetland of mosses and lichens/emerging wetland/shrub-forest wetland

C/b/h1. Rocky and unconsolidated bottom/water bed/rocky and unconsolidated littoral/emerging wetland

C/b/h2. Rocky and unconsolidated bottom/water bed/rocky and unconsolidated littoral/emerging wetland

C/b/h3. Rocky and unconsolidated bottom/water bed/rocky and unconsolidated littoral/emerging wetland

C/b/h4. Rocky and unconsolidated bottom/water bed/littoral unconsolidated/wetland of mosses and lichens/emerging wetland/shrub-forest wetland

Table 1 contd. ...

...Table 1 contd.

C/c/h2. Rocky and unconsolidated bottom/water bed/littoral unconsolidated/wetland of mosses and lichens/emerging wetland/shrub-forest wetland

C/c/h3. Rocky and unconsolidated bottom/water bed/littoral unconsolidated/wetland of mosses and lichens/emerging wetland/shrub-forest wetland

C/c/h4. Rocky and unconsolidated bottom/water bed/littoral unconsolidated/wetland of mosses and lichens/emerging wetland/shrub-forest wetland

C/e/h4. Rocky and unconsolidated bottom/water bed/littoral unconsolidated/wetland of mosses and lichens/emerging wetland/shrub-forest wetland

C/f/h4. Rocky and unconsolidated bottom/water bed/littoral unconsolidated/wetland of mosses and lichens/emerging wetland/shrub-forest wetland

The interactions of physical, chemical, and biological components of a wetland as part of the planet's natural infrastructure, such as soils, water, plants, and animals, enable the wetland to perform vital functions. The role of wetlands in the hydrological cycle, such as regulators of water flows and storage of water in the recharge and discharge of groundwater, as well as in the treatment of contaminated water is a very important one. They play a role in biogeochemical cycles. Due to its damping effect these ecosystems can receive, process, and/or retain nutrients, organic matter, sediments, and pollutants. They participate in mitigation and adaptation to climate change, mainly by contributing to the reduction of emissions of gases into the atmosphere and the increase of the amount of carbon stored in natural sinks (carbon sinks), which is adsorbed through natural biochemical processes by various organisms such as aquatic and terrestrial plants, trees, soil microorganisms, and even marine species such as snails and clams in structures as diverse as leaves, stems, roots, and shells. Besides, these ecosystems can be considered regulators of the processes of disturbance, both natural and anthropic, by contributing to the stabilization and coastal protection, control soil erosion, protecting against climatic effects such as storms and hurricanes, mitigating floods and cushion drought, as well as stabilize local climatic conditions, particularly rainfall and temperature (López et al. 2010; CONANP 2013, 2015; Ramsar Convention Secretariat 2016).

Apart from this, wetlands often provide enormous economic benefits, including water supply (quantity and quality), fishing (more than two thirds of the world's fish harvest is linked to the health of wetland areas), agriculture (by maintaining groundwater and nutrient retention in alluvial plains), wood and other building materials, energy resources (such as peat and plant material), wildlife resources, transportation, plus a wide range of other wetland products, including herbal medicines, and last but not least less important, recreation and tourism opportunities (CONANP 2013; Ramsar Convention Secretariat 2016).

Convention on Wetlands *(Ramsar, Iran, 1971)*

Ramsar is the oldest of the modern global intergovernmental environmental agreements. The treaty has its origins in the decade of the 60's of the 20th century, when at the Conference Sea (from Marshes, Marécages, Marismas) organized by Dr. Luc Hoffmann in Les Saintes Maries-de-la-Mer in the French Camargue, governments of various countries, non-governmental organizations (NGOs) and wetland experts, concerned about the increasing loss and degradation of wetland habitat for migratory waterbirds, call for an international treaty on wetlands and for a list of internationally important wetlands. Later, Iran's Game and Fish Department organizes and holds a conference at the Caspian seaside resort of Ramsar, Iran, where the Convention on Wetlands of International Importance especially as Waterfowl Habitat is agreed by representatives of 18 nations. The Treaty is signed on 3 February 1971. However, it was until 1 December 1975, when the Convention enters into force after the United Nations Educational, Scientific and Cultural Organization (UNESCO), the Convention's depositary, receives from Greece an instrument of accession to become the Convention's 7th Contracting Party (Secretaría de la Convención de Ramsar 2013; Ramsar 2017a).

In general, the Convention on Wetlands held in 1971, although today the name commonly used to designate the Convention is "Convention on Wetlands (Ramsar, Iran, 1971)", is commonly known as the "Ramsar Convention" and is a multilateral treaty that serves as a framework for national action and international cooperation for the conservation and wise use of wetlands and their resources. Ramsar is the first of the modern multilateral treaties on the conservation and sustainable use of natural resources in a particular ecosystem and has the secretariat in Gland, Switzerland. Currently, it has 169 Contracting Parties or Member States, which cover all the geographic regions of the planet (Secretaría de la Convención de Ramsar 2013; Ramsar Convention Secretariat 2016).

The Convention's mission is "the conservation and wise use of all wetlands through local and national actions and international cooperation, as a contribution towards achieving sustainable development throughout the world" (Ramsar 2017b). Ramsar's philosophy revolves around the concept of rational use. The rational use of wetlands is defined as "the maintenance of their ecological characteristics, achieved through the implementation of ecosystem approaches, within the context of sustainable development". Consequently, wetland conservation, as well as its sustainable use and that of its resources, are at the center of rational use for the benefit of mankind. Given that wetlands are important for maintaining fundamental ecological processes and because of their rich flora and fauna and the benefits they bring to local communities and to human society in general, the general objectives of the Convention are to ensure its conservation and suitable use. States adhering to the Convention accept four main commitments (Secretaría de la Convención de Ramsar 2013; Ramsar Convention Secretariat 2016):

1. "*Listed sites*. The first obligation under the Convention is for a Party to designate at least one wetland at the time of accession for inclusion in the List of Wetlands of International Importance, the Ramsar List, to promote its conservation, and in addition to continue to designate suitable wetlands within its territory for the List."

2. "*Wise use*. Under the Convention there is a general obligation for the Contracting Parties to include wetland conservation considerations in national planning, this might include, for example, land use planning, water resource management planning, or development planning. They have committed themselves to formulate and implement their planning so as to promote, as far as possible, the wise use of wetlands in their territory."

3. *"Reserves and training.* Contracting Parties have also undertaken to establish nature reserves in wetlands, whether or not they are considered to be internationally important and included in the Ramsar List, and they also endeavor to promote training in the fields of wetland research and wetland management."

4. *"International cooperation.* Contracting Parties have agreed to consult with other Contracting Parties about implementation of the Convention, especially in regard to transboundary wetlands, shared water systems, and shared species."

Though the central Ramsar message is the need for the sustainable use of all wetlands, the flagship of the Convention is the List of Wetlands of International Importance (the Ramsar List). Presently, the Parties have designated for this List more than 2,260 wetlands for special protection as Ramsar Sites, covering 215,276,293 hectares (ha) (20,152,762.93 km^2), larger than the surface area of Mexico (Ramsar 2017e).

Table 2 shows the List of Wetlands of International Importance. Figure 1 show how the Ramsar Sites are distributed around the world (Ramsar 2017c).

Mexican Wetlands

According to the List of Wetlands of International Importance (Ramsar List), Mexico is the country (second only to the United Kingdom (which has 174 Ramsar Sites)), with the largest number of designated Ramsar Sites, that is, a total of 142 Ramsar Sites, with a surface of 8,643,579 ha (86,435.79 km^2) equivalent to 17 percent of the total country's surface (INEGI 2016b; Ramsar 2017e).

Figure 2 shows the distribution of Ramsar Sites in the national territory (Ramsar 2017d), of which, 58 are part of a Natural Protected Area (*ANP*, in Spanish, for *Área Natural Protegida*) and 37 of them have a management program published in the Official Journal of the Federation (*DOF*, in Spanish, for *Diario Oficial de la Federación*) (CONANP 2013). With the objective of having cartographic, environmental, and statistical information on the country's wetland ecosystems to guide decision-making and support management in terms of its sustainable use and preservation, it Mexico's National Inventory of Wetlands, a versatile tool, was created. It allows researchers to locate the country's wetlands, establish their characteristics, and provide baselines to monitor their dynamics of change, in addition to providing data on the biodiversity of the country's wetland ecosystems as it includes information on characteristic, endemic, and at risk species (CONANP 2013).

The Mexico's National Water Commission (*CONAGUA* in Spanish, for *Comisión Nacional del Agua*) is responsible for carrying and maintaining the *INH*, as well as to delimit them, classify them, propose standards for their protection, restoration and exploitation, promote and, if necessary, carry out the necessary actions and measures to rehabilitate or restore wetlands, as well as to establish a natural environment or perimeter of protection of the wetlands, in order to preserve their hydrological conditions and the ecosystem (CONAGUA 2017a).*

---

* Text continued on Page 12

Table 2. List of Wetlands of International Importance (Ramsar 2017e).

Country	Number of designated Ramsar sites	Surface (ha)
AFRICA		
Algeria	50	2,991,013
Benin	4	1,179,354
Botswana	1	5,537,400
Burkina Faso	19	677,722
Burundi	4	78,515
Cabo Verde	4	2,300
Cameroon	7	827,060
Central African Republic	2	376,300
Chad	6	12,405,068
Comoros	3	16,030
Congo	13	13,758,741
Côte d'Ivoire	6	127,344
Democratic Republic of Congo	3	7,435,624
Djibouti	1	3,000
Egypt	4	415,532
Equatorial Guinea	3	136,000
Gabon	9	2,480,086
Gambia	3	31,244
Ghana	6	176,134
Guinea	16	6,422,361
Guinea-Bissau	4	1,189,633.2
Kenya	6	265,449
Lesotho	1	434
Liberia	5	95,879
Libya	2	83
Madagascar	15	1.526,888
Malawi	1	224,800
Mali	4	4,204,640
Mauritania	4	1,240,600
Mauritius	3	401
Morocco	24	272,010
Mozambique	2	4,534,872
Namibia	5	676,564
Niger	12	4,317,869
Nigeria	11	1,076,728
Rwanda	1	6,736
Sao Tome and Principe	1	23
Senegal	5	99,993
Seychelles	3	44,025
Sierra Leone	1	295,000
South Africa	23	555,678
South Sudan	1	5,700,000
Sudan	3	2,489,600
Swaziland	3	1,183
Togo	4	1,210,400
Tunisia	41	840,363
Uganda	12	454,303
United Republic of Tanzania	4	4,868,424
Zambia	8	4030,500
Zimbabwe	7	453,828

Table 2 contd. ...

...Table 2 contd.

Country	Number of designated Ramsar sites	Surface (ha)
ASIA		
Bahrain	2	6,810
Bangladesh	2	611,200
Bhutan	3	1,226
Cambodia	4	75,942
China	49	4,112,424
India	26	689,131
Indonesia	7	1,372,976
Iran (Islamic Republic of)	24	1,486,438
Iraq	4	537,900
Japan	50	148,002
Jordan	1	7,372
Kazakhstan	10	3,188,557
Kuwait	1	50,948
Kyrgyzstan	3	679,408
Lao People's Democratic Republic Popular	2	14,760
Lebanon	4	1,075
Malaysia	7	134,158
Mongolia	11	1,439,530
Myanmar	3	108,243
Nepal	10	60,561
Oman	1	107
Pakistan	19	1,343,807
Philippines	7	244,017
Republic of Korea	22	19,164
Sri Lanka	6	198,172
Syrian Arab Republic	1	10,000
Tajikistan	5	94,600
Thailand	14	399,714
Turkmenistan	1	267,124
United Arab Emirates	5	20,278
Uzbekistan	2	558,400
Viet Nam	8	117,813
Yemen	1	580

Table 2 contd. ...

...Table 2 contd.

Country	Number of designated Ramsar sites	Surface (ha)
EUROPE		
Albania	4	98,181
Andorra	3	6,870
Armenia	3	493,511
Austria	23	124,968
Azerbaijan	2	99,560
Belarus	26	778,303
Belgium	9	46,944
Bosnia and Herzegovina	3	56,779
Bulgaria	11	49,913
Croatia	5	94,358
Cyprus	1	1,107
Czech Republic	14	60,207
Denmark	43	2,315,395
Estonia	17	304,778
Finland	49	799,518
France	45	3,623,749
Georgia	2	34,480
Germany	34	868,226
Greece	10	163,501
Hungary	29	261,422
Iceland	6	128,666
Ireland	45	66,994
Israel	2	366
Italy	53	60,759
Latvia	6	150,318
Liechtenstein	1	101
Lithuania	7	65,581
Luxembourg	2	17,213
Malta	2	16
Monaco	1	23
Montenegro	2	20,150
Netherlands	54	905,601
Norway	63	889,452
Poland	13	145,075
Portugal	31	132,487
Republic of Moldova	3	94,705
Romania	19	1,156,448
Russian Federation	35	10,323,676
Serbia	10	63,919
Slovakia	14	40,697
Slovenia	3	8,205
Spain	74	303,090
Sweden	68	666,625
Switzerland	11	14,668
The former Yugoslav Republic of Macedonia	2	21,616
Turkey	14	184,487
Ukraine	39	786,321
United Kingdom	174	1,281,989

Table 2 contd. ...

...Table 2 contd.

Country	Number of designated Ramsar sites	Surface (ha)
LATIN AMERICA AND THE CARIBBEAN		
Antigua and Barbuda	1	3,600
Argentina	22	5,625,407
Bahamas	1	32,600
Barbados	1	33
Belize	2	23,592
Bolivia (Plurinational State of)	11	14,842,405
Brazil	13	7,260,873
Chile	13	361,761
Colombia	6	708,683.9
Costa Rica	12	569,742
Cuba	6	1,188,411
Dominican Republic	4	135,097
Ecuador	18	290,815
El Salvador	7	207, 387
Grenada	1	518
Guatemala	7	628,592
Honduras	9	270,224
Jamaica	4	37,847
Nicaragua	9	406,852
Panama	5	183,992
Paraguay	6	785,970
Peru	13	6,784,042
Saint Lucia	2	85
Suriname	1	12,000
Trinidad and Tobago	3	15,919
Uruguay	3	435,837
Venezuela (Bolivarian Republic of)	5	263,636
NORTH AMERICA		
Canada	37	13,086,767
MEXICO	**142**	**8,643,580**
The United States of America	38	1,860,761
OCEANIA		
Australia	65	8,319,663
Fiji	1	615
Kiribati	1	1,033
Marshall Islands	2	70,119
New Zealand	6	56,639
Palau	1	500
Papua New Guinea	2	594,924
Samoa	1	470

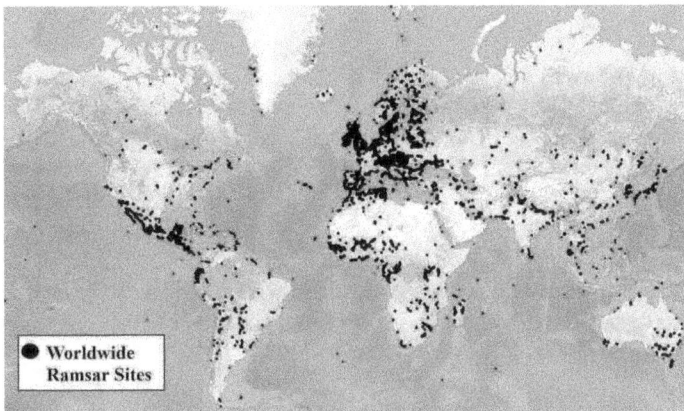

Fig. 1. Worldwide Ramsar Sites (Ramsar 2017c).

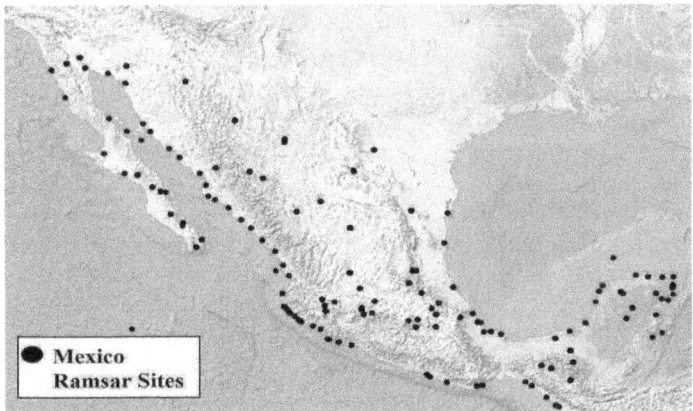

Fig. 2. Mexico Ramsar Sites (Ramsar 2017d).

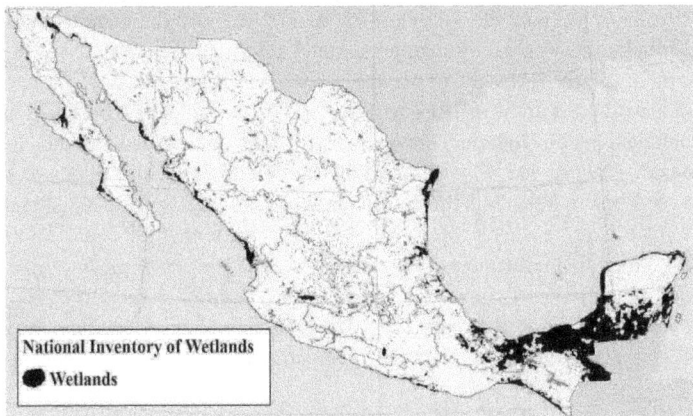

Fig. 3. Distribution of wetlands in Mexico according to the National Inventory of Wetlands (*INH* in Spanish for *Inventario Nacional de Humedales*) (CONANP 2013, CONAGUA 2017b).

According to this *INH*, in Mexico there are 6,331 wetland and wetland complexes in more than 9,924,624 ha (99,246.24 km²), which represent approximately five percent of the national territory. In Fig. 3, the distribution of wetlands at the national level can be observed according to the *INH*. Of this total, 38 percent are palustrine, 31 percent fluvial, 15 percent estuarine, eight percent lacustrine, and another eight percent created. Veracruz de Ignacio de la Llave is the state with the highest number of wetlands (664), followed by Chiapas (476), Tabasco (387), Oaxaca (381), Chihuahua (375), Sonora (332), Sinaloa (310), and Jalisco (303). On the other hand, of the wetlands at national level, the states that have the largest surface area of them are Campeche and Tabasco with 26 and 16 percent respectively, followed by the states of Chiapas and Veracruz de Ignacio de la Llave with nine percent (CONANP 2013; CONAGUA 2017b). This information is essential to formulate the necessary public policies aimed at the sustainable management of different systems.

Xochimilco, a Case Study

In the *Xochimilco* demarcation, one of the 16 political entities in which Mexico City is divided (Ortiz-Ordoñez et al. 2016), there exists the called *Sistema Lacustre Ejidos de Xochimilco y San Gregorio Atlapulco* (in Spanish) or *Xochimilco* Lake System *Ejidos*[1] and *San Gregorio Atlapulco*, formed by wetlands, channels, and *chinampas* areas with a surface of or 2,657 ha (26.57 km²) (Ramsar 2004).

Ejido is a Spanish word from Latin (*exītus*, for *exĭtus* 'outlet'), but the Spanish dictionary meaning does not truly depicts its meaning in Mexico where it was defined as a common piece of land of a community that was not used for private cultivation but for common people's activities and, until 1988, it could not be sold because this form of property was protected by the Mexican constitution. In 1988 this article of the constitution was changed and it is one of the causes why the use of the land in *Xochimilco* has deeply changed transforming it into an urban area with dwellings in a zone that has no conditions for this use (Orozco-Garibay 2010).

Xochimilco is a name that comes from the *Nahuatl* or Aztec language, *xochitl*=flower, *milli*=fertile land or sowing and *co*=place or site (locative termination), according to Cabrera (2002) and, commonly, it means a place where flowers grow ("*sementera de flores*" in Spanish), has a territorial extension of 118 km² and concentrates a population of 415,933 inhabitants, representing approximately eight and five percent of the area and the total population, respectively, of Mexico City (INEGI 2016a).

This demarcation is in the southeastern part of Mexico Valley and bordered to the north with the demarcations of *Tlalpan, Coyoacan, Iztapalapa,* and *Tláhuac*, to the east with the demarcations of *Tláhuac* and *Milpa Alta*, to the south with the demarcations of *Milpa Alta* and *Tlalpan* and, to the west, with the demarcation of *Tlalpan* (INEGI 2008, 2016a).

The *Xochimilco* area is made up of 105 colonies and communities (*pueblos* and *barrios* in Spanish). Among them, the most widely known are the *barrios* of the *Santísima Trinidad Chililico, Belem de Acampa, Asunción Colhuacatzinco, San Francisco Caltongo, San Pedro Tlanáhuac,* and the *pueblos* of *Santa María Nativitas, San Gregorio Atlapulco, San Mateo Xalpa, Santa Cruz Acalpixca,* and *Santiago Tepalcatlalpan*.[2]

---

[1] Spanish Word for common land.

[2] After the conquest in 1521, the Spanish priests encouraged the civil authorities to maintain the *Nahuatl* or Aztec name for most existing locations, adding the name of a Saint that was to be celebrated on the day the population of such town had its pre-Columbian local celebrations in order to introduce the catholic religion through these celebrations.

Its geographic location is North 19°19', South 19°09' of North latitude, East 98°58' and West 99°10' of the West longitude. It is located in the Transversal Neovolcanic Axis (*Eje Neovolcánico Transversal* in Spanish), and its average altitude from its historic downtown to *Tulyehualco* is 2,240 meters above the sea level, whereas, in the mountainous zone its elevations are more important, as the volcanoes *Zompole* and *Teuhtli* that rise above the 2,600 m above the sea level, while the *Tlacualleli* and *Xochitepec* hills have 2,420 and 2,500 m above the sea level, respectively (INEGI 2008, 2016a). It is an important agricultural area of the Mexico Valley with at least 500 years of history in Mesoamerica, since, according to 2005 data, 77 km² (65 percent of its surface) of them are used for agricultural purposes (INEGI 2016a).

Among the great variety of products grown in the area, vegetables such as celery, chard, broccoli, zucchini, spring onion, peas, *cilantro*, cabbage, cauliflower, spinach, peppermint, lettuce, corn, parsley, radish, and carrot, among others, as well as various types of fruit trees such as plum, peach, fig, apple, and pear, in the same area, in addition, there is a large production of *quelites*, *romeritos*, and *verdolagas*, which are edible deep green plants that are wild and have always been part of the environment of the places known as *chinampas*[3] and *milpas*[4], the latter being maize sowing or planting areas. So, the addition of herbicides in these *milpas* is somewhat absurd as these edible plants that grow along with maize improve the diet with deep- or dark-green-leave plants which are necessary for its iron and vitamins content.

The largest production is currently in the area of greenhouses, where they cultivate a large variety of flowers and ornamental plants, which, according to 2015 data, highlight the so called Christmas flower (*nochebuena* in Spanish), a plant originally from Mexico and now associated with Christmas in all the countries of the world. Its *Nahuatl* name is *cuitlaxóchitl* or flower that grows in soils rich in organic or humic residues, from *cuítlatl*=organic or humic residues or "manure" (a term that did not exist before the Spaniards brought domestic farm animals to Mexico), and *xóchitl*=flower (Cabrera 2002), as well as geraniums and roses, with an annual output of 2,142,085; 2,424,000 and 1,128,675 tons respectively (INEGI 2008; SAGARPA 2016; SIAP 2017).

Lake System *Ejidos de Xochimilco* and *San Gregorio Atlapulco*

Historically, Mexico Basin was made up of five large lakes were interconnected: *Chalco*, *Texcoco*, *Xaltocan*, *Xochimilco*, and *Zumpango*, which covered an area of approximately 920 km² (Fig. 4) (Valek 2000; Ramsar 2004).

This basin was supplied with water by means of 45 rivers, among which the *Cuautitlán*, *Magdalena*, and *Tepotzotlán* rivers stood out and about 10 springs located in the region of *Chalco-Xochimilco* (Legorreta 2004; Santoyo-Villa et al. 2005). Nowadays, only the vestige of what was once the great Mexico Basin remains and the three small bodies of water belonging to the wetlands of *Zumpango*, *Chalco*, and *Xochimilco* remain, the latter being the largest (Bojórquez and Villa 1997; Legorreta 2004).

---

[3] From *chinámitl*, grid of branches or canes, and *pan*=over, these were also called *tepechtle*=rug, a support made of *Phragmites* sp. or *Typha* sp. canes covered with sediments from the lake to form that is floating on the lake or channels surface, that are fixed using trees or poles to the sediment (Cabrera 2002).

[4] From the *Nahuatl* language *milli*=sowing and *pan*=on or above, according to Cabrera (2002).

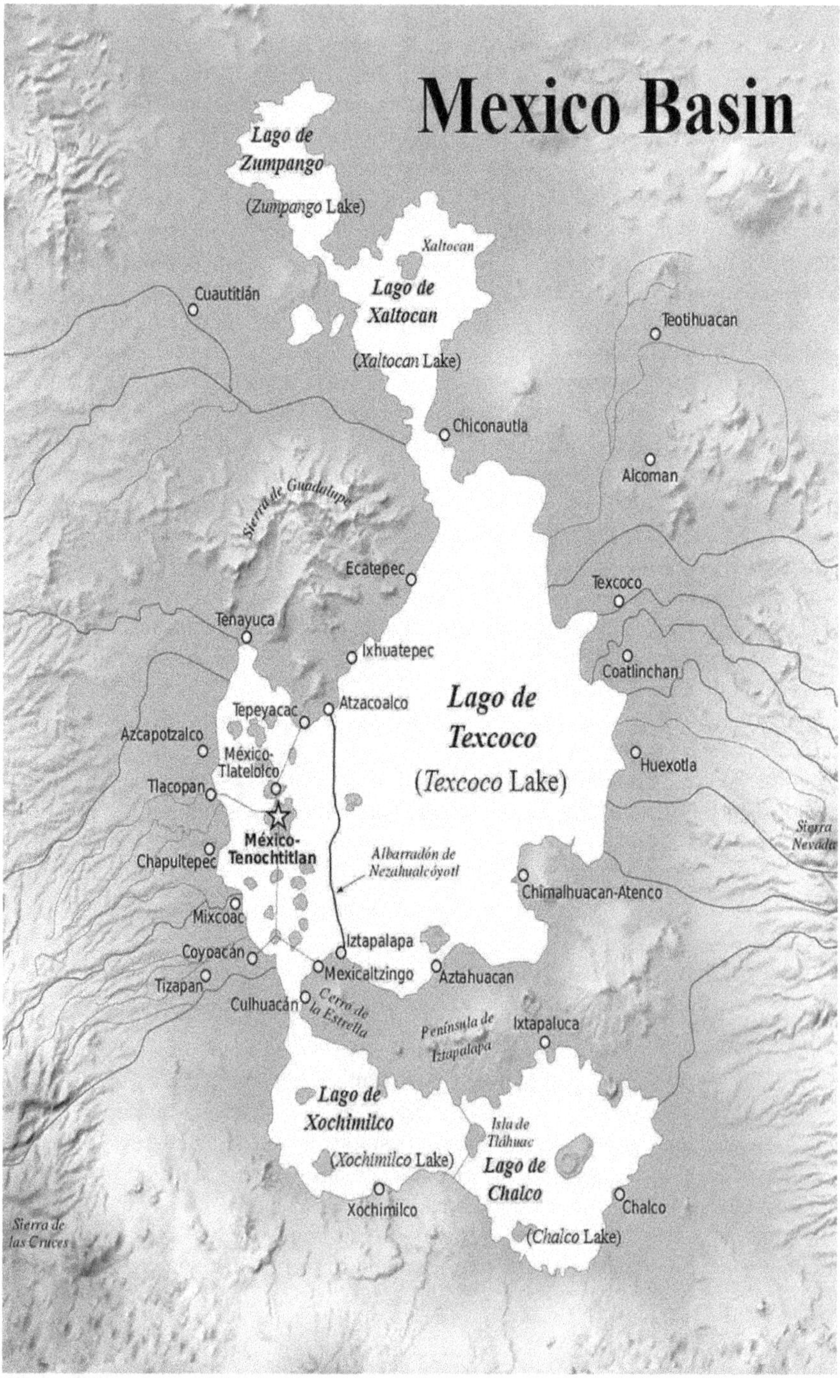

Fig. 4. Original lakes and rivers system of the Mexico Basin viewed by the Spanish conquerors upon their arrival in the 16th century (modified from internet: clio-mexico-luiselli.blogspot.mx/2013/09/mapa-de-la-cuenca-de-mexico-hacia-1519.html).

The Lake System of *Xochimilco* is located in the *Panuco* River region, in the sub-basin of Lakes *Texcoco* and *Zumpango*, belonging to the hydrological basin of the *Moctezuma* River. Its extreme geographic locations are 19°15'11" and 19°19'15" North latitude and 99°00'58" and 99°07"08" West longitude. The climate is temperate sub-humid with rains in summer (of average humidity), the average annual precipitation is 700 mm, and has an average annual temperature of 16°C (Ramsar 2004; INEGI 2008, 2016a).

At present, this system is reduced to a series of channels, *apantles,*[5] and lagoons (some permanent and others temporary), which form a natural area of discharge of the underground flow, which provides water through an aquifer and, in addition, operates as a regulator of flows at local and regional level. It is estimated that this system has an approximate length of 203 km of canals connected to each other. Among the most important channels are *Ampampilco, Apatlaco, Caltongo, Chalco, Cuemanco, Japón, Del Bordo, Nacional, San Sebastián, Santa Cruz, Texhuilo,* and *Zacapa*. Besides, this system is also made up of three main lagoons, *Caltongo, El Toro,* and the Lake of Flora, Fauna, and Aquaculture Conservation of *San Gregorio Atlapulco* (*Lago de Conservación de Flora, Fauna y Acuacultura de San Gregio Atlapulco* in Spanish) (Ramsar 2004).

The sources that supplied water naturally to the Lake System of *Xochimilco* have practically disappeared, causing the canals and lagoons to depend on artificial feeding through the supply of treated water from the wastewater treatment plants (WWTPs) *Cerro de la Estrella, San Lorenzo Tezonco,* and *San Luis Tlaxialtemalco,* which contribute to the channels system with 1.0, 0.08, and 0.065 $m^3 s^{-1}$, respectively, besides the contribution of the runoff from the *Sierra Ajusco-Cuautzin*. Another source of supply is the surface runoff originated by the *San Gregorio, San Lucas,* and *Santiago* or *Parres* rivers, which form in the foothills of the *Sierra Ajusco-Cuautzin*. The *Parres* River descends from the western slopes of *Cuautzin* to *San Lucas Xochimanca* dam, which has a capacity to store 850,000 m^3. The *San Lucas* and *San Gregorio* rivers conduct runoff and capture the sewage and rainwater of the towns of *San Francisco Tlalnepantla, San Lucas Xochimanca, San Mateo Xalpa, San Miguel Topilejo,* and *Santiago Tepalcatlalpan* (Ramsar 2004; Aguilar et al. 2006).

Since pre-Columbian times, the Lake System of *Xochimilco* has been an important place for the center of the country, due to the realization of agricultural activities, through the creation of the structures for cultivation called *chinampas,* also a name coming from the *Nahuatl* language, as mentioned before. This form of cultivation is considered unique in the world (Fig. 5a) and the use of *trajineras* (a hybrid word between Spanish and *Nahuatl* to name these particular canoes used as transport for regional trade within the *Xochimilco* channels) (Fig. 5b) (Ramsar 2004). The *chinampas* literally mean "floating terrains of short extension in the lagoons neighboring Mexico City, where flowers and vegetables are grown".

In general, they are rectangular islands, used for the cultivation of herbs, flowers, and vegetables, constructed from layers of aquatic vegetation, mainly *Typha* sp. and *Phragmites* sp., and mud from the bottom of the lake, forming a carpet of interwoven

---

[5] In *Nahuatl* language means "irrigation water ditch", *atl*=water and *pantli*=row or pile, according to Cabrera (2002).

Fig. 5. (a) A *chinampa* (modified from internet: www.xataka.com.mx/ciencia/desechos-organicos-podrian-ser-la-clave-para-rehabilitar-chinampas-de-xochimilco)), and (b) A *trajinera* (modified from internet (www.diariodf.mx/delegaciones/en-riesgo-zona-patrimonial-de-xochimilco-tlahuac-y-milpa-alta).

sticks, floating in shallow waters, subject to a fencing of poles and/or *ahuejote's* trees (*Salix lasialepis* Benth)[6] to avoid erosion.

These systems were able to sustain the population of what it is now Mexico City roughly estimated in several millions when the Spaniards arrived to continental America in 1519 that, in very few years later, was decimated by the diseases brought by the Spaniards, particularly the black or smallpox and the very poor conditions to which the indigenous population was subjected (Sahagún 1577, version in Spanish and *Nahuatl*, available in 2017).

Chinampas emerged as an alternative to develop agriculture, as well as to expand the territory on the surface of lakes and lagoons in the Mexico Valley. Also known as floating gardens (Chávez-López et al. 2011), the *chinampas* are nowadays mainly used for floriculture and for vegetables planting. The main features as per FAO 2017 are:

1. "They are made up of a diversified agriculture, which includes horticulture, floriculture, and the production of basic crops, both for consumption within and outside the region."

2. "They have a great biodiversity, because besides of the agricultural products, they offer ecologic niches for the aquatic fauna and for the endemic or transitory bird populations."

3. "The landscape follows a special pattern of elongated islands with a prevailing direction that shows a slight deflection of 15° from North to East. This direction has a coincidence with other pre-Columbian civil and religious construction models, among them, the *Teotihuacan's*[7] urban pattern. The accumulation of thousands of cultivated *chinampas*, *ahuejote* trees, and hundreds of kilometers of water channels in a huge territory not only make up an impressive cultural and productive site itself, but it also represents an esthetic landscape articulated by the water, the soil, the trees, the fauna, and the natural environment, which can be enjoyed by this great megacity's people with a population of more than 20 million inhabitants."

4. "The *chinampa* system is officially administrated by the local authorities but its operation and agriculture productivity are still in the original farmers' hands, which are the 'landowners' or *ejidatarios*."

5. "They provide a history with sustainable paths of economic and viability capacities to face process of changes, such as urbanization with resilience."

6. "The cultural diversity and the natural environment offer different ecosystem products and services."

In response to the global trends that undermine family agriculture and traditional agricultural systems, in 2002, the Food and Agriculture Organization (FAO) of

---

[6] *Ahuejote* comes from *Nahuatl* language, *atl*=water and *huexótl*=willow and means "water willow" (Cabrera 2002).

[7] *Teotihuacan* is a very important archaeological zone in the State of Mexico, north from Mexico City, where there are two very imposing pyramids. It is a *Nahuatl* word, from *téotl*=god and *huacan*=surrounds plus the phonetic union *ti* that means "place where gods gather or place where the gods are or city of the gods", according to Cabrera (2002).

the United Nations launched a Global Partnership Initiative on conservation and adaptive management of Globally Important Agricultural Heritage Systems (GIAHS). Currently, there are 36 recognized GIAHS in 15 countries and the *Chinampera Agriculture System in the Patrimonial Zone of Xochimilco, Tláhuac and Milpa Alta, Mexico City*, is a candidate system to be a GIAHS and it's in the second of three stages, evaluation by the Scientific Advisory Group (SAG), to be declared as such (FAO 2017).

Nowadays, strong anthropic pressure has led to uses other than *chinampas*, beyond its agricultural tradition. Thus, four large areas can be defined within the Lake *Xochimilco* according to the current main use of the *chinampas* (Zambrano et al. 2009; Merlo-Galeazzi 2014):

1. *Chinampa* area: The main activity is the traditional pre-Columbian biotechnology to produce vegetables and flowers.

2. Urban area: The main activity is for dwellings that unfortunately have no sewage systems provided by the local government.

3. Tourist area: It is also considered within the urban zone but its main activity is related with the tourist sector with rides along the channels using a traditional type of wooden canoes decorated with flowers known as trajineras where people can eat Mexican traditional food and listen to Mexican traditional music.

4. Mixed area: In this zone there is a mixture of urban dwellings and agricultural areas.

In Fig. 6, the zones of the Lake System of *Xochimilco* can be seen according to their main use. In addition to these areas, along the Lake *Xochimilco*, there are three areas that belong to other towns or *ejidos*, such as *San Gregorio Atlapulco, San Luis Tlaxialtemalco*, and the wetland region belonging to the *San Gregorio ejido*.

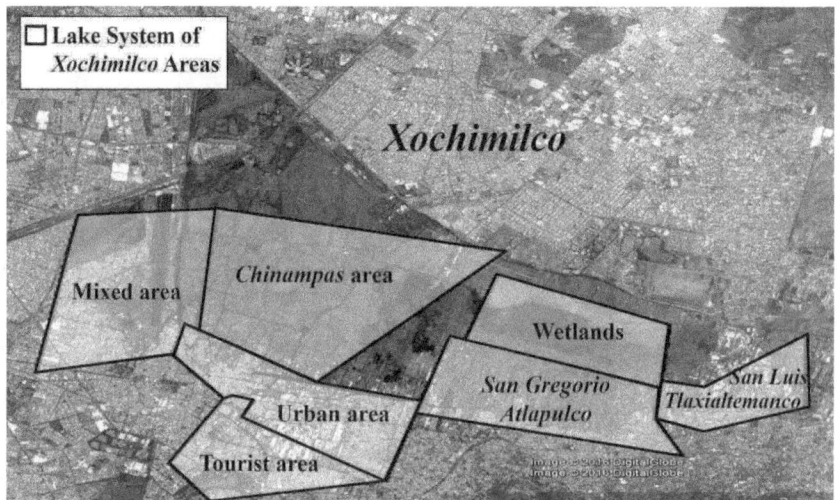

Fig. 6. Lake System of *Xochimilco* main areas (modified from Zambrano et al. 2009, Merlo-Galeazzi 2014).

Ecological Importance

The Lake System of *Xochimilco* represents an important ecosystem for the maintenance of the biodiversity of the City and/or the Mexico Valley, since they host a great amount of endemic aquatic and terrestrial species of flora and fauna, some of them vulnerable and of very restricted distribution (Ramsar 2004; INEGI 2016a). With respect to terrestrial flora, there are 146 species distributed in 101 genera and 46 families, while the aquatic vegetation is represented by 115 species, distributed in 63 genera, and the group of Chlorophytas is the most representative. It emphasizes the presence of *Nymphaea mexicana* (ninfa), a species under the threatened category according to the Mexican Norm NOM-059-SEMARNAT-2010, and the *tulares* from the *Nahuatl tulli* (*Typha latifolia*), and also *Scirpus americanus, Cyperus* sp., *Juncus* sp., among the different aquatic wetlands plants. This system is hosting 139 species, 21 of which are fish, 6 amphibians, 10 reptiles, 79 birds, and 23 mammals. In this area there are populations that have been reduced, such as the two species of *ajolote* (*Proteus mexicanus* L., *Ambystoma mexicanum*),[8] the *Moctezuma* and *Xochimilco* frogs, and the socalled *acociles*,[9] small shrimp (*Cambarellus montezumae* and *Hyallela azteca*), are also menaced species. *Axolotl* or *ajolote* is the most known amphibian of the Mexican Basin due to its strange features. Nowadays, its presence is rare because of the deterioration of the environment and the excessive fishing to which it was subjected for decades (Ramsar 2004; Zambrano et al. 2009).

This lacustrine system is of great importance since, besides to being cataloged in Mexico as Protected Natural Area (*ANP* in Spanish) with the character of Zone Subject to Ecological Conservation *Ejidos* of *Xochimilco* and *San Gregorio Atlapulco*, worldwide, its included in the Ramsar List, as well as in the List of World Heritage Sites by the UNESCO (UNESCO 1988; DOF 1992; Ramsar 2004). Figure 7 shows the area of *Xochimilco* that is of greater ecological importance, both nationally and globally. However, environmental conditions are not optimal, leading to endemic species on the verge of extinction (Contreras et al. 2009).

Present Ecological Problems

In *Xochimilco*, population growth has been accelerated, which has led to an increase in urbanization (Starkl et al. 2013). Due to the structure of the lake system and its proximity to the urban zone, there is a strong pressure inside it, so there are irregular human settlements. The population settled in the lake area is estimated at around 24,102 inhabitants and, in the area of immediate influence, 121,131 inhabitants (Ramsar 2004).

The Lake System of *Xochimilco* is a permanent wetland, which was formed in a place of natural discharge, in a closed basin. There are no tides and has no connection with seawater or any other body of water. Its depth varies considerably according to the area. Some sites, such as channels and flooded areas, can be up to 60 cm deep, while in some lagoons, these can reach 3 to 6 m depth. This system has had gradual processes of artificial extraction for potabilization plants to provide water to the Mexico City population (more than 9 millions), both of surface water (springs) and groundwater

---

[8] From *Nahuatl atl*=water and *xólotl*=small toy, small aquatic toy.
[9] From *Nahuatl atl*=water and *cuitzilli*=bent, a bent aquatic animal, aquatic bent animal.

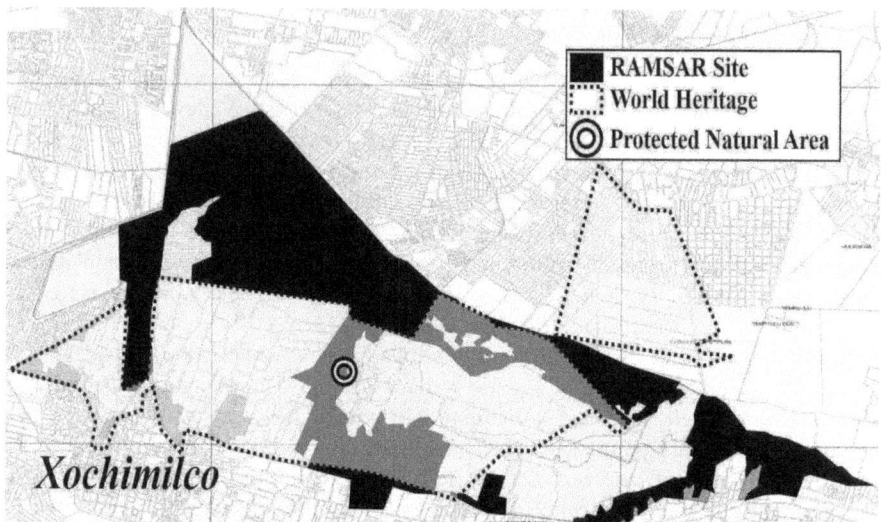

Fig. 7. *Xochimilco* ecological site of national and international importance (modified from UNESCO 1988, DOF 1992, Ramsar 2004).

(aquifers) to meet the needs of the Mexico City urban area (Ramsar 2004; Contreras et al. 2009). This has determined the drying up of the springs that fed it.

As mentioned before, the reduction of the spring surface water has been replaced by the Mexico City authorities with treated wastewater from the neighboring WWTPs. However, the overexploitation of the aquifers has led to a decrease in the level of the lake waters, channels, and ditches, promoting a gradual subsidence of the soil. These actions have led to a notable decrease in the productivity of *chinampas* and to the development of agricultural activities different from the traditional ones (Ramsar 2004; Enríquez-García 2009; González and Torres 2014).

It has been determined that the main problem that is present in the Lake System of *Xochimilco* is the association of the quality of the water supplied with the urban use of the area (Espinosa et al. 2009; Díaz-Torres et al. 2013; Sandoval 2013). Today, as stated in previous paragraphs, the lake system depends on artificial recharge using partially treated wastewater from two small WWTPs, *San Lorenzo Tezonco* and *San Luis Tlaxialtemalco*, and a large WWTP, known as *Cerro de la Estrella*, the latter being the most important source of contamination as it is the one that contributes the highest flow of water to the system (Ramsar 2004; González and Torres 2014). According to the Ramsar (2004) information, in *Xochimilco* may be found two types of water: natural (runoff) and of bad quality (WWTPs). The last one contain chemical compounds such as salts and heavy metals (cadmium, mercury, and lead), as well as hydrocarbons, fats and oils, herbicides, and chemical wastes from the neighboring industrial area of *Iztapalapa* that were not eliminated in the WWTP of *Cerro de la Estrella* (Bojórquez and Amaro 2003; Ramsar 2004). Agrochemicals as auxiliaries in the agricultural activities are carried out in some of the *chinampas* as well as the ever growing use of the area for dwellings have also originated an environmental pollution problem for the aquatic system of *Xochimilco* (Alcántara 2014).

Other factors that are associated with the loss of environmental quality are wastewater discharges and anthropic activities, such as those related to dwellings land use without sanitation facilities and non-traditional agriculture (Zambrano et al. 2009; González and Torres 2014; Merlo-Galeazzi 2014). There are some studies indicating the presence of organic pollutants in sediments, soil, and water of the *Xochimilco* channels. Chavéz-López et al. (2011) and Alcántara (2014) determined the presence of methyl-paration in soil of a *chinampa* located in the agricultural zone as well as organochlorinated and organophosphorous pesticides. These references indicate that in *Xochimilco* the use of pesticides is increasing, especially where Christmas and November flowers (for the November 1 and 2 festivities, this is also a Mexican flower called *cempasúchil*,[10] *Tagetes erecta* L., *Tagetes multiseta* Dc. and *cempasúchil chiquito, Adenophyllum prophyllum* Hemsl.) are being produced. All of the above have been leading to serious problems of pollution and environmental impact that are undermining the status of World Cultural and Natural Heritage and Ecological Conservation Zone of the Lake System of *Xochimilco*.

Final Remarks

This system has gone from being a natural wetland with unique characteristics in the world to a large artificial wetland purifier of poorly treated sewage of mixed origin, urban industrial, and dwellings origin. Therefore, from the scientific point of view to carry out its protection and to rescue this wetland, several measurements have to be taken by the local government of Mexico City:

1. Maintaining the quality and dynamics of the aquifer.
2. Conservation of the soil of the Mexican Basin.
3. Preservation of the endemism and the singularity of its natural communities and associated habitats.
4. Preservation of the regional ecological balance, and for the cultural value that it represents as a living natural museum, the support to the traditional agro-systems such as the *chinampas*.

It is necessary to carry out detailed studies such as the characterization of surface waters and sediments, but above all, to know how the water is flowing through the different channels and *apantles*. Since it is hypothesized that this is the reason that in areas where the use of agrochemicals is not carried out (organic areas), these are present as it will be seen in the next chapter.

As a final remark, the ecocide caused in the Mexico Basin (Fig. 4), now a valley full of buildings and streets and avenues, should be reverted with the support not only of the government but of the Mexico City's whole population if they want to survive. Resilience is the word (Durán-Domínguez-de-Bazúa et al. 2017).

Acknowledgements

The first author acknowledges his graduate student scholarship from the Mexican Science and Technology Council (*CONACYT*, in Spanish *Consejo Nacional de*

---

[10] *Cempasúchil* comes from *Nahuatl*: *cempoalli*, twenty and *xóchitl*, flower, twenty petals flower or many petals flowers, commonly used in the altars in honor of the beloved dead relatives and friends.

Ciencia y Tecnología). Materials were provided by *UNAM DGAPA* projects *PAPIME EN103704*, *PE101709*, and *PE100514* and to *UNAM Programa de Apoyo a la Investigación y el Posgrado de la Facultad de Química, PAIP 5000-9067*. The four authors appreciate the collegial support in the review of the first draft by Dr. José Manuel Barrera-Andrade from Mexico's National Polytechnical Institute (*IPN, Instituto Politécnico Nacional in Spanish*).

References Cited

Aguilar, A., A.C. Espinosa and C. Caraballo. 2006. El manejo del agua: Tema central en Xochimilco. pp. 183–200. *In*: VV.AA. [ed.]. Xochimilco, un proceso de gestión participativa. UNESCO. Ciudad de México, Mex. In Spanish

Alcántara, V. 2014. Caracterización y diagnóstico de la contaminación por plaguicidas en el lago de Xochimilco. Doctoral Thesis, UNAM, México. In Spanish.

Berlanga-Robles, C.A., A. Ruiz-Luna and G. de la Lanza-Espino. 2007. Esquema de clasificación de los humedales de México. Investigaciones Geográficas 66: 25–46. In Spanish.

Bojórquez, C.L. and E.J. Amaro. 2003. Caracterización múltiple de la calidad del agua de los canales de Xochimilco. pp. 281–302. *In*: Stephan-Otto, E. [ed.]. El Agua en la Cuenca de México. Sus problemas históricos y perspectivas de solución. UAM-Xochimilco—Patronato del Parque Ecológico de Xochimilco A.C. México. In Spanish.

Bojórquez, L. and F. Villa. 1997. La zona lacustre de Xochimilco: Reconstrucciones hipotéticas. pp. 468–493. *In*: Stephan-Otto, E. [ed.]. Primer Seminario Internacional de Investigadores de Xochimilco. Asociación Internacional de Investigadores de Xochimilco A.C. Ciudad de México, Mexico. In Spanish.

Cabrera, L. 2002. Diccionario de aztequismos. Puesto en orden y revisado por J. Ignacio Dávila-Garibi. Luis Reyes-García revisó los términos nahuas y Esteban Inciarte los que aparecen en latín. Ed. Colofón, S.A. 5ª ed. Mexico City, Mexico. In Spanish.

Chávez-López, C., A. Blanco-Jarvio, M. Luna-Guido, L. Dendooven and N. Cabirol. 2011. Removal of methyl parathion from a chinampa agricultural soil of Xochimilco Mexico: A laboratory study. Eur. J. Soil Biol. 47: 264–269.

CONABIO. 2017. What is a megadiverse country? http://www.biodiversidad.gob.mx/v_ingles/country/whatismegcountry.html.

CONAGUA. 2008. Catálogo tipológico de los humedales lacustres y costeros de Chiapas. Comisión Nacional del Agua. SEMARNAT, México. https://www.gob.mx/cms/uploads/attachment/file/110132/Humedales_Chiapas.pdf. In Spanish.

CONAGUA. 2017a. Humedales de la República Mexicana—Inventario Nacional de Humedales. Comisión Nacional del Agua. http://www.gob.mx/conagua/acciones-y-programas/inventario-nacional-de-humedales-inh. In Spanish.

CONAGUA. 2017b. Humedales de la República Mexicana—INH. Sistema de Información Geográfica del Agua. Comisión Nacional del Agua. http://sigagis.conagua.gob.mx/Humedales/. In Spanish.

CONANP. 2013. Política Nacional de Humedales. Comisión Nacional de Áreas Naturales Protegidas. http://ramsar.conanp.gob.mx/docs/PNH_SEMARNAT.pdf. In Spanish.

CONANP. 2015. Folleto: Las Áreas Naturales Protegidas, respuestas naturales frente al cambio climático. Comisión Nacional de Áreas Naturales Protegidas. http://www.gob.mx/conanp/documentos/folleto-las-areas-naturales-protegidas-respuestas-naturales-frente-al-cambio-climatico. In Spanish.

CONANP. 2016a. BROCHOURE: Áreas Naturales Protegidas. Comisión Nacional de Áreas Naturales Protegidas. http://www.gob.mx/cms/uploads/attachment/file/173571/BROCHOURE-cop13.pdf. In Spanish.

CONANP. 2016b. Programa Nacional de Áreas Naturales Protegidas 2013-2018. Comisión Nacional de Áreas Naturales Protegidas. http://entorno.conanp.gob.mx/documentos/PNANP.pdf. In Spanish.

Contreras, V., E. Martínez-Meyer, E. Valiente and L. Zambrano. 2009. Recent decline and potential distribution in the last remnant area of the microendemic Mexican axolotl (*Ambystoma mexicanum*). Biol. Cons. 142: 2881–2885.

Cowardin, L.M., V. Carter, F.C. Golet and E.T. LaRoe. 1979. Classification of wetlands and deepwater habitats of the United States. U.S. Department of the Interior, Fish and Wildlife Service, Washington, D.C. https://www.fws.gov/wetlands/Documents/Classification-of-Wetlands-and-Deepwater-Habitats-of-the-United-States.pdf.

Díaz-Torres, E., R. Gibson, F. González-Farías, A.E. Zarco-Arista and M. Mazari-Hiriart. 2013. Endocrine disruptors in the Xochimilco wetland, Mexico City. Water Air Soil Pollut. 224: 1586–1596.

DOF. 1992. Declaratoria que establece como zona prioritaria de preservación y conservación del equilibrio ecológico y se declara como área natural protegida, bajo la categoría de zona sujeta a conservación ecológica, la superficie que se indica de los ejidos de Xochimilco y San Gregorio Atlapulco, D.F. (Segunda publicación). In Diario Oficial de la Federación. Volume CDLXIV. No. 5. 1st section. pp. 53–58. Mexico. In Spanish.

Durán-Domínguez-de-Bazúa, M.-del-C., I. Salgado-Bernal, A.E. Navarro-Frómeta, U. Kappelmeyer, B. Espinosa-Aquino, N.J. Ruiz-Cárdenas et al. 2016. Polluted water threats to urban resilience: Case studies in Mexico and Havana Cities. Informe de la Red sobre el Uso de Humedales Artificiales para Reducir la Vulnerabilidad de las Comunidades con Escasez de Agua/*Network report on the use of artificial wetlands to reduce the vulnerability of communities with water scarcity* (Feb. 2015–Ene. 2016), presented in RESURBE III, INTERNATIONAL CONFERENCE ON URBAN AND REGIONAL RESILIENCE, Empowering Local Communities for Local Action. UNAM, Dirección de Cooperación e Internacionalización. With the support of INVENTEC Enterprise through S. A. Sánchez-Tovar. Electronic abstract (compact disk), Feb. 2016. 1st. Ed. 24 pages. Mexico City, Mexico.

Enríquez-García, E., S. Nandini and S.S.S. Sarma. 2009. Seasonal dynamics of zooplankton in Lake Huetzalin, Xochimilco (Mexico City, Mexico). Limnologica 39(4): 283–291.

Espinosa, A.C., C.F. Arias, S. Sánchez-Colón and M. Mazari-Hiriart. 2009. Comparative study of enteric viruses, coliphages and indicator bacteria for evaluating water quality in a tropical high-altitude system. Environ. Health 8(49): 1–10.

FAO. 2017. Sistemas Importantes del Patrimonio Agrícola Mundial (SIPAM). Organización de las Naciones Unidas para la Agricultura y la Alimentación. http://www.fao.org/giahs/es/. In Spanish.

González-Carmona, E. and C.I. Torres-Valladares. 2014. La sustentabilidad agrícola de las chinampas en el Valle de México: caso Xochimilco. Revista Mexicana de Agronegocios 34: 699–709. In Spanish.

Hu, S., Z. Niu, Y. Chen, L. Li and H. Zhang. 2017. Global wetlands: Potential distribution, wetland loss, and status. Sci. Total Environ. 586: 319–327.

INEGI. 2008. Xochimilco. Distrito Federal. Cuaderno estadístico delegacional 2008. Instituto Nacional de Geografía y Estadística. Aguascalientes, Ags. México. http://www.inegi.org.mx//est/contenidos/espanol/sistemas/cem08/estatal/df/m013/default.htm. (In Spanish).

INEGI. 2016a. Anuario estadístico y geográfico de la Ciudad de México 2016. Instituto Nacional de Geografía y Estadística. Aguascalientes, Ags. México. http://internet.contenidos.inegi.org.mx/contenidos/Productos/prod_serv/contenidos/espanol/bvinegi/productos/nueva_estruc/anuarios_2016/702825084318.pdf. In Spanish.

INEGI. 2016b. Anuario estadístico y geográfico de los Estados Unidos Mexicanos 2016. Instituto Nacional de Geografía y Estadística. Aguascalientes, Ags. México. http://internet.contenidos.inegi.org.mx/contenidos/Productos/prod_serv/contenidos/espanol/bvinegi/productos/nueva_estruc/AEGEUM_2016/702825087340.pdf. In Spanish.

LAN. 2014. Ley de Aguas Nacionales. Nueva Ley publicada en el Diario Oficial de la Federación el 1º de diciembre de 1992. Texto vigente, Última reforma publicada DOF 11-08-2014. Ciudad de Mexico, Mexico. In Spanish.

Legorreta, J. 2004. El agua de Xochimilco y su región lacustre. La Jornada daily newspaper. http://www.jornada.unam.mx/2004/05/31/eco-d.html. In Spanish.

López, J.A.P., V.M.R. Vásquez, L.R.A. Gómez and A.G.S. Priego. 2010. Humedales. pp. 227–248. *In*: Aguilar, I.M. [ed.]. Atlas del patrimonio natural, histórico y cultural de Veracruz. Tomo 1. Patrimonio natural. Comisión del Estado de Veracruz para la Conmemoración de la Independencia Nacional y la Revolución Mexicana. Xalapa, Veracruz, Mexico. In Spanish.

Merlo-Galeazzi, A. 2014. Efectos de la heterogeneidad espacial sobre la diversidad taxonómica y funcional de los macroinvertebrados acuáticos de Xochimilco. Master Thesis, UNAM, Mexico. In Spanish.

NOM-059-SEMARNAT-2010. Protección ambiental—Especies nativas de México de flora y fauna silvestres—Categorías de riesgo y especificaciones para su inclusión, exclusión o cambio—Lista de especies en riesgo. http://biblioteca.semarnat.gob.mx/janium/Documentos/Ciga/agenda/DOFsr/DO2454.pdf. In Spanish.

Orozco-Garibay, P.A. 2010. Naturaleza del ejido, de la propiedad ejidal, características y limitaciones. Revista Mexicana de Derecho (12): 163–193. In Spanish.

Ortiz-Ordoñez, E., E. López-López, J.E. Sedeño-Díaz, E. Uría, I.A. Morales, M.E. Pérez et al. 2016. Liver histological changes and lipid peroxidation in the amphibian *Ambystoma mexicanum* induced by sediments elutriates from the Lake Xochimilco. J. Environ. Sci. 46: 156–164.

RAE. 2017. Humedal. Real Academia Española. http://dle.rae.es/?id=KoF3wQ4.

Ramsar. 2004. Ficha Informativa de los Humedales Ramsar (FIR). Ramsar. https://rsis.ramsar.org/RISapp/files/RISrep/MX1363RIS.pdf. 23/10/2014. In Spanish.

Ramsar. 2017a. Historia de la convención de Ramsar. http://www.ramsar.org/es/acerca-de/historia-de-la-convención-de-ramsar. In Spanish.

Ramsar. 2017b. La convención de Ramsar y su misión. http://www.ramsar.org/es/acerca-de/la-convención-de-ramsar-y-su-misión. In Spanish.

Ramsar. 2017c. Servicio de información sobre Sitios Ramsar. https://rsis.ramsar.org/es/ris-search/?language=es. In Spanish.

Ramsar. 2017d. Servicio de información sobre Sitios Ramsar. https://rsis.ramsar.org/es/ris-search/?language=es&f[0]=regionCountry_es_ss%3AAmerica%20del%20Norte&f[1]=regionCountry_es_ss%3AM%C3%A9xico. In Spanish.

Ramsar. 2017e. The List of Wetlands of International Importance (The Ramsar List). Ramsar. http://www.ramsar.org/sites/default/files/documents/library/sitelist.pdf.

Ramsar Convention Secretariat. 2016. An Introduction to the Ramsar Convention on Wetlands (previously The Ramsar Convention Manual), 7ª edition, vol. 1. Ramsar Convention Secretariat, Gland, Switzerland. http://www.ramsar.org/sites/default/files/documents/library/handbook1_5ed_introductiontoconvention_final_e.pdf.

SAGARPA. 2016. La agricultura en chinampas. Secretaría de Agricultura, Ganadería, Desarrollo Rural, Pesca y Alimentación. https://www.gob.mx/sagarpa/articulos/la-agricultura-en-chinampas?idiom=es. In Spanish.

Sahagún, B. 2017. Historia general de las cosas de Nueva España. Written between 1540 and 1585. Also known as Códice Florentino (its original is in the Biblioteca Medicea Laurenciana, Florence, Italy). https://www.wdl.org/es/item/10096/. In Spanish.

Sandoval, J. 2013. Evaluación de la calidad del agua en los canales de Xochimilco para su recuperación ecológica. Master Thesis, UNAM, Ciudad de México, México. In Spanish.

Santoyo-Villa, E., E. Ovando Shelley, F. Mooser and E. León Plata. 2005. Síntesis geotécnica de la cuenca del valle de México. TGC Enterprise. Mexico City, Mexico. In Spanish.

Secretaría de la Convención de Ramsar. 2013. Manual de la Convención de Ramsar: Guía a la Convención sobre los Humedales (Ramsar, Irán, 1971), 6ª edición. Secretaría de la Convención de Ramsar, Gland, Suiza. http://ramsar.conanp.gob.mx/descargas/ManualConvenci%C3%B3nRamsar_6aEdicion.pdf. In Spanish.

Semeniuk, C.A. and V. Semeniuk. 1995. A geomorphic approach to global classification for inland wetland. Vegetation 118: 103–124.

SIAP. 2017. Anuario Estadístico de la Producción Agrícola. Servicio de Información Agroalimentaria y Pesquera. http://infosiap.siap.gob.mx/aagricola_siap_gb/icultivo/index.jsp. In Spanish.

Starkl, M., I. Bisschops, L. Essl, E. López, J.L. Martínez, D. Murillo et al. 2013. Opportunities and constraints for resource efficient environmental management in rapidly developing urban áreas: The example of Mexico City. Aquat. Procedia 1: 100–119.

UNESCO. 1988. Convention concerning the protection of the world cultural and natural heritage. Report of the world heritage committee. United Nations Educational, Scientific and Cultural Organization. Paris, France.

Valek, G. 2000. Agua: reflejo de un valle en el tiempo. Offset, S.A. de C.V. Mexico. In Spanish.

Zambrano, L., V. Contreras, M. Mazari-Hiriart and A.E. Zarco-Arista. 2009. Spatial heterogeneity of water quality in a highly degraded tropical freshwater ecosystem. Environ. Manage. 43: 249–263.

2

Rhizospheric Processes for Water Treatment

Background Principles, Existing Technology, and Future Use

Uwe Kappelmeyer[1], and *Lara A. Aylward[2]*

INTRODUCTION

Access to clean water and proper sanitation is considered a basic human right, as addressed in the United Nations (UN) resolution 64/292. There are a number of ways to fulfill this goal, but the focus here will be the implementation of green technology for the treatment of wastewater prior to its re-use or release into the natural ecosystem or municipal distribution network. The importance of this in terms of the protection of natural water resources and human health cannot be over-stressed.

Currently applied technologies range from highly sophisticated wastewater treatment plants with multiple technological steps to simple, lower cost, less technologically advanced methods utilizing natural filtration and purification processes. Large-scale treatment plants could rely on, for example, one or a combination of the following technologies:

- the activated sludge process

- oxidation processes (e.g., Fenton's process, ozonation, hydrogen peroxide treatment, advanced hybrid oxidation processes) (Gogate et al. 2004a,b)

[1] Group for Ecological Water Treatment Technologies, (EWaTT) – Constructed Wetlands, Department of Environmental Biotechnology, Helmholtz Centre for Environmental Research – UFZ, Permoserstrasse 15, 04318 Leipzig, Germany.

[2] Industrial and Mining Water Research Unit (IMWaRU), School of Chemical and Metallurgical Engineering & Centre in Water Research and Development (CiWaRD), University of the Witwatersrand, 1 Jan Smuts Avenue, Braamfontein, 2000, Johannesburg, South Africa.
Email: aylward.lara@gmail.com

*Corresponding author: uwe.kappelmeyer@ufz.de

- electrochemical and electro-oxidative processes (Chen 2004)
- membrane filtration

In contrast, reliable green technologies could include:

- wastewater ponds (Park and Craggs 2010; Rawat et al. 2011)
- slow sand filters (Seelaus et al. 1986; Wohanka 1995)
- treatment or constructed wetlands (CWs)

All of the above listed green technologies are biological treatment systems because they rely primarily on rhizospheric processes and microbial activity for cleaning wastewater (Stottmeister et al. 2003).

The focus in this chapter will be on CWs, which are artificial or man-made systems. Their appeal, over natural wetland systems, is that they can be built to provide a much higher degree of control and also aid in ensuring that natural ecosystems will be protected. CWs are complex systems, in which specialized plants (helophytes) and microbial communities interact with each other, and the filter matrix (soil or gravel) if present. In these systems, wastewater is remediated via physical, chemical, and biological processes and these could include sedimentation, precipitation, adsorption onto the biofilm or wetland matrix surface, microbial transformations and uptake by the plant roots (followed by phytovolatilization or internal metabolic processing) (Kumar and Zhao 2011). As wastewater passes through the wetland and the root-zone, permanent changes occur in local environmental conditions and these largely govern the contaminant degradation and transformation mechanisms. The treatment process is controlled by parameters such as wastewater composition, light intensity, and various other climatic factors.

In overview, this chapter provides information on the use of planted systems to treat various classes of wastewater. It begins with a definition and the general classification of CWs, before moving on to a description of the three key elements defining these systems and a discussion around the fundamental principles of wetland operation. Following on from this, the close link between wetland design and modeling is highlighted, before looking at the current, and widely implemented, modeling techniques and design recommendations. From this, the prevalence of the black-box approach to wetland design will be evident to the reader and this leads well into a description of a new methodology for CW experimental research. The motivation for this proposed methodology is two-fold. Firstly, it is a way of understanding the processes occurring within the CW black-box and, secondly, is a useful tool for the development of design guidelines which show lower reliance on local wetland conditions and would be more widely, or generically, applicable and more over, it may also permit the development of mechanistic models in the field of CWs.

Constructed or Artificial Wetlands for Wastewater Treatment

CWs or AWs are engineered systems, designed to utilize soil, vegetative, and microbial processes typically occurring in natural wetlands (Vymazal 2005). As such, they typically consist of a soil, gravel, or sand-filled bed planted with various types of macrophyte (Brix 1994). CWs are a widely accepted wastewater cleaning system and can handle inflows of different origins; including domestic and municipal sewage, industrial and agricultural wastewater, landfill leachate, acid mine drainage (AMD), storm-water run-off, and mining wastewater (Reed 1993). CWs are characterized by complex interactions between the three main elements of which they are composed: The plants (and the plant roots in particular), the soil/gravel matrix, and the microbial community. It is easy to understand that just the microbial community itself shows complex interactions because it is composed of different species all involved in manifold biochemical transformations. To begin, it is important to classify the planted systems suited to wastewater treatment.

Classification of Constructed Wetland Systems

Before the operational characteristics of these systems are discussed in more detail, it is useful to begin with an overview, or classification, of the different planted system configurations currently in use. A diagram of this general classification is provided in Fig. 1. As shown, CWs can be divided into two main classes: the floating plant mat system (Fig. 1A) and the matrix-based system. After this general classification, matrix-based CWs are further described by their vegetation. The wetland plants may be either completely submerged (Fig. 1B) or rooted beneath the water surface with the stems and leaves extending above it (Fig. 1C). Rooted systems are then defined by one of two principal types of flow; namely horizontal flow (HF) (Fig. 1D) and vertical flow (VF) (Fig. 1E). Finally, HF wetlands may be characterized by either surface flow (SF) or sub-surface flow (SSF). The CW classification will now be discussed in more detail, starting with floating plant mat systems.

Floating Plant Mat Wetland Systems

Floating plant mat wetlands (Fig. 1A) are a special case of CW. They do not contain a permanent supporting matrix and can be viewed as a further development of the existing wastewater pond systems. As the filter matrix is absent, the root surfaces provide the only surface for biofilm development. There are various advantages and disadvantages associated with these types of wetlands. The drawbacks include:

- A longer start-up phase while the root system matures or having to buy an already developed root mat.
- Flow irregularities and the development of preferential flow paths, particularly during the start-up phase, because the growing root zone provides a higher resistance to flow as compared to the free water zones. As a consequence, treatment efficiency is impaired.

On the other hand, the benefits of these systems include:

- Reduced construction costs because of the elimination of the soil/gravel matrix.
- Based on the initial investigations, improved treatment performance for special types of wastewater such as AMD (Richter et al. 2016).

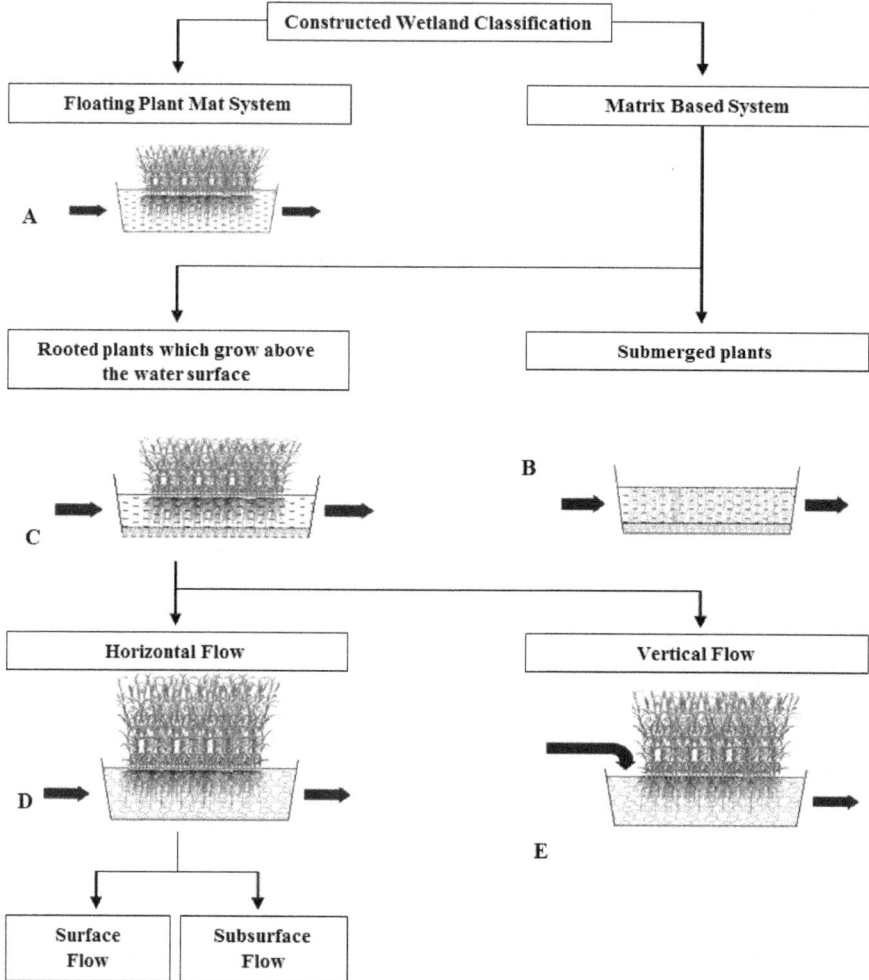

Fig. 1. Characterization of planted system technologies for wastewater treatment (the arrows in images A–E show the location of the inlet and outlet streams and the primary direction of flow).

Research into floating plant mat systems is still in the pilot stage and basic research questions regarding the influence of climatic conditions, system gradients, and the scale-up and implementation process are being addressed. The experimental and descriptive phase will finally end in the design of large scale treatment wetlands for ground water contamination.

Matrix-based Wetland Systems

In contrast, CWs may be constructed with a supporting matrix of soil, sand, or gravel which provides support to the plants, increased surface area for biofilm attachment and growth, and acts as a type of filter due to its porous nature. The first category of matrix system is the submerged CW in which the plants grow below the water surface entirely (Fig. 1B). In these systems, the stems and leaves provide additional surface area for biofilm growth, but the water can prevent penetration of sunlight and reduce the level of photosynthetic activity. Hence, the water must be shallow and pre-treated so that it is as clear as possible. Submerged systems are generally only used in special cases. Far more commonly implemented are the rooted or emergent CWs where the plants are rooted below the water surface in the soil/gravel and the stems and leaves extend above the surface (Brix 1994). One of the longest-established and most widely used is the horizontal flow constructed wetland HF CW (Fig. 1D). In contrast, vertical flow constructed wetlands VFCWs may also be chosen (Fig. 1E). The primary direction of flow—either vertical or horizontal—is a major characterizing feature of CWs and is the result of the inflow and outflow installations, as also demonstrated in Fig. 1. When the inflow is positioned above the surface of the supporting matrix, the result is a surface flow constructed wetland SFCW. On the other hand, when the inflow is positioned below the upper layer of supporting matrix surface, the CW will demonstrate sub-surface flow.

HF CWs are easy to apply and, on average, provide adequate treatment capacity. These systems are particularly effective for organics and suspended solids removal, but nutrient removal is generally low. For example, complete nitrification is poorly supported by HF systems because of their limited oxygen transfer capacity and phosphorus removal is restricted by the sorption capacity of the supporting matrix material. VF CWs show superior performance when nitrification or any other aerobic processes are favored (Vymazal 2005).

Hybrid Wetland Systems

It became evident that single-stage CW systems could not treat the increasing diversity of wastewaters to the desired level. This lead to the development of hybrid CWs. Hybrid CWs consist of two or more types of CW, connected together in various configurations of series and parallel units, to better utilize the treatment capabilities of the individual systems. For example, nitrogen removal efficiency may be poor in a single-stage HSSF CW, but in combination with a VF wetland, overall nitrogen efficiency can be improved (Vymazal 2005).

A more detailed discussion of the properties and operational characteristics of the main elements of a CW system, namely the matrix (for matrix-based systems), the vegetation, and the microbial community are presented in the following section.

The Three Elements of Constructed Wetland Systems

Wetland Support Matrix

The gravel, soil, or sand matrix is a characterizing feature of a CW, whose primary function is to support and stabilize the plant root systems. The root zone can be assumed to be the microbially-active region of the CW where the majority of contaminant transformation processes take place. Various studies have reported a root depth in the range of 50–60 cm. Furthermore, it provides an additional surface for hosting properties different to those inhabiting the plant root surface. For example, it might support a biofilm of different microbial composition as compared to the plant roots.

The wetland matrix has a characteristic sorption capacity, based on the specific material type and composition, and plays an important role in the nutrient buffering process as a result of sorption and desorption activity. More advanced CW design guidelines will recommend coating the matrix with a layer of material having enhanced adsorption and binding properties. This is often done to improve the removal of phosphorus from the wastewater stream (Molle et al. 2005). Reviews of phosphorus removal in CWs reported wide ranges of phosphorus sorption capacities up to 420 g P/kg of matrix material (Vianna et al. 2016; Vohla et al. 2011). With this, the versatility of CWs is demonstrated because their treatment potential spans from one of the simplest technologies all the way to specifically modified units capable of, for instance, removing increased quantities of phosphorous from livestock farming wastewaters.

From a design perspective, the matrix material is chosen to achieve a defined hydraulic residence time (HRT) and hydraulic conductivity. The hydraulic conductivity is a result of the matrix characteristics like porosity and particle size distribution. These are of particular importance within wetland systems because they describe flow behavior since treatment efficiency and the extent of pollutant removal are greatly impacted by how and how quickly water moves through the wetland. HRT time describes the average length of time that water spends in the wetland. The biogeochemical reactions responsible for contaminant removal require sufficient contact time between the pollutant species and the micro-organisms in the root zone, while sedimentation processes are closely linked to fluid velocity (Headley and Kadlec 2007). Related to flow behavior is hydraulic conductivity, which describes the ease with which water moves through the pore spaces of the soil/gravel matrix. It is recommended that porous media with a high hydraulic conductivity are chosen for the wetland substrate because poor material permeability results in clogging and greatly reduces hydraulic efficiency (Vymazal 2005). It has been found that the hydraulic conductivity of an established CW stabilizes in the range of 10^{-5} to 10^{-6} m s^{-1} during normal operation (Brix 1997).

Wetland Vegetation

To better understand the complexity of CW systems, a closer look at the wetland vegetation will be given. The root systems of the wetland plants are supported by the soil or gravel matrix. The most commonly used plant type are the helophytes, which are well adapted to survive under waterlogged and anaerobic conditions and are resistant to a reduced reduction-oxidation or redox environment, which can be caused by toxic components, such as hydrogen sulfide gas (H_2S).

The helophyte's resistance to reduced conditions can be explained by their anatomic and functional specialization. Helophytic plants have developed a specialized gas transport tissue, the aerenchyma, to transport oxygen from the leaves down to the root systems. In the root system, the aerenchyma also controls the partial release of oxygen into the surrounding area (Armstrong 1969). This is one of the major driving forces for the oxygen input into the constructed wetland soils. The result is an oxidized layer at the outer root walls and, overall, an oxidized region in the immediate vicinity of the root zone (Armstrong et al. 2000). The diffusion of oxygen out of the roots, and the resulting expansion of the oxidized layer, is dependent on the plant species and specific oxygen release rate; light conditions and the level of photosynthetic activity; temperature; redox conditions, and the amount of reduced elements in the surrounding (Jespersen et al. 1998; Wießner et al. 2002a,b, 2005).

The dependency of the rate of oxygen release into the root zone on the plant species gives rise to the question of plant selection for CWs. One can, of course, select plants with a high oxygen release rate like *Typha latifolia* or *Phragmites australis* (Wießner et al. 2002b) but a usefull selection of plant species is much more complex. Plant selection can also be guided by a need to create additional income for the local community, in which case ornamental plants like *Iris pseudocarus, Tulbaghia violacea, Cyperus papyrus*, etc., could be chosen. Otherwise, special adaptations, such as heavy-metal resistance in wetlands treating AMD, would mean that species from the family of metallophytes would be a more appropriate selection (Ali et al. 2013). Another point to consider during the selection of plant species is the amount of root exudates (Mucha et al. 2005; Koelbener et al. 2010). Various treatment processes, like denitrification and the treatment of AMD, are reliant on additional carbon. During the AMD treatment process, external carbon sources are usually added to initiate sulfate reduction and sulfide generation. Sulfide production is essential for heavy metal precipitation. A suitable choice of plant species could minimize the external carbon requirement because it could be substituted by root exudates (Richter et al. 2016).

In summary, although plant selection is a sensitive topic with regards to specialized applications, for many CWs treating municipal wastewater, the type of helophyte chosen is actually of minor importance. Planting a diversity of helophytes (including ornamental plants), rather than a single species, might also be favored to improve remediation and wetland value.

Wetland Microbial Community

Water passes through different redox-conditions as it flows through the wetland. The changes in the redox conditions are on the micro-scale. As explained in the previous section, the root surface and its close surroundings are in a more oxidized state than the regions with fewer or no roots. The co-localization of oxic–anoxic conditions in close proximity to the plant roots, due largely to the ability of macrophytes to transport oxygen to the rhizosphere, and is responsible for the distribution of the microbial community. More specifically, it allows for both aerobic and anaerobic communities to co-habit this narrow region (Bodelier and Dedysh 2013). If it were not for this phenomenon, aerobic microbial populations could not survive in an otherwise anaerobic environment.

Microbial communities are found within a specialized support matrix known as a biofilm. Biofilms are complex layers of microbial consortia surrounded by a self-produced, extracellular polymeric substance. It is valid to assume that aerobic microbial species, like nitrifying bacteria, will form a biofilm on the root-surface where more oxygen is available. Anaerobic microbial species, such as sulfate reducing bacteria, will more likely form a biofilm on the gravel at some distance away from the root surface or on older, lignified sections out of which oxygen cannot leak. In their 2002 investigation, Reddy et al. showed that aerobic microbial populations are found primarily in the water column, the outer regions of the plant detritus layer, and the periphyton mats, while anaerobic populations are found in biofilms on the soil/gravel particle surfaces and in the soil interstices. Furthermore, they determined that anaerobic populations are dominant in the subsurface region of the wetland (Reddy et al. 2002).

Microorganisms are the key role-players in CW functionality. In general, the microbial consortia show spatial heterogeneity even within small distances and the community structure and function indicate a strong correlation with electron acceptor availability. For example, the abundance of methanogens is negatively correlated with the abundance of nitrate- and sulfate-reducing bacteria and methane production is highest with low electron acceptor availability (He et al. 2015). The amount of microbes and the community structure are dependent on the type of wetland. This has been validated by microbial investigations carried out in operational CW systems (Ibekwe et al. 2003) and in wetland systems during restoration (He et al. 2015). For this reason, it is recommended that sufficient time is allowed for the start-up development and adaptation of both the plant stock and the microbial community.

Design, Modeling, and Usage of Treatment Constructed Wetlands

The design of CWs is based on multiple factors, including:

- National, regional, and local water quality legislation
- Source, type, and composition of wastewater
- Local climatic conditions
- Land area available and surrounding infrastructure

Wetland modeling and wetland design are closely related. Wetland models are created and validated using experimental data and used to optimize design criteria (Kumar and Zhao 2011). This process is often cyclical or iterative; requiring modeling to support design recommendations and then designed systems to be operational in order to confirm model reliability.

Several design criteria are derived from a statistical analysis of existing CW performance data combined with model experiments (Wu et al. 2014). Statistic models only allow for predictions in narrow boundaries. It is obvious that predictive results outside of the experimental data range are of questionable quality because the model would not be statistically validated.

The design strategy of the future would involve the application of more mechanistic, rather than static, models. Currently, there are a few studies outlining mechanistic modeling approaches (Langergraber 2008). For instance, Mitchel and McNevin (2001) describe a mechanistic model for BOD removal in SSF CWs. The model uses the BOD Monod removal kinetics and is applicable for the calculation of the minimal area of a CW to avoid under-sizing.

Target Contaminants

One of the primary objectives of a CW is the removal of target components, which could include biochemical oxygen demand (BOD), chemical oxygen demand (COD), nitrogenous compounds, and phosphorous-containing species, amongst others, depending on the origin of the wastewater. The extent of removal required is often defined by legal limits for discharge, such as the current German regulations for waste water (AbwV 2004). With national standards as a guideline, local water authorities will define treatment targets based on both legislation and the fate of the treated water. In broad terms, the fate of the remediated water could either be re-use or release into natural water bodies. Water re-use, in general, requires further treatment, such as reduction of the pathogens as well as general water sanitation, because of human health implications. The reaction rate constants for each individual process and the percentage contaminant removal are important experimental parameters used in the CW sizing and the prediction of the extents of pollutant removal in these systems (Kumar and Zhao 2011).

Nutrient Removal

The removal of nutrients from water systems is a growing global concern. Nitrogen and phosphorus are nutrients essential to plant health, but high concentrations of these species in natural water bodies lead to excessive plant growth and, eventually, eutrophication. High levels nitrogen and phosphorus in natural water bodies are often the consequence of the release of agricultural run-off, and untreated food processing wastewater and large quantities of nitrogenous species, in particular, are by-products of food manufacturing (Erisman et al. 2013). Hence, there is a strong need for nitrogen removal to avoid coastal and terrestrial eutrophication and nitrous oxide emission.

Nitrogen is removed from a water body via biological processes and is released as gaseous nitrogen. In contrast, phosphorus (and in particular phosphate) is eliminated via physical processes and focus is directed toward its removal via simple, green technological approaches. Nutrient elimination is also of importance in the case of water re-use in arid regions. Nitrogen and phosphorus are often added as fertilizers to improve crop growth. Therefore, nutrient removal from recycled agricultural run-off water is unnecessary if the water will be used for further irrigation of the field. However, seeing as the presence of other contaminants (pesticides, herbicides, etc.) is undesirable, a monitoring and water cycle modeling strategy would be the alternative. CWs are applicable in this filed and might also be used in a cascade with other low-tech systems of proven efficiency. In any event, a predictive model is needed to show the effect of combined technologies on overall nutrient removal. This is essential for nitrogen transformation and elimination because the multi-step microbial processes underlying nitrogen removal mean that a simple first-order or Monod kinetic model is of limited use (Gerke et al. 2001).

Optimization of Contact Time in the Root Zone

In principle, CW design is based on the optimization of contact time in the root zone. In other words, the HRT should to be long enough so as to allow for intensive contact between all wetland elements and the target components in the wastewater. The essential contact time is based on removal kinetics, which are specific to the particular component, climatic conditions, and the type of wetland.

Constructed Wetland Modeling

There are several papers describing various modeling approaches aimed at better understanding wastewater treatment in wetlands. These are detailed in the review of Kumar and Zhao (2011). Existing models range from simple, first order to dynamic, compartmental models of varying degrees of complexity.

There are two classes of wetland models; namely "Black Box" and mechanistic models. Black-box model are based on the systems theoretically approach of simplification. The simplification based on neglect the internal interaction of the appearing processes. This results in a system description only based on the measured in- and outflow values.

Black-box models, such as the *simple regression, first-order kinetic, retardation, Tanks-In-Series, Monod-kinetic*, etc., are created using only wetland input and output data. The most widely used CW modeling technique is still the first-order degradation model (Kadlec and Wallace 2009). However, it is known that the microbial transformation processes within biological systems, such as CWs, are better described by *Monod kinetics* and the Monod model (Mitchell and McNevin 2001). For many applications, these simple black-box models are a reasonable approximation of system performance.

For more detailed descriptions of CW behavior and, in particular internal behavior, sophisticated numerical models have been developed. In addition to accounting for inlet and outlet compositions, these mechanistic models simulate various internal wetland phenomena. This could include reactive transport through porous media, transformation and degradation of organic matter and nutrients, adsorption/desorption, clogging, and biomass growth to name a few. The mechanistic models make use of specialized software packages, such as *HYDRUS, STELLA, PHWAT,* or *RetrascoCodeBright*. Mechanistic CW modeling may still be considered as in its early stages (Kumar and Zhao 2011).

Recommendations for Constructed Wetland Design

Crites (1994) described the straight forward design criteria for SF and SSF CWs. The following design criteria have been recommended for SF systems:

- Hydraulic residence time: 5 to 14 days
- Hydraulic loading rate: 7 to 60 mm d^{-1}
- Maximum BOD loading rate: of 80 kg ha^{-1} d^{-1}
- Water depth: 0.1 to 0.5 m

One of the disadvantages of SF systems, particularly in the tropic and sub-tropic regions, is that the free water surface can become a breeding ground for mosquitos. A system without a free water surface and the potential for stagnating paddles is the best way to control mosquito populations and prevent spreading of disease. In this case, SSF CWs would be preferable. SSF systems have the further advantage of a remarkably shorter hydraulic residence time (2 to 7 days) and provide greater removal efficiency (Crites 1994). The simple explanation for this higher efficiency is more intensive contact between the microbial biofilm, the plant roots, and the entire wastewater stream. To illustrate additional aspects of CW wetland design and give some examples of CWs currently in use, some case studies from South America, Europe, and Asia are briefly described in the following sections.

Global Case Studies

Brazil. The use of a hybrid CW for the treatment of grey water in Brazil was described by Paulo et al. (2009). The hybrid system consisted of a horizontal and a vertical flow CW. Due to the particular type of grey water inflow, the system experienced a higher than usual hydraulic loading rate of 152 mm d^{-1}. This resulted in a low hydraulic residence time of about 1 day, in contrast to a full treatment system which would receive municipal wastewater and have a typical hydraulic residence time of about 7 days. A septic tank was installed for pre-treatment of the water before it entered the hybrid CW. As a result, no clogging issues were reported (Paulo et al. 2009). In general, first-stage septic tanks or other types of sedimentation units are strongly recommended in many sets of design guidelines.

France. "French systems" are two-stage VF CW systems, which receive raw sewage to the first stage without any form of pre-treatment, such as a septic or sedimentation tank. Molle et al. (2005) conducted a survey on the performance of these VF systems. At the time of the study, about 300 treatment plants servicing small communities in France (65% < 300 PE) were operating VF CWs of various sizes. The survey showed that this type of system is highly efficient in meeting French wastewater standards, despite variations in hydraulic and organic loads, and a fairly small area per person equivalent (PE) is required. Specifically, this is 1.2 m^2 PE^{-1} for the first stage and 0.8 m^2 PE^{-1} for the second. This is a relatively small per capita area due the tropical climate and elevated temperature gradient. A layer of sludge develops on the first stage as a result of the lack of pre-treatment. This layer greatly helps with feed water distribution, but must be removed every 10–15 years once is has reached 20 cm in depth (Molle et al. 2005).

Germany. The German CW guidelines define a specific area per inhabitant, or person equivalent, of 3–10 m^2 depending on the type of CW and the temperature conditions (DWA 2014). As seen in the previous section, VF CWs have a smaller area but greater technical infrastructure requirement than HF CWs. CWs operating in warm climates (temperature > 20°C) require a smaller land area than those in colder climates (temperature < 10°C). The German guidelines also specify the type of filter sand to be used and it is obligatory to pass the wastewater through a sedimentation tank prior to entry into the wetland system.

Denmark. The new Danish guidelines suggest VF CWs with a specific area of 3.2 m²
PE^{-1}, an effective filter depth of 1 m and a filter sand composition of d$_{10}$ 0.25–1.2 mm
and d$_{60}$ 1–4 mm. In addition, the uniformity coefficient (U = d$_{60}$/d$_{10}$) should be less than
3.5 (Brix and Arias 2005).

Nepal. VF systems have also been found to treat municipal wastewater more
efficiently than HF systems under the specific climatic conditions in Nepal (Shrestha
et al. 2001). This observation is supported by published studies and can be explained
by the general loading regime (Cooper 1999; Brix and Arias 2005). HF CWs usually
receive a steady water load and the wetland is usually filled with water. Flow is also
typically in the laminar range (supported by low flow velocities). As a result, diffusion
of oxygen through the surface is minimal in comparison to the oxygen released by the
plants. Hence, HF CWs support more anaerobic processes. In contrast, the water load
to a VF CW is distributed over the entire bed surface and permeates the bed under
the influence of gravity. Each feed in interval is followed by a drying cycle, which
sucks oxygen down into the filter and ensures good aeration within the system. The
alternate feed and drying cycles ensure good mixing of the fresh wastewater intensive
with the residual oxygen in the filter. Hence, VF systems favor aerobic transformation
processes, such as nitrification (Wu et al. 2014). As already alluded to, VF CWs may
have a lower area requirement but require a buffering tank, dosage pumps and control
systems to regulate the interval wastewater dosage.

These successful case studies highlight the great potential of CWs in treating
wastewaters of different origins. For example, domestic sewage (Kivaisi 2001; García
et al. 2010); industrial effluent (Calheiros et al. 2012); agricultural effluent (Millhollon
et al. 2009) and highway and stormwater run-off (Isteniç et al. 2012; Scholes et al.
1999). It should be noted that special design criteria are necessary while targeting
urban run-off and stormwater treatment wetlands because of the potential for large
surges in the volume of water that received by these wetlands in a short period of time
(Castelle et al. 1994; Shutes et al. 1997; Tshrintzis and Hamid 1997).

The Inadequacy of the "Black-box" Model

Many of the published case studies investigating full-scale treatment CWs describe
the systems by considering only the inflow and outflow concentrations of the target
components. This knowledge is then widely used to develop design guidelines for
both the construction and operation of CWs (Kadlec and Wallace 2009; DWA 2014;
Dotro et al. 2017). Moreover, it is also evident that there are many sets of local design
guidelines in the literature. Due to the strong influence of local environmental and
climatic conditions on CW performance, it has not been possible to generalize the data
derived from model experiments.

A closer analysis of the above cited and discussed papers, particularly with regards
to design guidelines, highlights the lack of internal process information gathered and
used for CW design. Solving the black-box dilemma is not a trivial task for a number
of reasons. The complexity of the processes taking place inside CWs is a large factor;
indicated the general lack of data showing and finding an explanation for the internal
behavior of CWs. In the past, it has been considered adequate to understand only
the overall system performance and know that wastewater fed into CW systems
will be treated to ensure compliance with local regulations. Now, the reason for (i)

the collection of just inflow and outflow data, (ii) the popularity of the black-box description, and (iii) the wide implementation of simple statistical performance analysis becomes clear.

It is clear that resource and time-intensive, active experiments involving the changing of various parameters are the only way to improve understanding of CW processes. However, it is also evident that these could be considered redundant when the primary function of a treatment wetland is cleaning wastewater and the concern is merely reaching a specified water quality. Here, solely descriptive experiments (collection of inflow and outflow data) are adequate. However, unless internal information is obtained, operators can never know whether the wetland is operating most efficiently and whether small adjustments could produce much improved performance.

Inside the "Black-box"

Designing active experiments around the components of interest, with the aim of developing a deeper process understanding, would be really valuable to investigate the ability of CWs to handle the increasingly diverse wastewaters. Before proceeding with this idea, it is acknowledged that it would be impractical to use full-scale CWs for active experiments, such as adding various components to describe maximum transformation activity or to reach plant toxicity levels. Nevertheless, sophisticated analytical tools producing negligible system disturbance are available and could be applicable.

The use of stable isotope probing is one of these monitoring tools. This technique traces the *in situ* flux of nutrients in biochemical cycles. The heavier stable isotopes (^2H, ^{13}C, or ^{15}N) can be enriched in the wastewater stream due to the preferential use of the lighter stable isotopes (^1H, ^{12}C, or ^{14}N) in bacterial transformations (Radajewski et al. 2000). The calculated enrichment factors can be linked with data from microbial transformations, in pure culture and under defined conditions, with known pathways. Therefore, it is possible to predict the primary degradation or transformation pathway. Vogt et al. (2008) combined carbon and hydrogen isotope fractionation to show toluene transformation under different aerobic and anaerobic pathways.

A conflict of sorts will always exist between descriptive experiments (which facilitate a better description of the overall wetland performance) versus active experiments (which facilitate a better understanding of the internal processes in full-scale CWs). To lend justification to the growing importance of the latter, a methodology will be proposed. This methodology will give specific design guidelines for CWs to meet local boundary conditions.

Future Perspectives in Constructed Wetland Research and Development

A New Methodology for Active Constructed Wetland Experiments

Laboratory-based experiments are the best, and most widely accepted, way to develop deeper process understanding. Model wetland systems are well suited to active experiments and can be used to describe transformations under multiple redox conditions, as shown in the papers of Wießner et al. (2013) and Wu et al. (2013a). The design of an ideal model system should be comparable with that of the particular full-scale CW under consideration, but with reduced size and complexity. The most important real-system parameter to replicate is the flow behavior and, in particular, the HRT of the full-scale wetland must be simulated in the model system. Tracer flow tests and tracer distribution monitoring are recommended to achieve this objective (Giraldi et al. 2010; Bonner et al. 2017a,b). Prior to commissioning for remediative purposes, full-scale treatment wetlands are investigated by tracer tests in order to identify flow irregularities, such as dead-zones and preferential flow (Levenspiel 1998). This is important because flow patterns greatly influence overall system performance.

There are numerous experimental set-up possibilities in the laboratory environment; spanning from simple single plant studies to advanced reactor-like arrangements. For this methodology, the objective is to implement a biotechnologically-driven scale-up procedure. Specifically, ideal flow behavior in the model system will be used as the control parameter because flow through porous media (particularly under laminar flow conditions as are often maintained in full-scale systems) is a well-described physical process. It is known that small units need careful attention during the design process to obtain an ideal flow pattern, but the attainment of a steady condition is not beyond the realm of possibility. The basis of the proposed methodology is the laboratory-scale "Planted Fixed Bed Reactor" (PFR). This model wetland system can be used to collect information about microbial processes, transformation kinetics, and phytovolatilization; amongst other processes.

In the proposed experimental methodology, three different small-scale systems are recommended; namely two idealized PFRs (Fig. 2) (Kappelmeyer et al. 2002; Wießner et al. 2005; Lunsmann et al. 2016a,b); a laboratory-scale gravel bed CW (Fig. 4) (Gonzalias et al. 2007; Wießner et al. 2010) and an outside pilot-scale wetland photographed in Fig. 5.

The Planted Fixed Bed Reactor

A photograph of the model PFR system is given in Fig. 2. The PFR is a finite volume element of a CW. It is composed of a 22L glass vessel into which a gravel-filled cage has been installed. In the centre of the gravel bed, there is a hollow, narrow flow cylinder. The vegetation is planted within the gravel. The glass vessel is covered by a custom-made teflon lid containing five opening through which the plants can emerge. The PFR has been designed for the following conditions:

- The fluid within the reactor is completely mixed and general flow behavior is comparable with that in a continuously stirred tank reactor (CSTR).
- A HRT similar to that of a typical full-scale treatment wetland can be attained.
- The wetland microcosm (Fig. 2a) is large enough to hold more than one plant, so as to overcome any single plant influence on the experiment.
- Circulation flow is achieved by pumping water out of the central cylinder using a peristaltic pump and dosing it back into the outer space between the glass vessel wall and the gravel cage. A schematic drawing is given in Fig. 2b for better visualization. The inflow is well-mixed and dosed, together with the circulation stream, into the outer ring. The PFR is also equipped with a level control unit.

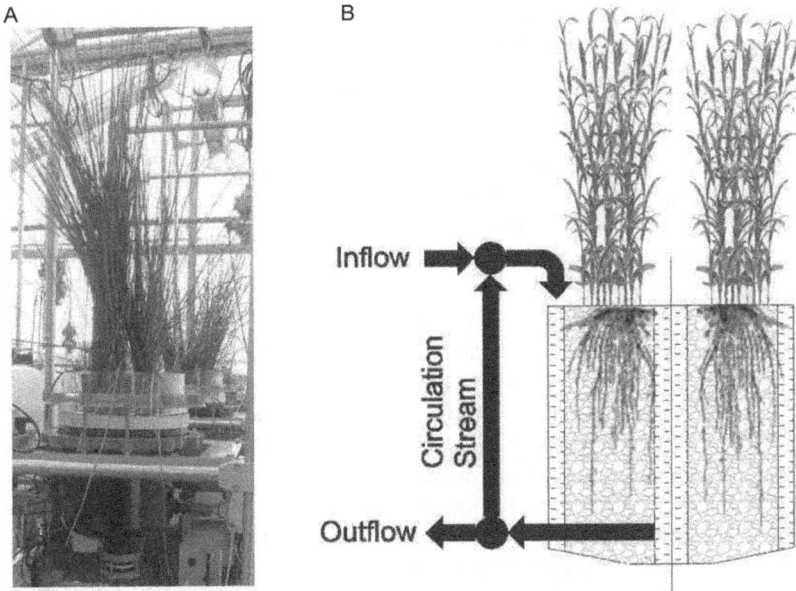

Fig. 2. The laboratory Planted Fixed Bed Reactor system (a) and a schematic diagram showing the flow regime inside the Planted Fixed Bed Reactor (b).

As shown in Fig. 2b, the circulation flow outside of the PFR passes via a pH-, redox-, and oxygen electrode and readings can be recorded at user-specified intervals. The advantage of the automated recording, internal mixing, and circulation flow system is that the day and night (diel *versus* diurnal) redox and oxygen conditions can be measured; an example of which is given in Fig. 3. Figure 3 indicates that oxygen release by the plant roots is the main oxygen source and, together with oxygen consumption owing to various microbial transformation processes, results in redox fluctuations—also detectable over larger distances (Fig. 3—redox oscillation dependent on the sunlight).

The PFR is an ideal experimental system for basic investigations into mass transfer and fluid flow pathways in the fields of microbiology, molecular biology, and biochemistry. It can also be used to give insight into the degradation and fate of new chemical species in wetlands. The PFR shows behavior similar to that in a full-scale CW experimental system, but its design eliminates macro gradients and, therefore, any spatial sampling dependency. This idealized behavior is realized by the mixing of the reactor liquid with a decoupled residence time.

Case Studies—Planted Fixed Bed Reactor

Determination of the Pathways for Microbial Toluene Degradation. The PFR was used in several case studies to investigate microbial transformations. In model PFR experiments using toluene as the sole additional external carbon source, microbial toluene degradation was investigated using DNA-based methods in combination with stable isotope fractionation analysis and characterization of toluene-degrading microbial isolates. Toluene was selected because the microbial transformation pathways and conditions for this hydrocarbon are well known. This study is a 'proof-of-principle' investigation to show the applicability of sophisticated analytical tools in understanding some of the internal dynamics of complex systems.

In the experimental period, which ran over more than two years, the overall redox conditions in two similar systems operated in parallel were hypoxic to anoxic. Phylogenetic analysis of 16S rRNA gene amplicons revealed the families *Xanthomonadaceae, Comamonadaceae,* and *Burkholderiaceae*; plus *Rhodospirillaceae* in just one of the reactor replicates. Stable isotope probing was successfully applied in this model toluene experiment. The investigation showed the dominance of mono-oxygenation under the reduced conditions in the PFR. The redox potential measured in the circulation flow reflects the average redox status of the whole reactor. As the roots are the main oxygen sources, particularly in this system, due to the limitation of other oxygen input paths, it could be concluded that the location of the observed aerobic transformations was the biofilm attached to the root system. Further metagenomic analysis (using a DNA microarray of catabolic potentials for aromatic compound degradation) suggested the presence of the ring mono-oxygenation pathway in both systems. The fact that the above described aerobic pathway are found in the hypoxic to anoxic environment of the PFR was a surprising result. These model experiments also lead to the detection of the anaerobic toluene

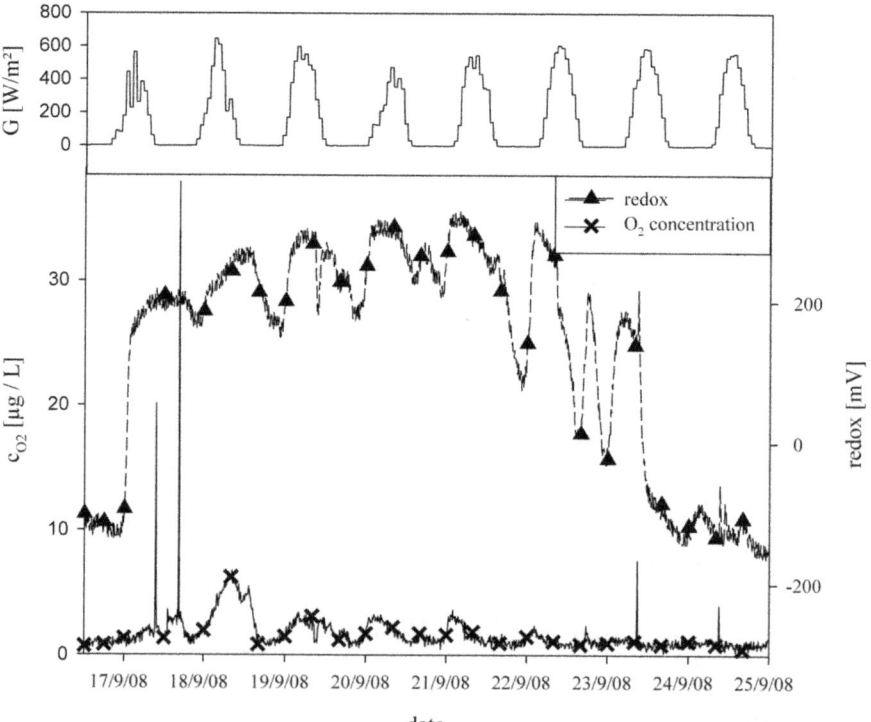

Fig. 3. Exemplary redox and dissolved oxygen values measured in the circulation flow of a Planted Fixed Bed Reactor.

pathway with a high abundance of *Rhodospirillaceae*. Mono-oxygenation and di-oxygenation were also confirmed by a stable isotope fractionation analysis which showed low-levels of carbon fractionation and minimal hydrogen fractionation in both PFRs. In summary, the combination of these different methods suggests that toluene degradation occurs predominantly via ring mono-oxygenation and acts as justification for the proposed methodology to be able to discover degradation pathways in a complex system (Martínez-Lavanchy et al. 2015).

Impact of Fluctuations in Root Exudation on Bacterial Communities. The execrated amount of oxygen, etc. may follow the day and night changes of the photosynthetic activity. Information about the influence of these chemical fluctuations on bacterial activity is scarce. As a follow-up study to the toluene degradation pathway description, trends in chemical fluctuations were investigated using metaproteomics. The diurnal expression patterns of genes involved in aerobic and anaerobic toluene degradation was investigated by quantitative real-time PCR. Thereafter, a stable daily and night time aerobic toluene turnover by *Burkholderiales* was observed. A day-night fluctuation was observed in polyhydroxyalkanoate synthesis was up regulated by these bacteria during the day. This suggests that the dominant bacterial community feed off organic root exudates during the day and re-use stored carbon compounds via the glyoxylate cycle during the night. Instrument detection limits and sample volume availability for the metaproteomic analysis hinders the proof of the relatively abundant mRNA copy numbers encoding the anaerobic enzyme benzyl succinate synthase (bssA) and its protein detection (Lunsmann et al. 2016b).

Application of PFR Principles in Full-scale Constructed Wetland Systems. Barreto et al. (2015) described the use of PFR principles in the field by installing this type of reactor type into a full-scale CW. Such a modification has been made to the pilot-scale horizontal subsurface flow (HSSF) CWs at the Helmholtz UFZ, Leipzig. Figure 4a is

Fig. 4. Pilot-scale constructed wetlands at the Helmholtz UFZ, Germany, showing (a) the locations of three Planted Fixed Bed reactor installations and (b) one of the Planted Fixed Bed reactor baskets when removed from the wetland showing a well-developed root system.

a photograph of the pilot wetland systems in use at the UFZ and Fig. 4b shows a PFR basket when it was removed from the wetland. The *in situ* PFRs provide an opportunity for active experiments, including plant toxicity level evaluation, in microcosms established within an operational full-scale system; assuming that the same microbial community will be present inside the PFR as in the surrounding CW. Further, if one uses the same inflow stream for both installations, experiments carried out in the PFR have essentially been performed under identical conditions to those experienced by the large CW. It is recommended that more than one PFR be installed at various distances along the flow path of the full-scale system (Fig. 4a) (Barreto et al. 2015). This is a perfect tool to investigate spatial and gradient variances in operational CWs.

Laboratory-Scale 1 m Constructed Wetland Systems

The next step in the proposed methodology involves increasing the complexity of the model system, but still under laboratory conditions. The new model system is a laboratory-scale HSSF wetland system (1 m x 0.2 m x 0.5 m) with a gravel bed height of 0.45 m, shown in Fig. 5. In contrast to the PFR, this system is marked by real macro gradients along the 1 m flow path. As the wastewater passes through the wetland, gradients in flow velocity can develop and the contaminants are transformed, resulting in the formation of concentration gradients. As in full-scale treatment CWs, the inflow is positioned at one end, the water runs through the gravel bed, and exits the system on the opposite end after a given residence time. The water level is adjustable by a siphon system installed at the system outlet on the left side of the 1 m CW in the picture.

Fig. 5. The 1 m laboratory constructed wetland systems.

Case Studies–1 m Model HSSF Constructed Wetlands

AMD Remediation. The 1 m laboratory scale wetlands are further applicable to preliminary testing of new wetland installations on a small scale and under controlled conditions. An AMD (acid mine drainage) remediation process was investigated in one of these 1 m model wetlands. The hypothesis was that plant root exudates would act as a carbon source for microbial sulfate reduction in the AMD wastewater which typically has a low-carbon concentration. It was shown that sulfate and metal removal from model AMD wastewater was the result of chemolithotrophic microbial sulfate reduction. The microbial transformation processes were supported by adding supplementary hydrogen to the model wetland via a bubble-free gas dosage system. In addition, the influence of rhizodeposition on chemo-organotrophic sulfate reduction was investigated. Microbially produced bicarbonate aided pH neutralization. The pH of the AMD wastewater was observed to rise from 3.3 to 8.2 (Richter et al. 2016).

Organic Carbon Degradation and Nitrification. In a separate case study, the removal kinetics of various organic components together with nitrification kinetics were studied in the 1 m model wetlands (Wu et al. 2013b).

These idealized and controlled laboratory set-ups have provided detailed information about transformation processes/pathways, reaction kinetics, and microbial community structures. The next step in the proposed methodology is scale-up to pilot wetland systems.

Pilot-Scale Constructed Wetland Systems

It is now necessary to gather information on a scale larger than what can be achieved in the laboratory, but still in a size that is practical for collecting relatively large data sets. For this, a pilot-scale CW is used. This type of system should be operating under natural climatic conditions for two reasons. Firstly, to validate the laboratory results and, secondly, to formulate more robust guidelines for CW design.

Case Studies–Pilot-Scale Constructed Wetlands

Fuel-grade Hydrocarbon Removal. Groundwater contaminated with benzene, toluene, ethylbenzene, xylene, BTEX, the fuel additive methyl *tert*-butyl ether, and ammonium was investigated in several pilot-scale CWs, including an unplanted gravel filter system and a hydroponic plant root mat system. Hydrodynamic behavior was evaluated by means of temporal moment analysis of outlet tracer breakthrough curves. The hydraulic indices were related to contaminant mass removal. This detailed knowledge of flow patterns within the pilot CWs allowed the estimation of local flow rates and contaminant loads. The highest mass removals were achieved by the plant root mat system at low water level with reductions of 98 percent (544 mg m^{-2} d^{-1}), 78 percent (54 mg m^{-2} d^{-1}), and 74 percent (893 mg m^{-2} d^{-1}) for benzene, MTBE, and ammonium-nitrogen, respectively. Within the CWs, flow behavior was depth-dependent and influenced most notably by the vegetation and the position of the outlet tube. Both an elevated flow rate and contaminant flux were observed immediately below the densely rooted, porous substrate region in the planted CW. The unplanted control system showed faster flow in the bottom region of the wetland (Seeger et al. 2013).

Chlorinated Hydrocarbon Removal. Low and high chlorinated hydrocarbon degradation performance in sulfate-rich groundwater in a pilot-scale CW planted with *Phragmites australis* was investigated. The aim was to explore the effect of the vegetation. The low chlorinated hydrocarbon, monochlorobenzene (MCB), was removed at a rate of about 208 mg m^{-2} d^{-1} and the high chlorinated hydrocarbon, perchloroethylene (PCE), was removed at a rate of about 49 mg m^{-2} d^{-1}. The removal rate of PCE was the same in both the planted and unplanted systems. MCB removal in the planted system, on the other hand, showed a large difference compared to that achieved in the unplanted system (58 m^{-2} d^{-1}). Due to the combination of the groundwater contaminants present and the plant derived organic carbon, sulfide formation of up to 15 mg/L in the planted wetland was observed. Aerobic MCB degrading bacteria benefited from the oxygen released by the plant roots and plant-derived organic carbon stimulated dissimilatory sulfate reduction, which had a negative effect on MCB removal. Hence, it is clear that the presence of the vegetation has a definite effect on the removal of MCB (Chen et al. 2012).

Pilot-scale CWs have also been used for an internal, spatial analysis of MCB and oxygen compositions during the groundwater treatment process. The upper layer was proven to be the active region for MCB removal in the investigated pilot-scale wetlands, with up to 77 percent removal efficiency. Oxygen concentrations were below 0.16 mg/L throughout the wetland transect. To avoid false interpretation, the low oxygen concentration doesn't mean there is now more oxygen available. The measured oxygen is the remaining oxygen from the dynamic process of oxygen release by the plants and oxygen consumption by aerobic microorganisms at the measured location. Stable isotope enrichment analysis of ^{13}C in the residual MCB fraction at increasing distance from the inflow indicated microbial MCB degradation throughout the wetland (Braeckevelt et al. 2007).

Nitrogen Transformations. The investigation of Coban et al. (2015) used ^{15}N isotopic techniques and molecular-biological methods to show that nitrogen transformation by the Anammox process is of minor importance in pilot-scale HSSF CWs treating ground water. This investigation was confirmed by a high throughput sequencing approach. The three systems investigated (a HSSF CW, an unplanted gravel bed, and a floating plant mat system) were sampled at three distances along the flow path. The planted and unplanted systems were further sampled at three depths for each distance. This approach shows that the diversity at family level, based on the Simpson Index, was on average 3 to 4 times higher for the planted wetland than for the unplanted wetland and the floating plant mat (Coban et al. 2015). This finding addresses the question regarding an evidence-based comparative system to ascertain the influence of the selected system physical properties on wetland treatment performance. In particular, the widely used unplanted CW is usable from a technological point of view, but it is doubtful as to whether this type of system is practical given the positive influence of the plants on wetland remediative processes.

Conclusions

This chapter presented exciting design guidelines; all of which are based on the black-box assumption. It has been shown that these general recommendations are specific, for instance, to the climatic conditions and type of wastewater to which the treatment wetland is exposed. To overcome these limitations and to develop more general guidelines, a new experimental methodology has been proposed and consists of three major parts:

1. Small-scale, tightly controlled laboratory experiments using specially designed equipment; namely the planted fixed bed reactor.

2. Additional laboratory experiments, in 1 m model CWs, in which real wetland conditions are more closely simulated.

3. Experiments run under minimal control in outdoor pilot-scale CWs which are exposed to real climatic conditions. These are for validation of the results of parts 1 and 2.

This new methodology provides a systematic approach to data collection and analysis for the compilation of improved constructed wetland models and design guidelines.

Acknowledgements

We would like to thank Dr. Otoniel Carranza-Díaz for the review of the chapter. Special thanks are directed to Dra. María del Carmen Durán-Domínguez-de-Bazúa for her wonderful editorial work during the finalizing of this chapter.

References Cited

AbwV. 2004. Ordinance on Requirements for the Discharge of Waste Water into Waters. BMU Germany [ed.]. Waste Water Ordinance. Federal Minister for the Environment. Federal Law Gazette I. Germany.

Ali, H., E. Khan and M.A. Sajad. 2013. Phytoremediation of heavy metals—Concepts and applications. Chemosphere 91(7): 869–881.

Armstrong, W. 1969. Rhizosphere oxidation in rice—An analysis of intervarietal differences in oxygen flux from roots. Physiol. Plantarum. 22(2): 296–303.

Armstrong, W., D. Cousins, J. Armstrong, D.W. Turner and P.M. Beckett. 2000. Oxygen distribution in wetland plant roots and permeability barriers to gas-exchange with the rhizosphere: A microelectrode and modelling study with *Phragmites australis*. Ann. Bot-London 86(3): 687–703.

Barreto, A.B., G.R. Vasconcellos, M. von Sperling, P. Kuschk, U. Kappelmeyer and J.L. Vasel. 2015. Field application of a planted fixed bed reactor (PFR) for support media and rhizosphere investigation using undisturbed samples from full-scale constructed wetlands. Water Sci. Technol. 72(4): 553–560. doi: 10.2166/wst.2015.238.

Bodelier, P.L.E. and S.N. Dedysh. 2013. Microbiology of wetlands. Front. Microbiol. 4: 79 (4 pages).

Bonner, R., L. Aylward, C. Harley, U. Kappelmeyer and C.M. Sheridan. 2017a. Heat as a hydraulic tracer for horizontal subsurface flow constructed wetlands. J. Water Proc. Eng. 16: 183–192.

Bonner, R., L. Aylward, U. Kappelmeyer and C.M. Sheridan. 2017b. A comparison of three different residence time distribution modelling methodologies for horizontal subsurface flow constructed wetlands. Ecol. Eng. 99: 99–113.

Braeckevelt, M., H. Rokadia, G. Mirschel, S. Weber, G. Imfeld, N. Stelzer et al. 2007. Biodegradation of chlorobenzene in a constructed wetland treating contaminated groundwater. Water Sci. Technol. 56(3): 57–62.

Brix, H. 1994. Use of constructed wetlands in water pollution control: Historical development, present status and future perspectives. Water Science and Technology 30(8): 209–223.

Brix, H. 1997. Do macrophytes play a role in constructed treatment wetlands? Water Sci. Technol. 35(5): 11–17.

Brix, H. and C.A. Arias. 2005. The use of vertical flow constructed wetlands for on-site treatment of domestic wastewater: New Danish guidelines. Ecol. Eng. 25(5): 491–500.

Calheiros, C.S.C., P.V.B. Quitério, G. Silva, L.F.C. Crispim, H. Brix, S.C. Moura et al. 2012. Use of constructed wetland systems with *Arundo* and *Sarcocornia* for polishing high salinity tannery wastewater. J. Environ. Manage. 95(1): 66–71.

Castelle, A.J., A.W. Johnson and C. Conolly. 1994. Wetland and stream buffer size requirements—A review. J. Environ. Qual 23(5): 878–882.

Chen, G. 2004. Electrochemical technologies in wastewater treatment. Sep. Purif. Technol. 38(1): 11–41.

Chen, Z., S. Wu, M. Braeckevelt, H. Paschke, M. Kästner, H. Köser et al. 2012. Effect of vegetation in pilot-scale horizontal subsurface flow constructed wetlands treating sulphate rich groundwater contaminated with a low and high chlorinated hydrocarbon. Chemosphere 89(6): 724–731.

Coban, O., P. Kuschk, U. Kappelmeyer, O. Spott, M. Martienssen, M.S.M. Jetten et al. 2015. Nitrogen transforming community in a horizontal subsurface-flow constructed wetland. Water Res. 74: 203–212.

Cooper, P. 1999. A review of the design and performance of vertical-flow and hybrid reed bed treatment systems. Water Sci. Technol. 40(3): 1–9.

Crites, R.W. 1994. Design criteria and practice for constructed wetlands. Water Sci. Technol. 29(4): 1–6.

Dotro, G., G. Langergraber, P. Molle, J. Nivala, J. Puigagut, O. Stein et al. 2017. Treatment Wetlands, Biological Wastewater Treatment Series (Volume 7). IWA Publishing. London.

DWA. 2014. Grundsätze für Bemessung, Bau und Betrieb von Pflanzenkläranlagen mit bepflanzten Bodenfiltern zur biologischen Reinigung kommunalen Abwassers, 03/2006. Germany. In German.

Erisman, J.W., J.N. Galloway, S. Seitzinger, A. Bleeker, N.B. Dise, A.M.R. Petrescu et al. 2013. Consequences of human modification of the global nitrogen cycle. Philos. Trans. R. Soc. Lond., B, Biol. Sci. 368(1621): 20130116.

García, J., D.P.L. Rousseau, J. Morató, E. Lesage, V. Matamoros and J.M. Bayona. 2010. Contaminant removal processes in subsurface-flow constructed wetlands: A review. Crit. Rev. Env. Sci. Technol. 40(7): 561–661.

Gerke, S., L.A. Baker and Y. Xu. 2001. Nitrogen transformations in a wetland receiving lagoon effluent: sequential model and implications for water reuse. Water Res. 35(16): 3857–3866.

Giraldi, D., M. de Michieli Vitturi and R. Iannelli. 2010. FITOVERT: A dynamic numerical model of subsurface vertical flow constructed wetlands. Environ. Modell. Softw. 25(5): 633–640.

Gogate, P.R. and A.B. Pandit. 2004a. A review of imperative technologies for wastewater treatment I: oxidation technologies at ambient conditions. Adv. Environ. Res. 8(3-4): 501–551.

Gogate, P.R. and A.B. Pandit. 2004b. A review of imperative technologies for wastewater treatment II: Hybrid methods. Adv. Environ. Res. 8(3-4): 553–597.

Gonzalias, A.E., P. Kuschk, A. Wiessner, M. Jank, M. Kästner and H. Koser. 2007. Treatment of an artificial sulphide containing wastewater in subsurface horizontal flow laboratory-scale constructed wetlands. Ecol. Eng. 31(4): 259–268.

He, S.M., S.A. Malfatti, J.W. McFarland, F.E. Anderson, A. Pati, M. Huntemann et al. 2015. Patterns in wetland microbial community composition and functional gene repertoire associated with methane emissions. Mbio 6(3): 1–15.

Headley, T.R. and R.H. Kadlec. 2007. Conducting hydraulic tracer studies of constructed wetlands: A practical guide. J. Environ. Qual. 7(3-4): 269–282.

Ibekwe, A.M., C.M. Grieve and S.R. Lyon. 2003. Characterization of microbial communities and composition in constructed dairy wetland wastewater effluent. Appl. Environ. Microbiol. 69(9): 5060–5069.

Isteniç, D., C.A. Arias, J. Vollertsen, A.H. Nielsen, T. Wium-Andersen, T. Hvitved-Jacobsen et al. 2012. Improved urban stormwater treatment and pollutant removal pathways in amended wet detention ponds. J. Environ. Sci. Health A Tox. Hazard Subst. Environ. Eng. 47(10): 1466–1477.

Jespersen, D.N., B.K. Sorrell and H. Brix. 1998. Growth and root oxygen release by *Typha latifolia* and its effects on sediment methanogenesis. Aquat. Bot. 61: 165–180.

Kadlec, R.H. and S.D. Wallace. 2009. Treatment Wetlands. 2nd Ed. CRC Press. Boca Raton, FL.

Kappelmeyer, U., A. Wießner, P. Kuschk and M. Kästner. 2002. Operation of a universal test unit for planted soil filters—planted fixed bed reactor. Eng. Life Sci. 2(10): 311–315.

Kivaisi, A.K. 2001. The potential for constructed wetlands for wastewater treatment and reuse in developing countries—A review. Ecol. Eng. 16: 545–560.

Koelbener, A., L. Ström, P.J. Edwards and H. Olde Venterink. 2010. Plant species from mesotrophic wetlands cause relatively high methane emissions from peat soil. Plant Soil 326(1): 147–158.

Kumar, J.L.G. and Y.Q. Zhao. 2011. A review on numerous modeling approaches for effective, economical and ecological treatment wetlands. J. Environ. Manage. 92(3): 400–406.

Langergraber, G. 2008. Modeling of processes in subsurface flow constructed wetlands: A review. Vadose Zone J. 7(2): 830–842.

Levenspiel, O. 1998. Chemical Reaction Engineering. 3rd Ed. Wiley. US.

Lunsmann, V., U. Kappelmeyer, R. Benndorf, P.M. Martinez-Lavanchy, A. Taubert, L. Adrian et al. 2016a. *In situ* protein-SIP highlights *Burkholderiaceae* as key players degrading toluene by para ring hydroxylation in a constructed wetland model. Environ. Microbiol. 18(4): 1176–1186.

Lunsmann, V., U. Kappelmeyer, A. Taubert, I. Nijenhuis, M. von Bergen, H.J. Heipieper et al. 2016b. Aerobic toluene degraders in the rhizosphere of a constructed wetland model show diurnal polyhydroxyalkanoate metabolism. Appl. Environ. Microbiol. 82(14): 4126–4132.

Martínez-Lavanchy, P., Z. Chen, V. Lünsmann, V. Marin-Cevada, R. Vilchez-Vargas, D. Pieper et al. 2015. Microbial toluene removal in hypoxic model constructed wetlands occurs predominantly via the ring monooxygenation pathway. Appl. Environ. Microbiol. 81: 6241–6252.

Millhollon, E.P., P.B. Rodrigue, J.L. Rabb, D.F. Martin, R.A. Anderson and D.R. Dans. 2009. Designing a constructed wetland for the detention of agricultural runoff for water quality improvement. J. Environ. Qual. 38(6): 2458–2467.

Mitchell, C. and D. McNevin. 2001. Alternative analysis of BOD removal in subsurface flow constructed wetlands employing Monod kinetics. Water Res. 35(5): 1295–1303.

Molle, P., A. Lienard, C. Boutin, G. Merlin and A. Iwema. 2005. How to treat raw sewage with constructed wetlands: An overview of the French systems. Water Sci. Technol. 51(9): 11–21.

Mucha, A.P., C.M.R. Almeida, A.A. Bordalo and M.T.S.D. Vasconcelos. 2005. Exudation of organic acids by a marsh plant and implications on trace metal availability in the rhizosphere of estuarine sediments. Estuar. Coast. Shelf S. 65(1-2): 191–198.

Park, J.B.K. and R.J. Craggs. 2010. Wastewater treatment and algal production in high rate algal ponds with carbon dioxide addition. Water Sci. Technol. 61(3): 633–639.

Paulo, P.L., L. Begosso, N. Pansonato, R.R. Shrestha and M.A. Boncz. 2009. Design and configuration criteria for wetland systems treating greywater. Water Sci. Technol. 60(8): 2001–2007.

Radajewski, S., P. Ineson, P.R. Nisha and C. Murrell. 2000. Stable-isotope probing as a tool in microbial ecology. Nature 403(6770): 646–649.

Rawat, I., R. Ranjith Kumar, T. Mutanda and F. Bux. 2011. Dual role of microalgae: Phycoremediation of domestic wastewater and biomass production for sustainable biofuels production. Appl. Energy 88(10): 3411–3424.

Reddy, K.R., A. Wright, A. Ogram, W.F. Debusk and S. Newman. 2002. Microbial processes regulating carbon cycling in subtropical wetlands. Paper read at 17th WCSS, Symposium no. 11, at Thailand.

Reed, S.C. 1993. Subsurface flow constructed wetlands for wastewater treatment: A technology review. U.S. EPA Office of Water. Washington DC.

Richter, J., A. Wiessner, A. Zehnsdorf, J.A. Müller and P. Kuschk. 2016. Injection of hydrogen gas stimulates acid mine drainage treatment in laboratory-scale hydroponic root mats. Eng. Life Sci. 16(8): 769–776.

Scholes, L.N.L., R.B.E. Shutes, D.M. Revitt, D. Purchase and M. Forshaw. 1999. The removal of urban pollutants by constructed wetlands during wet weather. Water Sci. Technol. 40(3): 333–340.

Seeger, E.M., U. Maier, P. Grathwohl, P. Kuschk and M. Kästner. 2013. Performance evaluation of different horizontal subsurface flow wetland types by characterization of flow behavior, mass removal and depth-dependent contaminant load. Water Res. 47(2): 769–780.

Seelaus, T.J., D.W. Hendricks and B.A. Janonis. 1986. Design and operation of a slow sand filter. J. Am. Water Works Assoc. 78(12): 35–41.

Shrestha, R.R., R. Haberl and J. Laber. 2001. Constructed wetland technology transfer to Nepal. Water Science and Technology 43(11): 345–360.

Shutes, R.B.E., D.M. Revitt, A.S. Mungur and L.N.L. Scholes. 1997. The design of wetland systems for the treatment of urban run off. Water Science and Technology 35(5): 19–25.

Stottmeister, U., A. Wießner, P. Kuschk, U. Kappelmeyer, M. Kästner, O. Bederski et al. 2003. Effects of plants and microorganisms in constructed wetlands for wastewater treatment. Biotechnol. Adv. 22(1-2): 93–117.

Tshrintzis, V.A. and R. Hamid. 1997. Modeling and management of urban stormwater runoff quality: A review. Water Resour. Manage. 11(2): 136–164.

Vianna, M.T.G., M. Marques and L.C. Bertolino. 2016. Sun coral powder as adsorbent: Evaluation of phosphorus removal in synthetic and real wastewater. Ecol. Eng. 97: 13–22.

Vogt, C., E. Cyrus, I. Herklotz, D. Schlosser, A. Bahr, S. Herrmann et al. 2008. Evaluation of toluene degradation pathways by two-dimensional stable isotope fractionation. Environ. Sci. Technol. 42(21): 7793–7800.

Vohla, C., M. Kõiv, H.J. Bavor, F. Chazarenc and Ü. Mander. 2011. Filter materials for phosphorus removal from wastewater in treatment wetlands—A review. Ecol. Eng. 37(1): 70–89.

Vymazal, J. 2005. Horizontal sub-surface flow and hybrid constructed wetlands systems for wastewater treatment. Ecol. Eng. 25(5): 478–490.

Wießner, A., P. Kuschk, M. Kästner and U. Stottmeister. 2002a. Abilities of helophyte species to release oxygen into rhizosphere with varying redox conditions in laboratory-scale hydroponic systems. Int. J. Phytoremediation 4(1): 1–15.

Wießner, A., P. Kuschk and U. Stottmeister. 2002b. Oxygen release by roots of *Typha latifolia* and *Juncus effusus* in laboratory hydroponic systems. Acta Biotechnol. 22: 209–216.

Wießner, A., U. Kappelmeyer, P. Kuschk and M. Kästner. 2005. Influence of the redox condition dynamics on the removal efficiency of a laboratory-scale constructed wetland. Water Res. 39(1): 248–256.

Wießner, A., K.Z. Rahman, P. Kuschk, M. Kästner and M. Jechorek. 2010. Dynamics of sulphur compounds in horizontal sub-surface flow laboratory-scale constructed wetlands treating artificial sewage. Water Res. 44(20): 6175–6185.

Wießner, A., U. Kappelmeyer, M. Kästner, L. Schultze-Nobre and P. Kuschk. 2013. Response of ammonium removal to growth and transpiration of *Juncus effusus* during the treatment of artificial sewage in laboratory-scale wetlands. Water Res. 47: 4265–4273.

Wohanka, W. 1995. Disinfection of recirculating nutrient solutions by slow sand filtration. Sci. Hortic. 382: 246–255.

Wu, S., P. Kuschk, A. Wiessner, J. Müller, R.A.B. Saad and R. Dong. 2013a. Sulphur transformations in constructed wetlands for wastewater treatment: A review. Ecol. Eng. 52: 278–289.

Wu, S., A. Wiessner, M. Braeckevelt, U. Kappelmeyer, R. Dong, J.A. Müller et al. 2013b. Influence of nitrate load on sulfur transformations in the rhizosphere of *Juncus effusus* in laboratory-scale constructed wetlands treating artificial domestic wastewater. Environ. Eng. Manag. J. 12(3): 565–573.

Wu, S., P. Kuschk, H. Brix, J. Vymazal and R. Dong. 2014. Development of constructed wetlands in performance intensifications for wastewater treatment: A nitrogen and organic matter targeted review. Water Res. 57: 40–55.

3

Use of Artificial Wetlands as Degradation Systems of Recalcitrant Contaminants and Transformers of Energy to Electricity

Aranys del Carmen Borja-Urzola,* Citlaly Marisol Hernández-Arriaga, Oscar Hugo Miranda-Méndez, María Guadalupe Salinas-Juárez, Marisela Bernal-González and *María del Carmen Durán-Domínguez-de-Bazúa*

INTRODUCTION

The rapid development in the area after the change in the constitutional protection of the *Xochimilco* land property in the 1980s has resulted in the inadequate use of water for the wetlands existing there (INEGI 2008; Zambrano et al. 2009; INEGI 2016a,b; García-Luna et al. 2017).

Barceló and López (2008) in Spain have a similar study on this type of problem. The presence of biologically active substances known as emerging pollutants, from different origins and having different chemical natures, apparently have a non-significant presence in the environment due to its distribution and/or concentration, have the potential to negatively impact the environment, as well as the health of all biosystems (Stuart et al. 2012). It has been established that these compounds are incorporated into the water through sewage and industrial effluents (Daughton 2004;

Universidad Nacional Autónoma de México, Facultad de Química, Departamento de Ingeniería Química, Laboratorios de Ingeniería Química Ambiental y de Química Ambiental (UNAM-FQ-DIQ-LIQAyQA), Ciudad Universitaria, Ciudad de México, Mexico.
* Corresponding author: araborurz@hotmail.com

Fent et al. 2006), hospitals effluents (Kummerer 2001; Velázquez-González et al. 2005), the agricultural and livestock activities (Watanabe et al. 2010), and the septic tanks (Swartz et al. 2006).

Emerging pollutants include a wide variety of chemical compounds, pharmaceuticals, chemical substances used for personal care, detergents, plasticizers, industrial additives, and pesticides that are not necessarily included within the follow up on water quality (Herrero et al. 2012). There is still limited information available on the possible effects on human health and ecology (Baltazar et al. 2014; Zuñiga et al. 2012). Due to the fact that these pollutants are present in different concentrations in surface water, environmental quality criteria are still being specified (Eggen et al. 2010). Besides this, conventional wastewater treatment plants are not designed to eliminate these compounds (García-Mercado et al. 2011). Therefore, these pollutants are source of concern for the scientific community and for the environmental regulatory agencies. Thus, an important part of the research presently conducted is centered in these pollutants found particularly in surface water because this water source is more susceptible to containing higher concentrations of pollutants than subterranean or groundwater (Lapworth et al. 2012).

A group of emerging pollutants of high importance at short and long terms are pesticides, a group of substances or mixtures aimed to prevent, destroy, repel, or mitigate pests and its effects (Gil et al. 2012). Due to the strict regulation derived of the known effects of organochlorine pesticides, these compounds have been studied for decades and there is a reasonable knowledge of their presence and fate in the aquatic environment. However, in the last years, a new concern is present: its metabolites or degradation products that were ignored and that little by little have shown its toxicity that, in some cases, is even higher than that from the compounds from which these metabolites came from (Sinclair and Boxall 2003).

This type of emerging pollutants may be eliminated from water using physicochemical methods such as hydrolysis, photolysis, and activated carbon adsorption (Derylo-Marczewska et al. 2017), advanced oxidation methods with oxygen reactive species (Khan et al. 2015), Fenton and Foto-Fenton reactions (Benzaquén et al. 2009; Affam and Chaudhuri 2013), and nanophotocatalysis (Sharma et al. 2016) but they are costly and are difficult to implement in real systems.

An alternative option through which most compounds are transformed into simpler chemical molecules with variable toxicity is the microbiological degradation using bacteria, actinomycetes, yeasts, microalgae, fungi, protozoa, among others, since these microorganisms use these pollutants as an energy source (Hansen et al. 2013) to carry out their habitual biochemical reactions.

Natural and artificial wetlands have demonstrated to be efficient systems for the biological degradation of pesticides. Environmental conditions associated with water interaction with the support or sediment, vegetal matter, and microorganisms present in these systems promote synergic mechanisms of depuration, reducing pesticides concentration, particularly some organochlorine ones: DDT, lindane, eldrin, etc. (Li et al. 2014); urea based: isoproturon; carboxamide: boscalid; triazolic ones: tebuconazol (Vallée et al. 2016); triazinic ones: atrazine (Vandermaesen et al. 2016).

A particular example for the biological elimination of an emerging pollutant would be the degradation of the chemical herbicide atrazine using a microbial consortium isolated within the natural wetlands represented by the *Xochimilco* channels. The atrazine is a widely-used herbicide due to its effect on the Photosystem II of the photosynthetic organisms eliminating competitive plants with wide leaves that may affect space, nutrients, and sunlight for the cultivated species in spite of its toxicity for amphibians and mammals. Atrazine biodegradation may produce metabolites such as deethylatrazine (DEA), deisopropylatrazine (DIA), cyanuric acid (CA), and hydroxiatrazine (HA) (Hansen et al. 2013).

The quantitative determination of these pollutants in the environmental matrices is possibly due to the existence of extraction techniques such as Solid Phase Extraction (SPE) and instrumental techniques such as Gas Chromatography (GC). Solid phase extraction coupled with gas chromatography and a flame ionization detector (GC-FID) to quantify ATZ, DIA, and DEA, and CA and HA were the methods used in this investigation. The procedure will be described in the next part of this chapter.

Finally, it should be mentioned that artificial wetlands systems aim to clean wastewaters plus produce electricity from bacteria that liberate electrons in the anaerobic zone of the artificial wetlands (Salinas-Juárez 2016). Considering the problems that *Xochimilco* is facing with the change of use of soil, this option might be attractive to be put into use for some of the areas that have neither sanitation nor electricity in spite of being located in the Mexico City area.

Although most of the results are still at research level, this technology may be an option especially for far located communities where no conventional sources of energy are available. This combination artificial wetland-microbial fuel cell, AW-MFC, has the advantage to simultaneously clean the wastewaters and produce electric energy (Aelterman et al. 2006; Chiranjeevi et al. 2013).

In the next paragraphs, three research approaches are presented.

Use of Native Bacterial Consortia to Remove Atrazine and Some by-products

Samples of channels water were collected during the months of January, February, March, and May 2016. These were analyzed using field parameters such as pH, electric conductivity, and dissolved oxygen concentration. At the labs, organic matter was assessed using chemical and biochemical oxygen demand, COD and BOD_5, nitrites, nitrates, nitrogen (total, organic, ammonia), and total phosphorus, following the recommended techniques by APHA-AWWA-WPCF (1998a).

A "synthetic water" was formulated to emulate the results obtained from sampling to have a homogeneous medium to work with: Organic matter measured as COD, between 71.7/227.1 mg L^{-1}; total phosphorus between 3.0–7.6 mg L^{-1}; nitrates between 0.9–22.3 mg L^{-1}; nitrites 0.01–1.02 mg L^{-1}; total nitrogen between 1.6–21.1 mg L^{-1}, and ammonia nitrogen between 0.4–10.3 mg L^{-1} (Hernández-Arriaga 2017).

Additionally, the water samples collected were tested using microbiological media: agar eosine-methylene blue, agar mannitol salt, agar potato dextrose with chloramphenicol, nutritive, and MacConkey's to reach bacterial strains isolation that, according to literature, were capable of degrading atrazine at laboratory level following the conditions found in other places of study with a similar situation to *Xochimilco* (Da Cunha et al. 2013).

Isolated strains were identified as *Nocardia* sp., *Bacillus megaterium*, *Morococcus* sp., and *Agrobacterium* sp. These microorganisms were mixed equimolecularly and adapted to atrazine concentrations of 15 and 20 mg L^{-1} during 15 days to avoid a reduction of growth in the experimental stages to be performed. The biodegradation stage lasted 30 days, testing three different atrazine concentrations: 15, 10, 5 mg L^{-1}. A follow up procedure was established using the colonies forming units accounting per milliliter (CFU mL^{-1}) to assess the growth of bacteria during the experimental period.

During the studied lapse, it was observed that the CFU mL^{-1} values in the bioreactors that contained atrazine were significantly higher with respect to the blank with no herbicide (0.05%). The highest values were 2.6E + 08 for 15 mg L^{-1}; 2.9E + 08 for 10 mg L^{-1}; and 2.6E + 08 for 5 mg L^{-1}, whereas for the blank it was 8.0E + 07. Therefore, bacteria in the consortium utilize atrazine as a source of nitrogen increasing its growth (Sene et al. 2010; Hansen et al. 2013).

Along the experimental time samples were taken every three days that were extracted through Oasis HLB cartridges (200 mg) with a 6 mL capacity, using methanol and acetonitrile as eluents to recover atrazine and its possible metabolites. Using HPLC and GC it was possible to determine the biodegradation of atrazine and the formation of metabolic byproducts: The consortium degraded atrazine in 33.73, 40.11, and 51.30 percent for the concentrations of 15, 10, and 5 mg L^{-1}, respectively. The resulting metabolic products were DEA, CA, and HA in concentrations between 0.54 – 1.25 mg L^{-1} for DEA, 1.66 mg L^{-1} for AC, and between 0.21–0.34 mg L^{-1} for HA. It is important to mention that these metabolites and particularly DEA were only found in the first day of the experiments. It might be assumed that afterwards this compound was also biodegraded to simpler molecules perhaps with lesser toxicity such as deisopropyl-deethylatrazine (DIDEATZ) or even simpler as urea, allophanate, CO_2, and NH_3 (Sene et al. 2010; Hansen et al. 2013; Fang et al. 2015a; La Cecilia and Maggi 2016) but these were neither identified nor quantified in this stage of the experiments.

Considering the kinetic models proposed by Monod and Michaelis-Menten and using the strategy presented by La Cecilia and Maggi (2016) and Khan et al. (2015), kinetic and analytical parameters were calculated to evaluate the biodegradation, obtaining the growth constant value, the generation time (d), and the reproduction rate (d^{-1}) for 15, 10, and 5 mg atrazine/L: 16.52, 0.0419, 23.85; 16.62, 0.0417, 23.99 d; 16.67, 0.0416 and 24.05 d^{-1}, respectively. The analytical parameters calculated were the degradation constant (k') and the half life time $t_{1/2}$ (d) of the pollutant obtaining the following values: 0.013, 51 d; 0.017, 41 d; 0.024, 29 d, for 15, 10, and 5 mg/L, respectively, whereas for the blank with no microorganisms, affected only by the hydrolysis and photolysis the values were 0.0009 and 736 d. Consequently, the presence of the microbial consortium significantly increases the atrazine degradation obtaining half-life times of 29 days whereas for abiotic processes is more than two years (765 days).

Finally, toxicity of the compounds studied was obtained after the biodegradation stage using the dry biomass obtained for a wild algae *Chlorococcum* sp., isolated from a lab scale artificial wetland operating as a vegetal fuel cell (Miranda-Méndez 2017). Fixed compounds concentrations, 5 mg L^{-1}, for ATZ, DEA, CA, and HA were added, following the recommended procedure of APHA-AWWA-WPCF (1998b). Results from this bioassay indicated that ATZ significantly affects the algal growth with respect to a blank with no toxic compound. The same results were found for DEA. However, for CA and HA, algal growth was significantly augmented, probably because no chlorine atoms are found in these compounds being used as nitrogen source promoting the microalgae growth. If it was found at lab scale is probably that it may be present in *Xochimilco*. This result is relevant because there is an ecotoxicological importance of these two compounds. Its presence may be promoting indirectly the increase of algal biomass in the *Xochimilco* channels and in this way, eutrofication phenomena (DOF 1992).

These results indicate how a former natural wetlands system may be an important source of microorganisms with the ability to substantially degrade toxic compounds at laboratory scale. In natural conditions, this type of ecosystems can self-depurate, constituting a natural source for eliminating, and/or transforming toxic emerging pollutants of environmental and public health importance.

Application of the Solid Phase Extraction to Determine Emerging Pollutants

Analytical chemistry is a very important subject to determine emerging pollutants in trace concentrations when these compounds are in environmental matrices. Presently, it is possible to find highly selective analytically sensitive techniques, easy to manipulate to identify and quantify these compounds in a specific sample. Among these techniques gas and liquid chromatography (GC and HPLC) using different detectors (flame ionization detector, FID, thermal conductivity detector, TCD, or coupled with other detection techniques such as mass spectrometry (MS), UV-Visible spectroscopy, and nuclear magnetic resonance (NMR)). However, all of them require samples that have been pretreated to eliminate impurities that may affect the results as well as to increase the analyte concentration.

Since environmental samples have trace concentrations of the emerging pollutants, preparation techniques are needed to obtain concentrated extracts free of interferences, improving the sensitivity of the analytical instruments (Pawliszyn 2012). Among the samples preparation techniques, the ones with more applicability in the environmental laboratories are: solid phase extraction (SPE), matrix solid phase dispersion (MSPD), and the magnetic stir bar sorption extraction (MSBSE). These techniques allow the extraction procedure with reduced volumes of sample and organic solvents greatly improving the separation operation and reducing its environmental impact (Arias et al. 2014; Stashenko and Martínez 2011).

These techniques for samples preparation are based in sorption processes where analytes interact with a solid phase yielding good retention percentages through chemical interactions (dipole–dipole, ionic exchange, hydrogen bonds). Once the retention of the analytes occurs, they should be eluted with a suitable solvent— different polarity to increase analyte solubility (Andrade-Eiroa et al. 2016). This sorption process yields an extract where the analytes are concentrated and free of interfering compounds. Thus, the chromatographic analyses may be carried out with the assurance that good results will be obtained.

Solid phase extraction (SPE) is one of the most widely known techniques in the environmental labs particularly for aqueous samples, concentrating and cleaning the extracts before its analysis (Hurtado-Sánchez et al. 2013; Pawliszyn 2012). SPE consists in charging a solution (sample) in a very low amount of a chemical compound in solid phase generally found in very small columns or discs known as cartridges, capable to adsorb or retain compounds of a specific polarity. Then the analyte of interest is eluted from the solid adsorbent using a specific solvent according to the analyte polarity. Then the dissolved analyte is sent to its identification and quantification analysis by gas or liquid chromatography (Andrade-Eiroa et al. 2015).

SPE efficiency depends upon some parameters that have to be optimized to reach the highest recovery of the analytes (Moldoveanu 2015). However, the development of a good analytical method using SPE implies the correct selection of the adsorbent material as well as a suitable protocol for the treatment and elution of the sample (Andrade-Eiroa et al. 2016). Some important parameters are the following:

- Type and quantity of the adsorbent phase: the selection of the phase is fundamental for the recovery of the analytes. SPE may be a selective technique for a specific group of compounds if the suitable conditions are chosen (Andrade-Eiroa et al. 2016).

- Cartridge capacity (leakage volume): It is needed to know the sample volume that may be applied through a cartridge without loses of the compounds of interest (Bielicka-Daszkiewicz and Voelkel 2009; Poole et al. 2000). In other words, it is the maximum sample volume that can be fed to the cartridge without causing the saturation of the active sites of the adsorbent material obtaining yields near to 100%.

- Volume and type of solvent to elute the analytes of interest: The choice of the solvent selection is important if a suitable method for a family of compound need to be established. The solvent selected must break the chemical bonds between the analyte and the solid phase and be strong enough to elute all the analyte adsorbed or retained, using the lowest possible amount to obtain a concentrated extract; a special consideration when trace concentrations are to be found in the sample.

- SPE operational aspects: Among the operational aspects concerning SPE, the most important one is the flow rate (mL min^{-1}) of the liquid sample through the cartridge. Analytes adsorption on the solid phase requires an equilibrium time, and thus, it is recommended to work at a constant flow between 5 and 10 mL min^{-1}. Another important aspect is the pretreatment of the sample. In some cases, it is necessary to adjust the pH value, for example.

SPE is versatile in its applications. Thus, it is common to find studies where it is employed as the extraction method to determine pollutants in surface and groundwater (Hurtado-Sánchez et al. 2013; Min et al. 2008), in biological samples (Tamrakar et al. 2009), and in the follow up of the efficiency of wastewater treatment (Farajzadeh et al. 2017; Fuentes et al. 2017).

In the next paragraphs, an example of this procedure carried out by Borja-Urzola (2017) is presented. It includes the development of a solid phase extraction technique followed by gas chromatography using a flame ionization detector (SPE-GC-FID) applicable to the identification of an herbicide pertaining to the triazine family, known as atrazina (ATZ), as well as two of its degradation metabolites: deethylatrazine (DEA) and deisoprophylatrazine (DIA) in samples of surface water from the Xochimilco channels system. The steps to implement the analytical method and the implications to use this SPE technique are shown.

Development of the Technique

Implementation of the Detection Method in the Analytical Instrument

If identification and quantification of a specific compound in an environmental matrix is to be performed it is needed a certified standard of the compound of interest. With it, the calibration curves are built in the analytical equipment, gas and/or liquid chromatograph. For this case of study, the standards for atrazine, deethylatrazine, and deisopropylatrazine were proportioned by ChemService with purities of 98.6, 98.4, and 98.9%, respectively.

Optimal operating conditions implemented in an Agilent 7890 A gas chromatograph with flame ionization detector are shown in Table 1. Figure 1 shows the resulting chromatogram for a mixture of 200 µg/L of the three compounds. Calibration data are shown in Table 2 together with some other analytical parameters evaluated to corroborate the correct operation of the instrumental system.

Table 1. Optimal conditions of operation for the Agilent 7890 A chromatograph for the separation of the analytes of interest.

INJECTOR	OVEN	DETECTOR
Injection: Splitless	**Column:** ZB5 (30 m*	**T:** 250°C
Conditions	0.25 mm* 0.25 µm)	**Frequence:** 50 Hz/0.04 min
T: 280°C	**Flujo He:** 1.127 mL	**Gases Ratio**
P: 101.863 kPa (14.774 psi)	**Temperature Program**	H$_2$: 35 mL/min
Injection Volume: 1 µL	(1) 80°C 2 min	Air: 350 mL/min
	(2) 48°C/min 176°C 0 min	He (make up): 35 mL/min
	(3) 6°C/min 212°C 1 min	

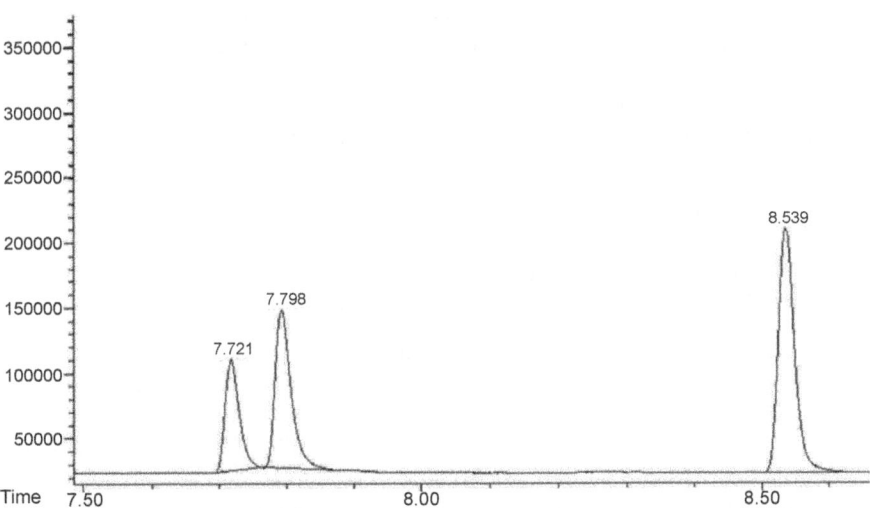

Fig. 1. Chromatogram corresponding to the 1 µL injection of standard mixture for ATZ, DEA, and DIA at a concentration of 200 µg/L. Column ZB-5 (5% phenyl-polysiloxane – 30 m long * 0.25 mm internal diameter* 0.25 µm phase thickness).

Table 2. Evaluation of the linearity of the instrumental system in a concentrations range of 15–60 μg/L, n = 12, confidence 95%.

Compound	Slope (m)	Interception (b)	r	Equation y = mx+ b	LD (μg/L)	LQ (μg/L)	F_{Fisher}	IC for the slope
DIA	677	−3299	0.993	y = 677x − 3299	2.77	6.02	720	± 56
DEA	956	−2746	0.994	y = 956x − 2746	3.05	5.00	930	± 70
ATZ	1752	−532	0.987	y = 1752x − 532	2.70	4.12	387	± 198

r, Pearson correlation coefficient; LD, limit of detection; LQ, limit of quantification; IC, interval of confidence

Table 3. Optimization of the technique for SPE applied to ATZ, DEA, and DIA.

Phase	Oasis®HLB	**Conditioning**: 6 mL methanol and 6 mL deionized water during 5 min each one
Sample Volume	200 mL	**Sample Volume**: 200 mL at a flow between 5–10 mL/min
Solvent Type	Mezcla 50:50 (ACN:MeOH)	**Washing:** 10 mL deionized water
		Drying: Centrifugation during 10 min at 3500 rpm
Solvent Volume	4 mL	**Elution:** 4 mL of a mixture 50:50 acetonitrile:methanol

Optimization of the extraction procedure. After the instrumental method is optimized, it is necessary to optimize the solid phase extraction technique: selecting the adsorbent phase, sample quantity, and solvent type and volume needed for the elution of the compounds.

In the cited example, to select the phase the extraction technique was studied comparing phase C_{18} (octadecyl bound silica) and the Oasis®HLB solid-phase extraction sorbent [poly(divinylbenzene-co-N-vinylpyrrolidone)]. The amount of sample was between 100 and 200 mL. To choose the suitable solvent some tests were performed using acetonitrile, dichloromethane, methanol, and a mixture 50:50 acetonitrile:methanol, with two volumes, 2 and 4 mL. Operating conditions that gave the best percentage of recovery for the extraction technique are presented in Table 3.

Technique validation. Every time that a new technique is developed for a specific problem it is necessary to demonstrate that it is suitable (Eurachem 2016). This procedure is known as method validation. Results for the validation of the technique developed for determining the presence of ATZ, DEA, and DIA in the Xochimilco channels points of sampling are presented in Table 4.

Application to real samples. For the application of these techniques to determine the quality of the water samples from *Xochimilco* channels considering atrazine and its metabolites, a sampling procedure was carried out following the official norm applicable in Mexico (PROY-NMX-AA-003/3-SCFI-2008) to the four zones: urban, a zone where there is the use of agrochemicals, transition (urban-with agrochemicals use), and the so-called organic zone where the traditional *chinampa* pre-Columbian method without any agrochemical is employed. Samples were stored in amber glass containers previously treated according to the norm. Samples were maintained at 4°C.

Table 4. Validation of the proposed technique for the determination of ATZ, DEA, and DIA in the Xochimilco channels system taken samples: 15–60 µg/L, n = 12, confidence 95%.

Compounds \ Parameters	DIA	DEA	ATZ
Equation	y = 456x – 2819	y = 926x – 15	y = 1707x – 4385
r	0.992	0.995	0.996
Slope CI	± 74	± 183	± 226
%Recovery	97–109	90–102	87–100
Accuracy (%CV)	1.5	1.85	1.37
Reproducibility	1.22	1.36	1.14
LD (µg/L)	2.8	2.24	2.39
LQ (µg/L)	5.4	2.82	2.78

r, Pearson correlation coefficient; LD, limit of detection; LQ, limit of quantification; CV, coefficient of variation

Table 5. Results for the determination of ATZ, DEA, and DIA in Xochimilco samples without the standards addition.

Point	Zone	DIA			DEA			ATZ		
		Tr	Area	[µg/L]	Tr	Area	[µg/L]	Tr	Area	[µg/L]
1	Urban	7.693	2663	**12.02**	-	-	-	-	-	-
2		-	-	-	-	-	-	-	-	-
3	Use of	7.694	3865	**14.65**	-	-	-	-	-	-
4	agrochemicals	-	-		-	-	-	-	-	-
5	Transition	-	-	-	-	-	-	-	-	-
6		-	-	-	-	-	-	-	-	-
7	Organic	7.695	1360	**9.16**	7.781	3378	**3.66**	-	-	-
8		-	-	-	-	-	-	-	-	-

A volume of 200 mL of sample were taken. These samples were filtered to eliminate suspended solids. Filtered samples of 200 mL were processed in the Oasis®HLB using the indicated procedure (Table 3).

The first analysis was done without the addition of a known amount of standards. Results indicated DIA presence in three of the sampled points corresponding to the urban zone, the zone with the use of agrochemicals, and the organic zone. DEA presence was also found in the organic zone. Table 5 shows the concentrations obtained for each of the sampling points.

With the application of SPE with GC-FID it is possible to design extraction and quantification methods for pollutants in natural wetlands in trace concentrations, especially when the wetlands are as important from the ecological point of view as *Xochimilco*, Mexico.

Energy Production in Artificial Wetlands used for Wastewaters Sanitation

Artificial wetlands (AW) are a very versatile ecotechnology that may be widely use for the treatment of wastewaters due to its very low cost, easy operation, and minimal maintenance, compared with conventional WWT systems. Its use is ample, from domestic sewage to mining effluents (Liu et al. 2015; García-Mercado et al. 2017). Some disadvantages are also present, such as the clogging of the support media (Ruiz et al. 2010), low nitrogen removal efficiencies (Rodríguez-Monroy and Durán-de-Bazúa 2006; Durán-de-Bazúa et al. 2008; Orduña-Bustamante et al. 2013; Wu et al. 2014), and low elimination yields of heavy metals and recalcitrant pollutants (Melian et al. 2008; Ruiz-López et al. 2010; Salgado-Bernal et al. 2012). In the last decade, an important number of research publications on the combination of artificial wetlands (AW) with microbial fuel cells (MFC) have appeared. These treatment systems aim to clean wastewaters as well as produce electricity from bacteria that liberate electrons in the anaerobic zone of the artificial wetlands (Salinas-Juárez 2016).

Although most of the results are still at research level, this technology may be an option especially for far located communities where no conventional sources of energy are available. This combination AW-MFC has the advantage to simultaneously clean the wastewaters and produce electric energy (Aelterman et al. 2006; Chiranjeevi et al. 2013).

Naturally, it is necessary to include in the design of the AW an electrochemical system so that liberated electrons may be captured in order to obtain a suitable electric current (Salinas-Juárez 2016).

In Xochimilco, channels water also contains dissolved organic matter that should be removed to avoid eutrofication processes. Although there are many conventional processes to do this, for this particular case, the use of this type of novel technology might be applicable. Considering the vegetal microbial fuel cell, VMFC, as it is now known, it would be possible to remove this organic matter and to produce electricity. As this technology is still in its research phase (Helder et al. 2013; Strik et al. 2008), it is important to consider all its aspects to increase its sustainability (Yadav et al. 2012; Doherty et al. 2015a; Salinas-Juárez 2017, the following chapter in this book).

Logan et al. (2006) schematically present the principle of operation of these microbial fuel cells (MFC) where bacteria act as catalysts oxidizing both organic and inorganic matter and producing electric current.

Electrons produced by the bacteria from these substrates are transferred to the anode and flow towards the cathode since these are bonded together by a conductive material. The main elements of a MFC are:

1) Anode: Construction materials should be able to conduct electricity and be compatible with the microorganisms that interact to transfer the electrons. Anode should also be chemically stable (He et al. 2005).

2) Cathode: Here the reduction processes, where oxygen is the electron acceptor for a MFC due to its high oxidation potential, availability, low cost, sustainability,

and because with these reactions do not produce chemical residues (water is the only final product), are occurring. Cathode material selection is an important factor for its performance and variety of applications (Cheng et al. 2006).

3) Some MFC designs use cathode and anode compartmentalization through an electrons exchange membrane. Those configurations that exclude the use of membranes are the naturally separated systems or the ones especially designed to be of one compartment only (Liu and Logan 2004).

Presently, technology and operation of the MFC has been improved with different configurations. However, a common factor and one that has great relevance is the formation of a microbial film on the anode (Kim et al. 2006). Although there are studies for the microbial ecology of these biofilms, the relationships among the members of the microbial communities and its contribution to electron flow from anode to cathode is still under study, as well as the mechanisms favoring its formation and the metabolic processes that take place during its establishment (Lee et al. 2003).

On the other hand, the most important factor for a MFC to produce an efficient electron current is the type of microorganisms present to carry out the degradation process of the organic matter to CO_2 and H_2O and the electrons liberation to the system. To obtain these microorganisms, an inoculum is required: anaerobic sludge, sewage, industrial wastewaters, marine or aquatic sediments, etc. (Falcón et al. 2009).

Vertical flow artificial wetlands systems may develop a significant oxidation-reduction gradient along its height, with an aerobic environment in the upper part, and an anaerobic one in the bottom (Zhao et al. 2013). This distribution redox configuration offers the possibility to integrate the MFC technology to an artificial wetland. Thanks to this technology power has been produced at lab scale vertical flow systems (Erazo-Cortés 2015; Salinas-Juárez 2016; Salinas-Juárez et al. 2016).

Power production estimates based on the results obtained indicate that these systems may become an alternative for communities in the rural areas that have neither sewage treatment facilities nor electricity due to the high costs associated to the attention of these distant and disperse communities (Salinas-Juárez 2016). Among the first studies presenting the combination of these technologies are Salinas et al. (2011) and Yadav et al. (2012) obtaining power densities of 18.2 and 15.73 mW m^{-2} and maximum current density of 39.83 and 69.75 mA m^{-2}, respectively.

Although it is an innovative and growing technology, it is possible to find studies (Table 6) where the authors suggest improvements to make MFC-AW more efficient. Operacional parameters, different configurations, and different types of wastewaters, among other approaches, have been evaluated.

Most of the reactors operated are of vertical flow to maximize anaerobic conditions in the anode and the aerobic ones in the cathode as mentioned above (Zhao et al. 2013; Miranda-Méndez 2017).

However, yield comparisons between vertical and horizontal flow systems have demonstrated that no significant differences, in terms of organic matter elimination measured as chemical oxygen demand, voltage production, and power density calculated considering dissolved oxygen and redox potential, have been found (Liu et al. 2015).

Table 6. Comparison between organic matter removal measured as chemical oxygen demand (COD) and power density in integrated microbial fuel cells and artificial wetlands.

Reference	Initial COD (mg/L)/ Removal (%)	Power Density (mW/m^2)
Venkata et al. 2011	1350/86.67	69.72
Yadav et al. 2012	1500/74.90	15.73
Chiranjeevi et al. 2013	121/82.88	179.78
Zhao et al. 2013	1058/76.50	9.40
Fang et al. 2013	180/85.70	30.20
Liu et al. 2013	193–205/94.80	12.42
Villaseñor et al. 2013	250/80–100	43.00
Villaseñor-Camacho et al. 2014	230/86	8.00
Doherty et al. 2015b	411–854/64	48.71
Fang et al. 2015b	300/72.50	85.20
Oon et al. 2015	314.8/100	6.12
Corbella et al. 2015	232/60	36.00
Srivastava et al. 2015	770–887/84	76.99
Oon et al. 2016	624/99	9.30
Xu et al. 2016	484/90	87.79

Most of these results have been obtained in laboratory scale experiments, and thus, to scale it up some challenges are still to be considered:

- Possible reduction of the power density produced when the working volume is increased (Ge et al. 2014).
- Reduction in the electrodes space whereas maintaining an anodic milieu thermodynamically and kinetically favorable for the electrogenic bacteria (An et al. 2013).
- Study of the power current ratio produced among the rhyzodeposits and wastewaters, that is, the amount of organic matter from wastewaters is transformed into electricity (Doherty et al. 2015a,b).
- Limitation of the nitrification or denitrification (Vymazal 2013; Hu et al. 2014).

In spite of the fact that artificial wetlands and microbial fuel cells have been exhaustively studied in an individual form, its combination is fairly recent. Net Energy Recovery (NER) may be an appropriate parameter for the cross comparison among studies, since it is based on the wastewaters characteristics and less dependent on the AW-MFC dimensions. The highest NER found is of 0.047 kWh kg^{-1} COD (Liu et al. 2013).

These AW-MFC systems may be a promising technology for the treatment of the *Xochimilco* channels having as an added value the production of electricity.

However, with the results obtained up to date, the highest advantage of combining these two technologies is for improving the wastewater treatment due to the favoring of the anaerobic conditions in the bottom of the wetland. To enhance the electricity production, it is necessary to investigate new parameters that would favor the redox required conditions besides improving the electrogenic microorganisms environment for the transferring of electrons from the anode to the cathode.

Final Remarks

A worrying result from these experimental studies is that *Xochimilco*, an originally natural wetlands area recognized by UNESCO (1988) as a cultural heritage of humanity, instead of receiving water from springs and rivers, is now receiving badly treated wastewaters to maintain the water level since clean water from springs and rivers is being used as a water source by a megalopolis where authorities instead of protecting one of the world cultural and natural heritage sites are allowing multifamily constructions where there were before single family dwellings with garden areas (Valek 2000; Chávez-López et al. 2011; Díaz-Torres et al. 2013; Alcántara 2014; Zabaleta-Solís 2017).

Acknowledgements

The first four authors acknowledge their graduate students scholarships from the Mexican Science and Technology Council (*CONACYT*, in Spanish *Consejo Nacional de Ciencia y Tecnología*). Materials and reagents were provided by *UNAM DGAPA* projects *PAPIME EN103704, PE101709*, and *PE100514* and to *UNAM Programa de Apoyo a la Investigación y el Posgrado de la Facultad de Química, PAIP 5000–9067*. The five authors appreciate the collegial support in the review of the first draft by Dr. José Manuel Barrera-Andrade from Mexico's National Polytechnical Institute (*IPN, Instituto Politécnico Nacional in Spanish*).

References Cited

Aelterman, P., K. Rabaey, P. Clauwaert and W. Verstraete. 2006. Microbial fuel cell for wastewater treatment. Water Sci. Technol. 54: 9–15.
Affam, A.C. and M. Chaudhuri. 2013. Degradation of pesticides chlorpyrifos, cypermethrin and chlorothalonil in aqueous solution by TiO_2 photocatalysis. J. Environ. Manage. 130: 160–165.
Alcántara, V. 2014. Caracterización y diagnóstico de la contaminación por plaguicidas en el lago de Xochimilco. Doctoral Thesis, UNAM, México.
An, J., B. Kim, J. Nam, H.Y. Ng and I.S. Chang. 2013. Comparison in performance of sediment microbial fuel cells according to depth of embedded anode. Bioresour. Technol. 127: 138–142.
Andrade-Eiroa, A., M. Canle, V. Leroy-Cancellieru and V. Cerdá. 2015. Solid phase extraction of organic compounds: A critical review. Part I. Trends Analyt. Chem. 80: 1–42.
Andrade-Eiroa, A., M. Canle, V. Leroy-Cancellieru and V. Cerdá. 2016. Solid phase extraction of organic compounds: A critical review. Part II. Trends in Analytical Chemistry 80: 655–667.
APHA-AWWA-WPCF. 1998a. Standard methods for the examination of water and wastewater. American Public Health Association, American Water Works Association, Water Pollution Control Federation. 19th Edition. Washington, DC, US. pp. 4-149,151,195, 197 and 5-4,5,19, 20.

APHA-AWWA-WPCF. 1998b. Standard methods for the examination of water and wastewater. American Public Health Association, American Water Works Association, Water Pollution Control Federation. 19th Edition. Washington, DC, US. pp. 8-49–8-59.

Arias, J.L., C. Rombaldi, S.S. Caldas and E.G. Primel. 2014. Alternative sorbents for the dispersive solid-phase extraction step in quick, easy, cheap, effective, rugged and safe method for extraction of pesticides from rice paddy soils with determination by liquid chromatography tandem mass spectrometry. J. Chromatogr. A. 1360: 66–75.

Baltazar, M.t., R.J. Dinis, M.L. Bastos, A.M. Tsatsakis, J.A. Duarte and F. Carvalho. 2014. Pesticides exposure as etiological factors of Parkinson's disease and other neurodegenerative diseases—a mechanistic approach. Toxicology Letter 230(2): 85–103.

Barceló, D. and M.J.L. López. 2008. Contaminación y calidad química del agua: el problema de los contaminantes emergentes. pp. 2–4. *In*: Jornadas de presentación de resultados: El estado ecológico de las masas de agua. Panel científico-técnico de seguimiento de la política de aguas. Sevilla, Spain. In Spanish.

Benzaquén, T.B., I. Curcio, M.A. Isla and O.M. Alfano. 2009. Degradación del herbicida atrazina en agua mediante los procesos Fenton y foto-Fenton. Desarrollo Tecnológico para la Industria Química 23(1): 1–10.

Bielicka-Daszkiewicz, K. and A. Voelkel. 2009. Theoretical and experimental methods of determination of the breakthrough volume of SPE sorbents. Talanta 80: 614–621.

Borja-Urzola, A.C. 2017. Influencia de la materia orgánica en la extracción en fase sólida de la atrazina y dos de sus metabolitos de degradación. Caso de estudio Canales de Xochimilco. M.S. Thesis,UNAM, México.

Chávez-López, C., A. Blanco-Jarvio, M. Luna-Guido, L. Dendooven and N. Cabirol. 2011. Removal of methyl parathion from a chinampa agricultural soil of Xochimilco Mexico: A laboratory study. Eur. J. Soil Biol. 47: 264–269.

Cheng, S., H. Liu and B.E. Logan. 2006. Power densities using different cathode catalysts (Pt and CoTMPP) and polymer binders nafion and PTFE in single chamber microbial fuel cells. Environ. Sci. Technol. 40: 364–369.

Chiranjeevi, P., R. Chandra and S.V. Mohan. 2013. Ecologically engineered submerged and emergent macrophyte based system: An integrated eco-electrogenic design for harnessing power with simultaneous wastewater treatment. Ecological Engineering 5: 181–190.

Corbella, C., M. Guivernau, M. Viñas and J. Puigagut. 2015. Operational, design and microbial aspects related to power production with microbial fuel cells implemented in constructed wetlands. Water Research 84: 232–242.

Da Cunha, J., L. Pinelli, L.M.I. Bellini, D. Davyt and S.A. Fernández. 2013. Determinación de atrazina e intermediarios de biodegradación en enriquecimientos bacterianos provenientes de cursos de agua superficial de Uruguay. Innotec. 8: 23–29. In Spanish.

Daughton, C. 2004. Non-regulated water contaminants: emerging research. Environ. Impact Asses. 24(7): 711–732.

Derylo-Marczewska, A., M. Blachnio, A.W. Marczewski, A. Swiatkowski and B. Buczek. 2017. Adsorption of chlorophenoxy pesticides on activated carbon with gradually removed external particle layers. Chem. Eng. J. 308: 408–418.

Díaz-Torres, E., R. Gibson, F. González-Farías, A.E. Zarco-Arista and M. Mazari-Hiriart. 2013. Endocrine disruptors in the Xochimilco wetland, Mexico City. Water Air Soil Pollut. 224: 1586–1588.

DOF. 1992. Declaratoria que establece como zona prioritaria de preservación y conservación del equilibrio ecológico y se declara como área natural protegida, bajo la categoría de zona sujeta a conservación ecológica, la superficie que se indica de los ejidos de Xochimilco y San Gregorio Atlapulco, D.F. (Segunda publicación). In Diario Oficial de la Federación. Volume. No. 5. 1st section. pp. 53–58. Mexico. In Spanish.

Doherty, L., Y. Zhao, X. Zhao, Y. Hu, X. Hao, L. Xu et al. 2015a. A review of a recently emerged technology: Constructed wetland—Microbial fuel cells. Water Res. 85: 38–45.

Doherty, L., X. Zhao, Y. Zhao and W. Wang. 2015b. The effects of electrode spacing and flow direction on the performance of microbial fuel cell-constructed wetland. Ecological Engineering 79: 8–14.

Durán-de-Bazúa, C., A. Guido-Zárate, T. Huanosta, R.M. Padrón-López and J. Rodríguez-Monroy. 2008. Artificial wetlands performance: nitrogen removal. Water Sci. Technol. 58(7): 1357–1360.

Eggen, T., M. Moeder and A. Arukwe. 2010. Municipal landfill leachates: A significant source for new and emerging pollutants. Sci. Total Environ. 408(21): 5147–5157.

Erazo-Cortés, G.I. 2015. Evaluación de medios de empaque para un reactor que simula un humedal artificial asistido electroquímicamente. Professional Thesis in Chemical Engineering, UNAM, México.

Eurachem. 2016. Guía Eurachem. La adecuación al uso de los métodos analíticos. Una guía de laboratorio para la validación de métodos y temas relacionados. Eurolab, España. Primera Edición Española. In line: https://www.eurachem.org/images/stories/Guides/pdf/MV_guide_2nd_ed_ES.pdf.

Falcón, A., E. Lozano and K. Juárez. 2009. Bioelectricidad. BioTecnología. 13(3): 62–78. In Spanish.

Fang, H., J. Lian, H. Wang, L. Cai and Y. Yu. 2015a. Exploring bacterial community structure and function associated with atrazine biodegradation in repeatedly treated soils. J. Hazard. Mater. 286: 457–465.

Fang, Z., H.L. Song, N. Cang and X.N. Li. 2013. Performance of microbial fuel cell coupled constructed wetland system for decolorization of azo dye and bioelectricity generation. Bioresour. Technol. 144: 165–171.

Fang, Z., H. Song, N. Cang and X. Li. 2015b. Electricity production from Azo dye wastewater using a microbial fuel cell coupled constructed wetland operating under different operating conditions. Biosensors & Bioelectronics 68: 135–41.

Farajzadeh, M.A., A. Yadegharia and L. Khoshmaramb. 2017. Combination of dispersive solid phase extraction and dispersive liquid–liquid microextraction for extraction of some aryloxy pesticides prior to their determination by gas chromatography. Microchemical Journal 131: 182–191.

Fent, K., A. Weston and D. Caminada. 2006. Ecotoxicology of human pharmaceuticals. Aquat. Toxicol. 76: 122–159.

Fuentes, M.S., E.E. Raimondo, M.J. Amoroso and C.S. Benimeli. 2017. Removal of a mixture of pesticides by Streptomyces consortium: influence of different soil systems. Chemosphere 173: 359–367.

García, G.C., M.P. Gortáres and P. Drogui. 2011. Contaminantes emergentes: efectos y tratamientos de remoción. Química Viva. 2(10): 96–105. In Spanish.

García-Luna, V.J., M. Bernal-González, F.A. García-Jiménez and M.C. Durán-Domínguez-de-Bazúa. 2017. Xochimilco, Mexico, a natural wetland or an artificial one? pp. 1–24. *In*: M.C. Durán-Domínguez-de-Bazúa, A.E. Navarro-Frómeta and J.M. Bayona-Termens [eds.]. Artificial or Constructed Wetlands: A Suitable Technology for Sustainable Water Management. Science Publishers/CRC Press, Boca Raton, FL, US.

García-Mercado, H.D., G. Fernández-Villagómez, M.A. Garzón-Zúñiga and M.C. Durán-Domínguez-de-Bazúa. 2017. Remediation of mercury polluted soils using artificial wetlands. Int. J. Phytoremediation 19(1): 3–13.

Ge, Z., J. Li, L. Xiao, Y. Tong and Z. He. 2014. Recovery of electrical energy in microbial fuel cells: Brief review. Environ. Sci. Technol. 1: 137–141.

Gil, M.J., M.A. Soto, J.I. Usma and D.O. Gutiérrez. 2012. Contaminantes emergentes en aguas, efectos y posibles tratamientos. Producción Limpia 7(2): 52–73. In Spanish.

Hansen, A., G. Treviño, H. Márquez, M. Villada, C. González and A. Guillén 2013. Atrazina: Un herbicida polémico. Revista Internacional de Contaminación Ambiental 29: 65–84. In Spanish.

He, Z., S.D. Minteer and L.T. Angenent. 2005. Electricity generation from artificial wastewater using an upflow microbial fuel cell. Environ. Sci. Technol. 39: 5262–5267.

Helder, M., W.S. Chen, E.J.M. Van der Harst, D.P.B.T.B. Strik, H.V.M. Hamelers and C.J.N. Buisman 2013. Electricity production with living plants on a green roof: Environmental performance of the plant-microbial fuel cell. Biofuels, Bioproducts and Biorefining 7: 52–64.

Hernández-Arriaga, C.M. 2017. Biodegradación del herbicida atrazina por un consorcio microbiano aislado de los canales de Xochimilco. M.S. Thesis, UNAM, Mexico.

Herrero, O., P. Pérez, J. Fernández, L. Carvajal, A. Pedropadre and M. Hazen. 2012. Toxicological evaluation of three contaminants of emerging concern by use of the *Allium cepa* test. Mutation Research 743: 20–24.

Hu, Y.S., Y.Q. Zhao and A. Rymszewicz. 2014. Robust biological nitrogen removal by creating multiple tides in a single bed tidal flow constructed wetland. Sci. Total Environ. 471: 1197–1204.

Hurtado-Sánchez, M.C., R. Romero-González, M.I. Rodríguez-Cáceres, I. Durán-Merásand and A. Garrido-Frenich. 2013. Rapid and sensitive on-line solid phase extraction-ultra high performance liquid chromatography-electrospray-tandem mass spectrometry analysis of pesticide in surface waters. J. Chromatogr. A. 1305: 193–202.

INEGI. 2008. Xochimilco. Distrito Federal. Cuaderno estadístico delegacional 2008. Instituto Nacional de Geografía y Estadística. Aguascalientes, Ags. Mexico. In Spanish.

INEGI. 2016a. México en Cifras. Instituto Nacional de Estadística y Geografía. http://www.beta.inegi.org.mx/app/areasgeograficas/?ag=09. 14/09/2016. In Spanish.

INEGI. 2016b. Anuario estadístico y geográfico de los Estados Unidos Mexicanos 2016. Instituto Nacional de Geografía y Estadística. Aguascalientes, Ags. México. In Spanish.

Khan, J.A., N.S. Shah and H.M. Khan. 2015. Decomposition of atrazine by ionizing radiation: kinetics, degradation pathways and influence of radical scavengers. Sep. Purif. Technol. 156: 140–147.

Kim, G.T., G. Webster, W.T. Wimpenny, B.H. Kim, H.J. Kim and A.J. Weighman. 2006. Bacterial community structure, compartmentalization and activity in a microbial fuel cell. Appl. Microbiol. 101: 698–710.

Kummerer, K. 2001. Drugs in the environment: Emission of drugs, diagnostic aids and disinfectants into wastewater by hospitals in relation to other sources. A review. Chemosphere 45: 957–969.

La Cecilia, D. and F. Maggi. 2016. Kinetics of atrazine, deisopropylatrazine, and deethylatrazine soil biodecomposers. J. Environ. Manage. 183: 673–686.

Lapworth, D., N. Baran, M. Stuart and R. Ward. 2012. Emerging organic contaminants in groundwater: A review of sources, fate and occurrence. Environ. Pollution 163: 287–303.

Lee, J.N., T. Phung, I.S. Chang, B.H. Kim and H.C. Sung. 2003. Use of acetate for enrichment of electrochemically active microorganism and their 16s rDNA analyses. FEMS Microbiol. Lett. 223: 185–191.

Legorreta, J. 2004. El agua de Xochimilco y su región lacustre. La Jornada daily newspaper. http://www.jornada.unam.mx/2004/05/31/eco-d.html. In Spanish.

Li, H., H.H. Zeng and Y.P. Liang. 2014. Removal of organochlorine pesticides in constructed wetlands. Applied Mechanics and Materials 692: 40–43.

Liu, H. and B.E. Logan. 2004. Electricity generation using an air-cathode single chamber microbial fuel cell in the presence and absence of a proton exchange membrane. Environ. Sci. Technol. 38: 4040–4046.

Liu, R., Y. Zhao, L. Doherty, Y. Hu and H. Hao. 2015. A review of incorporation of constructed wetland with other treatment processes. Chem. Eng. J. 279: 220–230.

Liu, S., H. Song, X. Li and F. Yang. 2013. Power generation enhancement by utilizing plant photosynthate in microbial fuel cell coupled constructed wetland system. International Journal of Photoenergy. Article ID 172010, 10 pages.

Logan, B., B. Hamelers, R. Rozendal, U. Schröder, J. Keller, S. Freguia et al. 2006. Microbial fuel cells: Methodology and technology. Environ. Sci. Technol. 40(17): 5181–5192.

Melian, J.A.H., J. Arana, J.A. Ortega, F.M. Munoz, E.T. Rendon and J.P. Pena. 2008. Comparative study of phenolics degradation between biological and photocatalytic systems. Energy Eng. 130: 1–7.

Min, G., S. Wang, H. Zhu, G. Fang and Y. Zhang. 2008. Multi-walled carbon nanotubes as solid phase extraction adsorbents for determination of atrazines and its principal metabolites in water and soil samples by gas chromatography-mass spectrometry. Sci. Total Environ. 396: 79–85.

Miranda-Méndez, O.H. 2017. Producción de electricidad en sistemas de humedales artificiales asistidos electroquímicamente: flujo continuo versus flujo intermitente. M.S. Thesis, UNAM, Mexico.

Moldoveanu, S. 2015. Modern sample preparation for chromatography. 1st ed. Elsevier. pp. 191–286.

Oon, Y.L., S.A. Ong, L.N. Ho, Y.S. Wong and F.A. Dahalan. 2016. Synergistic effect of up-flow constructed wetland and microbial fuel cell for simultaneous wastewater treatment and energy recovery. Bioresour. Technol. 203: 190–197.

Oon, Y.L., S.A. Ong, L.N. Ho, Y.S. Wong, Y.S. Oon and H.K. Lehl. 2015. Hybrid system up-flow constructed wetland integrated with microbial fuel cell for simultaneous wastewater treatment and electricity generation. Bioresour. Technol. 186: 270–275.

Orduña-Bustamante, M.A., M. Vaca-Mier, J.A. Escalante-Estrada and C. Durán-Domínguez-de-Bazúa. 2013. Nitrogen and potassium variation on contaminant removal for a vertical subsurface flow lab scale constructed wetland. Bioresour. Technol. 102: 7745–7754.

Pawliszyn, J. 2012. Comprehensive sampling and sample preparation. Vol. 2. Elsevier. US. pp. 273–297.

Poole, C.F., A.D. Gunatilleka and R. Sethuraman. 2000. Contributions of theory to method development in solid phase extraction. J. Chromatogr A 885(1-2): 17–39.

PROY-NMX-AA-003/3-SCFI-2008. Proyecto de Norma Mexicana. WATER ANALYSIS.- RESIDUAL WATER SAMPLING. On line: http://www.economia-nmx.gob.mx/normas/nmx/2009/proy-nmx-aa-003-1-scfi008.pdf.

Rodríguez-Monroy, J. and C. Durán-de-Bazúa. 2006. Remoción de nitrógeno en un sistema de tratamiento de aguas residuales usando humedales artificiales de flujo vertical a escala de banco/Nitrogen removal from wastewaters on a bench scale vertical flow artificial wetlands system. Tecnol. Ciencia Ed. (IMIQ). 21(1): 25–33. In Spanish.

Ruiz, I., M.A. Diaz, B. Crujeiras, J. Garcia and M. Soto. 2010. Solids hydrolysis and accumulation in a hybrid anaerobic digester-constructed wetlands system. Ecol. Eng. 36: 1007–1016.

Ruiz-López, V., M.R. González-Sandoval, J.A. Barrera-Godínez, G. Moeller-Chávez, E. Ramírez-Camperos and M.C. Durán-Domínguez-de-Bazúa. 2010. Remoción de Cd y Zn de una corriente acuosa de una empresa minera usando humedales artificiales/Cadmium and zinc removal from a mining reprocessing aqueous stream using artificial wetlands. Tecnol. Ciencia Ed. (IMIQ). 25(1): 27–34. In Spanish.

Salgado-Bernal, I., M.E. Carballo-Valdés, A. Martínez-Sardiñas, M. Cruz-Arias and C. Durán-Domínguez. 2012. Interacción de aislados bacterianos rizosféricos con metales de importancia ambiental. Rev. Tecnol. Ciencias Agua. 3(3): 83–95. In Spanish.

Salinas-Juárez, M.G. 2016. Estudio de la generación de electricidad en un humedal artificial asistido electroquímicamente. Doctoral Thesis, UNAM, Mexico.

Salinas-Juárez, M.G. 2017. Electrochemically assisted constructed wetlands: Generating electricity from wastewater treatment, a review. pp. 241–261. *In*: M.C. Durán-Domínguez-de-Bazúa, A.E. Navarro-Frómeta and J.M. Bayona-Termens [eds.]. Artificial or Constructed Wetlands: A Suitable Technology for Sustainable Water Management. Science Publishers/CRC Press, Boca Raton, FL, US.

Salinas-Juárez, M.G., P. Roquero and M.C. Durán-Domínguez-de-Bazúa. 2011. Estudio de la generación de energía eléctrica con una celda de combustible microbiana con plantas en forma acoplada con el tratamiento de agua residual en un humedal artificial a escala de laboratorio. Ph.D. Thesis Protocole, UNAM, Mexico.

Salinas-Juárez, M.G., P. Roquero and M.C. Durán-Domínguez-de-Bazúa. 2016. Plant and microorganisms support media for electricity generation in biological fuel cells whit living hydrophytes. Bioelectrochemistry 112: 145–152.

Sene, L., A. Converti, G. Secchi and R. Simão. 2010. New aspects on atrazine biodegradation. Braz Arch .Biol. Technol. 53(2): 487–496.

Sharma, A.K., R.K. Tiwari and M.S. Gaur. 2016. Nanophotocatalytic UV degradation system for organophosphorus pesticides in water samples and analysis by Kubista model. Arab. J. Chem. 9: S1755–S1764.

Sinclair, C. and A. Boxall. 2003. Assessing the ecotoxicity of pesticide transformation products. Sci. Total Environ. 37: 4617–4625.

Srivastava, P., A.K. Yadav and B.K. Mishra. 2015. The effects of microbial fuel cell integration into constructed wetland on the performance of constructed wetland. Bioresour. Technol. 195: 223–230.

Stashenko, E and J.R. Martínez. 2011. Preparación de la muestra: Un paso crucial para el análisis por GC-MS. Scientia Chromatographical 3(1): 25–49. In Spanish.

Strik, D.P.B.T.B., H.V.M. Hamelers, J.F.H. Snel and C.J.N. Buisman. 2008. Green electricity production with living plants and bacteria in a fuel cell. International Journal of Energy Research 32: 870–876.

Stuart, M., D. Lapworth, E. Crane and A. Hart. 2012. Review of risk from potential emerging contaminants in UK groundwater. Sci. Total Environ. 416: 1–21.

Swartz, C., S. Reddy, J. Benotti, H. Yin, L. Barber and B. Brownawell. 2006. Steroid estrogens, nonylphenolethoxylate metabolites, and other wastewater contaminants in groundwater affected by a residential septic system on Cape Cod, MA. Environ. Sci. Technol. 40(16): 4894–4902.

Tamrakar, U., S.B. Mathew, V.K. Gupta and A.K. Pillai. 2009. Determination of atrazine in environmental and biological samples using solid phase extraction and spectrophotometry. J. Analyt. Chem. 64(4): 386–389.

UNESCO. 1988. Convention concerning the protection of the world cultural and natural heritage. Report of the World Heritage Committee. United Nations Educational, Scientific and Cultural Organization. Paris, France.

Valek, G. 2000. Agua: reflejo de un valle en el tiempo. Offset, S.A. de C.V. Mexico. In Spanish.

Vallée, R., S. Dousset and D. Billet. 2016. Influence of substrate water saturation on pesticide dissipation in constructed wetlands. Environ. Sci. Pollut. Res. 23(1): 109–119.

Vandermaesen, J., B. Horemans, K. Bers, P. Vandermeeren, S. Herrmann and A. Sekhar. 2016. Application of biodegradation in mitigating and remediating pesticide contamination of freshwater resources: state of the art and challenges for optimization. Appl. Microbiol. Biotechnol. 100(17): 7361–7376.

Velázquez-González, A., G. Solórzano-Ochoa, C. Gutiérrez-Palacios, J. Villalba-Caloca, J. Pérez-Neria, L. Alfaro-Ramos et al. 2005. Estrategia para la generación de una base de datos sobre residuos peligrosos corrosivos, reactivos, explosivos, tóxicos e inflamables, CRETI, en unidades hospitalarias/Strategy for generating a data base on hazardous residues, corrosive, reactive, explosive, toxic, and flammable, CRETF, in hospitals. Tecnol. Ciencia Ed. (IMIQ) 20(2): 45–56.

Venkata, M., G. Mohanakrishna and P. Chiranjeevi. 2011. Sustainable power generation from floating macrophytes based ecological microenvironment through embedded fuel cells along with simultaneous wastewater treatment. Bioresour. Technol. 102: 7036–7042.

Villaseñor-Camacho, J., M.C. Montaño-Vico, M. Andres, R. Rodrigo, F.J. Fernández-Morales and P. Cañizares-Cañizares. 2014. Energy production from wastewater using horizontal and vertical subsurface flow constructed wetlands. Environ. Eng. Manage. J. 13(10): 2517–2523.

Villaseñor, J., P. Capilla, M.A. Rodrigo, P. Cañizares and F.J. Fernández. 2013. Operation of a horizontal subsurface flow constructed wetland – Microbial fuel cell treating wastewater under different organic loading rates. Water Research 47(17): 6731–6738.

Vymazal, J. 2013. The use of hybrid constructed wetlands for wastewater treatment with special attention to nitrogen removal: a review of a recent development. Water Res. 47: 4795–4811.

Watanabe, N., B. Bergamaschi, K. Loftin, M. Meyer and V. Harter. 2010. Use and environmental occurrence of antibiotics in freestall dairy farms with manure forage fields. Environ. Sci. Technol. 44(17): 6591–6600.

Wu, S., P. Kuschk, H. Brix, J. Vymazal and R. Dong. 2014. Development of constructed wetlands in performance intensifications for wastewater treatment: A nitrogen and organic matter targeted review. Water Res. 57: 40–55.

Xu, L., Y. Zhao, L. Doherty, Y. Hu and X. Hao. 2016. Promoting the bio-cathode formation of a constructed wetland-microbial fuel cell by using powder activated carbon modified alum sludge in anode chamber. Scientific Reports. 6: Article 26514.

Yadav, A.K., P. Dash, A. Mohanty, R. Abbassi and B.K. Mishra. 2012. Performance assessment of innovative constructed wetland-microbial fuel cell for electricity production and dye removal. Ecological Engineering 47: 126–131.

Zabaleta-Solís, D. 2017. El Proyecto Unesco-Xochimilco (PUX), en la Ciudad de México. Alcances y límites de la gobernanza democrática en iniciativas propuestas por gobiernos locales con institucionalidad débil. Instituto de Investigación y Debate sobre la Gobernanza, IIDG. NGO. http://www.institut-gouvernance.org/es/experienca/fiche-experienca-27.html.

Zambrano, L., V. Contreras, M. Mazari-Hiriart and A.E. Zarco-Arista. 2009. Spatial heterogeneity of water quality in a highly degraded tropical freshwater ecosystem. Environ. Manage. 43: 249–263.

Zhao, Y.S., Q. Collum, S. Phelan, M. Goodbody, T. Doherty and L. Hu. 2013. Preliminary investigation of constructed wetland incorporating microbial fuel cell: batch and continuous flow trials. Chem. Eng. 229: 364–370.

Zuñiga, E., E. Arellano, L. Camarena, W. Daesslé, C. Von-Glascoe, J.C. Leyva and B. Ruíz. 2012. Daño genético y exposición a plaguicidas en trabajadores agrícolas del Valle de San Quintín, Baja California, México. Rev. Salud. Ambiental. 12(2): 93–101.

4

Constructed Wetlands for the Tertiary Treatment of Municipal Wastewaters

Case Studies in Mexico at Mesocosm Level

Jorge Antonio Herrera-Cárdenas and
*Amado Enrique Navarro-Frómeta**

INTRODUCTION

The demand for water increases proportionally to population growth. About 70% of the freshwater is used in agriculture (a higher percent in the least developed countries). Energy production, industry, and other domestic purposes use about 15, 5, and 10% respectively and only less than 1% is used for drinking. Urbanization and improvements in sanitation will increase the demand for water for domestic uses by about 30% by 2050. To face all these needs, an appropriate integrated water management that considers the challenges imposed by climate change to guarantee the water supply and its associated energy consumption is a crucial aspect for solving actual development problems in a sustainable way (WWAP 2012, 2014, 2017).

During its use, water is polluted and needs treatment to be safely disposed in the environment or reused. About 56% of the used water is released as wastewater from municipal and industrial activities or drainage water from agriculture. High income, upper, middle, and low-income countries treat around 70, 38, 28, and 8% of the wastewater, respectively. The disposal of untreated wastewater in the environment poses severe threats to human and ecosystems health. Water pollution accounts for

Universidad Tecnológica de Izúcar de Matamoros, Prolongación Reforma 168, Barrio de Santiago Mihuacán, 74420 Izúcar de Matamoros, Puebla, México.
* Corresponding author: navarro4899@gmail.com

a high fraction of the global burden of disease. For example, diarrheal diseases are responsible for about 20% of all deaths in children under five years. A large proportion of these diarrheal diseases is caused by pathogens present in polluted drinking water. In 2012, an estimated of 842,000 deaths in the middle- and low-income countries were caused by contaminated drinking water and inappropriate or inadequate sanitation services (Prüss-Üstün et al. 2016; WWAP 2017).

Considering the increasing water needs, the decrease of the sources of clean water, and the increase in the volume of wastewater, it is evident that the perception of wastewater has to be changed from a burden to be disposed of or a nuisance to be ignored to a potentially safe, affordable, and sustainable source of water, energy, other useful by-products, considering the reduction of pollution at the source, removing of contaminants, the treated wastewater reuse, and the recovery of valuable materials and energy (the "four R′s"). This means that treatment of wastewater is very necessary. Obviously, the extension of treatment will be conditioned by the use that will be given later to the treated water, available resources, and social issues.

At the present time is considered that low-cost decentralized systems are a better option than the conventional treatment plants, especially for countries with warm climates (Ávila et al. 2014). It has been estimated that the investment requirements for low-cost treatment systems represent only 20–50% of that of centralized systems, with lower operation and maintenance costs. The dissemination of knowledge and lessons about the available low-cost technologies for water quality improvement and the replication of best practices require that both authorities and the stakeholders become involved in the process of sewage appropriation as a valuable good (Angelakis and Snyder 2015; UN-Water 2015; WWAP 2017). In a preeminent place between the green technologies are the constructed wetlands. Some considerations about its use for tertiary treatment of wastewater are discussed below.

Tertiary Treatment: Methods and Costs

Tertiary treatment, also known as advanced treatment, has the purpose of providing a final polishing stage for improving water quality in order to meet more stringent or specific requirements before it is reused, recycled, or discharged to the environment. The above is achieved through the removal of inorganic compounds, nutrients such as the nitrogen and phosphorus, suspended and colloidal solids, specific persistent, priority or emerging pollutants (pesticides, detergents, fragrances, drugs, etc.), as well as pathogens (bacteria, viruses and parasites) remaining from the previous primary and secondary treatments.

Tertiary purification techniques include different physical, chemical, biological processes and/or their combinations such as: chlorination, ozonation, ultraviolet irradiation; reverse osmosis, micro- and ultrafiltration with membrane technologies; treatment with granular activated carbon and ion-exchange with resins, slow filtration with sand or diatomaceous earth; physicochemical coagulation and precipitation; air or steam entrainment processes; electrochemical, oxidation and advance oxidation technologies, biological processes, and eco-friendly green technologies based on natural processes (e.g., different kind of ponds, constructed wetlands), each of them targeting specific pollutants and achieving reasonable high specific removals for them, and thus producing different water qualities for different uses (Topare et al. 2011;

Abdel-Raouf et al. 2012; Gupta et al. 2012; Lee et al. 2015; Mohammed and ElBably 2016).

The decision about the tertiary polishing method to use depends on different variables: available financial resources (one of the determining factors), characteristics of the water to treat and the effluent quality needed for a specific use, the cost/benefit ratio, etc. Frequently, the methods used to assess the economic feasibility tend to focus on internal costs, while the external benefits are not considered disregarding the important role they play in the overall economic balance of the treatment process evaluation. Often, not enough attention is paid to the sustainability of the tertiary treatment project during its analysis. In this sense, multi-criteria approaches allow the examination of all the dimensions of the problem, including life-cycle assessment, considering the greenhouse emissions and keeping in mind that wastewater management is always a socially desirable and economically rewarding option, which must be considered in the decision-making process (Molinos-Senante et al. 2011; Remy et al. 2014; Hernández-Sancho et al. 2015; Plakas et al. 2016).

The costs of water purification encompass different items, including disposal, maintenance, personnel, and energy, which represents more than 80% of overall costs and tertiary treatment represents around 40% (Mas-Ortega 2016). There is sufficient information related to costs regarding wastewater treatment, however, it is complicated to carry out comparative studies because the approaches are very diverse. One of the limitations of the techniques of tertiary treatment is the high energy and economic consumption; for example, with an ionizing disinfection treatment up to 90% of elimination of bacteria is achieved. However, it has an energy expenditure of 0.12 Wh L^{-1} d^{-1}, and 36.8 and 96.53 Wh L^{-1} d^{-1} with ozone and UV radiation, respectively (Lee et al. 2015); the activated sludge process requires approximately 0.6 kWh to treat an m^3 of water (Mc Carty et al. 2011). Advanced oxidation treatments with hydrogen peroxide (commercial grade) range from \$390.00 to \$500.00 per ton of oxidizing agent (Zhu and Logan 2013). Ozone has shown a high removal of pathogenic microorganisms, but not of certain pollutants of interest and their byproducts (Zimmermann et al. 2011).

The goal of achieving disinfection and reducing the consumption of disinfectants and operational costs remains a challenge in wastewater treatment (Shah et al. 2012; Li et al. 2017). Disinfectants inactivate pathogens in water sources (Stalter et al. 2016). There are several methods of disinfection (Victoria 2002), such as ozonation, chlorination, and UV radiation; however, chlorination is one of the most economical methods which has been widely used, has allowed a reduction of hydrotransmissible diseases (Sedlak and Gunten 2011), and represents a possible solution to minimize the dispersion of resistance to antibiotics by bacteria. Among these preoxidants, chlorine, chlorine dioxide, and ozone form disinfection byproducts chlorite and bromate ion, respectively. The use of chlorine in disinfection processes, besides representing a viable option, also has some disadvantages such as toxicity, corrosivity, and risks during transportation and storage (USEPA 1999). The less toxic species are chlorinated lime and sodium hypochlorite, in addition to being the ones that need less equipment. Another problem of disinfectants is that they react with organic matter and bromide/iodide to form disinfection byproducts (Stalter et al. 2016). Chlorination of water can generate a variety of chlorinated and brominated disinfection byproducts because bromide is found naturally in various water sources (Phan and Zhang 2013).

The main groups of compounds formed are trihalomethanes, halo-acetic acids, and halo-acetonitriles (Kanchanamayoon 2015). Organic matter and some organic pollutants may react to form various disinfection byproducts (Richardson and Postigo 2016). The pH value is one of the factors favoring the formation of these compounds. For example, the triclosan contaminant, which is used in many antibacterial hand soaps, can be transformed into the presence of chlorine to form six DBPs, including chloroform, three chlorophenoxyphenols, and two chlorophenols.

Natural Wetlands and their Role in Tertiary Polishing

Natural wetlands are areas of transition between the terrestrial and aquatic environment. They are characterized by having two or more of the following attributes: the soil or substrate must be temporarily or permanently saturated with water, presenting a layer of shallow water or groundwater near the surface, whether permanent or temporary and at least periodically, the land must predominantly maintain aquatic or hydrophilic vegetation (Landgrave and Moreno-Casasola 2012). Having been considered as unprofitable areas and sources of diseases, now these areas are understood to provide important ecological functions and services: fish and wildlife habitats, pollution abatement and water quality improvement, protection against floods, the source of cheap useful natural products, sites for recreation, etc. The productivity of wetlands as ecosystems is comparable to that of the rain forests and coral reefs. Natural wetlands historically have played an important role in the natural purification of polluted waters and are the habitat of various species of fauna and flora (Ramsar 2013; USEPA 2016).

Traditionally, natural wetlands have been used to dispose wastewaters, including those from waste water treatment plants (WWTP) (Kinyua 2014). In these systems occur a series of complex physical, chemical, and microbiological processes that create conditions for the development of microorganisms communities that use the pollutants as nutrient and energy sources, being responsible for changing or stabilizing the characteristics of the wastewaters that are discharged in these systems (Haarstad et al. 2012), allowing the removal of pollutants loads, for which they have been described as the kidneys of the natural environment or "water living filters" (Dordio et al. 2008), because of the functions they can perform in the hydrological and chemical cycles (Batenganya et al. 2015), and as biological supermarkets, due to the extensive food webs and the rich biological diversity that sustain, being used even in very cold climate (Chouinard et al. 2014).

During the last decades, there have been important changes in the hydrology of the tropics, which have been caused by natural causes and also as a consequence of human activities. Wetlands are intimately linked to watersheds that receive water runoff, with suspended solids, and nutrients. This makes wetlands very vulnerable to inadequate management of the basin. Changes in sedimentation rates, freshwater runoff, and intertidal floods as a result of human activities at the local and regional level, as well as the direct removal of some natural barriers from human settlements, industry, agriculture, and the subsequent clogging, have directly impacted these ecosystems and natural wetlands which continue to decline both in area and in quality, a problem that is worsening with climate change. As a result, the ecosystem services that wetlands provide to society are diminished (Junk et al. 2013; Bassi et al. 2014; Davidson 2014; Ramsar 2015). However, the elimination of wastewater through proper management (secondary treatment) may favor natural wetlands, increase their productivity and raise the level of water that can serve to mitigate the intrusion of seawater in coastal areas (Day et al. 2004).

Constructed Wetlands

Constructed wetlands (CW), also called artificial or treatment wetlands are engineered systems that mimic natural wetlands for the treatment of different kinds of wastewater as domestic sewage, agricultural, industrial effluents, mining drainage, landfill leachate, storm water, polluted riverine waters, and urban runoff at different scales (Wallace and Knight 2006; Vymazal 2009, 2011; Calheiros et al. 2012). They are mainly composed of vegetation (generally vascular plants), a supporting medium or substrate (soils, porous media), microorganisms, and water. CW remove water pollutants improving water quality such as organic load, total suspended solids, nutrients, fecal indicators, heavy metals, organic compounds and nanomaterials among others. CW have certain advantages compared to other technologies applied during tertiary polishing due to low operating costs and energy consumption, therefore, representing a good alternative to small communities (less than 2000 inhabitants) or where exists budgetary restrictions for the installation of conventional systems and enough land is available (Kadlec and Wallace 2009; Vymazal 2010; Westerhoff et al. 2014; Wu et al. 2015a; Wu et al. 2015b).

In CW occur complex physical, chemical, and biological processes that remove the pollutants: microbial degradation or transformation of pollutants in the biofilms; sedimentation of solids; adsorption mediated by the supporting media; evaporation; direct oxidation and photooxidation (where light access the system), of the pollutants; bacterial die-off and predation; phytoremediation processes and in particular, the interactions between both processes. Most biochemical reactions that transform pollutants occur at the solid-liquid interfaces present in water, granular medium, rhizosphere, and especially on the bacterial biofilms. Each of the components that make up the system has an important role in its functioning; water is the vehicle for the entry and distribution of nutrients within the system, as well as being a support for plant material and establishment of bacterial colonies. Its main function is to provide the necessary nutrients for the development and metabolism of bacteria, which have an important influence on the processes of removal, biodegradation, and biotransformation of the substances that enter it. The supporting media provide the site for biofilms development and also take part in pollutants removal through surface phenomena. Finally the plants perform various functions such as: support for some bacterial colonies, oxygen injection, and supply of biomass through the dry leaves that are deposited on the surface of the system in addition to the regulation of temperature by acting as a barrier against the sunlight (García et al. 2010; Haarstad et al. 2012; Saeed and Sun 2012; Idris et al. 2014; Reddy et al. 2014; Stefanakis et al. 2014; Gökalp et al. 2016; Qomariyah et al. 2017).

The efficiency of these systems depends on the combination of the main parameters such as hydraulic design and residence time, porous media granulometry, operation mode, etc. In terms of the hydraulic design, these systems are classified as surface flow and subsurface flow wetlands. The seconds are subdivided into horizontal and vertical

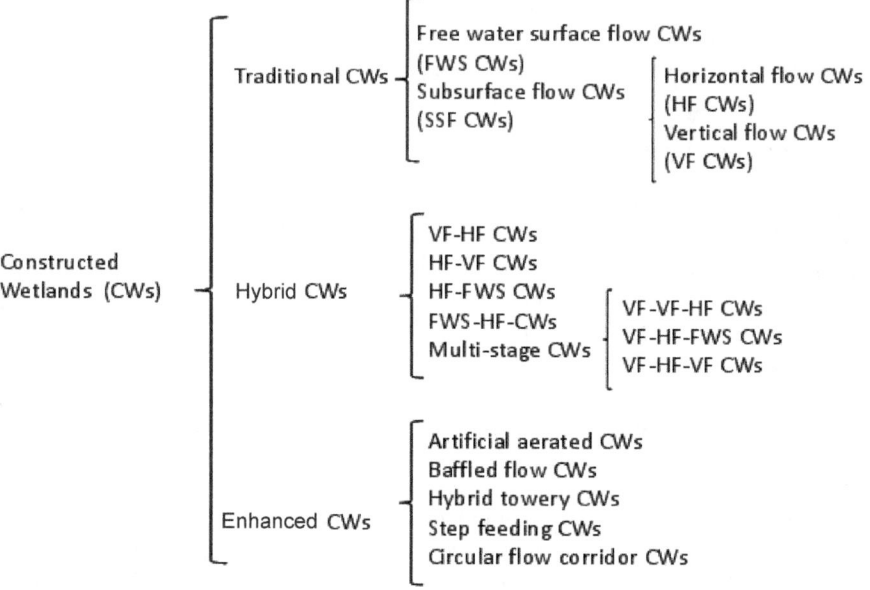

Fig. 1. Classification of constructed wetlands (Wu et al. 2015b).

subsurface flow systems (Fig. 1). Different types of wetlands have different prevailing conditions and mechanisms for pollutants removal, for example, the predominance of aerobic and anaerobic processes and bacteria. The more critical aspect of the CW is the gradual clogging of the porous media as a result of the processes that take place in the system. Therefore, a pretreatment process is essential to enhance the performance and lengthen the service life of the system. Considering the major drawbacks of CW, their combination with other treatment technology is a promising technological approach (Nivala et al. 2012; Gikas et al. 2013; Adrados et al. 2014; Liu et al. 2015; Aiello et al. 2016; Badejo et al. 2017).

Although constructed wetlands represent a cost-effective alternative for the tertiary treatment of sewage of various types, they also present some limitations and aspects that have not yet been understood. These limitations can be grouped according to the stage of system development, such as design, construction, and operation. Some of them are indicated in Table 1.

The low content of organic matter in secondary treated waters may be an obstacle in microbial biodegradation efficiency, for example in hybrid CWs (Wu et al. 2017b). However, the use of plant species (ornamental) that do not require high concentrations of nutrients, not only counteract the problem but also contributes to the improvement of the landscape.

It is well known that the treatment plants are fixed sources of pollutants to surface water bodies. In the case of organic microcontaminants, there is ample evidence that conventional treatment processes are insufficient for their removal, which constitutes a potential risk given the pseudo persistence of these compounds and their effects on the endocrine systems of living organisms (Kosma et al. 2014; Nam et al. 2014; Verlicchi and Zambello 2014). On the other hand, other tertiary treatments such as advanced oxidation processes and membrane filtration, though promising, need

Table 1. Advantages and disadvantages of CWs.

VF CWs	Advantages	• Lower surface area (Abou-Elela et al. 2013) • Efficient odor elimination • Higher efficiency in BOD removal • Better removal performance for TSS • Greater ability to oxidize ammoniacal nitrogen (Vymazal 2010) • Greenhouse gas emissions are lower compared to HF CW (Mander et al. 2014)
	Disadvantages	• Complex flow control • Low organic matter content may result in a low nitrification-denitrification activity • Such systems are more susceptible to clogging compared to horizontal systems (García-Serrano and Corzo-Hernández 2009) • The increase of the hydraulic load decreases the capacity of removal of some priority pollutants as: ammonium, total phosphorus, COD, and BOD_5 (Zhang et al. 2007) • These systems do not have the capacity to reduce environmental risk, due to the resistance of some analgesics and antibiotics (Verlicchi and Sambello 2014). Something that usually does not happen with systems of FWS CWs (Berglund et al. 2014)
HF CWs	Advantages	• Simple flow control • Higher efficiency of removal of total phosphorus
	Disadvantages	• High surface areas, from 2 to 5 m^2 per person (Kadlec et al. 2000) • Long residence times for the elimination of odors • Due to the low presence of oxygen, can result in low denitrification (Hu et al. 2012; Fan et al. 2013) • The recirculation of the effluent can cause problems in this type of system due to the increase of the hydraulic load (Stefanakis and Tsihrintzis 2009) • Emissions of greenhouse gases are higher compared to SFCW (Mander et al. 2014) • Despite advances in the characterization of microbial communities, some aspects of bacterial spatial dynamics are still unknown; The knowledge of these aspects, would allow to optimize the processes of removal (Weber and Legge 2009)

investment and have operating costs, that in most cases in developing countries, exceed the financial possibilities of the local authorities in charge of the WWTP. This remarks the convenience of CW for tertiary polishing of wastewater. Another aspect that has an impact on the selection of CW as an option for tertiary treatment is their relatively low consumption of energy (Yeh and Wu 2009; Plappally and Lienhard 2012). It has been reported that CWs have costs up to 4–5 times less than conventional treatments (USEPA 2000).

The cost of installing these systems is variable and include costs of: design, geotechnical tests, legal requirements, and other unexpected expenses, which are not usually included in the reports. The uncertainty of these estimates could come from data related to the project and factors such as complexity of design details, variation of local legal requirements, undeclared soil conditions, and other specific aspects of the site (BARR 2011). The cost also depends on the location (Woltemade 2000), the area's agricultural potential (Iowa CREP Program Overview 2016), and the amount of nutrients removed (Gachango et al. 2015). According to the CONAGUA catalogs, the cost of a wetland of 1000 m² is approximately $12,000 USD, equivalent to 120 USD/m² considering the rustic and geomembrane work. One third of the amount mentioned above corresponds to the compensation of the owner of the *ejido*[1] where the wetland would be built. According to data from studies in North America, the cost of installing a wetland in a site with high potential for maize cultivation is approximately $2,500.00 USD, and considering a 40 year lifespan the annual cost would be $200.00 USD (Iowa CREP 2016). The cost of these systems has also been reported based on the amount of phosphorus removed (data for 2014), ranging from 44,313 to 165,126 $USD per kilogram of phosphorus removed (20,100 to 74,900 $USD per pound removed) (Nobles et al. 2017). Finally, it should be noted that maintenance and freight costs for total suspended solids range from $7,830.00 USD ha⁻¹ yr⁻¹ and 11.00 to 21.00 $USD per kg of TSS, respectively.

Tertiary Treatment with Free Water Surface Wetlands

Free Water Surface (FWS) flow CWs are the systems that most resemble the natural swamps, with their advantages and drawbacks and are well described in the literature (Vymazal 2010). They are considered to be most aesthetically designed of CWs, attracting wildlife as they provide a valuable habitat for nesting birds and animals because water is exposed to atmosphere and sunlight. However, at the same time, this is a disadvantage because the exposed water body attracts insects such as mosquitoes and increases water losses from direct evaporation. In this sense, they are useful as a water reservoir during the dry season and as tertiary polishing systems in the rainy season in tropical countries (Sun et al. 2015). FWS CWs, through different mechanisms, remove microorganisms and parasites, such as Cryptosporidium, and control the dissemination of antimicrobial resistant bacteria. The study of the abundance, metabolic activity, and functional diversity of the bacterial communities in the inflow and outflow of a FWS CWs polishing secondary treated water, allowed posing the hypothesis that CWs are capable of transforming the original anthropological bacterial community in the WWTP effluent to a bacterial community resembling those found in surface waters. These facts account for the reduction of risks due to pathogens using FWS CWs as an advanced treatment (Mulling et al. 2014; Nasser 2016; Vivant et al. 2016; Wu et al. 2016). FWS CWs have shown their capacity to act as temperature regulators for tertiary effluents, allowing dissolved oxygen availability in hotter months in temperate climates, although they have shown little efficiency in the removal of phosphorus and ammonia (Beutel 2012).

Tertiary Treatment with Horizontal Subsurface Flow Constructed Wetlands

The use HF CWs for different kinds of wastewater treatment, including municipal ones, has been described in the literature (Vymazal 2009, 2010). They are able to

---

[1] See Chapter 1 of this book for the definition of this land property type in Mexico (Note of first editor).

Fig. 2. HF CWs plots and fluvial park in the Besós, River, Barcelona. Map adapted from the leaflet of the park (Diputación de Barcelona 2017). Photos from the authors.

remove a wide range of contaminants such as BOD, COD, TSS, total nitrogen, and total phosphorus and can supply water for irrigation of "no-food" crops with an attractive biomass yield and a complete restitution of evapotranspiration losses, which constitutes a contribution to energy production from renewable sources. Their effluents may be used with caution for irrigation of vegetable crops (a certain persistence of microbial contamination in the irrigated soil has been observed). Besides their removal efficiencies remain unchanged after more than a decade of use (Saeed and Sun 2011; Barbagallo et al. 2014; Castorina et al. 2016). In these systems, the plant biomass plays an important role in the bacterial community present in the system and both are the responsible for most of the removal processes (Chen et al. 2015). As other types of CWs, HF CWs efficiently remove organic micropollutants, disinfection by-products, and pharmaceuticals, a process that can be improved with active aeration (Chen et al. 2014; Auvinen et al. 2017; Vymazal et al. 2017). It has been demonstrated that the physical design of the HF CW can give high operational reliability against clogging (Pozo-Morales et al. 2014).

Although constructed wetland technology is more suitable for rather small residual water flows, the need to augment the streamflow of a highly deteriorated urban stream, the Besós River in Catalonia, Spain, led to use this technology to give tertiary treatment to the effluent water from the Montcadai Reixac activated sludge WWTP. A series of plots were distributed along 3.2 km of the basin (Fig. 2). The water from the tertiary treatment showed the following percentage of removal: 40% for suspended solids; 62% for COD; 20% for NH4-N; 58% for P; and 1.1 log CFU $(100$ mL$)^{-1}$ for fecal coliforms. This allowed creating a fluvial park, for the recreation of the inhabitants of Barcelona as shown in Fig. 4 (Huertas et al. 2006). Moreover, these wetlands have shown their capability to remove emerging contaminants with

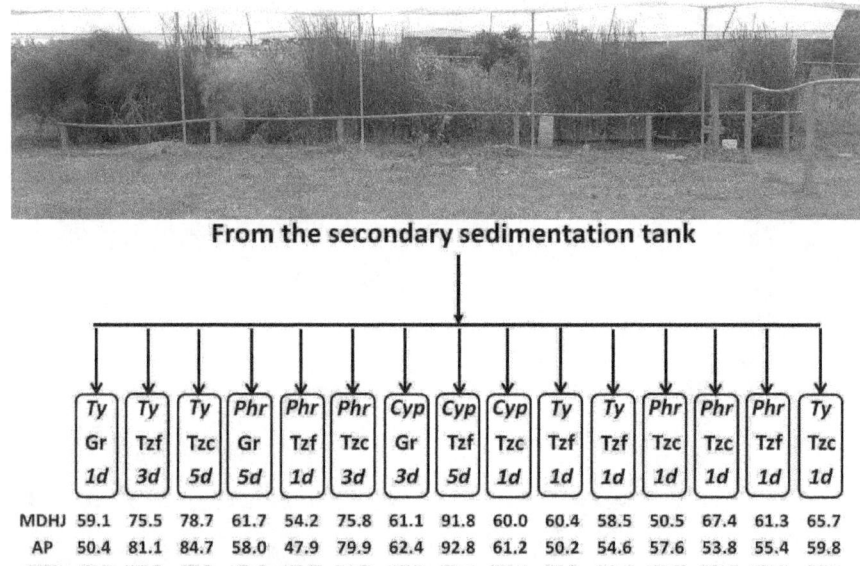

	Ty	Ty	Ty	Phr	Phr	Phr	Cyp	Cyp	Cyp	Ty	Ty	Phr	Phr	Phr	Ty
	Gr	Tzf	Tzc	Gr	Tzf	Tzc	Gr	Tzf	Tzc	Tzf	Tzf	Tzc	Tzc	Tzf	Tzc
	1d	3d	5d	5d	1d	3d	3d	5d	1d	1d	1d	1d	1d	1d	1d
MDHJ	59.1	75.5	78.7	61.7	54.2	75.8	61.1	91.8	60.0	60.4	58.5	50.5	67.4	61.3	65.7
AP	50.4	81.1	84.7	58.0	47.9	79.9	62.4	92.8	61.2	50.2	54.6	57.6	53.8	55.4	59.8
BOD	69.6	82.6	85.2	76.9	75.9	84.2	79.6	91.5	78.0	65.1	74.4	65.2	72.1	77.8	82.3

Fig. 3. General view of the HSSF-CW in the WWTP of Izúcar de Matamoros, experimental design and mean values of the removal percentages of methyl dihydrojasmonate (MDHJ), alkylphenols and BOD$_5$. Ty—*Typha latiffolia*; Phr—*Phragmites australis*; Cyp—*Cyperus papyrus*; Gr—gravel; Tzf—fine volcanic gravel; Tzc—coarse volcanic gravel. 1d, 3d, and 5d—hydraulic residence times in days.

some preference for more hydrophobic and higher molecular mass compounds (Matamoros et al. 2017).

Considering the Besós experience, the tertiary treatment of the effluent of the WWTP of the Izúcar de Matamoros City (Puebla State, México) with HF CWs, was evaluated at mesocosms level using a modified Latin square experimental design with the variables the porous medium, PM (river gravel, G; fine volcanic gravel known as tezontle, Tzf; coarse volcanic gravel, Tzc), macrophyte type, M (*Thypa latifolia*, Ty; *Phragmites australis*, Phr; *Cyperus papyrus*, Cy), and hydraulic residence time, HRT (1, 3, 5 days) (Fig. 3). Obtained results (in Fig. 3 mean removal percentages are shown for selected pollutants), show that the variable having the greatest effect is the HRT, followed by the type of porous media with better results for Tzf, which is related to a larger surface area favoring the establishment of a higher number of bacterial colonies, and, finally, the macrophytes that better removed were *Ty* and *Cy*. However, *Phr* also showed considerable removal, which gives versatility to the systems, an important characteristic for installing these systems by using the local resources (Herrera et al. 2016).

Vertical Subsurface Flow Artificial Wetlands for Tertiary Polishing

The design, pollutants removal mechanisms, and other characteristics of VF CWs have been described in the literature (Vymazal 2010; Stefanakis et al. 2014). These systems (Fig. 4) are loaded intermittently. Therefore, saturation conditions with water

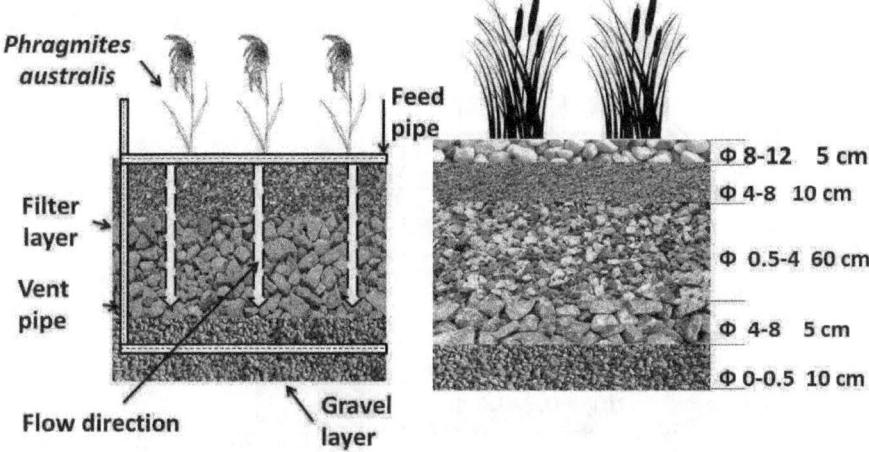

Fig. 4. Structural characteristics and particle sizes in VF CWs.

in the matrix bed are followed by periods of unsaturation, stimulating the oxygen supply. The introduction of atmospheric oxygen to the substrate allows the increase of the rate of oxygen transfer enhancing processes such as oxidation of organic matter and nitrification (Butterworth et al. 2016). Besides, it makes possible the use of a clay-rich soil for water infiltration. Nevertheless, it has been demonstrated that they can be operated in continuous and semi-continuous mode (Fan et al. 2013; Nivala et al. 2013; Bai et al. 2014; Bisone et al. 2017; Kumar and Singh 2017). Such systems are more susceptible to clogging compared to horizontal systems (García-Serrano and Corzo-Hernández 2008). However, the problem is mitigated by the appropriate choice of the filter medium (Stefanakis and Tsihrintzis 2012) and an appropriate distribution of particle size (Fig. 4) which also has a significance influence on the system performance (Bruch 2014).

Vertical Flow (VF) constructed wetlands CWs are highly efficient in the removal of suspended solids, biological oxygen demand (BOD_5), the nitrification of ammonium, as well as in the removal of significant levels of metals, organic compounds, and pathogens. Due to the presence of a greater amount of oxygen, there are no significant anaerobic reactions, which inhibit denitrification processes and the inability to eliminate nitrates is evident since aerobic processes are the most important route (Yang et al. 2011; Azcoitia-Toribio 2012). They efficiency in nutrients removal explain their use in the tertiary polishing of secondary effluents and combined sewer overflows which make them a good option for decentralized water treatment. Recirculation enhances performance and odors removal (Sharma and Brighu 2014; Kumar and Singh 2017; Masi et al. 2017; Weedon 2017). VF CWs have shown to be efficient in tropical climates as warm temperatures enhance removal processes (Molle et al. 2015).

To evaluate the effectiveness of vertical systems, in the WWTP of Izúcar de Matamoros were installed two mesocosms of 1 m², one planted with reeds and the other with bulrush. The scheme of the systems and the particle size of the porous media used (from a rocky material in the bottom to coarse sand in the top layer) are shown in Fig. 5. Daily, during an adaptation period of three months and the five-week

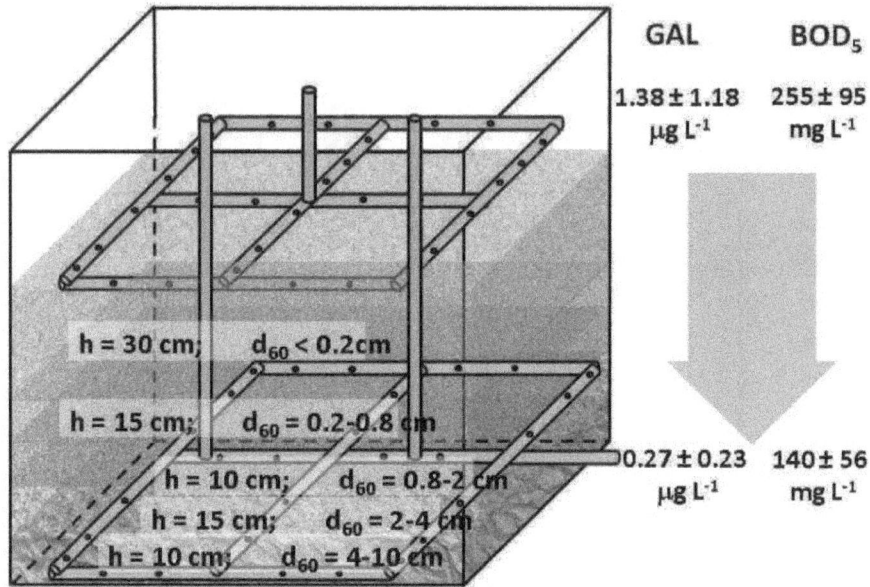

Fig. 5. Scheme and performance of the VFCW modules in the WWTP of Izúcar de Matamoros, Puebla, Mexico.

experiment, four 40L pulses of the final effluent from the WWTP were applied every 2 hours, with a resting interval of 16 hours, to adjust to a working day of the plant's personnel. The performance of the system was evaluated through the determination of the BOD_5 and the concentration of some emerging contaminants (caffeine, galaxolide, tonalide, methyl dihydrojasmonate, sunscreen UV-15, Parsol MCX, alkylphenols, their mono- and diethoxylates and naproxen, by gas chromatography/mass spectrometry). No significant differences in removal percentages between the systems planted with the two macrophytes were found. The removal of the organic microcontaminants was from a 53% for methyldihidrojasmonate, up to 85% for the caffeine. Figure 5 shows the input and output concentrations of galaxolide (GAL) and of the BOD_5, equivalent to a removal of 80.0 and 45.8%, respectively.

Hybrid Wetlands for Tertiary Treatment

Hybrid constructed wetlands (HCW) are systems that combine different types of CW arranged sequentially. This allows the combination of the best characteristics of each particular system (redox potential, pH conditions, communities of microorganisms present, etc.), which complements the weak points of each individual system and results in very efficient systems for the removal of wastewater contaminants, in particular when the treatment objective is the re-use of water, and there is a need for a proper management of nitrogen compounds. There is a wide range of combination possibilities of the individual systems. Horizontal (HF) and vertical (VF) flow systems, are interchangeable: VF–HF HF–VF, observing a certain preference with the first combination. Also, multistage systems have been designed for these two types of CW and have been combined with surface flow wetlands (SF). There are examples of stacked systems in which the VF is superimposed to the HF. All these systems are effective for the treatment of wastewater of different types (Sayadi et al. 2012; Vymazal 2013; Wu et al. 2014; Zhang et al. 2015; Dai et al. 2017; Wu et al. 2017b). It has been reported the high efficiency of hybrid systems in the removal of human pathogens and fecal indicators (Wu et al. 2016).

Like all CW, the weak point is the possibility of clogging. Therefore, attaching of pretreatments that reduce the amount of solids facilitates their operation and enables the reduction of the area needed to achieve good removals. Coupled with septic tanks, Imhoff filters, trickling filters, anaerobic reactors (digesters), and even vermifiltration systems have proven to be an attractive alternative for the treatment of domestic wastewater from small communities as they also remove organic microcontaminants with efficiency (Hudcová et al. 2013; Kim et al. 2014; Ávila et al. 2015; Ayaz et al. 2015; Samal et al. 2017). The availability of dissolved oxygen for the processes of organic matter removal and nitrification is an aspect that has a great impact on the performance of the HCW so different strategies of aeration are currently used, as the physical design of the systems and effluent recirculation, to increase the removal of contaminants and reduce the area needed for the treatment, with values close to 1 m^2 PE^{-1} (Wu et al. 2014; Ávila et al. 2017; Ilyas and Masih 2017a,b).

Given that the primary and secondary treatments remove a considerable amount of solids, the possibility of using the hybrid systems for tertiary treatment of municipal wastewater with secondary treatment has been used with success, showing high percentages of organic micropollutants removal (Verlicchi and Zambello 2014; Zhu and Chen 2014; Haghshenas-Adarmanabadi et al. 2016). In this case, it should be noted that there is a need for a considerable area by the volume of effluent from the WWTP to treat.

For the tertiary polishing of the effluent of the WWTP of the city of Izúcar de Matamoros, a pilot system of HCW with the VF–VF–HF sequence was designed, considering two arrays of macrophytes: reed-reed-bulrush and reed-bulrush-bulrush (Fig. 6). A hydraulic load of 20 cm d^{-1} (considering the area of the first VF), in 4 pulses every 2 hours and a rest of 16 hours was used. The system's ability to improve the quality of the treated water and remove the organic load and the organic microcontaminants was evaluated through the determination of multiple physical-chemical and microbiological parameters in the inputs and outputs of the different cells, as well as the concentrations, by gas chromatography/mass spectrometry, of a group of organic microcontaminants. Figure 6 illustrates the system performance in the removal of the chemical oxygen demand and loads of galaxolide and alkylphenols. In general, the concentrations of contaminants decreased by 3–4 orders of magnitude, from the ppb level to ppt and traces, with removals between 88.7 and 99.6% of the load in the effluent from the WWTP. Among the aspects relevant to the improvement of the quality of water that discharges into the Nexapa River, adjacent to the plant, it can be noted that the effluent showed pH values close to neutrality and a very low turbidity (3.15 ± 0.66 NTU), a value that is lower than the established value in the Mexican regulations for urban public water usage (10 NTU). Also, there was the absence of the typical smell of municipal wastewater, which is attributed to the high removal of the compounds responsible for the odor, as for example the methylindoles, whose concentration decreased from 137 ± 133 mg L^{-1} to 20 ± 30 ng L^{-1}. Although the removal of fecal coliforms was 94%, this only meant little more than an order

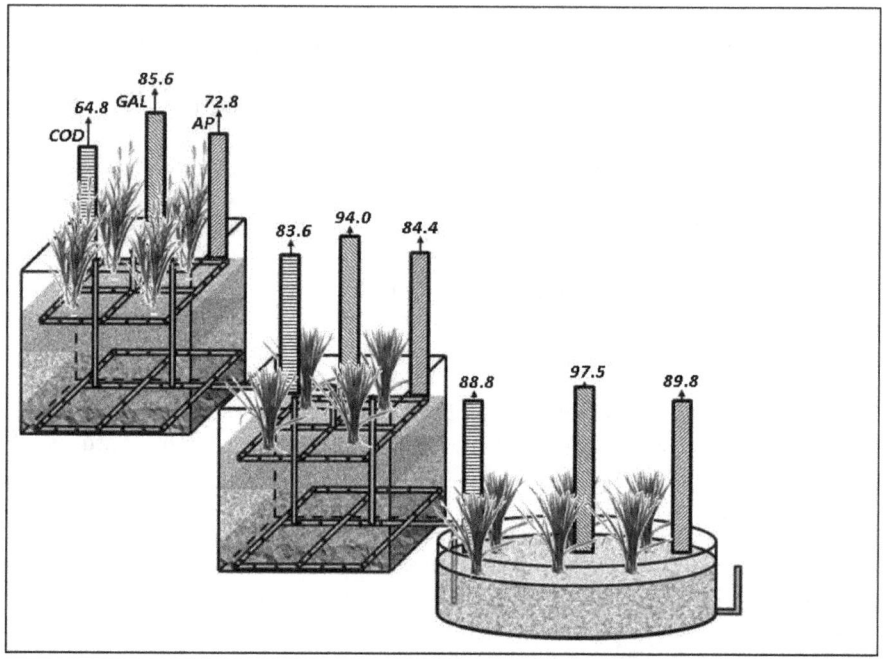

Fig. 6. Scheme and performance of hybrid systems in the WWTP of Izúcar de Matamoros. Puebla, Mexico, COD—Chemical oxygen demand; GAL—galaxolide; AP—alkylphenols.

of magnitude (from 6.3 to 5.0 $\log_{10}$ [CFU (100 mL)$^{-1}$]), so it is advisable to apply other disinfection technology after the wetlands to reduce microbial contamination. No significant differences were found between the results of the second vertical cells with reeds and bulrush ($p = 0.05$), which permits to be flexible in the design with these macrophytes. The system showed to be robust, since final concentrations of contaminants were obtained in narrow ranges, despite the significant differences in their concentrations in the effluent from the plant.

Comparison of the Performance of the CWs Studied in the Izúcar de Matamoros WWTP

In Table 2 are compared the mean input (I), output (O) and removed loads (R), of caffeine (CAF), galaxolide (GAL), naproxen (NAP), and alkyl phenols in the three types of CWs studied in the Izúcar de Matamoros WWTP (mean ± standard error of the mean). For the HCWs, all the figures were divided by the total 3 m^2 of the vertical and horizontal cells. For the HF CWs, the results for one day of hydraulic residence time were selected to better compare with the other systems.

As expected in a real uncontrolled system, the input loads for each pollutant may vary in a wide range, making the comparison difficult.

At first glance, it seems that HF and VF systems achieve better load removals, but the high quality of the effluent in the output of the HCWs shows that there still is the possibility to increase the input loads and obtaining better removals per total area

Table 2. Removal of loads (mg/m^2 d; mean ± standard error of the mean) of selected organic micropollutants in the studied CWs.

	CAF Caffeine	GAL Galaxolide	NAP Naproxen	AP Alkylphenols
HF CWs (n = 135)				
I	1096 ± 64	522 ± 23	2527 ± 111	2380 ± 142
O	111 ± 10	121 ± 8	1445 ± 177	1093 ± 80
R	996 ± 64	389 ± 21	1333 ± 93	1363 ± 125
VF CWs (n = 11)				
I	382 ± 65	221 ± 57	119 ± 23	6121 ± 1612
O	63 ± 10	53 ± 14	55 ± 8	2382 ± 642
R	319 ± 55	168 ± 49	64 ± 23	3739 ± 1454
HCW (n = 6)				
I	837 ± 316	147 ± 63	319 ± 105	1605 ± 516
O	4 ± 1	2 ± 0.4	27 ± 6	126 ± 14
R	833 ± 199	145 ± 39	292 ± 49	1479 ± 319

I: mean input, O: output, and R: removed loads

of the system. Turbidity results were better also for the hybrid systems (system–input turbidity/output turbidity, mean values): HF CWs – 85.6/17.6, VF CWs – 67.0/13.5, HCWs – 81.1/3.1. The turbidity mainly affects the penetration of light into water bodies (Larsen et al. 2017) is associated with the presence of viruses or parasites (Kale 2016), affects the photosynthetic processes (Piepho 2017), and decreases the productivity of water bodies (Ruiz et al. 2017).

It may be concluded that HCWs are the most promising option to give a tertiary polishing to the effluent of the Izúcar de Matamoros WWTP previous its disposal to the Nexapa River. This is favored by an adequate slope to attach the vertical systems and a suitable subsequent space for the horizontal systems, draining the whole set directly to the river without the need for other engineering works or pumping systems.

Future Research Needs

The most important aspects in which more research is needed are mainly related to the increase in the effectiveness and efficiency of constructed wetlands for tertiary treatment effective, which is very closely linked with the overall needs of studies to achieve better results with these systems, whatever the stage of treatment in which they are used. This includes the following:

The increase in the efficiency and the reduction of the area per population equivalent. The use of new designs and management of wetlands, for example, systems of multilayer filter elements (Nakamura et al. 2017), the use of bioaugmentation strategies, for example by using barriers of iron scraps (Kumar and Singh 2017), as well as the conversion to hybrid systems (Jóźwiakowski et al. 2017), combinations with other stages of treatment with plants (Boat and Borin 2017), or the use of nonconventional materials as a means of support which even reduces the risk of clogging (Tatoulis et al. 2017).

The evaluation of the long-term performance of the systems, including the clogging and its causes, as well as the socio-economic aspects of these systems including its life cycle assessment (Carballeira et al. 2017; Garfí et al. 2017; Matos et al. 2017; Wu et al. 2017).

The better understanding of the mechanisms of pollutant removal as well as studies that cover removal of more compounds including those that are incorporated into the lists of priority pollutants or of interest (Gorito et al. 2017; Lv et al. 2017; Vymazal et al. 2017a).

Conclusions

The problem of water shortage and pollution must be addressed immediately because the effects on the environment are becoming more severe and irreversible. Wastewater should be considered as a valuable resource and therefore properly treated depending in its further use. This is critical in regions where financial and technological resources are not available for the implementation of technologically complex techniques. Therefore constructed wetlands represent a viable, economical, technically simple, and efficient option for the purification of water of different nature at the local level. The experiences documented in the literature mentioned throughout this chapter ratify the benefits of using wetlands to efficiently address water issues not only in small communities but also for the recovery of larger water bodies and areas with high degree of urbanization. The selection of the specific system depends on the local conditions and resources. The case studies of the effectiveness of HF, VF, and HCWs in the Izúcar de Matamoros WWTP show that HCWs are the most promising option for the tertiary treatment of the effluent of the WWTP before its disposal into the Nexapa River.

Acknowledgements

Authors gratefully acknowledge Dr. Víctor Matamoros-Mercadal for his manuscript revision.

References Cited

Abdel-Raouf, N., A.A. Al-Homaidan and I.B.M. Ibraheem. 2012. Microalgae and wastewater treatment. Saudi Journal of Biological Sciences 19(3): 257–275.
Abou-Elela, S.I., G. Golinielli, E.M. Abou-Taleb and M.S. Hellal. 2013. Municipal wastewater treatment in horizontal and vertical flows constructed wetlands. Ecol. Eng. 61: 460–468.
Adrados, B., O. Sánchez, C.A. Arias, E. Becares, L. Garrido, J. Mas et al. 2014. Microbial communities from different types of natural wastewater treatment systems: vertical and horizontal flow constructed wetlands and biofilters. Water Res. 55: 304–312.
Aiello, R., V. Bagarello, S. Barbagallo, M. Iovino, A. Marzo and A. Toscano. 2016. Evaluation of clogging in full-scale subsurface flow constructed wetlands. Ecol. Eng. 95: 505–513.
Angelakis, A.N. and S.A. Snyder. 2015. Wastewater treatment and reuse: Past, present, and future. Water (Switzerland) 7(9): 4887–4895.

Auvinen, H., I. Havran, L. Hubau, L. Vanseveren, W. Gebhardt, V. Linnemann et al. 2017. Removal of pharmaceuticals by a pilot aerated subsurface flow constructed wetland treating municipal and hospital wastewater. Ecol. Eng. 100: 157–164.

Ávila, C., J. Nivala, L. Olsson, K. Kassa, T. Headley, R. Mueller et al. 2014. Emerging organic contaminants in vertical subsurface flow constructed wetlands: Influence of media size, loading frequency and use of active aeration. Sci. Total Environ. 494: 211–217.

Ávila, C., J.M. Bayona, I. Martín, J.J. Salas and J. García. 2015. Emerging organic contaminant removal in a full-scale hybrid constructed wetland system for wastewater treatment and reuse. Ecol. Eng. 80: 108–116.

Ávila, C., C. Pelissari, P.H. Sezerino, M. Sgroi, P. Roccaro and J. García. 2017. Enhancement of total nitrogen removal through effluent recirculation and fate of PPCPs in a hybrid constructed wetland system treating urban wastewater. Sci. Total Environ. 584-585: 414-425.

Ayaz, S.Ç., Ö. Aktaş, L. Akça and N. Fındık. 2015. Effluent quality and reuse potential of domestic wastewater treated in a pilot-scale hybrid constructed wetland system. J. Environ. Manage. 1(156): 115–120.

Azcoitia-Toribio, E.D. 2012. Efecto de sobrecargas hidráulicas en el rendimiento de humedales construidos para la depuración de aguas. Tesina de especialidad. Universidad Politecnica de Cataluña, Spain. In Spanish.

Badejo, A.A., D.O. Omole, J.M. Ndambuki and W.K. Kupolati. 2017. Municipal wastewater treatment using sequential activated sludge reactor and vegetated submerged bed constructed wetland planted with *Vetiveria zizanioides*. Ecol. Eng. 99: 525–529.

Bai, L., C. Wang, C. Huang, L. He and Y. Pei. 2014. Reuse of drinking water treatment residuals as a substrate in constructed wetlands for sewage tertiary treatment. Ecol. Eng. 70: 295–303.

Barbagallo, S., A.C. Barbera, G.L. Cirelli, M. Milani and A. Toscano. 2014. Reuse of constructed wetland effluents for irrigation of energy crops. Water Sci. Technol. 70(9): 1465–1472.

Barco, A. and M. Borin. 2017. Treatment performance and macrophytes growth in a restored hybrid constructed wetland for municipal wastewater treatment. Ecol. Eng. 107: 160–171.

BARR. 2011. NorthMet Project Baseline Wetland Type Evaluation Prepared for PolyMet Mining Inc. https://www.leg.state.mn.us/docs/2015/other/150681/PFEISref_1/Barr%202011d.pdf.

Bassi, N., M.D. Kumar, A. Sharma and P. Pardha-Saradhi. 2014. Status of wetlands in India: A review of extent, ecosystem benefits, threats and management strategies. J. Hydrol. Regional Studies 2: 1–19.

Bateganya, N.L., D. Nakalanzi, M. Babu and T. Hein. 2015. Buffering municipal wastewater pollution using urban wetlands in sub-Saharan Africa: a case of Masaka municipality, Uganda. Environ. Technol. 36(17): 2149–2160.

Berbeka, K., M. Czajkowski and A. Markowska. 2012. Municipal wastewater treatment in Poland— efficiency, costs and returns to scale. Water Sci. Technol. 66(2): 394–401.

Berglund, B., G.A. Khan, S.E. Weisner, P.M. Ehde, J. Fick and P.E. Lindgren. 2014. Efficient removal of antibiotics in surface-flow constructed wetlands, with no observed impact on antibiotic resistance genes. Sci. Total Environ. 476: 29–37.

Beutel, M.W. 2012. Water quality in a surface-flow constructed treatment wetland polishing tertiary effluent from a municipal wastewater treatment plant. Water Sci. Technol. 66(9): 1977–1983.

Bisone, S., M. Gautier, M. Masson and N. Forquet. 2017. Influence of loading rate and modes on infiltration of treated wastewater in soil-based constructed wetland. Environ. Technol. 38(2): 163–174.

Bruch, I., U. Alewell, A. Hahn, R. Hasselbach and C. Alewell. 2014. Influence of soil physical parameters on removal efficiency and hydraulic conductivity of vertical flow constructed wetlands. Ecol. Eng. 68: 124–132.

Butterworth, E., A. Richards, M. Jones, G. Dotro and B. Jefferson. 2016. Assessing the potential for tertiary nitrification in sub-surface flow constructed wetlands. Environ. Technol. 5(1): 68–77.

Calheiros, C.S., P.V. Quitério, G. Silva, L.F. Crispim, H. Brix, S. Moura et al. 2012. Use of constructed wetland systems with Arundo and Sarcocornia for polishing high salinity tannery wastewater. J. Environ. Manage. 95(1): 66–71.

Carballeira, T., I. Ruiz and M. Soto. 2017. Aerobic and anaerobic biodegradability of accumulated solids in horizontal subsurface flow constructed wetlands. International Biodeterioration and Biodegradation 119: 396–404.

Castorina, A., S. Consoli, S. Barbagallo, F. Branca, A. Farag, F. Licciardello et al. 2016. Assessing environmental impacts of constructed wetland effluents for vegetable crop irrigation. Int. J. Phytoremediat. 18(6): 626–633.

Chen, Y., Y. Wen, Z. Tang, L. Li, Y. Cai and Q. Zhou. 2014. Removal processes of disinfection byproducts in subsurface-flow constructed wetlands treating secondary effluent. Water Res. 51: 163–171.

Chen, Y., Y. Wen, Z. Tang, J. Huang, Q. Zhou and J. Vymazal. 2015. Effects of plant biomass on bacterial community structure in constructed wetlands used for tertiary wastewater treatment. Ecol. Eng. 84: 38–45.

Chouinard, A., C.N. Yates, G.C. Balch, S.E. Jørgensen, B.C. Wootton and B.C. Anderson. 2014. Management of tundra wastewater treatment wetlands within a lagoon/wetland hybridized treatment system using the SubWet 2.0 Wetland Model. Water 6(3): 439–454.

Dai, Y., R. Tao, Y. Tai, N.F. Tam, A. Dan and Y. Yang. 2017. Application of a full-scale newly developed stacked constructed wetland and an assembled bio-filter for reducing phenolic endocrine disrupting chemicals from secondary effluent. Ecol. Eng. 99: 496–503.

Davidson, N.C. 2014. How much wetland has the world lost? Long-term and recent trends in global wetland area. Mar. Freshwater Res. 65(10): 934–941.

Day, J.W., J.Y. Ko, J. Rybczyk, D. Sabins, R. Bean, G. Berthelot et al. 2004. The use of wetlands in the Mississippi Delta for wastewater assimilation: a review. Ocean. Coast. Manage. 47(11): 671–691.

Diputación de Barcelona. 2017. Desplegable del Parque Fluvial del Besós (English). http://parcs.diba.cat/c/document_library/get_file?uuid=0260693a-c81c-453f-93b3-21c2ead11c0c&groupId=5280469.

Dordio, A., A.J.P. Carvalho and A. Pinto. 2008. Wetlands: Water "living filters"? pp. 15–71. *In*: Russo, R.E. [ed.]. Wetlands Ecology Conservation and Restoration. Nova Science Publishers, Inc., Hauppauge, NY, US.

Fan, J., W. Wang, B. Zhang, Y. Guo, H.H. Ngo, W. Guo et al. 2013. Nitrogen removal in intermittently aerated vertical flow constructed wetlands: impact of influent COD/N ratios. Bioresour. Technol. 143: 461–466.

Gachango, F.G., S.M. Pedersen and C. Kjærgaard. 2015. Cost-effectiveness analysis of surface flow constructed wetlands (SFCW) for nutrient reduction in drainage discharge from agricultural fields in Denmark. Environ. Manage. 56(6): 1478–1486.

García, J., D.P.L. Rousseau, J. Morató, E. Lesage, V. Matamoros and J.M. Bayona. 2010. Contaminant removal processes in subsurface-flow constructed wetlands: A review. Critical Reviews in Environ. Sci. Technol. 40(7): 561–661.

García-Serrano, J. and A. Corzo-Hernández. 2008. Guía práctica de diseño, construcción y explotación de sistemas de humedales de flujo subsuperficial. http://hdl.handle.net/2117/2474. http://upcommons.upc.edu/bitstream/handle/2117/2474/JGarcia_and_ACorzo.pdf;sequence=1.

Garfí, M., L. Flores and I. Ferrer. 2017. Life Cycle Assessment of wastewater treatment systems for small communities: Activated sludge, constructed wetlands and high rate algal ponds. J. Clean Prod. 161: 211–219.

Gikas, P., E. Ranieri and G. Tchobanoglous. 2013. Removal of iron, chromium and lead from waste water by horizontal subsurface flow constructed wetlands. J. Chem. Technol. Biotechnol. 88(10): 1906–1912.

Gökalp, Z., S. Sedat Karaman, I. Taş and H. Kirnak. 2016. Constructed wetland technology to prevent water resources pollution. Curr. T. Natur. Sci. 5(9): 125–132.

Gorito, A.M., A.R. Ribeiro, C.M.R. Almeida and A.M.T. Silva. 2017. A review on the application of constructed wetlands for the removal of priority substances and contaminants of emerging concern listed in recently launched EU legislation. Environ. Pollut. 227: 428–443.

Gupta, V.K., I. Ali, T.A. Saleh, A. Nayak and S. Shilpi Agarwal. 2012. Chemical treatment technologies for waste-water recycling—an overview. RSC Adv. 2: 6380–6388.

Haarstad, K., H.J. Bavor and T. Mæhlum. 2012. Organic and metallic pollutants in water treatment and natural wetlands: a review. Water Sci. Technol. 65(1): 76–99.

Haghshenas-Adarmanabadi, A., M. Heidarpour and S. Tarkesh-Esfahani. 2016. Evaluation of horizontal–vertical subsurface hybrid constructed wetlands for tertiary treatment of conventional treatment facilities effluents in developing countries. Water Air Soil. Pollut. 227: 28 (18 pages).

Hernández-Sancho, F., B. Lamizana-Diallo, J. Mateo-Sagasta and M. Qadir. 2015. Economic Valuation of Wastewater: The cost of action and the cost of no action. UNEP Report. https://wedocs. unep.org/bitstream/handle/20.500.11822/7465/-Economic_Valuation_of_Wastewater_The_ Cost_of_Action_and_the_Cost_of_No_Action-2015Wastewater_Evaluation_Report_Mail.pdf. pdf?sequence=3&isAllowed=y.

Herrera, J.A., A.E. Navarro and E. Torres. 2016. Effects of porous media, macrophyte type and hydraulic retention time on the removal of organic load and micropollutants in constructed wetlands. J. Environ. Sci. Heal. A 51(5): 380–388.

Hu, Y., Y. Zhao, X. Zhao and J.L. Kumar. 2012. High rate nitrogen removal in an alum sludge-based intermittent aeration constructed wetland. Environ. Sci. Technol. 46(8): 4583–4590.

Hudcová, T., J. Vymazal and M.K. Dunajský. 2013. Reconstruction of a constructed wetland with horizontal subsurface flow after 18 years of operation. Water Sci. Technol. 68(5): 1195–1202.

Huertas, E., M. Folch, M. Salgot, I. Gonzalvo and C. Passarell. 2006. Constructed wetlands effluent for streamflow augmentation in the Besòs River (Spain). Desal. 188(1): 141–147.

Idris, A., A.G.L. Abdullah, Y. Hung and L.K. Wang. 2014. Wetlands for wastewater treatment and water reuse. pp. 643–680. *In*: Wang, L.K. and C.T. Yang [eds.]. Handbook of Environmental Engineering, Modern Water Resources Engineering. Volume 15. Humana Press, Totowa, NJ, US.

Ilyas, H. and I. Masih. 2017a. The performance of the intensified constructed wetlands for organic matter and nitrogen removal: A review. J. Environ. Manage. 198(1): 372–383.

Ilyas, H. and I. Masih. 2017b. Intensification of constructed wetlands for land area reduction: A review. Environ. Sci. Pollut. Res. 24: 12081–12091.

Iowa CREP Program Overview. 2016. Conservation Reserve Enhancement Program (CREP) Iowa State. Fact Sheet. Feb. 2011. UNITED STATES DEPARTMENT OF AGRICULTURE. FARM SERVICE AGENCY. http://www.fsa.usda.gov/Assets/USDA-FSA-Public/usdafiles/ Conservation/PDF/crepiafactsheet.pdf.

Jóźwiakowski, K., P. Bugajski, Z. Mucha, W. Wójcik, A. Jucherski, M. Nastawny et al. 2017. Reliability and efficiency of pollution removal during long-term operation of a one-stage constructed wetland system with horizontal flow. Sep. Pur. Tech. 187: 60–66.

Junk, W.J., S. An, C.M. Finlayson, B. Gopa, J. Kvĕt, S.A. Mitchell et al. 2013. Current state of knowledge regarding the world's wetlands and their future under global climate change: a synthesis. Aquat. Sci. 75(1): 151–167.

Kadlec, R.H., R.L. Knight, J. Vymazal, H. Brix, P. Cooper and R. Haberl. 2000. Constructed Wetlands for Pollution Control. Processes, Performance, Design and Operation. IWA Specialist Group on the Use of Macrophytes in Water Pollution Control, IWA Scientific and Technical Report No. 8, IWA Publishing, London.

Kadlec, R.H. and S.D. Wallace. 2009. Treatment Wetlands, 2nd. Edition. Boca Raton, CRC Press, 1016 p.

Kale, V.S. 2016. Consequence of temperature, pH, turbidity and dissolved oxygen water quality parameters. IARJSET 3(8): 186–190.

Kanchanamayoon, W. 2015. Sample preparation methods for the determination of chlorination disinfection byproducts in water samples. Chromatographia 78(17-18): 1135–1142.

Kim, B., M. Gautier, P. Molle, P. Michel and R. Gourdon. 2014. Influence of the water saturation level on phosphorus retention and treatment performances of vertical flow constructed wetland combined with trickling filter and FeCl$_3$ injection. Ecol. Eng. 80: 53–61.

Kinyua, G.J. 2014. Effectiveness of natural wetland in waste water treatment: a case study of Tibia Wetland, Limuru Municipality, Kenya. Master Thesis. School of Environmental Studies of Kenyatta University.

Kosma, C.I., D.A. Lambropoulou and T.A. Albanis. 2014. Investigation of PPCPs in wastewater treatment plants in Greece: occurrence, removal and environmental risk assessment. Sci. Total Environ. 466-467: 421–438.

Kumar, M. and R. Singh. 2017. Performance evaluation of semi continuous vertical flow constructed wetlands (SC-VF CWs) for municipal wastewater treatment. Bioresour. Technol. 232: 321–330.

Landgrave, R. and P. Moreno-Casasola. 2012. Evaluación cuantitativa de la pérdida de humedales en México. Investigación Ambiental 4(1): 19–35.

Larsen, T.C., N.K. Browne, A.C. Erichsen, K. Tun and P.A. Todd. 2017. Modelling for management: Coral photo-physiology and growth potential under varying turbidity regimes. Ecol. Model. 362: 1–12.

Lee, O.M., H.Y. Kim, W. Park, T.H. Kim and S. Yu. 2015. A comparative study of disinfection efficiency and regrowth control of microorganism in secondary wastewater effluent using UV, ozone, and ionizing irradiation process. J. Hazard. Mater. 295: 201–208.

Li, Y., M. Yang, X. Zhang, J. Jiang, J. Liu, C.F. Yau et al. 2017. Two-step chlorination: A new approach to disinfection of a primary sewage effluent. Water Res. 108: 339–347.

Liu, R., Y. Zhao, L. Doherty, Y. Hu and X. Hao. 2015. A review of incorporation of constructed wetland with other treatment processes. Chem. Eng. J. 279: 220–230.

Lv, T., P.N. Carvalho, L. Zhang, Y. Zhang, M. Button, C. Arias et al. 2017. Functionality of microbial communities in constructed wetlands used for pesticide remediation: Influence of system design and sampling strategy. Wat. Res. 110: 241–251.

Mander, Ü., G. Dotro, Y. Ebie, S. Towprayoon, C. Chiemchaisri, S.F. Nogueira et al. 2014. Greenhouse gas emission in constructed wetlands for wastewater treatment: a review. Ecol. Eng. 66: 19–35.

Masi, F., A. Rizzo, R. Bresciani and G. Conte. 2017. Constructed wetlands for combined sewer overflow treatment: Ecosystem services at Gorla Maggiore, Italy. Ecol. Eng. 98: 427–438.

Mas-Ortega, J.G. 2016. Análisis coste/beneficio aplicado a los procesos de depuración y reutilización. Master Thesis, University of Alicante, Spain.

Matamoros, V., Y. Rodríguez and J.M. Bayona. 2017. Mitigation of emerging contaminants by full-scale horizontal flow constructed wetlands fed with secondary treated wastewater. Ecol. Eng. 99: 222–227.

Matos, M.P., M. von Sperling, A.T. Matos, S.T. Miranda, T.D. Souza and L.M. Costa. 2017. Key factors in the clogging process of horizontal subsurface flow constructed wetlands receiving anaerobically treated sewage. Ecol. Eng. 106: 588–596.

McCarty, P.L., J. Bae and J. Kim. 2011. Domestic wastewater treatment as a net energy producer–can this be achieved? Environ. Sci. Technol. 45(17): 7100–7106.

Mohammed, A.N. and M.A. ElBably. 2016. Technologies of domestic wastewater treatment and reuse: options of application in developing countries. JSM Environ. Sci. Ecol. 4(3): 1033.

Molinos-Senante, M., F. Hernández-Sancho and R. Sala-Garrido. 2011. Cost–benefit analysis of water-reuse projects for environmental purposes: A case study for Spanish wastewater treatment plants. J. Environ. Manage. 92(12): 3091–3097.

Molle, P., R.L. Latune, C. Riegel, G. Lacombe, D. Esser and L. Mangeot. 2015. French vertical-flow constructed wetland design: adaptations for tropical climates. Water Sci. Technol. 71(10): 1516–1523.

Mulling, B.T., A.M. Soeter, H.G. Van Der Geest and W. Admiraal. 2014. Changes in the planktonic microbial community during residence in a surface flow constructed wetland used for tertiary wastewater treatment. Sci. Total Environ. 466: 881–887.

Nakamura, K., R. Hatakeyama, N. Tanaka, K. Takisawa, C. Tada and K. Nakano. 2017. A novel design for a compact constructed wetland introducing multi-filtration layers coupled with subsurface superficial space. Ecol. Eng. 100: 99–106.

Nam, S.W., B.I. Jo, Y. Yoon and K.D. Zoh. 2014. Occurrence and removal of selected micropollutants in a water treatment plant. Chemosphere 95: 156–165.

Nasser, A.M. 2016. Removal of Cryptosporidium by wastewater treatment processes: A review. J. Water Health 14(1): 1–13.

Nivala, J., P. Knowles, G. Dotro, J. García and S. Wallace. 2012. Clogging in subsurface-flow treatment wetlands, Measurement, modeling and management. Water Res. 46: 1625–1640.

Nivala, J., S. Wallace, T. Headley, K. Kassa, H. Brix, M. van Afferden et al. 2013. Oxygen transfer and consumption in subsurface flow treatment wetlands. Ecol. Eng. 61: 544–554.

Nobles, A.L., J.L. Goodall and G.M. Fitch. 2017. Comparing costs of onsite best management practices to nutrient credits for stormwater management: A case study in Virginia. JAWRA J. Am. Water Resour. Assoc. 53(1): 131–143.

Phan, Y. and X. Zhang. 2013. Four groups of new aromatic halogenated disinfection byproducts: effect of bromide concentration on their formation and speciation in chlorinated drinking water. Environ. Sci. Technol. 47(3): 1265–1273.

Piepho, M. 2017. Assessing maximum depth distribution, vegetated area, and production of submerged macrophytes in shallow, turbid coastal lagoons of the southern Baltic Sea. Hydrobiologia 794(1): 303–316.

Plakas, K.V., A.A. Georgiadis and A.J. Karabelas. 2016. Sustainability assessment of tertiary wastewater treatment technologies: a multi-criteria analysis. Water Sci. Technol. 73(7): 1532–1540.

Plappally, A.K. and V.J.H. Lienhard. 2012. Energy requirements for water production, treatment, end use, reclamation, and disposal. Renew. Sust. Energ. Rev. 16(7): 4818–4848.

Pozo-Morales, L., M. Franco, D. Garvi and J. Lebrato. 2014. Experimental basis for the design of horizontal subsurface-flow treatment wetlands in naturally aerated channels with an anti-clogging stone layout. Ecol. Eng. 70: 68–81.

Prüss-Üstün, A., J. Wolf, C. Corvalán, R. Bos and M. Neira. 2016. Preventing disease through healthy environments: A global assessment of the burden of disease from environmental risks. World Health Organization, Geneva, Switzerland.

Qomariyah, S., A.H. Ramelan and P. Setyono. 2017. Use of macrophyte plants, sand & gravel materials in constructed wetlands for greywater treatment. IOP Conf. Series: Materials Science and Engineering 176 012018. http://iopscience.iop.org/article/10.1088/1757-899X/176/1/012018/pdf.

Ramsar. 2013. The Ramsar Convention Manual: a guide to the Convention on Wetlands (Ramsar, Iran, 1971), 6th ed. Ramsar Convention Secretariat, Gland, Switzerland.

Ramsar. 2015. State of the World's Wetlands and their Services to People: A compilation of recent analyses. Ramsar Convention. http://www.ramsar.org/sites/default/files/documents/library/bn7e_0.pdf.

Reddy, L., D. Kumar and S.R. Asolekar. 2014. Typologies for successful operation and maintenance of horizontal sub-surface flow constructed wetlands. Sciences 6(12): 157–164.

Remy, C., U. Miehe, B. Lesjean and C. Bartholomäus. 2014. Comparing environmental impacts of tertiary wastewater treatment technologies for advanced phosphorus removal and disinfection with life cycle assessment. Water Sci. Technol. 69(8): 1742–1750.

Richardson, S.D. and C. Postigo. 2016. Discovery of new emerging DBPs by high-resolution mass spectrometry. pp. 335–356. *In*: Pérez, S., P. Eichhorn and D. Barceló [eds.]. Comprehensive Analytical Chemistry, Volume 71. Chapter 11. Elsevier, Amsterdam.

Ruiz, J., D. Macías and G. Navarro. 2017. Natural forcings on a transformed territory overshoot thresholds of primary productivity in the Guadalquivir Estuary. Cont. Shelf Res. 148: 199–207.

Saeed, T. and G. Sun. 2011. A comparative study on the removal of nutrients and organic matter in wetland reactors employing organic media. Chem. Eng. J. 171(2): 439–447.

Saeed, T. and G. Sun. 2012. A review on nitrogen and organics removal mechanisms in subsurface flow constructed wetlands: dependency on environmental parameters, operating conditions and supporting media. J. Environ. Manage. 112: 429–448.

Samal, K., R.R. Dash and P. Bhunia. 2017. Treatment of wastewater by vermifiltration integrated with macrophyte filter: A review. J. Environ. Chem. Eng. 5(3): 2274–2289.

Sayadi, M.H., R. Kargar, M.R. Doosti and H. Salehi. 2012. Hybrid constructed wetlands for wastewater treatment: A worldwide review. Proc. Int. Acad. Ecol. Environ. Sci. 2(4): 204–222.

Sedlak, D.L. and U. von Gunten. 2011. The chlorine dilemma. Science 331(6013): 42–43.

Shah, A.D., S.W. Krasner, C.F.T. Lee, U. von Gunten and W.A. Mitch. 2012. Trade-offs in disinfection byproduct formation associated with precursor preoxidation for control of N-nitrosodimethylamine formation. Environ. Sci. Technol. 46(9): 4809–4818.

Sharma, G. and U. Brighu. 2014. Performance analysis of vertical up-flow constructed wetlands for secondary treated effluent. APCBEE Procedia 10: 110–114.

Stalter, D., E. O'Malley, U. Von Gunten and B.I. Escher. 2016. Point-of-use water filters can effectively remove disinfection by-products and toxicity from chlorinated and chloraminated tap water. Environ. Sci.: Water Res. Tech. 2(5): 875–883.

Stefanakis, A.I. and V.A. Tsihrintzis. 2009. Effect of outlet water level raising and effluent recirculation on removal efficiency of pilot-scale, horizontal subsurface flow constructed wetlands. Desalination 248(1-3): 961–976.

Stefanakis, A.I. and V.A. Tsihrintzis. 2012. Effects of loading, resting period, temperature, porous media, vegetation and aeration on performance of pilot-scale vertical flow constructed wetlands. Chem. Eng. J. 181: 416–430.

Stefanakis, A., C.S. Akratos and V.A. Tsihrintzis. 2014. Vertical Flow Constructed Wetlands, Ecoengineering Systems for Wastewater and Sludge Treatment. Elsevier, Boston, 392 pp.

Sun, G., T. Saeed, G. Zhang and N. Sivakugan. 2015. Water quantity and quality assessment on a tertiary treatment wetland in a tropical climate. Water Sci. Technol. 71(4): 511–517.

Tatoulis, T., C.S. Akratos, A.G. Tekerlekopoulou, D.V. Vayenas and A.I. Stefanakis. 2017. A novel horizontal subsurface flow constructed wetland: Reducing area requirements and clogging risk. Chemosphere 186: 257–268.

Topare, N.S., S.J. Attar and M.M. Manfe. 2011. Sewage/wastewater treatment technologies: a review. Sci. Revs. Chem. Commun. 1(1): 18–24.

UN-Water. 2015. Compendium of Water Quality Regulatory Frameworks: Which Water for Which Use? https://es.search.yahoo.com/search?p=Compendium%20of%20Water%20Quality%20Regulatory%20Frameworks%3A%20Which%20Water%20for%20Which%20Use%3F&fr=yset_ff_syc_oracle&type=newtab.

USEPA. 1999. Folleto informativo de tecnología de aguas residuales. United States Environmental Protection Agency. https://www3.epa.gov/npdes/pubs/cs-99-062.pdf.

USEPA. 2000. Wastewater Technology Fact Sheet Wetlands: Subsurface Flow. Fact sheet EPA 832-F-00-023. United States Environmental Protection Agency. https://www3.epa.gov/npdes/pubs/wetlands-subsurface_flow.pdf.

USEPA. 2016. Wetland function and values. United States Environmental Protection Agency. Watershed Academy Web. https://www.epa.gov/sites/production/files/2016-02/documents/wetlandfunctionsvalues.pdf.

Verlicchi, P. and E. Zambello. 2014. How efficient are constructed wetlands in removing pharmaceuticals from untreated and treated urban wastewaters? A review. Sci. Total Environ. 470-471: 1281–1306.

Victoria, E.P.A. 2002. Guidelines for Environmental Management: Disinfection of Treated Wastewater. Environmental Protection Authority Victoria. http://www.epa.vic.gov.au/~/media/Publications/730.pdf.

Vivant, A.L., C. Boutin, S. Prost-Boucle, S. Papias, A. Hartmann, G. Depret et al. 2016. Free water surface constructed wetlands limit the dissemination of extended-spectrum beta-lactamase producing *Escherichia coli* in the natural environment. Water Res. 104: 178–188.

Vymazal, J. 2009. The use constructed wetlands with horizontal sub-surface flow for various types of wastewater. Ecol. Eng. 35(1): 1–17.

Vymazal, J. 2010. Constructed wetlands for wastewater treatment. Water 2(3): 530–549.

Vymazal, J. 2011. Constructed wetlands for wastewater treatment: Five decades of experience. Environ. Sci. Technol. 45(1): 61–69.

Vymazal, J. 2013. The use of hybrid constructed wetlands for wastewater treatment with special attention to nitrogen removal: a review of a recent development. Water Res. 47(14): 4795–4811.

Vymazal, J., T.D. Březinová, M. Koželuh and L. Kule. 2017. Occurrence and removal of pharmaceuticals in four full-scale constructed wetlands in the Czech Republic–the first year of monitoring. Ecol. Eng. 98: 354–364.

Wallace, S.D. and R.L. Knight. 2006. Small-Scale Constructed Wetland Treatment Systems: Feasibility, Design Criteria and O & M Requirements. Water Environ Res. Foundation (WERF), Alexandria, Virginia, US.

Weber, K.P. and R.L. Legge. 2009. One-dimensional metric for tracking bacterial community divergence using sole carbon source utilization patterns. J. Microbiol. Methods 79(1): 55–61.

Weedon, C.M. 2017. Tertiary sewage treatment by a full-scale compact vertical flow constructed wetland. Environ. Technol. 38(2): 140–153.

Westerhoff, P., F. Sharif, R. Halden, P. Herckes and R. Krajmalnik-Brown. 2014. Constructed Wetlands for Treatment of Organic and Engineered Nanomaterial Contaminants of Emerging Concerns. Web Report #4334. Water Research Foundation. http://www.waterrf.org/PublicReportLibrary/4334.pdf.

Woltemade, C.J. 2000. Ability of restored wetlands to reduce nitrogen and phosphorus concentrations in agricultural drainage water. J. Soil Water Conserv. 55(3): 303–309.

Wu, H., J. Zhang, H.H. Ngo, W. Guo, Z. Hu, S. Liang et al. 2015a. A review on the sustainability of constructed wetlands for wastewater treatment: design and operation. Bioresour. Technol. 175: 594–601.

Wu, H., X. Wang, X. He, S. Zhang, R. Liang and J. Shen. 2017a. Effects of root exudates on denitrifier gene abundance, community structure and activity in a micro-polluted constructed wetland. Sci. Tot. Environ. 598: 697–703.

Wu, S., P. Kuschk, H. Brix, J. Vymazal and R. Dong. 2014. Development of constructed wetlands in performance intensifications for wastewater treatment: A nitrogen and organic matter targeted review. Water Res. 57: 40–55.

Wu, S., S. Wallace, H. Brix, P. Kuschk, W.K. Kirui, F. Masi et al. 2015b. Treatment of industrial effluents in constructed wetlands: challenges, operational strategies and overall performance. Environ Pollut. 201: 107–120.

Wu, S., P.N. Carvalho, J.A. Müller, V.R. Manoj and R. Dong. 2016. Sanitation in constructed wetlands: a review on the removal of human pathogens and fecal indicators. Sci. Total Environ. 541: 8–22.

Wu, Y., R. Han, X. Yang, Y. Zhang and R. Zhang. 2017b. Long-term performance of an integrated constructed wetland for advanced treatment of mixed wastewater. Ecol. Eng. 99: 91–98.

WWAP. 2012. The United Nations World Water Development Report 4: Managing Water under Uncertainty and Risk. World Water Assessment Programme. Paris, UNESCO.

WWAP. 2014. The United Nations World Water Development Report 2014: Water and Energy. United Nations World Water Assessment Programme. Paris, UNESCO.

WWAP. 2017. The United Nations World Water Development Report 2017: Wastewater, The Untapped Resource. United Nations World Water Assessment Programme. Paris, UNESCO.

Yang, Y., L. Zhang, Y.Q. Zhao, S.P. Wang, X.C. Guo, Y. Guo et al. 2011. Towards the development of a novel construction solid waste (CSW) based constructed wetland system for tertiary treatment of secondary sewage effluents. J. Environ. Sci. Health. Part A 46(7): 758–763.

Yeh, T.Y. and C.H. Wu. 2009. Pollutant removal within hybrid constructed wetland systems in tropical regions. Water Sci. Technol. 59(2): 233–240.

Zhang, D.Q., K.B. Jinadasa, R.M. Gersberg, Y. Liu, S.K. Tan and W.J. Ng. 2015. Application of constructed wetlands for wastewater treatment in tropical and subtropical regions (2000–2013). J. Environ. Sci. 30: 30–46.

Zhu, S. and H. Chen. 2014. The fate and risk of selected pharmaceutical and personal care products in wastewater treatment plants and a pilot-scale multistage constructed wetland system. Environ. Sci. Pollut. Res. Int. 21(2): 1466–1479.

Zhu, X. and B.E. Logan. 2013. Using single-chamber microbial fuel cells as renewable power sources of electro-Fenton reactors for organic pollutant treatment. J. Hazard. Mater. 252: 198–203.

Zimmermann, S.G., M. Wittenwiler, J. Hollender, M. Krauss, C. Ort, H. Siegrist et al. 2011. Kinetic assessment and modeling of an ozonation step for full-scale municipal wastewater treatment: micropollutant oxidation, by-product formation and disinfection. Water Res. 45(2): 605–617.

5

Effect of Support Media on Heavy Metals Removal in Constructed Wetlands Inoculated with Metallotolerant Strains

Leonel Ernesto Amabilis-Sosa,[1,]* *Marcela Arroyo-Ginez,*[2]
Ruth Pérez-González,[2] *Adriana Roé-Sosa,*[1] *Landy Irene Ramírez-Burgos*[2] and *María del Carmen Durán-Domínguez-de-Bazúa*[2]

INTRODUCTION

Heavy metals are known as a group of pollutants which generate great concern about the environmental and public health because of their inorganic and toxic nature that causes their persistence in the environment. Those characteristics allow heavy metals to be transferred by bioaccumulation and biomagnification processes to trophic chains, through environment compartments' migration to higher organisms including humans, where teratogenic, mutagenic, and carcinogenic effects are magnified (Dickson 2006; Amuda et al. 2007).

Unfortunately, in the last two decades, industrial growth in developed and developing countries has accelerated, which has increased the use of heavy metals and the need for their secure final disposal. This has increased the flow of metals to the ecosystems and therefore, the risk to human health. Given this, there are currently technologies for heavy metals removal from wastewater, whose operation imply physical and/or chemical separation processes.

[1] Universidad Politécnica del Estado de Morelos, Jiutepec, Morelos, México.
[2] Facultad de Química, Universidad Nacional Autónoma de México, Ciudad de México, México.
* Corresponding author: leoamabilis@yahoo.com.mx

The most commonly processes that had been used are ion exchange resins, activated carbon adsorption, strong alkali chemical precipitation (as hydroxides), and reverse osmosis. Nevertheless, all of these technologies are not accessible for most populations with heavy metal contamination problems. These technologies are only used in some industrial processes. In addition, some technologies generate large volumes of hazardous solid wastes, such as precipitation (Barakat 2011).

Ion exchange resins only change the heavy metals phase, so a secondary treatment is required for its by-products. Besides, reverse osmosis is extremely costly in spite of being very efficient and resulting in low volumes of waste (Kumari et al. 2015).

Based on the above, studies have been carried out around the world focused on heavy metals removal with sustainable and approachable technologies for the general population without compromising removal efficiency.

Biofilters and constructed wetlands systems have been reported as the most successful cases. Biofilters systems work with ion exchange between positive charges of metals and negative charges of organic matter. Although design parameters are referenced and even implemented at real scale, the design criteria denote empiricism when lifetimes and removal efficiencies of biofiltration systems are analyzed (Hatt et al. 2011; Dhokpande and Kaware 2013).

On the other hand, constructed wetland systems compete with secondary treatments such as activated sludge, trickle filters, upflow anaerobic sludge blanket reactors, and oxidation ditches, among others. These systems have defined design, sizing, and scaling parameters with robust engineering bases (Brix and Arias 2005).

These parameters had permitted to realize numerous studies, in which one of the research topics is heavy metals removal through phytoremediation (transfer of contaminants present in soil and/or water to vegetation organs previously selected for this purpose). These studies have shown heavy metals selectivity in either the aerial part (epigea) or the underground (hypogea) part of the plant, which leads to translocation phenomena that suggest a subsequent metal recovery (Cheng et al. 2002; Dhir 2010).

In addition, recent research contributions have been elucidated the role of microorganisms in heavy metals removal inside constructed wetlands. As these systems are a secondary treatment system with immobilized biomass, they behave as a packed bed bioreactor, in which after a certain hydraulic residence time, it is observed that heavy metals could be adsorbed, oxidized, or reduced enzymatically (Salgado-Bernal et al. 2012a,b, 2015; Amabilis-Sosa et al. 2015).

The above-mentioned components of constructed wetlands (vegetation and microorganisms) have supported in a greater or lower scale a large number of research studies concerning the potential of constructed wetlands for heavy metals removal. For example, Cd, Pb, and Zn removal was studied in 19 constructed wetlands, each one with a different plant species, getting a translocation between 40.5 and 90% (Liu et al. 2015). For their part, Galletti et al. (2010) investigated Cu, Ni, and Zn accumulation in horizontal subsurface flow constructed wetlands and its removal from the water flow. Average percentage removal rates were low for Cu (3–9%) and higher for Zn and Ni (25–35%). Besides that, when a higher Zn concentration was present in the influent, a 78–87% removal was found. Yadav (2010) investigated chromium and nickel removal from aqueous solutions in constructed wetlands using *Canna indica*. The highest chromium and nickel removal was 98.3 ± 0.32 and $96.2 \pm 1.52\%$ respectively, with a 10 mg L^{-1} initial concentration, 48 hr of HRT and 0.95 m gravel bed depth.

In addition to the studies previously mentioned, and others, there is a need to study the effect of support media on heavy metals removal in constructed wetland, due to the fact that for many years this support media has been considered an inert material only used for the bacterial consortia establishment or for vegetation support (Pérez-González 2017).

Actually, it is now known that depending on the contaminants to be treated and the geological composition of the support media, adsorption and absorption phenomena are present in these media. Similarly, it is known that during this adsorption microorganisms are adhered and have developed biofilms that constitute a new material support media that combines inorganic characteristics media with organic microbial biomass. Support media provide ion exchange and microbial biomass as its own cellular composition and excrete polymers that are adhered as biofilm (extracellular polymers) (Pérez-González 2017).

When microorganisms previously adapted to heavy metals or if these are metal tolerant-bacteria by genetic adaptation and transcription, its use is even more important to consider. In fact, Amábilis et al. (2015) found (through a heavy metals mass balance in constructed wetlands) that support media was responsible of 40 percent of lead and chromium removal.

Based on all previous information exposed, the present research objective is to study the effect of support media inoculated with tolerant strains in heavy metals removal at waste water industries concentrations in constructed wetlands. This knowledge will contribute to complement parameters design that currently exist for heavy metals removal in constructed wetlands, because presently the effect of support media is not currently considered for this purpose.

Materials and Methods

Heavy Metals Solution

The adsorption of total chromium and total lead was studied, due to their high toxicity and their usage in large volumes, and consequently, large discharges to the environment. For these reasons, those two metals have a high environmental and public health concern (Barakat 2011).

Synthetic wastewater was used in heavy metals adsorption tests with *tezontle*. This aqueous solution simulated a typical secondary municipal effluent with a theoretical chemical oxygen demand value of 400 mg L^{-1} (Masters and Ela 2008), which was derived from sucrose. In addition to carbon, synthetic wastewater contained 30 mg L^{-1} of nitrogen, 6 mg L^{-1} of phosphorus, and 30 mg L^{-1} of potassium, derived from ammonium sulphate, $(NH_4)_2SO_4$, sodium phosphate, Na_3PO_4, and potassium nitrate, KNO_3, respectively (Orduña-Bustamante et al. 2011). Likewise, the synthetic water was contaminated with analytic Merck reagents, $Pb(NO_3)_2$ and $Cr(NO_3)_6$, at concentrations of 24 and 18.74 mg L^{-1}, respectively. These values correspond to industrial effluents related with lead and chromium activities such as mining, tanning, chrome plating, electroplating, and battery production and disposal (Naja and Volesky 2009). Tap water was used to prepare synthetic water that was used in experimental tests with constructed wetlands and in controlled adsorption tests. It is worth mentioning that tap water was previously characterized by ionic chromatography to know N, P, and K concentration, which were considered for synthetic water preparation.

Constructed Wetlands Experimental Systems

Twelve artificial or constructed wetlands at laboratory scale (AW or CW) were built with polyvinyl chloride cylinders with dimensions of 39 cm in height and 20 cm in diameter. The containers were uniformly filled with fragments of volcanic rock known as "tezontle", *an Aztec or Nahuatl word meaning stone as light as hair (from tetl=stone and tzontli=hair)* (Cabrera 2002), with a 3.8–5 mm diameter, 1.22 g cm^{-3} density, and 38 percent porosity (EPA 2000). In theoretical and experimental studies, it was shown that 38 percent of porosity helps to avoid clogging phenomena (Masi and Martinuzzi 2007; Peruzzi et al. 2009). In addition, *tezontle* was washed and sterilized by autoclave before to being placed in the reactors, due to the objectives project related to metallotolerant strains used.

For all AW, the water level was adjusted to 5 cm below the surface of the support media, exhibiting a free pore water volume of 4.06 L and a height/diameter ratio of 1.7, according to the recommended standards (Winter and Goetz 2003; Caselles-Osorio et al. 2006; Puigagut et al. 2008). Six of the 12 AW were planted with *Phragmites australis* (CWP), three of them were inoculated with tolerant strains (CWP_I) described below. The remaining six CW were used as controls. Three were neither planted nor inoculated (CWC) and the other three were inoculated but not planted (CWI). Table 1 shows the experimental design with the configuration of the different lab scale wetlands systems.

Before the start of the experiment, the vegetation was propagated in a hydroponic system in the same location where the experiment was implemented. One hundred days before starting the experiments, the plants were transferred into the constructed wetlands, using the procedures described by Kadlec (2009). In the case of the three planted and inoculated systems (CWP_I), the rhizosphere was previously sterilized using NaClO (10 percent) and ethanol (70 percent) to inactivate microorganisms associated with the rhizome (De Souza et al. 1999).

Once vegetation was placed into the systems, the three sterilized CWP_I were inoculated with a bacterial consortium conformed by the following five strains which were designated, studied in detail, and classified in GenBank by Salgado-Bernal et al. (2012a): T117, T119, T1113, T1115, and T217, all of the genus *Bacillus*, Gram-

Table 1. Configuration of experimental design for evaluate adsorption of Pb and Cr in support media from constructed wetlands. All support media were sterilized before the treatment.

Constructed Wetland	Inoculated	Planted
CWP		XXX
CWP_I	XXX	XXX
CWI	XXX	
CWC		XXX

positives, and endospore formation with exception of T1113. Besides bacillus of T117 are large, thin, large chains, subterminal and oval without cell deformation. T119 bacillus are short, thick, medium chains, subterminal and terminal endospore, oval without cell deformation. T1113 bacillus are short, thin, without aggrupation. T1115 and T217 bacillus are short, thick, medium chains, with subterminal endospore, oval and manifests cell deformation.

For each bacterial isolate identification, Salgado-Bernal et al. (2012a) determined the 16S ribosomal DNA sequence (16S rDNA), amplifying by PCR by using the Accuprime Tag DNA Polymerase System (Invitrogen). These microorganisms showed resistance to heavy metals, including lead and chromium, and were even able to remove them in a certain percentage of the aqueous phase. The 12 CW operated for 151 days with discontinuous feeding at four-day and two-day intervals.

Concerning hydraulic residence time (HRT), two values within the limits of those suggested values for AW or CW by several authors (Brix and Arias 2005; Caselles-Osorio et al. 2006; Chong et al. 2009; Kadlec 2009; Orduña-Bustamante et al. 2011) were chosen. Considering that there are not established HRT values for heavy metals removal, in this work a high (4 d) and a low (2 d) HRTs were evaluated. After 151 operations days, *tezontle* was collected from each CW and properly washed to remove vegetation debris or other particles. Each support media was dried in an oven at 70°C for 48 hours to achieve constant mass. The dry mass was determined and the dried samples were powdered and subsequently digested with 5 mL of concentrated HNO_3:HCl (3:1) at 160°C until a colorless solution was obtained. After cooling down, the suspensions were filtered and the filtrate was adjusted to 50 mL with de-ionized water for heavy metals analysis (Allen 1989). A reagent blank reference and high purity grade of HNO_3 and HCl were used, plant (if applicable) and support media were included to verify the accuracy and precision of the digestion procedure and subsequent analysis.

Total chromium (Cr) and total lead (Pb) concentration of all samples were determined by the flame method of atomic absorption spectrometer Perkin-Elmer Optima 4300DV. The working volume for each lab scale wetland system (v) was uniform and Cr and Pb concentrations in effluents (c) were 24 and 18.74 mg L^{-1}, respectively. Thus, heavy metal mass in influent was calculated using equation (1).

$$MMI = v \sum_{t_i}^{t_f} ci \qquad (1)$$

where

MMI = total mass of metal in influent, mg
v   = constructed wetland volume, L
t_i = initial operation day, d
t_f = total days of feeding, d
ci   = concentration of heavy metal studied

Total mass of Pb and Cr in support media was calculated multiplying concentration, in mg kg^{-1}, by the total mass of dried support media. In this sense, the heavy metal percentage adsorbed by the support media was calculated by difference, considering total mass influent as 100 percent.

Carbon and Nitrogen Quantification in Support Media

Tezontle samples were analyzed for C, H, O, N, P, S content using an elemental analyzer Leco 832, in order to correlate the amount of metal adsorbed with the elemental molecular content of the substrate. The above, after 151 days of operation and based on the hypothesis that heavy metal adsorption is proportional to the nutrients content developed (by bacterial metabolisms) in constructed wetlands support media. The experimental design was shown in Table 1. The treatments were analyzed in triplicate. Also, negative and positive controls were analyzed.

Heavy Metals Adsorption Controlled Test

Because the removal of heavy metals is carried out directly by the adsorption on the support media in constructed wetlands, heavy metal removal, and controlled adsorption tests using *tezontle* and river stone (pebbles) were made at the same time. In controlled adsorption tests, the performed treatments were inoculated tezontle (IT), sterile no-inoculated tezontle (CT), inoculated river stone (RSI) and sterile no-inoculated river stone (RSC) all of them without vegetation. Adsorption tests for isotherms construction were done in closed batch reactors. Tests were done in reactors at 25°C. This arrangement system was done to avoid an exchange of liquid or gaseous substances with the surroundings.

River stones are another kind of the material mostly used for support media in constructed wetlands, especially when human population lives near to rivers. Studying them is important because they has geological properties that are completely different from *tezontle*, which are related to ion exchange capacity with heavy metals, besides offering different adhesion material for bacterial colonization in biofilm form. Moreover, considering that this was a controlled test, heavy metals solutions were evaluated together and separately, considering three assays (Pb adsorption, Cr adsorption, and Pb and Cr combined adsorption).

All *tezontle* and river stones utilized for filling or packing each reactor were autoclaved at 120°C during 30 minutes. After that, support media was prepared under sterile conditions in a laminar flow hood. The CT and RSC systems were also exposed to UV light (254 nm) to ensure that no microorganisms had grown. In the case of systems containing tolerant bacteria (TI and RSI), inoculation was performed under asepsis conditions.

In general terms, adsorption tests consisted in pouring different concentrations of interest solution in vials with the support media. Considering that concentration of interest for lead is 24 mg L^{-1} and that of Cr 17 mg L^{-1}, concentrations shown in Table 2 were used for all tests (different treatments). Each treatment was done in triplicate, so the experimental design, for each test (heavy metal to evaluate), consisted of 12 flasks, six with sterile *tezontle*, six with inoculated *tezontle*, six with sterile river stone, and six with inoculated river stone (Table 3). Each support media had the same physical characteristics as those used in experimental constructed wetland systems. These characteristics are 3.8–5 mm diameter, 38 percent porosity, and 1.22 g cm^{-3} density for *tezontle* and 1.32 g cm^{-3} for river stone.

Table 2. Pb and/or Cr concentrations used in controlled adsorption tests.

Flask		Cr, mg L^{-1}	Pb, mg L^{-1}
1a	1b	0	0
2a	2b	2.5	5
3a	3b	5.0	15
4a	4b	15	30
5a	5b	30	40
6a	6b	40	55

Table 3. Experimental design for controlled heavy metals adsorption tests.

Support media	Lead		Chromium		Lead and chromium	
	Sterile	Sterile and inoculated	Sterile	Sterile and inoculated	Sterile	Sterile and inoculated
Tezontle	XXX		XXX		XXX	
River stone	XXX		XXX		XXX	
Tezontle		XXX		XXX		XXX
River stone		XXX		XXX		XXX

After reactors were assembled with different support media, solutions with heavy metal were added for each test (Table 2), then reactors were hermetically sealed and placed on an orbital shaker (Orbit 1900) during three days at 30 revolutions per minute at a constant temperature of 25°C as already mentioned.

After three days, heavy metals concentration was quantified in the aqueous solution called the equilibrium concentration (c_e) and concentration adsorbed was calculated through equation (2) (Klewer et al. 1992). After that, the adsorption models were built according to the linear, Freundlich and Langmuir equations. The last two were linearized and the three correlation coefficients were compared to find the best adjust model.

$$C_s = \frac{(C_0 - C_e)V}{M} \qquad (2)$$

where

c_s = Adsorbed concentration, mg L^{-1}
c_0 = Concentration added before adsorption test, mg L^{-1}
c_e = Concentration quantified in the aqueous solution called the equilibrium concentration, mg L^{-1}
M = *Tezontle* or river stone dry matter inside reactor, mg
V = Solution volume added, L

Statistical Analysis

A one-way ANOVA followed by a Tukey test *post-hoc* analysis ($\alpha = 0.95$), was used to ascertain whether the amount of metals was significantly different among the four types of experimental units used. All environmental compartments previously described were considered. The normality and homogeneity of data were evaluated by the Kolmogorov-Smirnov test.

As for elemental analysis results in correlation with the adsorbed heavy metal percentage, a discriminant analysis was made in conjunction with a principal components analysis to know which molecules (C, N, or P) had more statistical influence on each metal adsorption. All statistical analyses of data were performed using Minitab 15 package for Windows.

Results and Discussion

Percentages of lead and chromium quantified in *tezontle* were similar between both HRT (4-day and 2-day); results are shown in Table 4. In artificial or constructed wetlands operation with two different HRT, CWC was the only CW that presented significant differences in both conditions ($F_{7,16} = 36.82$; $P = 0.0001$). In the case of lead, 1.17 g was accumulated in *tezontle* with 4 days of HRT, equivalent to 30.3 percent; whereas 1.06 g were accumulated in 2-day of HRT, which was only 14.01 percent of the total admission.

In the case of chromium, metal accumulation percentages in *tezontle* were similar to those exhibited for lead, although for CWC, the means and standard deviations of the two HRT were similar. In fact, it is reflected in the statistical test with a P value that tends to zero $P = 0.01 \times 10^{-8}$ (Table 4), which is related to the great affinity for specific area that has the *tezontle* and, in this case, chromium adsorption by the biofilm as mentioned by Ferraz et al. (2015). In addition, CWC only contains *tezontle* and a small amount of microorganisms, compared to the inoculated systems, which have been incorporated from the surrounding conditions since the reactors were located in open conditions and subjected to weathering.

There were particularities in comparing the removal of heavy metal between the reactors for both metals in both HRT. This behavior is described in detail in the next paragraphs.

Table 4. Heavy metals mass adsorbed by support media (*tezontle*) in artificial or constructed wetlands at the end of the operation.

		Lead mass, g			Chromium mass, g		
		Influent	Effluent	*Tezontle*	Influent	Effluent	*Tezontle*
CWP_1	4 d	3.86	1.34 ± 0.32	1.04 ± 0.34	2.89	1.13 ± 0.28	0.82 ± 0.20
	2 d	7.72	2.94 ± 0.58	2.14 ± 0.52	5.78	2.53 ± 0.33	1.36 ± 0.32
CWP	4 d	3.86	2.03 ± 0.45	0.56 ± 0.15	2.89	1.38 ± 0.27	0.41 ± 0.10
	2 d	7.72	4.26 ± 0.73	0.93 ± 0.22	5.78	2.88 ± 0.51	1.15 ± 0.23
CWC	4 d	3.86	2.68 ± 0.46	1.17 ± 0.29	2.89	1.85 ± 0.22	0.99 ± 0.32
	2 d	7.72	6.63 ± 0.93	1.06 ± 0.25	5.78	4.64 ± 0.45	1.11 ± 0.38
CWI	4 d	3.86	2.06 ± 0.29	1.58 ± 0.26	2.89	1.49 ± 0.26	1.39 ± 0.28
	2 d	7.72	5.48 ± 0.72	2.16 ± 0.31	5.78	3.83 ± 0.53	1.94 ± 0.33

In Table 4, results on lead are shown. Considering a 4-day HRT, these results indicate that CWI showed the highest metal removal by adsorption in *tezontle*. This CW was initially inoculated but without vegetation and 1.58 g of the total 3.86 g of lead were removed, which is equivalent to 40.9 percent, being significantly higher than in other three CWs ($F_{7,16}$ = 29.11; P = 0.002). The CWC was the second reactor that removed the most lead with 1.17 g equivalent to 30.3 percent with values very similar to CWP_I that removed 1.04 corresponding to 27 percent, without showing significant differences between these two systems ($F_{7,16}$ = 15,32; P = 0.11), but with significant differences with the other CW ($F_{7,16}$ = 36.82; P = 0.004). Finally, CWP had the significantly lowest lead accumulation in the *tezontle* (F7,16 = 37.61; P = 0.0013).

On the other hand, when the CW were operated with 2 days of HRT, the lead amount in the influent was 7.72 g, inoculated CWs were capable of significantly removing the most lead with 2.14 g CWP_I and 2.16 g CWI that are equivalent to 27.7 and 27.9 percent respectively, with virtually the same removal between them. A significantly lower amount of lead was removed by non-inoculated CW ($F_{7,16}$ = 33.42; P = 0.004), which was about 1.0 g of lead corresponding to 12.9 percent.

In the case of 2 days of HRT, lead adsorption is related to biofilm presence in the medium in which the systems were inoculated with metallotolerant bacteria, which is manifested in a greater amount of biofilm because of them. These bacteria develop their metabolic functions in presence of heavy metals compared to non-tolerant microorganisms (Treviño-Cordero et al. 2013). This is further supported by the results obtained from the elemental analysis, in which Fig. 1 shows that *tezontle* from the inoculated CW had a significantly higher amount of C and N compared to the non-inoculated systems (CWP and CWC). A highest amount of C and N was found in inoculated systems than in not inoculated systems. In general terms, bacterial density was higher in inoculated CW systems. This phenomenon was observed in a previous study of Amabilis-Sosa et al. (2015).

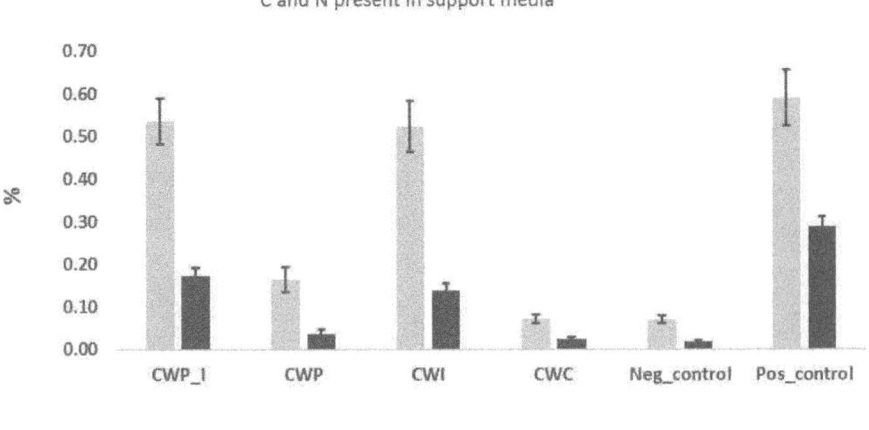

Fig. 1. Evaluation of carbon and nitrogen present in support media from different CW systems.

The above mentioned in mainly true with 2-day HRT, because with 4 days CWI and CWC (unplanted CW) had the highest lead removal. This effect could be explained because vegetation reduces the amount of metal available to be adsorbed by the packaging medium. This behavior is related to the fact that with enough HRT, vegetation begins to accumulate heavy metals by mechanisms of metal transportation and accumulation, independently of the metal translocation capacity to the aerial part (Clemens 2002).

On the other hand, Table 4 indicates chromium removal by the support media in the evaluated CW. Regarding 4 days of HRT, statistically different values between the four CW were observed ($F_{7,16} = 25.82$; $P = 0.0006$).

Chromium was adsorbed by *tezontle* in the highest amount by the CWI, removing 48 percent from the total (1.39 g of the 2.89 g). It was followed by the CWC with 0.99 g corresponding to 34.25 percent. Both CW with vegetation, CWP_I removed 0.82 g (28.37 percent) and CWP with 0.41 g (14.18 percent) had the lowest removal. These data have been discussed for lead and supported by Treviño-Cordero et al. (2013) and Amábilis-Sosa et al. (2015), with respect to the adsorption capacity increase by an inert material which contains an organic "coating" called biofilm. A higher percentage of carbon and nitrogen in the composition of these systems are shown in Fig. 1.

Finally, in heavy metals adsorption by *tezontle*, with 2 days of HRT the largest amount of chromium removed by the support media was 1.94 g, with 33.56 percent of the total input (5.78 g) in CWI. The second one was the other inoculated system with vegetation (CWP_I) that adsorbed 1.36 g equivalent to 23.53 percent, with significant differences between the two inoculated systems ($F_{7,16} = 28.76$; $P = 0.004$). On the other case, both non-inoculated CW presented no significant differences between them in chromium adsorption with 2 days HRT ($F_{7,16} = 12.33$; $P = 0.126$), and both removed chromium values around 1.12 g, which is 19 percent of the total entered.

From inoculated CW (CWP_I and CWI), the absence of vegetation allowed for a larger amount of the metal to be captured by support media because it did not have the presence of roots which also experience metal incorporation in their vegetal tissues. This could be interpreted as a kind of competition for analytes to adsorb as was suggested by LeDuc and Terry (2005), Clemens et al. (2006), and Liu et al. (2007). On the other hand, Gavrilescu (2004) and Dhir (2010) mention that roots produce exudates under stress conditions, which are usually organic acids (citric, malic, aspartic, and glycine) that when combined with metals (chelation), give rise to mobile complexes that are difficult to adsorb by *tezontle* but assimilable by the plant (Navarro-Noya et al. 2010).

In addition, of the two CW without vegetation, the one containing tolerant strains removed significantly more chromium related to more content of carbon and nitrogen, as showed in the next section.

As it was mentioned, because of its composition, *tezontle* does not exhibit ion exchange with heavy metals. However, biofilm bacterial growth occurs when there is an adequate support media such as the interstices of volcanic rock, resulting in an effective bioadsorbent (Chong et al. 2009).

In relation to the above, the elemental analysis described in the following lines indicates a greater amount of carbon and nitrogen (as cellular debris) precisely in the CW that adsorbed greater heavy metals amount.

This higher proportion of carbon and nitrogen is due to the fact that bacteria tolerant of heavy metals were not inhibited (or inhibited to a lesser extent), and therefore, developed their metabolic functions in a better way. As it is in this case, extracellular polymers generation for its adhesion to the support media were probably exuded (Nies 2003; Kosolapov et al. 2004; Odukuma and Emedolu 2005; Silver and Phung 2005).

Based on the above, it seems that adsorption by the presence of bacterial biofilm would not occur in reactors without inoculation CWP and CWC, especially in the last one that was sterile (sterilized support media) at the beginning of the experiment and did not contain vegetation that could contribute the presence of rizospheric bacteria.

This event could be explained by the entry of microorganisms into the system, which, despite not being able to survive because they are not tolerant, could attach their cellular material to *tezontle*, resulting in a bioadsorbent from which good heavy metals removal efficiencies have been reported (Gavrilescu 2004; Odukuma and Emedolu 2005).

Finally, the lead and chromium amount quantified in *tezontle* of CWP (conventional constructed wetlands) systems is similar to that reported by Matagi et al. (1998) and Wood and McAtamney (1996) who worked with different support media, but with similar constructive characteristics to the present study. In this sense, the heavy metals amount quantified in CWP_I systems is greater than that reported in similar studies, in which they did not use metal tolerant bacteria.

Carbon and Nitrogen in Support Media

Contents of total carbon and nitrogen in support media in four different CW is showed in Fig. 1. Also, a negative control (sterile *tezontle*) and positive control are included (*tezontle* inoculated with the consortia evaluated, but without heavy metals addition). Hitherto, both inoculated CW (CWI and CWP_I) contained practically the same amount of nitrogen as the positive control with mean values around 0.58 percent without presenting statistically significant differences ($F_{6,4}$ = 2.6; P = 0.098). This clearly shows the adaptation and tolerance of the selected microorganisms to grow in the inoculated CW in the presence of heavy metals during the entire operation of the CW. In fact, CWC with no tolerant bacteria (CWC and CWP) have a significantly lower amount than those inoculated ($F_{6,4}$ = 9.23; P = 0.000012) and the one with CWP vegetation is significantly higher than CWC with negative control. The relative abundance of nitrogen in the inoculated systems is due to the absence of inhibition in the presence of heavy metals, as mentioned by Obarska-Pempkowiak and Klimkowska (1999) and Yeh et al. (2009). Specific enzymatic aspects of oxidation and reduction reactions in synergy with the adsorption derived from the content of murein in the cell wall are related.

On the other hand, Fig. 1 shows significantly more content of carbon in inoculated CW (CWP_I and CWI) ($F_{6,4}$ = 8.3; P = 0.0003) but also, significantly, the least amount of the element in comparison with the positive control ($F_{6,4}$ = 9.2; P = 0.0001). Unlike the nitrogen content, the amount of carbon was affected by the presence of heavy metals, which is related to cellular lysis (Navarro-Noya et al. 2010). Notwithstanding, adsorption capacity of systems was not affected (Table 4), since the mix with extracellular polymers typical for the biofilm, exerted an effect of bioadsorbent, which means that even though the bacterial consortium was no longer totally viable, the bioadsorbent was effective in the adsorption of metals (Nithya et al. 2011). As it is observed in the results of Table 4, it is worth mentioning that the bioadsorbent capacity of inoculated CW is derived from the tolerance of the bacterial consortium towards heavy metals, since cellular lysis could only have appeared in cells that had already been developed in the beginning and this development was derived from having realized its metabolic functions despite the presence of heavy metals.

Controlled Adsorption Tests

Chromium

Table 5 indicates that, in almost all cases, *tezontle* was able to adsorb significantly more chromium than the river stone ($F_{7,3} = 16.2$, $P = 0.0002$). The last material at its two end points shows a certain saturation tendency, when maximum concentration adsorbed (10 mg L^{-1}) is repeated unlike the *tezontle*, in which a linear behavior is observed. That is, at the higher exposed concentration the greater the amount of the adsorbed metal. Table 5 also shows that *tezontle* in the presence of lead in solution, adsorbed more quantity than when only chrome was contained. Even in the last concentration levels the difference is maximized, reaching to adsorb up to 30 mg L^{-1}, which is equivalent to 75 percent of the initial total concentration. This is probably related to the effect of the pH that favors the presence of lead at high concentrations. It is a metal that by its nature consumes alkalinity, which has an influence on the adsorption capacity of chromium, indicating an inversely proportional relation between pH and percentage of adsorption (Alemayehu et al. 2011). In fact, this is supported with pH measurements. When heavy metals salts were added (at the beginning), pH values of chromium, lead, and mix chromium-lead treatments were 6.4, 6.0, and 5.2, respectively. At the end of the experiments there was a pH decrease between 0.5 and 0.7 units in all cases. These effects are probably related to the chemical nature of nitrate-based salts [Pb(NO$_3$)$_2$ and Cr(NO$_3$)$_6$], since by promoting an oxidizing environment pH is reduced, as it does during nitrification (Dickson 2006).

Based on the above, the researchers proceeded to apply the adsorption models to the *tezontle* for Cr alone and chromium combined with lead, since in them the amount of removal is considerable, in comparison to river stone. Within this context,

Table 5. Initial, C_0, and final, C_s, chromium concentrations after the accelerated adsorption process.

C_0, mg L^{-1}	C_s, mg L^{-1}		
	Tezontle	River stone	*Tezontle* in presence of lead (II)
0	0	0	0
2.5	2 ± 0.15	0.012 ± 0.002	2.35 ± 0.19
5	2.7 ± 0.2	0.2 ± 0.18	3.1 ± 0.3
15	5.03 ± 0.35	1.9 ± 0.12	12.1 ± 0.86
30	8.1 ± 0.42	10 ± 0.87	19.35 ± 1.1
40	19.12 ± 1.1	10 ± 0.89	30 ± 2.3

Fig. 2 shows the linear model applied for both cases. Figure 2a shows a correlation coefficient of 0.7028 which is acceptable for the adsorption models. In fact, for the equilibrium concentrations (C_e) above 3.0 mg L^{-1}, the model will present a better fit. Figure 2b shows a correlation coefficient of 0.9772 for the linear model in the adsorption of chromium and lead in the *tezontle*. This is a very high coefficient in comparison to those found in other studies where the removal of these metals (Malkoc et al. 2006; Di Natale et al. 2007).

It is worth mentioning that from kinetic models which are accepted for contaminants adsorption (linear, Freundlich, and Langmuir), the lineal model had the best fit, but with correlation coefficients lower than 0.8, which is not an ideal fit. This is because, kinetically, a sigmoidal behavior could be observed (Figs. 2a and 3a). Thus, this suggests that in future studies, sigmoidal models could be evaluated, at least for chromium, such as Brody, Richards, logistic, and Gompertz models (Malhado et al. 2008).

Figure 3 shows the result of applying the Freundlich model to chromium adsorption alone and with lead in *tezontle*. In this respect, Fig. 3a indicates the adjustment to chromium alone which is practically 0.80, slightly better than the linear one (Fig. 2a), which is why it is well accepted for adsorption isotherms. The main finding is shown in Fig. 3b, in which the adsorption of lead and chromium in *tezontle* shows a correlation coefficient of 0.9866, which is even better than the linear model, so that the Freundlich model can be incorporated to the design parameters of the constructed wetlands, to pilot scale, for chromium and lead removal. In particular, the good fit of

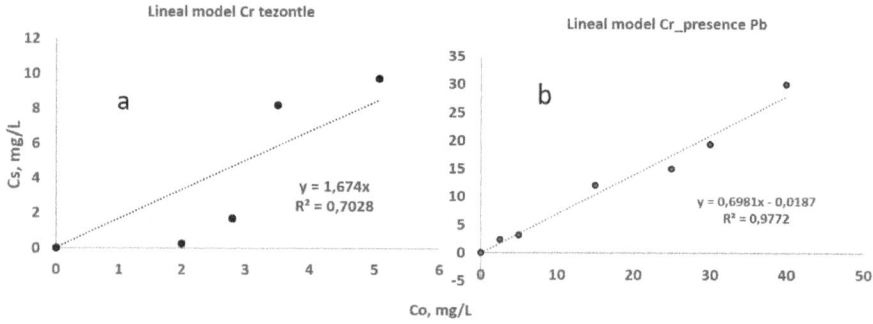

Fig. 2. Lineal model for adsorption of (a) chromium in *tezontle* and (b) chromium and lead in *tezontle*.

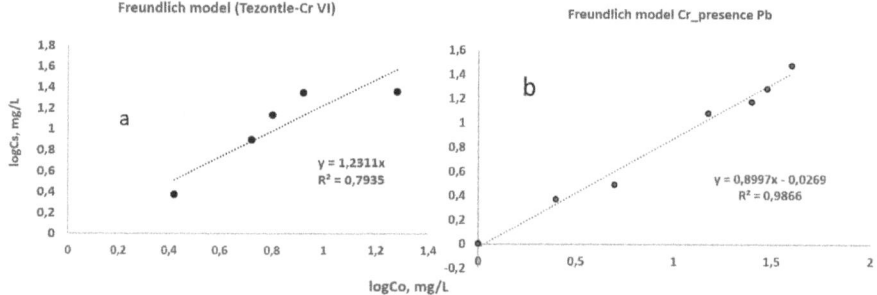

Fig. 3. Freundlich model for adsorption of (a) chromium in *tezontle* and (b) chromium and lead in *tezontle*.

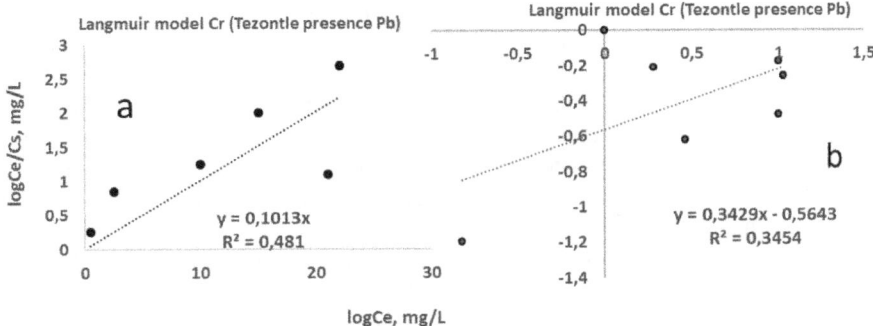

Fig. 4. Langmuir model for adsorption of (a) chromium in *tezontle* and (b) chromium and lead in *tezontle*.

the Freundlich model indicates that the adsorption capacity increases as the amount of the heavy metals in the aqueous medium increases, which refers to a number of "infinite" active sites with concentrations in this order of values (Kwon et al. 2005).

Finally, the Langmuir adsorption model presented the lowest correlation coefficients of the three models evaluated (Fig. 4), with no correlation between the values of $logC_e$ and $logC_e/C_s$. Given this, it is not feasible to predict the heavy metals of interest concentrations of adsorption through this model that theoretically considers continuous reactivations.

Lead

Unlike chromium adsorption results, Table 6 indicates that for lead removal, river stone was more efficient in all cases ($F_{7,3}$ = 17.1; P = 0.0002). Both materials show much similarity in their last two values, which means that they could not continue removing more heavy metal. Even *tezontle* adsorbed less with the initial concentration of 55 mg L^{-1} than with 44 mg L^{-1}. The tendency to saturation is probably related to the chemical or mineral conformation of river stone, since it is known to have twice as much silica as *tezontle* (Pérez-González 2017). This chemical element offers a certain capacity of ion exchange with lead, in addition to the fact that the river stone does not contain iron, an element that is 30 percent of the content of the *tezontle*

Table 6. Initial C_o, and final C_s, lead concentrations after the accelerated adsorption process.

C_o, mg L^{-1}	C_s, mg L^{-1}	
	Tezontle	River stone
0	0	0
5	0.85 ± 0.70	4.9 ± 0.40
15	4.4 ± 0.23	12.6 ± 1.07
30	5.5 ± 0.35	21.3 ± 1.80
40	18 ± 1.20	25.3 ± 1.90
55	16 ± 1.10	25.910 ± 1.80

and precisely the chemical element which does not offer ionic interchange with the lead (Pérez-González 2017). There is a displacement by molecular mass (Pacini and Gächter 1999).

Next, the application of the results of the adsorption models is presented. It is possible to mention that for lead one does not count the data of *tezontle* with the presence of another metal different from lead, since it is known that this cation is a displacement object when a component of similar molecular mass with a higher valence such as chromium is found (Vacca et al. 2005).

Figure 5 refers to lineal model applied to lead adsorption in *tezontle* (a) and river stone (b). The coefficient correlation in the first case is 0.8437, which is a high value, but it can be seen that the points above an initial concentration of 30 mg L^{-1} are very far from the trend line, so that only would adjust for concentrations below that value, with which the coefficient of adjustment of the model would be extremely high. On the other hand, in river stone, the coefficient correlation is 0.915, which could suggest the use of linear model for removal of lead with river stone.

Figure 6 shows that for both supports, the correlation coefficient of the Freundlich model is extremely high with 0.9334 for *tezontle* (Fig. 6a) and 0.977 (Fig. 6b) for river stone. This is especially true for the latter support media, since it is acknowledged that all points are very close to the trend line. Only considering that approximately 50 mg L^{-1} of initial lead value, the behavior of the adsorption capacity is asymptotic.

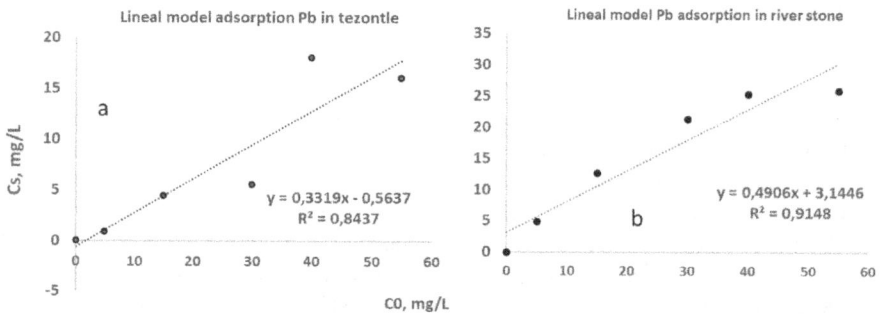

Fig. 5. Lineal model for adsorption of (a) lead in *tezontle* and (b) lead in river stone.

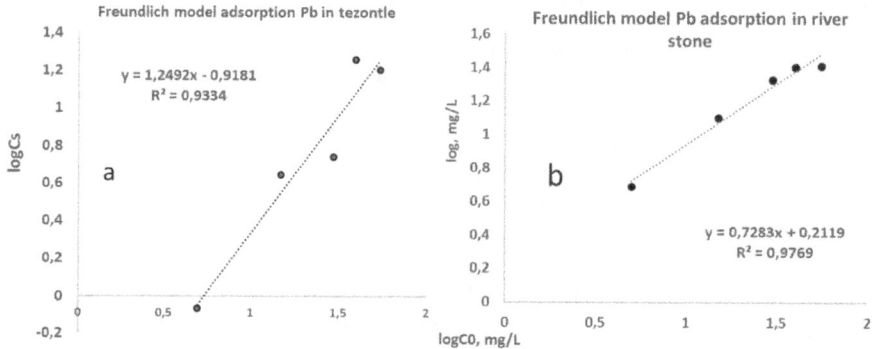

Fig. 6. Freundlich model for adsorption of (a) lead in *tezontle* and (b) lead in river stone.

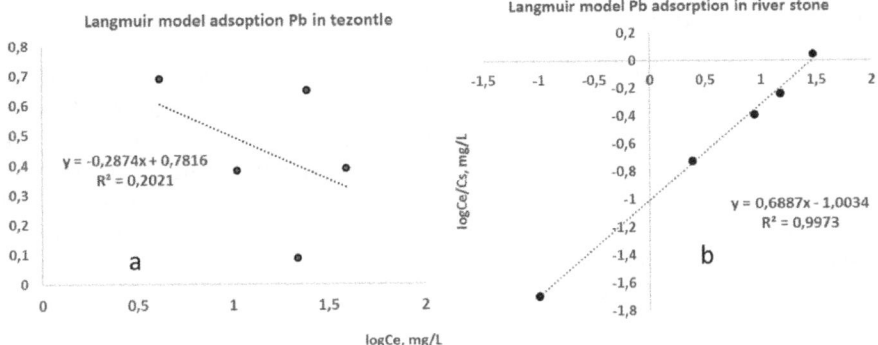

Fig. 7. Langmuir model for adsorption of (a) lead in *tezontle* and (b) lead in river stone.

Finally, the application of the Langmuir model in Fig. 7 indicated a very slow statistical correlation for adsorption in *tezontle* (0.2021). Figure 7b indicates a very high correlation coefficient of 0.9973, which seems to be an excellent fit of the model. Nevertheless, mathematically, the slope of this model should be negative by the application of a log function, which is finally interpreted as a very good fit, but only valid in points that are positive (Leyva 2010), so the Langmuir model is not applicable for lead adsorption in river stone.

The previous results of lead removal by adsorbents derived from volcanic slag are above those reported by Kwon et al. (2005) and very similar to those indicated by Pacini and Gächter (1999). The best fit of the models in the adsorption of lead in river stone is possibly related to the high silica content of the river stone (64 percent), as mentioned above, since this chemical element exhibits a high ion exchange capacity with metals such as lead and cadmium. In fact, zeolites have demonstrated efficient results in this respect due to their high content of silica being used in industrial processes (Kwon et al. 2005).

With the above results, it is worth mentioning that, regardless of the amount of lead adsorbed by the *tezontle* and river stone, it is remarkable the good adjustment to the linear and Freundlich models, so that the amount of metal to adsorb by the support could be calculated if the initial concentration of the metal is known.

Conclusions

In experiments related with adsorption, support media with metallotolerant microorganisms were capable of accumulating a considerable amount of heavy metals, probably due to biofilm development. This was experimentally supported by elemental analysis results.

Additional to recent research publications, the overall results of these experiments demonstrate that AW or CW, inoculated with metal tolerant bacteria, eventually could be used as an effective treatment of wastewaters containing lead and/or chromium.

Also, depending on the geographical area, river stone or *tezontle* could be used as support media, only considering the respective hydraulic calculations.

Acknowledgments

The authors are grateful for the pre-final draft version revision by Ph.D. Cand. Víctor de Jesús García-Luna belonging to Universidad Nacional Autónoma de México, Facultad de Química, Conjunto E. Also, the authors acknowledge the valuable academic advice and metallotolerant organisms provided by Mrs. Prof. Dr. Irina Salgado-Bernal from the Universidad de La Habana, Cuba, Head of the Department of Microbiology and Virology of the Faculty of Biology. Financial support from UNAM, Facultad de Química PAIP (Programme for Support to Research and Graduate Students Formation, in Spanish) is greatly acknowledged.

References Cited

Alemayehu, E., S. Thiele-Bruhn and B. Lennartz. 2011. Adsorption behaviour of Cr (VI) onto macro and micro-vesicular volcanic rocks from water. Sep. Purif. Technol. 78(1): 55–61.

Allen, S.E. 1989. Analysis of vegetation and other organic materials. pp. 46–61. *In*: Allen, S.E. [ed.]. Chemical Analysis of Ecological Materials Blackwell Scientific Publications. Oxford, UK.

Amabilis-Sosa, L.E., C. Siebe, G. Moeller-Chávez and M.d.C. Durán-Domínguez-de-Bazúa. 2015. Accumulation and distribution of lead and chromium in laboratory-scale constructed wetlands inoculated with metal-tolerant bacteria. Int. J. Phytoremediat. 17(11): 1090–1096.

Amuda, O.S., A.A. Giwa and I.A. Bello. 2007. Removal of heavy metals from industrial wastewater using modified activated coconut shell carbon. Biochem. Eng. J. 36(2): 174–181.

Barakat, M.A. 2011. New trends in removing heavy metals from industrial wastewater. Arab. J. Chem. 4(4): 361–367.

Brix, H. and C.A. Arias. 2005. The use of vertical flow constructed wetlands for one-site treatment of domestic wastewater: New Danish Guidelines. Ecol. Eng. 25(5): 491–500.

Cabrera, L. 2002. Diccionario de Aztequismos. Revision: J. Ignacio Dávila-Garibi. Nahuatl terms: Luis Reyes-García. Latin terms (botanical and zoological classifications): Esteban Inciarte. Ed. Colofón S.A. 5th edition. ISBN 968-867-038-3. Mexico City, Mexico. In Spanish.

Caselles-Osorio, A., J. Puigagut, E. Segú, N. Vaello, F. Granés, D. García et al. 2006. Solids accumulation in six full-scale subsurface flow constructed wetlands. Water Res. 41(6): 1388–1398.

Cheng, S., W. Grosse, F. Karrenbrock and M. Thoennessen. 2002. Efficiency of constructed wetlands in descontamination of water polluted by heavy metals. Ecol. Eng. 18(3): 317–325.

Chong, H.L.H., M.N. Ahmad and P.E. Lim. 2009. Growth of *Typha angustifolia* and media biofilm formation in constructed wetlands with different media. Born. Sci. 25(1): 317–325.

Clemens, S., G. Palmgren and U. Krämer. 2002. A long way ahead: Understanding and engineering plant metal accumulation. Plant Sci. 7(1): 309–315.

De Souza, M.P., C.P.A. Huang, N. Che and N. Terry. 1999. Rhizosphere bacteria enhance the accumulation of selenium and mercury in wetland plants. Plan 209(2): 259–263.

Dhir, B. 2010. Use of aquatic plants in removing heavy metals from wastewater. Int. Env. Eng. 2(1-3): 185–201.

Di Natale, F., A. Lancia, A. Molino and D. Musmarra. 2007. Removal of chromium ions form aqueous solutions by adsorption on activated carbon and char. J. Hazard. Mat. 145(3): 381–390.

Dickson, T.R. 2006. Química, un enfoque ecológico. Limusa Wiley. México, D.F. In Spanish.

Dhokpande, S.R. and J.P. Kaware. 2013. Biological methods for heavy metal removal: A review. Int. J. Eng. Sci. Innov. Tech. 2: 304–309.

EPA. 2000. Manual of constructed wetlands treatment of municipal wastewaters. Environmental Protection Agency. Cincinnati, Ohio, US.

Ferraz, A.I., F. Costa, M.T. Tavares and J.A. Teixeira. 2014. Mechanisms of Cr (III) biosorption onto residual brewer's yeast. *In*: IWA World Water Congress & Exhibition. Lisboa, Portugal. https://repositorium.sdum.uminho.pt/handle/1822/36037.

Galletti, A., P. Verlicchi and E. Ranieri. 2010. Removal and accumulation of Cu, Ni and Zn in horizontal subsurface flow constructed wetlands: contribution of vegetation and filling medium. Science of the Total Environment 408(21): 5097–5105.

Gavrilescu, M. 2004. Removal of heavy metals from the environment by biosorption. Eng. Life Sci. 4(3): 219–232.

Hatt, B.E., A. Steinel, A. Deletic and T.D. Fletcher. 2011. Retention of heavy metals by stormwater filtration systems: Breakthrough analysis. Water Science and Technology 64(9): 1913–1919.

Kadlec, R.H. 2009. Comparison of free water and horizontal subsurface treatment wetlands. Ecol. Eng. 35: 159–174.

Klewer, H.J. Langhoff and W. Weissenfels. 1992. Adsorption of polycyclic aromatic hydrocarbons (PAH's) by soils particles-influence on biodegrability and biotoxicity. Applied Microb. Biotechnol. 36: 689–696.

Kosolapov, D.B., P. Kusch, M.B. Vainshtein, A.V. Vatsourina, A. Wießner, M. Kästner et al. 2004. Microbial processes of heavy metals removal from carbon-deficient effluents in constructed wetlands. Eng. Life Sci. 4(5): 403–411.

Kumari, M., C.U. Pittman and D. Mohan. 2015. Heavy metals [chromium (VI) and lead (II)] removal from water using mesoporous magnetite (Fe_3O_4) nanospheres. J. Coll. Interf. Sci. 442: 120–132.

Kwon, J.S., S.T. Yun, S.O. Kim, B. Mayer and I. Hutcheon. 2005. Sorption of Zn (II) in aqueous solutions by scoria. Chemosphere 60: 1416–1426.

LeDuc, D.L. and N. Terry. 2005. Phytoremediation of toxic trace elements in soil and water. J. Ind. Microbiol. Biot. 32: 514–520.

Leyva, R. 2010. Fundamentos de adsorción. Centro de Investigación y Estudios de Posgrado, Facultad de Ciencias Químicas. Universidad Autónoma de San Luis Potosí. San Luis Potosí, México. In Spanish. http://www.cnea.gov.ar/xxi/ambiental/iberoarsen/docs/taller08/Presentaciones/4c%20 Fundamentos%20de%20Adsorcion%20Roberto%20Leyva.pdf.

Liu, J., Y. Dong, H. Xu, D. Wang and J. Xu. 2007. Accumulation of Cd, Pb and Zn by 19 wetland plant species in constructed wetland. J. Hazard. Mat. 147(3): 947–953.

Malhado, C.H.M., A.A. Ramos, P.L.S. Carneiro, J.C. Souza, F. Wechsler, J.P. Eler et al. 2008. Nonlinear models to describe the growth of the buffaloes of the murrah breed. Arch. Zoot. 57(220): 497–503.

Malkoc, E., Y. Nuhoglu and M. Dundar. 2006. Adsorption of chromium (VI) on pomace—an olive oil industry waste: Batch and column studies. J. Hazard. Mat. 138(1): 142–151.

Masi, F. and N. Martinuzzi. 2007. Constructed wetlands for the Mediterranean countries: Hybrid systems for water reuse and sustainable sanitation. Desalination 215(1): 44–55.

Masters, G. and W. Ela. 2008. Introduction to Environmental Engineering. Prentice Hall. US.

Matagi, S.V., D. Swai and R. Mugabe. 1998. A review of heavy metal removal mechanism in wetlands. Afr. J. Trop. Hydrobiology Fisheries 8(1): 13–25.

Naja, G.M. and B. Volesky. 2009. Toxicity and sources of Pb, Cd, Hg, Cr, As, and radionuclides in the environment. Heavy Metals in the Environment 8: 16–18.

Navarro-Noya, Y.E., J. Jan-Roblero, M.delC. González-Chávez, R. Hernández-Gama and C. Hernández-Rodríguez. 2010. Bacterial communities associated with the rhizosphere of pioneer plants (*Bahia xylopoda* and *Viguiera linearis*) growing on heavy metals-contaminated soils. Anton van Leeuw 97(4): 335–349.

Nies, D.H. 2003. Efflux-mediated heavy metals resistance in prokaryotes. FEMS Microbiol. Rev. 27(2-3): 313–339.

Nithya, C., B. Gnanalakshmi and S.K. Pandian. 2011. Assessment and characterization of heavy metal resistance in Palk Bay sediment bacteria. Mar. Environ. Res. 71(4): 283–294.

Obarska-Pempkowiak, H. and K. Klimkowska. 1999. Distribution of nutrients and heavy metals in a constructed wetland system. Chemosphere 39(2): 303–312.

Odukuma, L.O. and S.N. Emedolu. 2005. Bacterial sorbents of heavy metals associated with two Nigerian crude oils. Glob. J. Pure Appl. Sci. 11(3): 343–351.

Orduña-Bustamante, M.A., M. Vaca-Mier, J.A. Escalante-Estrada and C. Durán-Domínguez. 2011. Nitrogen and potassium variation on contaminant removal for a vertical subsurface flow lab scale constructed wetland. Bioresour. Technol. 102(17): 7745–7754.

Pacini, N. and R. Gächter. 1999. Speciation of riverine particulate phosphorus during rain events. Biogeochemistry 47(1): 87–109.

Pérez-González, R. 2017. Desempeño del medio de soporte de humedales artificiales en la remoción de plomo y cromo presentes en aguas residuales. Professional (B.S.) Thesis, Universidad Nacional Autónoma de México, Mexico City, Mexico. In Spanish.

Peruzzi, E., C. Macci, S. Doni, G. Maciandaro, P. Peruzzi, M. Aiello et al. 2009. *Phragmites australis* for sludge stabilization. Desalination 246(3): 110–119.

Puigagut, J., A. Caselles-Osorio, N. Vaello and J. García. 2008. Fractionation biodegradability and particle-size distribution of organic matter in horizontal subsurface-flow constructed wetland. pp. 289–297. Chapter 25. *In*: Vymazal, J. [ed.]. Wastewater Treatment Plant Dynamics and Management in Constructed and Natural Wetlands. Springer Netherlands.

Salgado-Bernal, I., J.E. Pérez-Ortega, M.E. Carballo-Valdés, A. Martínez and M. Cruz-Arias. 2015. Aplicación de rizobacterias en la biorremediación del cromo hexavalente presente en aguas residuales. Cuban Journal of Biological Sciences/Revista Cubana de Ciencias Biológicas 4(2): 20–34. In Spanish.

Salgado-Bernal, I., M.E. Carballo-Valdés, A. Martínez-Sardiñas, M. Cruz-Arias and C. Durán-Domínguez. 2012a. Interacción de aislados bacterianos rizosféricos con metales de importancia ambiental. Tecnol Cienc Agua. 3(3): 83–95. In Spanish.

Salgado-Bernal, I., A. Martínez-Sardiñas, M.E. Carballo-Valdés, M. Cruz-Arias and M.C. Durán-Domínguez. 2012b. Diversidad de las bacterias rizosféricas asociadas a plantas de *Typha dominguensis* en humedales del río Almendares. Revista CENIC Ciencias Biológicas 43(3): 1–7. In Spanish.

Silver, S. and L.T. Phung. 2005. A bacterial view of the periodic table: gene and proteins for toxic inorganic ions. J. Ind. Microbiol. Biot. 32(11-12): 587–605.

Treviño-Cordero, H., L.G. Juárez-Aguilar, D.I. Mendoza-Castillo, V. Hernández-Montoya, A. Bonilla-Petriciolet and M.A. Montes-Morán. 2013. Synthesis and adsorption properties of activated carbons from biomass of *Prunus domestica* and *Jacaranda mimosifolia* for the removal of heavy metals and dyes from water. Ind. Crop Prod. 42: 315–323.

Vacca, G., H. Wand, M. Nikolausz, P. Kusch and M. Kästner. 2005. Effect of plants and filter materials on bacteria removal in pilot scale constructed wetlands. Water Res. 39(7): 1361–1373.

Winter, K.J. and D. Goetz. 2003. The impact of sewage composition on the soil clogging phenomena of vertical flow constructed wetlands. Water Sci. Tech. 48(5): 9–14.

Wood, R.B. and C.F. McAtamney. 1996. Constructed wetlands for waste water treatment: the use of laterite in the bed medium in phosphorus and heavy metal removal. Hydrobiologia 340(1): 323–331.

Yadav, S.K. 2010. Heavy metals toxicity in plants: an overview on the role of glutathione and phytochelatins in heavy metal stress tolerance of plants. South African Journal of Botany 76(2): 167–179.

Yeh, T.Y., C.C. Chou and C.T. Pan. 2009. Heavy metal removal within pilot-scale constructed wetlands receiving river water contaminated by confined swine operations. Desalination 249(1): 368–373.

6

Full-Scale Applications of Constructed Wetlands in Africa

Diederik P.L. Rousseau

INTRODUCTION

The sanitation crisis in Africa is still high on the international agenda, as data from UNICEF/WHO showed that by 2015, 11% of the North-African population and 70% of the sub-Saharan African population were still using unimproved sanitation facilities, with the rural population lagging behind on the urban population in both cases (UNICEF and WHO 2015).

Constructed wetlands (CW), known as relatively cheap, robust, and easy-to-maintain wastewater treatment technologies, have been reported by various authors as suitable and sustainable sanitation technologies for developing countries (Kivaisi 2001; Zhang et al. 2014). The climate in those countries, especially the high temperatures, usually plays a conducive role in obtaining good treatment efficiencies through enhanced activity of microbiota and macrophytes.

In a country-by-country literature search, 49 peer-reviewed publications were identified which considered constructed wetlands for wastewater treatment in one or another African country. The earliest retained publication dates from 1985 (Chale 1985), the bulk of papers however dates from 2000 and later. Figure 1 gives an overview by country, which shows that Egypt, Kenya, and Tanzania are front runners in terms of constructed wetland research.

Ghent University Campus Kortrijk, Department of Green Chemistry and Technology, Graaf Karel de Goedelaan 5, 8500 Kortrijk, Belgium.
Email: diederik.rousseau@ugent.be

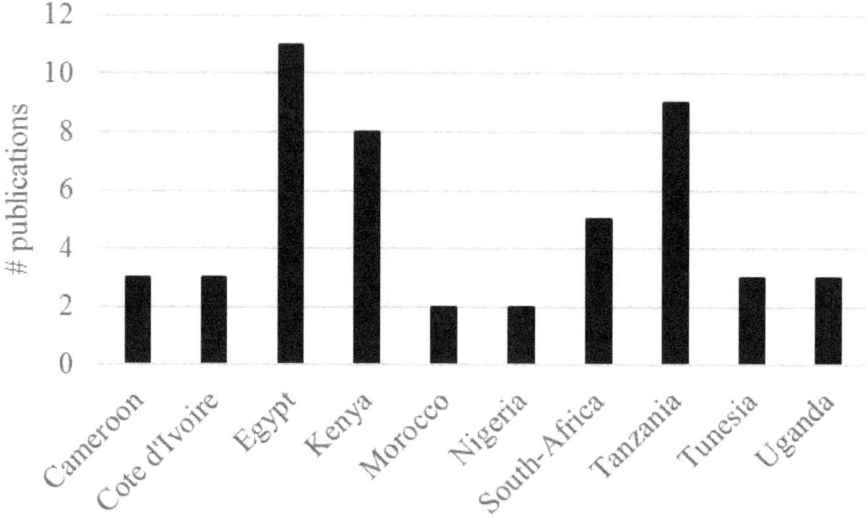

Fig. 1. Constructed wetland studies by African country, based on peer-reviewed publications.

From a search of the literature, it appeared that the majority of publications concerns lab-scale or pilot-scale systems. For more information on those systems, the reader is referred to Mekonnen et al. (2015) for a recent review.

In this chapter, the main attention is given to case studies of full-scale constructed wetland systems, which have been pragmatically defined as constructed wetlands with a (combined) surface area exceeding 100 m². These are further discussed below on a country-by-country basis.

Cameroon

Nya et al. (2002) and Kengne et al. (2003) reported on a floating macrophyte-based system located in Yaoundé (Cameroon). The treatment plant was already built in 1986, and consists of one anaerobic pond, followed by seven ponds in series, covered with water lettuce (*Pistia stratiotes*). The ponds have a total area of 720 m² and treat approximately 45 m³d⁻¹ of sewage from a residential quarter, with a hydraulic residence time varying between 9–16 days. Past treatment results (COD, BOD_5, and TSS removal all exceeding 80%) were confirmed by a new measurement campaign spanning the period November 1997–October 1998. A summary of the design and some of the obtained treatment results is given in Table 1.

One can see a clear exponentially decreasing trend of BOD_5 and TSS as wastewater travels through the system. In other words, the first ponds ensure the largest part of the removal, whereas the final ponds have a polishing role. Surprisingly though, the BOD_5 final effluent concentration remains rather high, despite the high oxygen concentrations reported. Nya et al. (2002) also observed good removal for several other parameters. Based on a bar chart shown in the latter publication, the derived approximate removal efficiencies were: 87% turbidity, 52% TP, 51% NH_4^+, 90% COD, 2.15 log units fecal coliforms (FC), and 2.00 log units fecal streptococci

(FS). Conditions were indeed beneficial for biodegradation: a fairly constant water temperature of around 26.5°C and a near-neutral pH of 7.0–7.6.

In addition to this normal operation, two other interesting questions were studied in the abovementioned papers: (1) does this type of wastewater treatment plant contributes to the proliferation of mosquitos, and (2) how does this macrophyte-based system compare with an algae-based (waste stabilization pond (WSP)) system?

To answer the first question, traps were set during the same time span as the water quality monitoring, with the purpose of catching adult mosquitos. In addition, water and roots were regularly sampled to capture immature stages (larvae and nymphs) (Kengne et al. 2003). An important proliferation of mosquitoes was indeed observed, with on average 43 imagoes per m² and per day (an insect in its final, adult, sexually mature, and typically winged state) emerging, of which about 54% were female. They belonged mainly to two genera, *Mansonia* and *Culex*. Fortunately, the potentially malaria transmitting *Anopheles* mosquito only constituted 0.02% of the captured imagoes. In the first couple of ponds, the high organic loads prevented the good root development of *Pistia*, thus favoring the growth of the free-living *Culex*. In contrast, the final set of ponds, with their better water quality, allowed a better root development and thus stimulated the growth of *Mansonia* which prefers to fixate itself onto these roots. Successful trials were also done in the same study with the mosquito fish *Gambusia*, which succeeded to significantly reduce the number of immature mosquitos.

For the second question, all ponds were stripped of their macrophyte coverage to allow the spontaneous development of microalgae. Subsequently, a new monitoring campaign was carried out between October 1999 and September 2000. In general, the macrophyte-based system performed better in terms of TSS, turbidity, BOD_5, and COD, whereas the algae-based system showed higher removals of nutrients and pathogens. The authors state that the latter especially might be of importance in Africa, where the abatement of waterborne diseases is still priority number one. Mburu et al. (2013) also compared the horizontal subsurface flow constructed wetlands and waste stabilization

Table 1. Pond design characteristics and water quality data of a floating macrophyte-based system in Yaoundé (Cameroon). Compiled from Nya et al. (2002) and Kengne et al. (2003).

Description	Area (m²)	Volume (m³)	BOD_5 (mg L⁻¹)	TSS (mgL⁻¹)	DO (mgO₂L⁻¹)
Influent composition			709 ± 139	835 ± 572	0.3 ± 0.8
P0 (anaerobic)	23.6	47.3	n.d.	n.d.	n.d.
P1 (vegetated – *Pistia*)	96.8	67.9	308 ± 12	277 ± 13	0.3 ± 0.0
P2 (vegetated – *Pistia*)	94.6	75.2	212 ± 13	149 ± 9	0.8 ± 0.1
P3 (vegetated – *Pistia*)	96.8	86.5	173 ± 10	103 ± 7	1.7 ± 0.2
P4 (vegetated – *Pistia*)	94.6	76.0	173 ± 9	70 ± 13	2.2 ± 0.2
P5 (vegetated – *Pistia*)	94.6	85.5	135 ± 8	64 ± 4	3.6 ± 0.3
P6 (vegetated – *Pistia*)	96.8	86.4	101 ± 7	45 ± 3	4.4 ± 0.2
P7 (vegetated – *Pistia*)	118.8	59.5	83 ± 2	30 ± 2	9.6 ± 2.0
Total system removal			88.3%	96.4%	

BOD_5, biochemical oxygen demand; TSS, total suspended solids; DO, dissolved oxygen; n.d. no data ponds based on data from a Kenyan treatment plant, and came to similar conclusions in terms of organic matter removal (CW outperformed WSP) and nutrients (WSP outperformed CW). They, however, demonstrated that apart from treatment efficiency, other factors such as land requirement, capital costs, and operation costs are equally important when selecting a certain wastewater treatment system.

Egypt

Lake Manzala Engineered Wetland Project

The Lake Manzala Engineered Wetland Project (LMEWP) is a large constructed wetland system located in Egypt, approximately 15 km south-west of Port Said. It was set up by the UNDP and funded through the GEF International Waters focal area, in order to improve the strongly deteriorated water quality of Lake Manzala, and hence the livelihood of those that depend on Lake Manzala's natural resources, such as fish and drinking water.

One of the important influxes of pollutants to Lake Manzala occurs via the Bahr El-Baqar drain, which transports mainly agricultural drainage water, in addition to both treated and untreated wastewater from Cairo and other smaller cities and villages in its drainage area. Due to its origin and additional dilution along the way, concentrations are rather low (El-Sheikh et al. 2010; Table 2), but the volumes are significant and of the order of 9 million m^3 d^{-1}.

Table 2. The Lake Manzala Engineered Wetland project (Egypt). Concentrations of the Bahr El-Baqar drain (min-max), influent to the constructed wetland (averages), effluent of high-loaded (HFR) and low-loaded (LFR) free water surface constructed wetlands with removal efficiency between brackets, and Egyptian regulations for water quality in agricultural drains. Compiled from El-Sheikh et al. (2010).

Parameter	Drain	Influent to CW*	HFR out*	LFR out*	Limit (Egyptian Law 48/1982)
COD (mg L^{-1})	72–122	140	69 (51%)	56 (60%)	80
BOD_5 (mg L^{-1})	28–100	55	28 (50%)	22 (60%)	40
TSS (mg L^{-1})	50–130	105	29 (72%)	22 (80%)	50
DO (mg L^{-1})	1.0–2.1	1.5	7.2	6.5	4
NH_4-N (mg L^{-1})	1.8–4.5	7.4	3.4 (54%)	2.7 (63%)	3
NO_3-N (mg L^{-1})	n.s.	1.8	6.9	7.8	n.s.
PO_4^{3-} (mg L^{-1})	0.8–2.3	2.4	0.95 (52%)	0.95 (52%)	1
pH (-)	7.3–7.6	7.9	7.8	7.8	6–9
Cu (mg L^{-1})	n.s.	0.048	0.029 (39%)	0.024 (50%)	n.s.
Zn (mg L^{-1})	n.s.	0.036	0.025 (30%)	0.023 (37%)	n.s.

COD, chemical oxygen demand; BOD_5, biochemical oxygen demand; TSS, total suspended solids; DO, dissolved oxygen; NH_4-N, ammonia-nitrogen; NO_3-N, nitrate-nitrogen, PO_4^{3-} phosphates; pH, hydrogen ions concentration; Cu, copper ions; Zn, zinc ions; n.s. no standard specified in legislation for that particular parameter
* some values estimated by visual interpretation from figures in mentioned reference

In order to alleviate pollutant fluxes, an intake structure and screw pumps lift 25,000 m³ d⁻¹ of wastewater from the drain to two sedimentation cells, each 16,650 m² surface area × 1.5 m depth. After sedimentation, 24,500 m³ d⁻¹ continues its way through a set of 10 parallel free water surface constructed wetlands (FWS CW), of which five are high-loaded (treating approximately 21,000 m³ d⁻¹) and five are low-loaded (treating about 4,000 m³ d⁻¹). Each FWS CW has dimensions 250 m × 50 m × 0.4 m and is planted with *Phragmites communis*, though some smaller patches of *Typha* sp. and *Cyperus papyrus* sp. also occur. The remaining 500 m³ d⁻¹ is diverted to two TVA-technology (Tennessee Valley Authority) inspired reciprocating subsurface flow wetlands of 1050 m² each, and is then fed to a number of fish ponds. The total site area approaches 100 ha, and the total project cost was in the order of 11.13 million US\$. Figure 2 gives an overview of the project site.

El-Sheikh et al. (2010) evaluated the difference in removal between the high-loaded wetlands (termed HFR) (HLR of 34.4 cm d⁻¹, HRT 1.16 days) and low-loaded wetlands (termed LFR) (HLR of 4.8 cm d⁻¹, HRT 8.33 days) based on data collected between 2005–2008. A summary of the obtained results is presented in Table 2.

COD, BOD_5, TSS, and DO easily met the water quality standards for agricultural drains, whereas PO_4^{3-} concentrations where flirting with the limit, and NH_4-N

Fig. 2. Aerial picture of the Lake Manzala Engineered Wetland project (31°09'48.2"N 32°11'45.4"E). (A) Bahr El-Baqar drain; (B) pumping station; (C) two sedimentation basins; (D) 10 free water surface constructed wetlands; (E) two reciprocating subsurface flow wetlands; (F) four hatchery and two fingerling ponds; (G) 24 fish ponds. Source: Google Maps & Digital Globe.

concentrations exceeded the standard for the HFR. In general, the LFR had lower effluent concentrations than the HFR, most probably as a consequence of the significantly longer hydraulic residence time. Exceptions were pH and PO_4^{3-} with equal effluent concentrations, and DO concentrations which were higher in the effluent of the HFR. Increased turbulence and thus higher surface reaeration because of higher flow velocities could explain the latter. Some seasonal differences where observed with better removal during summer, but since water temperature never dropped below 17°C, temperature inhibition never occurred.

El-Refaie (2010) made a further study on the influence of temperature in the HFR, based on biweekly samples collected during 2008. Total phosphorus removal and fecal coliform removal proved to be temperature-insensitive, as is usually accepted when using the first-order k-C* model (Kadlec and Knight 1996). However, clear trends where obtained for TSS and BOD_5, with strongly decreasing removal efficiencies at higher temperatures (R^2 in both cases 0.71). This contrasts with the usual temperature coefficients $\theta = 1$ in the first-order k-C* model (meaning no influence of temperature) and seems illogical as biological processes such as those that contribute to BOD_5 removal are normally enhanced by higher temperatures (Kadlec and Knight 1996). No explanation is given by the author, but two processes may possibly explain these phenomena. Firstly, at higher water temperatures, one may expect more evapotranspiration, which may lead to higher effluent concentrations. Secondly, as hinted by El Sheikh et al. (2010), algae growth might contribute as an extra internal load to BOD_5 and TSS concentrations.

An additional study by Hegazy et al. (2013) on the LMEWP shed some light on the removal of enteroviruses. The authors used two different techniques for virus detection, that is, Real Time RT-PCR and an infectivity assay (IA) with green monkey kidney cells. Enteroviruses where detected in all influent samples, showed significant reduction after the sedimentation basins (56% based on IA; 97.72% based on RT-PCR) and virtually disappeared after the free water surface constructed wetlands (100% removal based on IA; 99.99% removal based on RT-PCR), thus demonstrating the effectiveness of constructed wetlands in reducing fecal pollution.

Other Studies from Egypt

Abou-Elela and Hellal (2012) report on another full-scale application of a vertical subsurface flow constructed wetland for treating municipal wastewater near Cairo. The system consisted of coarse screen, oil removal, primary settling tank, and a 457 m² vertical subsurface flow constructed wetland. The latter had a 60 cm treatment layer of 10 mm diameter gravel, and a 20 cm drainage layer of 20 mm diameter gravel. Its surface area was split in four quadrants, one left unplanted and the others planted with *Canna, Phragmites australis*, and *Cyperus papyrus,* respectively. The wetland was operated at a hydraulic loading rate of 4.4 cm d^{-1} and the surface loading rate varied between 26.2–7.65 kg BOD_5 ha^{-1} d^{-1}.

As may be expected, the system reached high COD, BOD_5 and TSS removal, all in the order of 90%. The system further achieved simultaneous nitrification and denitrification, the latter being less common in vertical flow wetlands. Unfortunately, information on intermittent feeding frequency is lacking and hence it is difficult to explain this behavior. Phosphate was removed at 62%, probably due to the young age of the system and thus, a still high availability of sorption sites. However, plant nutrient uptake could also play an important role in this. According to the (average) influent values and harvested plant nutrient contents specified by the authors, a vertical flow wetland fully planted with *Canna* (all 457 m²) would be able to store 13.6% of the incoming nitrogen, and 43.3% of the incoming phosphorus in its harvestable biomass. Likewise, *Phragmites* would seem able to store 9.7% of incoming N, and 38.5% of incoming P. These values seem rather high when compared to those mentioned in the review paper by Vymazal (2007), but may be explained by the relatively low influent nutrient concentrations (31.3 mg $TKNL^{-1}$ and 4.7 mg TPL^{-1}) and the fact that the vegetation takes up more nutrients than usual while establishing itself on a newly built constructed wetland. Finally, the geometric mean of the bacterial indicators count, that is, total coliforms, fecal coliforms, and *E. coli*, in the treated effluent were all just above the permissible Egyptian limit of 1000 $(100 mL)^{-1}$ for restricted irrigation.

In a follow-up paper (Abou-Elela et al. 2013), the same treatment system was compared with a 654 m² horizontal subsurface flow constructed wetland (L × W × d 37.87 × 17.3 × 0.85 m), likewise filled with 20 mm diameter gravel and planted with *Canna, Phragmites australis,* and *Cyperus papyrus*. The horizontal flow wetland received the same 20 m³ d^{-1} flow rate, resulting in a hydraulic residence time of 11 days. Removal efficiencies of both systems were very similar and consistently above 90% for COD, BOD_5, and TSS. Surprisingly, differences in TKN and NH_4 removal were rather small, although one would expect more nitrification in the aerobic environment that vertical flow wetlands normally are. The rather long residence time of the horizontal flow wetland may explain this observation. An interesting feature of this study was the detailed monitoring of evapotranspiration. It was found that water losses in Egypt's arid climate rose from 35% of the influent hydraulic load during winter to 60% during summer with water losses of the vertical flow wetland on average about 5% lower than those of the horizontal flow wetland. These huge losses are obviously a drawback when aiming for water reclamation. In addition, residual pathogen levels of both systems were exceeding the local requirements for wastewater reuse in irrigation, thus additional disinfection was required.

A particularly interesting and robust technology for warmer climates are upflow anaerobic sludge blanket reactors (UASB) which can convert organic load into biogas. The effluent however needs further treatment before discharge or reuse, hence the use of constructed wetlands as post-treatment technology, as studied by Morsy et al. (2007) and El Khateeb et al. (2009). Both studied the same system located in Cairo, fed by municipal wastewater, and consisting firstly of a 1.3 m³ UASB reactor operating at an HRT of 8 hours and in a temperature range of 19–38°C. The effluent of the UASB was then continuously fed to a 2.4 × 1 × 0.25 m free water surface constructed wetland planted with *Typha latifolia*, operated at a 13.8 cm d^{-1} hydraulic loading rate. Finally, a 2.4 × 1 × 0.4 m horizontal subsurface flow constructed wetland filled with sand (ø 3–4 mm), planted with the same vegetation and operated at the same hydraulic loading rate, provided final treatment. The idea behind this particular wetland configuration was to optimize nitrogen removal via nitrification-denitrification and to ensure good solids retention via the fine texture media of the subsurface flow system.

El Khateeb et al. (2009) report a total removal for COD of 95.7%, for BOD_5 of 97.2%, and for TSS of 98.3%. Although the UASB removed most of the load (in the order of 70% for the three abovementioned parameters), the wetlands were necessary to obtain satisfactory effluent concentrations. In terms of pathogen densities (total coliforms, fecal coliforms, fecal streptococci, and *E. coli*), the relative contribution of each technology was inversed, with the UASB only removing 1.3–1.9 log units, whereas the combined wetlands were removed in the order of 6.35–7.55 log units. The combined units also achieved a 100% removal of all monitored helminth eggs and protozoan (oo) cysts. In the study by Morsy et al. (2007) for the same system, it was confirmed that *Cryptosporidium* was also completely removed by the combined system. In conclusion, this system achieved an effluent quality fully compliant with WHO guidelines for unrestricted irrigation.

Kenya

Engineering Natural Wetlands for Wastewater Treatment— The Kahawa Case

One of the oldest studies on wastewater quality improvement by means of wetlands in Africa was done already in 1980–1981, as reported by Chale (1985). In this investigation, the effectiveness of a semi-natural papyrus swamp (man-made impoundment transformed into a swamp) in removing nutrients from primary treated domestic wastewater was evaluated. The swamp is located in the north-east of Nairobi, covering an area of about 3.7 ha and with a maximum reported depth of 1.5 m. About 10% of the swamp is open water area, 65% is covered by *Cyperus papyrus*, and the remaining area is covered by a variety of other macrophytes. The general physico-chemical conditions were permissive for wastewater treatment (mean temperature between 20–24°C and pH between 7.0–7.5), leading to a 76.5% reduction in ammonium levels and a 79.5% reduction in orthophosphates, yielding effluent concentrations of around 1.9 mg NH_4-N L^{-1} and 0.33 mg P L^{-1}. Unfortunately, no data on flow rates or hydraulic residence times were specified to further interpret these data.

Splash-Carnivore Constructed Wetland

One of the better-known constructed wetlands in Africa is the Splash-Carnivore system, just south of Nairobi, treating about 80 m^3 d^{-1} of wastewater from the two eponymous restaurants. It is reputedly the first constructed wetland ever built in Kenya and was commissioned in 1994. The wastewater receives primary treatment in two septic tanks, and then flows through a baffled, horizontal subsurface flow constructed wetland of 1750 m^2 (70 × 25 m, 1.5 meter deep). The latter was purposely chosen to reduce odor nuisance. Further treatment is then achieved in three surface flow wetlands placed in series, with respective areas of 480, 1080, and 990 m^2. The total wetland area thus amounts to 4300 m^2 and yields a hydraulic residence time in the order of 11 days.

Different measurement campaigns have been executed, reported in various papers (Nzengy'a and Wishitemi 2001; Kelvin and Tole 2001; Kelvin et al. 2014). Results are summarized in Table 3.

Both studies by Kelvin et al. (2014) show that the treatment performance is not significantly affected by the season (dry or wet). In the cases of organic ammonia and

Table 3. Summary of wastewater quality data from the Splash-Carnivore constructed wetland system (Kenya). Compiled from Nzengy' and Wishitemi (2001): Data from both wet and dry seasons, given as mean ± standard deviation, Kelvin and Tole (2001): Data from dry season only, given as mean (min–max) and Kelvin et al. (2014): Data from wet season only, given as mean.

	Season	Influent (mg L⁻¹)	Effluent HSSF CW (mg L⁻¹)	Effluent third FWS CW (mg L⁻¹)	Removal efficiency (%)
BOD$_5$	Wet & dry	1603 ± 398	63 ± 11	15.1 ± 2.5	99.1
	Dry	286 (150–385)	n.s.	11 (9.4–13.5)	96.2
	Wet	472	n.s.	23	95.1
COD	Wet & dry	3750 ± 207	205 ± 102	95.5 ± 7.2	97.5
	Dry	2003	n.s.	47.5	97.6
	Wet	2174	n.s.	71.5	96.5
TSS	Wet & dry	195.4 ± 58.7	12.3 ± 3.3	4.7 ± 1.9	97.6
	Dry	102	n.s.	16	84.3
	Wet	116	n.s.	24	79.3
Ammonia	Wet & dry	14.6 ± 4.1	7.9 ± 2.5	BDL	/
Organic N	Dry	16.92	3.44	2.3	86.4
	Wet	16.92	3.44	2.3	86.4
Soluble phosphorus	Dry	13.8	n.s.	11.2	18.8
	Wet	13.8	n.s.	11.2	18.8

HSSF, Horizontal Sub-Surface Flow Constructed Wetland; FWS, Free Water Surface Constructed Wetland; BOD$_5$, biochemical oxygen demand; COD, chemical oxygen demand; TSS, total suspended solids; n.s., not specified; BDL, below detection limit (which was not quantified in the original paper)

soluble phosphorus, values are exactly the same, which rather suggests a mistake in the original publications. The authors attribute the similar treatment efficiencies to the relatively small temperature differences between both seasons in the tropics.

A similar statement is put forward by Nzengy'a and Wishitemi (2001), who observed very little temporal variation during their study.

Temperatures indeed never dropped below 13°C (daily mean minimum temperature), which may well explain the continuous good removal performances.

Treatment of Flower Farm Wastewater

Flower farming is an important economic sector in several East-African countries (Kenya, Ethiopia, and now also starting in Rwanda). One of the tools to control the environmental impact of the farms are constructed wetlands. Kimani et al. (2012) report on a hybrid system, the Kingfisher constructed wetland located on a flower farm near Lake Naivasha, which is almost identical in layout as the Splash-Carnivore one: a sedimentation chamber, then one baffled horizontal subsurface flow wetland followed by a buffer cell, and then three surface flow wetlands in series. The system was constructed in 2005, at a cost of approximately 40,000 US$, for a design flow rate of 45 m³ d⁻¹. Unfortunately, neither surface areas nor hydraulic residence times are specified. In the grey literature, a surface area of 4 hectares was found, which probably

Table 4. Wastewater treatment parameters of the Kingfisher wetland (Kenya) and relevant discharge standards. Compiled from Kimani et al. (2012).

	Influent	Final effluent	Removal efficiency (%)	Kenyan standard 2006
TSS (mg L^{-1})	233 ± 26.5	23 ± 3.9	90.1	30
BOD$_5$ (mg L^{-1})	138 ± 15.1	72 ± 5.5	47.8	30
COD (mg L^{-1})	569 ± 175	186 ± 62	67.3	50
TN (mg L^{-1})	5.1 ± 0.65	2.0 ± 0.34	67.3	n.s.
TP (mg L^{-1})	5.5 ± 0.82	2.6 ± 0.24	52.7	n.s.
Temperature (°C)	23.1 ± 0.35	18.3 ± 0.38	/	/
pH (-)	6.81 ± 0.17	6.65 ± 0.10	/	/
EC (μS/cm)	722 ± 632	514 ± 15	/	/

TSS, total suspended solids; BOD$_5$, biochemical oxygen demand; COD, chemical oxygen demand; TN, total nitrogen; TP, total phosphorus; pH, hydrogen ions concentration; EC, electrical conductivity; n.s. no discharge standard specified

also encompasses landscaping features around the wetland cells. The Kingfisher wetland treats a combined wastewater flow, not only from domestic sources like the staff canteen and wash rooms, but also from the pack houses and other technical facilities such as where spray gear washing occurs. Monitoring data reported in Kimani et al. (2012) are however limited to conventional wastewater parameters (e.g., no pesticides measured). An overview is given in Table 4.

Overall removal of organic matter (BOD$_5$ and COD) is rather low compared to other wetland systems. The authors mention frequent clogging of the horizontal subsurface flow wetland during their study. This may have a direct impact on removal efficiency, but indirectly also points to an overloaded system, which is also hinted at by the authors of this study. As a consequence, the effluent also fails to meet the specified Kenyan discharge standards. Nutrient removal, however, falls within normal reported ranges.

South-Africa

Both retained case studies from South-Africa are dealing with non-point source pollution, be it in completely different circumstances.

In the first study, Schulz and Peall (2001) report on a free water surface constructed wetland aimed at non-point source pollution control in the Lourens river near Cape Town. The system was built in 1991 and consists of a vegetated pond of dimensions 134 × 36 × 1.5 m. It was built at the outlet of a small sub-catchment, comprising some 15 ha of orchard, 10 ha of pasture land, and 18 ha of forest, and the system treats the total discharge of the stream (varying between 0.03 m³ sec^{-1} in the drier summer months to 0.32 m³ sec^{-1} in winter). The main aim was to reduce the suspended solids load.

At the time of study (1999), the pond had already lost much of its storage capacity due to silting up, with corresponding water depths only between 0.3–1.0 m. About one fifth of the wetland (near the inlet) was free of vegetation, the remainder was covered mainly by *Typha capensis* Rohrb, *Juncus kraussii* Hochst, and *Cyperus dives* Delile.

By means of a 6-month monitoring campaign, it was shown that the system fulfilled its purpose. During the dry periods (< 2 mm d^{-1} rainfall), TSS was reduced from 21.7 to 18.4 mg L^{-1} on average, whereas during the wet periods (> 2 mm d^{-1} rainfall), TSS was reduced from 105 to 23 mg L^{-1} on average. Higher loads were thus effectively neutralized. Similar observations were made for nitrate and orthophosphate. This proved that the wetland can (1) cope with higher loads and still yield low effluent concentrations, and (2) that lack of maintenance (i.e., desludging) did not affect long-term performance.

In addition, it was also shown by Schulz and Peall (2001) that the wetland was effective in retaining organophosphorus pesticides azinphos-methyl, chlorpyrifos, and endosulfan originating from the orchards, with effluent concentrations in most cases dropping below the detection limit of 0.01 μg L^{-1}. This was further backed up by an *in situ* exposure bioassay with *Chironomus*, which showed a mortality reduction up to 89% between inlet and outlet.

In the second study, du Toit and Campbell (2002) present an innovative project where a wetland system was built at the inlet of solar salt works near Port Elisabeth. Its aim was to reduce the nutrient load to the salt works and thus reduce eutrophication, which was negatively affecting the salt quality. This halotolerant wetland consisted of an initial pool of dimensions 25 × 10 × 1 m which was aimed at reducing the influent velocity. It was spontaneously colonized by species such as *Ulva rigida*, *Enteromorpha intestinali's* and *Enteromorpha flexuosa*. This initial pool was followed by a shallower 30 × 12 × 0.15 m pool, mainly covered by *Spartina maritima*. Salt works are fed depending on the weather conditions. In this case, the residence time of water in the wetland during feeding was only 2 hours. Despite this short residence time, significant nutrient reductions were obtained, with removals of ammonium, nitrate, and phosphate that on average were 56, 60, and 56%, respectively. This resulted in effluent levels which were claimed to be able to reduce the incidence of harmful algal blooms.

Tanzania

Banana Wine Wastewater, Arusha

Paschal et al. (2015) present an interesting case study on agro-industrial wastewater processing from a banana wine factory in Arusha, Tanzania. Their project clearly goes beyond the basic scope of environmental protection, as it includes energy, water, and nutrient recuperation. The system consists of four major stages, and was designed for treating 200 m^3 d^{-1}. Firstly, primary treatment and buffering is ensured through sequential bar screens, an equalization tank, and a primary clarifier. Secondly, a UASB reactor (ø 10 m, height 8 m) tackles organic pollution and converts it to biogas. The third stage, intended for tertiary treatment, consists of two sequential horizontal subsurface flow wetlands (each of dimensions 30 × 7.5 × 1 m), packed with 12.5–20 mm basaltic gravel and planted with papyrus. Finally, a sludge drying bed is foreseen to convert sludge into a suitable agricultural fertilizer.

Table 5. Results from a wastewater treatment plant in Arusha (Tanzania) treating banana wine wastewater. Concentrations in mg L^{-1} and given as mean ± standard deviation. R% = removal percentage. Compiled from Paschal et al. (2015).

	UASB influent	UASB effluent	R% UASB	CW effluent	R% CW	R% total system	Standard
COD	4959 ± 389	126 ±17	97.5	48 ± 6	62.0	99.0	60
BOD$_5$	1454 ± 110	36 ± 5	97.5	20 ± 4	45.0	98.6	30
TSS	2431 ± 191	114 ± 6	95.3	27 ± 3	76.7	96.0	100
NH$_4^+$	7.2 ± 1.1	66.5 ± 13.9	-	35.1 ± 4.9	43.5		n.s.
NO$_3^-$	23.4 ± 3.2	0.4 ± 0.1	98.2	2.6 ± 0.4	−585	88.7	50
PO$_4^{3-}$	5.1 ± 0.7	5.1 ± 1.1	-	2.5 ± 0.4	60.4	50.8	6

COD, chemical oxygen demand; BOD$_5$, biochemical oxygen demand; TSS, total suspended solids; NH$_4^+$, ammonium ions; NO$_3^-$, nitrates ions; PO$_4^{3-}$, phosphate ions; n.s. no discharge standard specified

A 17 month follow-up showed very encouraging results, with excellent removal efficiencies, and all applicable standards met. A summary can be found in Table 5.

Two factors should however be considered when interpreting these results. Firstly, it was found that the flow rate was far lower than expected, that is, 62.4 m³ d^{-1} versus the design flow of 200 m³ d^{-1}. This was attributed to the installation of a new, more water-efficient bottle-washing line. Secondly, the UASB was found to operate below its mesophilic design temperature of 35°C, that is, at 29.3 ± 1.7°C. The only problematic parameter was NH$_4^+$ which, because of lack of aerobic conditions throughout the system, accumulated. An intermediate aeration step between UASB and treatment wetland was suggested by the authors to solve this issue.

In addition, important energy savings where obtained by using the biogas (on average 163 m³ d^{-1}) in a boiler for steam production.

Paschal et al. (2015) mention a reduction of almost 3500 liter industrial diesel oil per month, equivalent to over 2000 US$ per month. Dried sludge was also analyzed, and was found to be suitable for agricultural purposes.

Constructed Wetlands for Tertiary Treatment of Municipal Wastewater in Arusha

Both Kipasika et al. (2014) and Kihila et al. (2014) report on a horizontal subsurface flow constructed wetland near Arusha, which treats municipal wastewater pre-treated by a waste stabilization pond system consisting of an anaerobic pond, two parallel facultative ponds, and two sequential maturation ponds. In both cases, the scope was on safe reuse of the effluent in agriculture, and in both cases a comparison was made between the wetland and four additional maturation ponds in series in terms of additional polishing.

The full system is designed to treat 4500 m³ d^{-1} whereas the wetland receives 200 m³ d^{-1}. At an estimated (no data given in said references) porosity of 40%, this would result in a hydraulic residence time in the order of 2.25 days within the wetland.

Kihila et al. (2014) report precise water balance details, showing firstly that the system is operating near its design flow (measured average inflow to the waste stabilization ponds 4,193 m³ d^{-1} and to the wetland 200 m³ d^{-1}) and secondly, that there are significant water losses as the combined pond-wetland outflow only amounts to 2587 m³ d⁻¹. In the constructed wetland, flow is reduced from 200 to 134 m³ d⁻¹. After passing the fish pond, the wetland effluent is used for irrigating 8,094 m² of mainly paddy farms and an additional 113,311 m² are irrigated by the effluent of the final maturation pond. When the irrigation demand is low, effluent is discharged into a nearby stream. An overview of the system is given in Fig. 3.

Water quality results of the Moshi system are summarized in Table 6 and Fig. 4, from which Kihila et al. (2014) conclude that the effluent is suitable for restricted irrigation only. The low removal efficiencies of the wetland were attributed to the fact that in cases of high demand, farmers appear to block the outlet of the wetland, thus

Fig. 3. Waste stabilization pond—constructed wetland system operated by Moshi Urban Water Supply and Sewerage Authority (MUWSA). AP: anaerobic pond (55 × 115 × 5 m), FP: facultative pond (67 × 147 × 2.5 m), MP: maturation pond (70 × 138 × 1.5 m), CW: constructed wetland (15 × 50 × 1.5 m), FiP: fish pond. Source: Google Maps and DigitalGlobe.

Table 6. Water quality results from the Moshi waste stabilization pond—constructed wetland system. MP: maturation pond, CW: constructed wetland. Compiled from Kihila et al. (2014).

	Influent (=MP2 effluent)	MP6 effluent	Removal efficiency M3-M6 (%)	CW Effluent	Removal efficiency CW (%)
pH (-)	8.0 ± 0.4	8.8 ± 1.1	-	7.7 ± 0.3	-
DO (mg L⁻¹)	4.6 ± 2.3	5.7 ± 2.2	-	3.6 ± 2.1	-
TDS (mg L⁻¹)	639 ± 36	587 ± 76	8.1	627 ± 32	1.9
Cl⁻ (mg L⁻¹)	969 ± 77	944 ± 88	2.6	929 ± 65	4.1
NO₃-N (mg L⁻¹)	2.53 ± 1.07	9.19 ± 6.83	−263	2.35 ± 0.64	7.1
PO₄-P (mg L⁻¹)	28.83 ± 16.94	29.89 ± 7.96	−3.7	32.45 ± 8.58	−12.6
FC (# (100 mL⁻¹)⁻¹)	7249 ± 3287	1001 ± 991	86.2	4627 ± 2003	36.2
COD (mg L⁻¹)	219 ± 61	160 ± 49	26.9	149 ± 59	32.0

pH, hydrogen ions; DO, dissolved oxygen; TDS, total dissolved solids; Cl-, chloride ions; NO_3-N nitrate-Nitrogen; PO_4^{3-}, phosphate-Phosphorus; FC (# (100 mL⁻¹)⁻¹), fecal coliforms number per 100 milliliters; COD, chemical oxygen demand

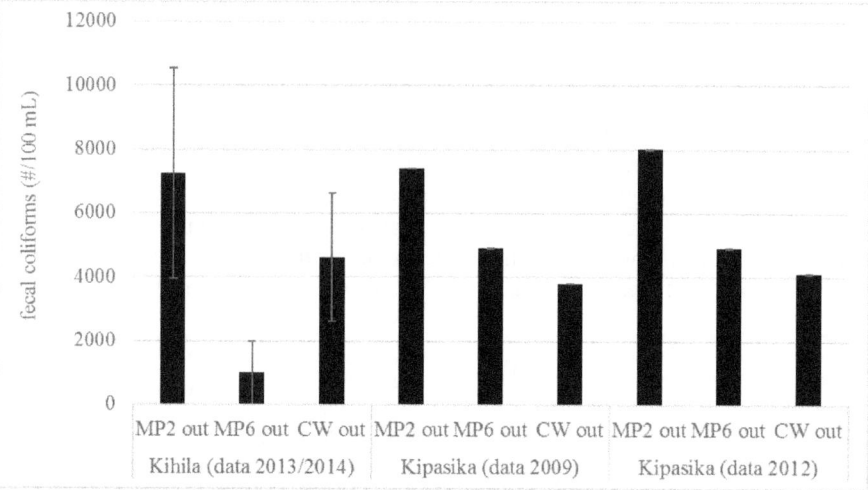

Fig. 4. Data on fecal coliform removal from the Moshi waste stabilization pond—constructed wetland system. MP: maturation pond, CW: constructed wetland. Compiled from Kihila et al. (2014) and Kipasika et al. (2014).

accumulating water until it overflows into nearby paddy fields. This type of operation may indeed negatively impact treatment performance.

One issue which is not further discussed by the authors is the apparent contradiction between the water balance results and the chloride concentrations. Chlorides are often considered as mostly conservative substances in natural treatment systems. Significant water losses through evapotranspiration would therefore result in an increasing chloride concentration, which is not the case here (Table 6).

An additional survey among farmers using the effluent indicated significant economic benefits. Firstly, those farmers were able to harvest twice per year, compared to one harvest per year for those practicing rain-fed agriculture. Secondly, a reduced or even non-existing use of artificial fertilizer compared to rain-fed agriculture was reported.

Tunisia

The reportedly first full-scale constructed wetland in Tunisia was built in 2004 in a small town called Jougar, about 80 km south-west of Tunis, which hosts about 1000 inhabitants. The system consists, in order, of a stormwater bypass, a bar screen, a primary treatment pond, a holding tank which allows batch-feeding of the subsequent 121 m² vertical flow constructed wetland, and finally, a 207 m² horizontal subsurface flow wetland. Average discharge reaching the system is reported to be in the order of 24 m³ d⁻¹. Both wetlands were planted with a mix of reeds and cattails.

Kouki et al. (2009) report on the performance of this system one year after it started (May–November 2005). A summary of their results is given in Table 7.

Despite the high concentrations and thus high (organic) loading rates, the system achieves good removal efficiencies for BOD$_5$, COD, and TSS, and lower nutrient removal efficiencies, which are however still in line with other reported results.

Table 7. Results from the hybrid constructed wetland system in Jougar (Tunisia). Compiled from Kouki et al. (2009).

	Influent	Effluent	Removal efficiency (%)	Tunisian standards for provision of wildlife habitat (N.T. 106.002)
TSS (mg L^{-1})	798 ± 302	18 ± 14	98	30
BOD$_5$ (mg L^{-1})	420 ± 144	30 ± 9	93	30
COD (mg L^{-1})	1339 ± 352	134 ± 25	90	90
TKN (mg L^{-1})	205 ± 70	117 ± 19	43	1
TP (mg L^{-1})	30.7 ± 12.4	7.17 ± 2.29	77	0.05
Temperature (°C)	23.4 ± 2.9	23.5 ± 3.5	/	25
pH (-)	8.03 ± 0.12	7.51 ± 0.11	/	6.5–9
FC (CFU (100 mL)$^{-1}$)	2.69 ± 4.14 x 10^7	1910 ± 2180	4.1 log units	2000
FS (CFU (100 mL)$^{-1}$)	9.35 ± 1.81 x 10^7	595 ± 611	5.2 log units	1000

TSS, total suspended solids; BOD$_5$, biochemical oxygen demand; COD, chemical oxygen demand; TKN, total Kjeldahl nitrogen; TP, total phosphorus; pH, hydrogen ions; FC (CFU (100 mL)$^{-1}$), fecal coliforms in colonies forming units per 100 milliliters; FS (CFU (100 mL)$^{-1}$), fecal streptococci in colonies forming units per 100 milliliters

Removal of pathogens is also excellent. Nonetheless, the effluent fails to meet all required effluent standards. Apart from the (too) high loading rates, clogging because of insufficient settlement pond maintenance and lack of resting periods for the vertical flow wetland are put forward by the authors to explain the lack of compliance.

Conclusions

About two thirds of the above-described case studies are dealing with domestic/ municipal wastewater, which conforms to the earlier-mentioned sanitation crisis which still exists in Africa. This is also clear from the fact that most studies purposely looked at removal of fecal coliforms and other fecal indicator organisms. In most cases, irrigation with effluent was proven possible, although in some cases only restricted irrigation was recommended. It also became clear that in those cases it was not so much the constructed wetland technology as such, but rather under-dimensioning (resulting in overloading) or wrong operation, which were causing the non-compliant effluent concentrations. This is a recurring point of attention for natural treatment technologies. However, the other case studies show that other applications of constructed wetlands also exist in Africa for treating agro-industrial wastewater, runoff water, and even nutrient-rich sea water.

An issue related to irrigation, which seems underexposed in all but one of the retained case studies, is the loss of water due to evapotranspiration. As Kihila et al. (2014) reported for the Arusha system, nearly 40% of the incoming water was lost and even higher losses are possible in arid areas. Apart from the loss of water, this also results in an accumulation of salts, which may negatively affect the irrigation potential of effluent.

Furthermore, it is interesting to point out that at the level of full-scale constructed wetlands in Africa, the reported studies are mainly limited to conventional wastewater parameters (COD, BOD_5, TSS, nutrients, etc.), in a few cases only complemented with data on heavy metals. Only one study was found which considered pesticides (Schulz and Peall 2001); other pollutants of emerging concern such as pharmaceuticals and personal care products or nanoparticles have not been yet studied.

In conclusion, good experience with full-scale constructed wetlands already exists in several African countries, and the multitude of other lab-scale and pilot-scale studies (Mekonnen et al. 2015) reported in the scientific literature suggests that a further expansion of this sustainable, green wastewater technology in Africa may be anticipated.

Acknowledgements

The author acknowledges Joan García from Barcelona Tech in Spain for reviewing the pre-final draft of this chapter.

References Cited

Abou-Elela, S.I. and M.S. Hellal. 2012. Municipal wastewater treatment using vertical flow constructed wetlands planted with Canna, Phragmites and Cyprus. Ecol. Eng. 47: 209–213.

Abou-Elela, S.I., G. Golinielli, E.M. Abou-Taleb and M.S. Hellal. 2013. Municipal wastewater treatment in horizontal and vertical flows constructed wetlands. Ecol. Eng. 61: 460–468.

Chale, F.M.M. 1985. Effect of a *Cyperus papyrus* L. swamp on domestic waste water. Aquat. Bot. 23: 185–189.

du Toit, S.R. and E.E. Campbell. 2002. An analysis of the performance of an artificial wetland for nutrient removal in solar saltworks. S. Afr. J. Bot. 68: 451–456.

El-Khateeb, M.A., A.Z. Al-Herrawy, M.M. Kamel and F.A. El-Gohary. 2009. Use of wetlands as post-treatment of anaerobically treated effluent. Desalination 245: 50–59.

El-Refaie, G. 2010. Temperature impact on operation and performance of Lake Manzala Engineered Wetland, Egypt. Ain Shams Eng. J. 1: 1–9.

El-Sheikh, M.A., H.I. Saleh, D.E. El-Quosy and A.A. Mahmoud. 2010. Improving water quality in polluted drains with free water surface constructed wetlands. Ecol. Eng. 36: 1478–1484.

Hegazy, A.M., A.H. El-Salakawy, M.M. Shaban, M.M. Yehia and M.S. AbuSalam. 2013. Eco-friendly management of enteroviruses in wastewater. Water Sci. 27: 19–29.

Kadlec, R.H. and R.L. Knight. 1996. Treatment Wetlands, CRC Press, 893 pp. Boca Raton, FL, US.

Kelvin, K. and M. Tole. 2011. The efficacy of a tropical constructed wetland for treating wastewater during the dry season: The Kenyan experience. Water Air Soil Pollut. 215: 137–143.

Kelvin, K., M. Tole, S.A. Obiero and S.W. Mwangi. 2014. The efficacy of a tropical constructed wetland for treating wastewater during the wet season: The Kenyan experience. J. Environ. Earth Sci. 4(15): 66–73.

Kengne, I.M., F. Brissaud, A. Akoa, R. AtanganaEteme, J. Nya, A. Ndikefor et al. 2003. Mosquito development in a macrophyte-based wastewater treatment plant in Cameroon (Central Africa). Ecol. Eng. 21: 53–61.

Kihila, J., K.M. Mtei and K.N. Njau. 2014. Wastewater treatment for reuse in urban agriculture; the case of Moshi Municipality, Tanzania. Phys. Chem. Earth 72–75: 104–110.

Kimani, R.W., B.M. Mwangi and C.M. Gichuki. 2012. Treatment of flower farm wastewater effluents using constructed wetlands in lake Naivasha, Kenya. Indian J. Sci. Technol. 5(1): 1870–1878.

Kipasika, H.J., J. Buza, B. Lyimo, W.A. Miller and K.N. Njau. 2014. Efficiency of a constructed wetland in removing microbial contaminants from pre-treated municipal wastewater. Phys. Chem. Earth 72–75: 68–72.

Kivaisi, A.K. 2001. The potential for constructed wetlands for wastewater treatment and reuse in developing countries: A review. Ecol. Eng. 16: 545–560.

Kouki, S., F. M'hirib, N. Saidi, S. Belaïd and A. Hassen. 2009. Performances of a constructed wetland treating domestic wastewaters during a macrophytes life cycle. Desalination 246: 452–467.

Mburu, N., S.M. Tebitendwa, J.J.A. van Bruggen, D.P.L. Rousseau and P.N.L. Lens. 2013. Performance comparison and economics analysis of waste stabilization ponds and horizontal subsurface flow constructed wetlands treating domestic wastewater: A case study of the Jujasewage treatment works. J. Environ. Manage. 128: 220–225.

Mekonnen, A., S. Leta and K.N. Njau. 2015. Wastewater treatment performance efficiency of constructed wetlands in African countries: A review. Wat. Sci. Technol. 71(1): 1–8.

Morsy, E.A., A.Z. Al-Herrawy and M.A. Ali. 2007. Assessment of Cryptosporidium removal from domestic wastewater via constructed wetland systems. Water Air Soil Pollut. 179: 207–215.

Nya, J., F. Brissaud, I.M. Kengne, C. Drakides, A. Amougou, R. AtanganaEteme et al. 2002. Sewage treatment in Cameroon: Comparison of the purifying efficiency of macrophytic and microphytic lagoon systems. pp. 726–736. *In*: Proceedings of International Symposium on Environmental Pollution Control and Waste Management (EPCOWM'2002). 7–10 January 2002. Tunis.

Nzengy'a, D.M. and B.E.L. Wishitemi. 2001. The performance of constructed wetlands for wastewater treatment: A case study of Splash wetland in Nairobi Kenya. Hydrol. Process. 15: 3239–3247.

Paschal, C., L. Gastory, J.H.Y. Katima and K.N. Njau. 2015. Application of up-flow anaerobic sludge blanket reactor integrated with constructed wetland for treatment of banana winery effluent. Proceedings 15th IWA International Conference on Wetland Systems for Water Pollution Control, 4–9 September 2016. Volume 2: 842–854. Gdańsk, Polska.

Schulz, R. and S.K.C. Peall. 2001. Effectiveness of a constructed wetland for retention of nonpoint-source pesticide pollution in the Lourens River Catchment, South Africa. Environ. Sci. Technol. 35: 422–426.

UNICEF and WHO. 2015. Progress on Sanitation and Drinking Water—2015 update and MDG assessment. ISBN 9 789241 509145. UNICEF and World Health Organization.

Vymazal, J. 2007. Removal of nutrients in various types of constructed wetlands. Sci. Total Environ. 380(1–3): 48–65.

Zhang, D.Q., K.B.S.N. Jinadasa, R.M. Gersberg, Y. Liu, W.J. Ng and S.K. Tan. 2014. Application of constructed wetlands for wastewater treatment in developing countries—A review of recent developments (2000–2013). J. Environ. Manage. 141: 116–131.

7

Application of a Multi-function Constructed Wetland for Stream Water Quality Improvement and Ecosystem Protection

A Case Study in Kaohsiung City, Taiwan

C.M. Kao,* W.H. Lin, P.J. Lien, Y.T. Sheu and Y.T. Tu

INTRODUCTION

Artificial or constructed wetlands (AW or CWs) have been successfully used for the treatment of surface runoff, landfill leachates, polluted river water, and wastewater (e.g., agricultural drainage, urban wastewater, industrial discharge) (Torrijos et al. 2016; Yi et al. 2016; Corbella et al. 2017).

CW is one of the natural treatment technologies which is a green, environmentally acceptable, and ecological technology. Due to the relatively lower construction as well as the operational and maintenance (O&M) costs, CW is especially an appropriate alternative treatment facility for small townships, communities, and industries that cannot afford traditional treatment systems (wastewater treatment plants with primary and secondary treatment units) with high installation and O&M costs (Collins and Gillies 2014; Zalewski 2014; Dunne et al. 2015; Wu et al. 2015).

In the practical application, CWs can be used to remove many types of pollutants including nutrients, organic chemicals, suspended solids (SS), trace elements,

Institute of Environmental Engineering, National Sun Yat-Sen University, Kaohsiung, Taiwan.
* Corresponding author: jkao@mail.msysu.edu.tw

pesticides, and even inorganic chemicals from different wastewaters and river water (Białowiec et al. 2014; Ávila et al. 2015; Vymazal and Březinová 2015; Leung et al. 2016).

Functions and Types of Constructed Wetlands

According to the major components and flow conditions, CWs can be divided to two different kinds of CWs including subsurface flow system (SSFS) and free water surface system (FWS) (Filoso et al. 2015; Lv et al. 2016). In FWS, water flows laterally through a shallow water basin with natural soils, gravels, or sands. Different wetland plants (e.g., emergent, floating, submerged, and floating-leaf plants) are planted in FWS, and plants in CWs are useful for water flow regulation (Vymazal and Březinová 2015; Lu et al. 2016). Wetland plants also serve as the location for bacteria growth and the contact of wastewater with reactive biological surfaces could effectively enhance the pollutants biodegradation rates. SSFS can be treated as a trickling filter with horizontal flow regime. It contains an underground basin with porous media (e.g., gravels, sands) mainly for biofilm establishment. The water level in SFS is below the media surface, and thus, mosquito breeding and odor problems can be minimized (Vymazal and Březinová 2015; Papaevangelou et al. 2017). As floating and submerged wetland plants cannot grow in SSFS, emergent plants are the only option for wetland plant selection. Emergent plants with extensive root systems can be grown in SSFS because the wetland plants can deliver oxygen from the atmosphere via the stems to the root zone area to enhance the biodegradation process under aerobic conditions. Moreover, root systems can maintain a higher permeability.

Although CWs can be used for water purification, the accumulated wetland sediments might become the sinks for heavy metals and organic compounds. The accumulated contaminants on sediments would result in sediment contamination. The release of heavy metals or organic contaminants from sediments into the environment would have adverse impacts on the ecosystem (Menon and Holland 2014; Saeedi and Jamshidi-Zanjani 2015; Li et al. 2016). Biological processes are major mechanisms causing the nutrients (N and P) cycles in CW ecosystems. Researchers have reported that similar microbial diversities were observed in wetland water and sediment environments (McGenity 2014; Valenzuela et al. 2017). This indicates that microbial processes also play important roles in contaminant removal the sediment environment (Ligi et al. 2014; Prasse et al. 2015; Kaplan et al. 2016).

Wetland Plants and Biodiversity

CWs are built to offer different ecological functions (Semeraro et al. 2015). Different species of macrophyte have been observed in CW systems worldwide, and many CW plants have been planted in CWs for nutrient removal (Vymazal 2013a). *Phragmites australis*, *Typha* spp., *Juncus* spp., *Scirpus (Schoenoplectus)* spp., and *Eleocharis* spp. are commonly used CW plants worldwide (Vymazal 2013b). Wetland vegetation can be also used as carbon sinks because CO_2 can be sequestered in wetland plants and sediments (de Klein and van der Werf 2014).

Commonly used CW plants include *Scirpus (Schoenoplectus)* spp., *P. australis, Juncus* spp., *Typha* spp., and *Eleocharis* spp. (Vymazal 2013a). CW plants can serve as sinks of carbon when CO_2 is sequestered in biomass (Audet et al. 2014; de Klein and van der Werf 2014). Although macrophyte species significantly contribute to the establishment of the wetland ecosystem, wetland biodiversity encompasses many different kinds of wildlife species, and these species contribute to the abundance of biodiversity (Syrbe et al. 2013). The following methods have been applied for biodiversity evaluation: Shannon Diversity, species abundance, Simpson Index, and species evenness (Chao et al. 2014; Duflot et al. 2015). Shannon Index and Simpson Index are also used for landscape composition identification. Variations due to the changes in landscape richness and evenness can be obtained by these indices. The Shannon Index can be used for landscape management and it is also a factor of the ecological framework. Simpson's Index can be also used for biodiversity (number and abundance of life form species) measurement (Chao et al. 2014; Wangchuk et al. 2014; Duflot et al. 2015).

Case Studies

CWs have been used for wastewater treatment and river purification successfully (Calheiros et al. 2014; Lai 2014; Morató et al. 2014). Zhai et al. (2016) applied the vertical-baffled flow wetlands (VBFWs) to treat sewage wastewater. The VBFWs were designed to treat approximately 300 m³/d of sewage from a community with 3,000 people. The system had a hydraulic residence time (HRT) of 2.65 d, and the wetland plant used in the system was *Cyperus alternifolius*. Results show that the removal efficiencies for chemical oxygen demand (COD), ammonium nitrogen (NH_3-N), total nitrogen (TN), and total phosphorus (TP) were 85, 76, 65, and 65%, respectively. Yan and Xu (2014) reported that higher removal rates were observed in the summer season with the higher temperature.

A full-scale partially saturated vertical subsurface flow (VF) CW planted with *C. papyrus* was applied to treat the effluent from urban wastewater in southern Brazil (Pelissari et al. 2017). The hydraulic loading rate (HLR) was 24.5 mm/d, and results show that more than 82% of total suspended solid (TSS), 79% of COD, 46% of TN, and 91% of NH_3-N could be removed through the VF system (Pelissari et al. 2017).

Labella et al. (2015) applied a horizontal SSFS to treat wastewater from urban sewer. Results indicated that about 69% of NH_3-N and 56% COD could be removed through system. Rossmann et al. (2013) applied VF to treat metal-contained wastewater from a paper mill. *Typha angustifolia, Erianthus arundinaceus,* and *Phragmites australis* were planted in the VF system for heavy metals removal. The result demonstrated that macrophytes played an important role in heavy metals removal from pulp and paper wastewater. The removal efficiencies for iron, copper, manganese, zinc, nickel, and cadmium were 74, 80, 60, 70, 71, and 70%, respectively in the VF system.

Tian et al. (2016) combined CWs and coal cinder-zeolite balls for aquaculture wastewater treatment. Their results demonstrated that a significant amount of COD and NH_3-N could be removed using the CW system (Tian et al. 2016).

Results from Xiong et al. (2015) showed that non-point source (NPS) pollutants from agricultural drainage caused the deterioration of the water quality of the HuaiRiver in Eastern China. Conventional Irrigation and Drainage System (CIDS) and Paddy Eco-ditch and Wetland System (PEDWS) were applied to reduce the NPS pollutant concentrations (Xiong et al. 2015). More than 70% of the total nitrogen TN and total phosphorus TP in NPS pollution could be reduced via the natural treatment system (Xiong et al. 2015).

Paing et al. (2015) used a vertical flow CW as a wastewater treatment plant to treat wastewater from a food-processing plant.

Wang et al. (2015) designed floating treatment wetlands (FTWs) to enhance the assimilative capacity and increase the biodiversity for a pond system in Fairfax, US. Results from their studies showed high treatment efficiencies (> 99%) for organic compounds and nutrients were observed.

Background

Taiwan has a population of 23.5 million and approximately half of the households are not hooked to the underground sewer system. Therefore, part of the improperly treated daily sewage is discharged into the surface water bodies (CPAMI 2017; MOI 2017). As most rivers in Taiwan have relatively steep slopes and short distances, they have relatively lower carrying capacities. The discharges of untreated domestic wastewater as well as the agricultural and industrial wastewaters into the water bodies result in contamination of more than 20% of rivers in Taiwan (TEPA 2017a). Thus, temporary measures need to be taken to minimize the impact of untreated wastewater on river water quality before the hookup rate of the sewer system can be increased.

In Taiwan, CWs have been applied for polluted rivers purification since the year of 2000 (Hsueh et al. 2014; Hsieh et al. 2015; Wu et al. 2015) because they have been recognized as the costly and green technology for wastewater treatment (Mander et al. 2014; He et al. 2015). To meet water reuse standards, CWs can be also used as the tertiary wastewater treatment facility for the polishing of treated wastewater after the conventional secondary wastewater treatment. The abovementioned CWs can purify wastewaters before they flow into the surface water bodies, and this is an expedient measure for wastewater treatment while sewer systems are under construction (TEPA 2017b). Most of these CWs are multi-functional wetlands, which are also used for environmental education, flood control, biodiversity, ecosystem protection, and recreation purposes.

Because of the inadequate discharges of industrial, domestic, and agricultural wastewaters into the stream, the Ju-Liao Stream is a highly polluted local river in Kaohsiung City. The stream water contains higher concentrations of organic contaminants, nutrients, and SS. The polluted stream water also result in the

Fig. 1. Photo of the KRRBCW located adjacent to the new (left) and old (new) Kaoping River Rail Bridges.

deterioration of the Kaoping River water quality, which is a receptor of the Ju-Liao Stream (Lin et al. 2015).

In the year 2010, the Taiwan Environmental Administration (TEPA), Kaohsiung City Government, and Taiwan Water Resources Agency initiated a CW full-scale CWs project by constructing a multi-function CW to improve the water quality of the Ju-Liao Stream and also protect the ecosystem of Dashu Region, Kaohsiung City, Taiwan. The CW is named the Kaoping River Rail Bridge Constructed Wetland (KRRBCW) because it is situated near the Kaoping River and is adjacent to the Kaoping River Rail Bridge. Figure 1 is a site map of the CW (KRRBCW). The KRRBCW has been operated for 13 years since its completion in 2004.

The CW is functioning well and is being operated and maintained by volunteers from a local environmental protection organization.

System Design and Performance Evaluation

The KRRBCW has two different treatment systems (Systems A and B). System A is composed of six basins (A1 to A6) and System B contains seven basins (B1 to B7). System A receives treated water (effluent of a final clarifier) from a secondary wastewater treatment plant of a local paper mill. System B receives polluted stream water from the Ju-Liao Stream.

The first basins (A1 or B1) are used to settle SS from both influents. In System A, Basins A2 to A5 are designed for wastewater polishing. In System B, Basins B2 to B6 are designed for stream water treatment. The last basins of both systems (A6 and B7) are used for wildlife habitats and ecological conservation. Figures 2 and 3 show the photos of wetland basins for Systems A and B, respectively.

Wetland water samples were collected and analyzed quarterly from 2008 to 2016 to assess the efficiency of using CWs for stream water quality improvement and wastewater polishing. For System A, water samples were collected from A1 inlet and

Fig. 2. Photos of wetland basins for System A.

outlets of A1, A2, A3, and A6. For System B, water samples were collected from B1 inlet and outlets of B1, B3, B4, and B7. Figure 4 shows the Inlets of Systems A and B. Figure 5 presents the water sampling locations.

For System A, the averaged total basin surface area, water depth, inflow rate, HRT, and HLR are 154,980 m², 1.1 m, 146,500 m³ d⁻¹, 10.1 d, and 0.48 m d⁻¹, respectively. For System B, the averaged total basin surface area, water depth, inflow rate, HRT, and HLR are 146,510 m², 0.7 m, 5,790 m³ d⁻¹, 14 d, and 0.04 m d⁻¹, respectively (Tu 2014; Lin et al. 2015). Different wetland plants including *T. latifolia* L., *P. australis* Cav., and *Scirpus* spp. are planted in both wetland basins.

Fig. 3. Photos of wetland basins for System B.

Fig. 4. Inlets of Systems A and B.

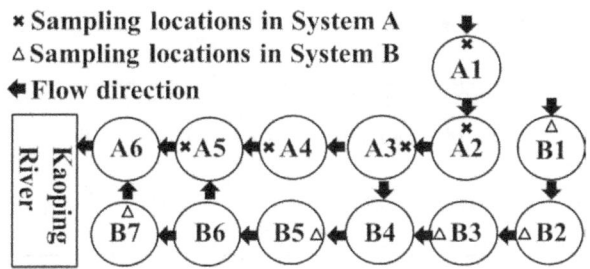

Fig. 5. Sampling locations of the two wetland systems.

Water Quality Evaluation

Tables 1 and 2 present the averaged influent and effluent concentrations and the removal efficiency, RE, for major water quality indicators of the wetland Systems A and B, respectively.

The first basins (A1 and B1) mainly planted with submerged plants (*Egeria densa*) were designed for SS sedimentation. Oxygen could be transferred by the submerged plants from the atmosphere to the water phases, and thus, the dissolved oxygen (DO) in the wetland water could be increased. The first basins in the two systems contain several different functions including flow adjustment, water treatment, sedimentation, and DO supplementation.

Emergent wetland plants (including *Phragmites australis*, *Phragmites communis*, *Typha orientalis*, and *Typha latifolia*) and floating wetland plants including *Pistia stratiotes* and *Ipomoea aquatica* are planted in the other basins (except the last basins in two systems). Nutrients and organic chemicals are biodegraded in these basins via biological mechanisms (e.g., biodegradation, plant uptakes). Figures 6a and 6b present the distributions of the influent (C_0) and effluent (C_e) concentrations for SS, BOD, chlorophyll a, and TC for Systems A and B, respectively. Figures 7a and 7b present the distributions of the influent and effluent DO, pH, and oxidation-reduction potential (ORP) measurements for Systems A and B.

Table 1. Influent and effluent concentrations and RE for major water quality indicators of the wetland System A.

items	unit	BOD	SS	NH₃-N	TN	TP	Chlorophyll a
C_0	mg/L	16 ± 12	7 ± 5[1]	2.2 ± 1.8	7 ± 4.4	1.6 ± 1.3	33 ± 16
C_e	mg/L	5 ± 3	13 ± 10	0.6 ± 0.2	1.7 ± 1.1	0.1 ± 0.2	37 ± 19
RE[3]	%	69	–[2]	73	76	94	–
Standard[4]	mg/L	2	25	0.3	–	0.05	–

[1]mean ± standard deviation; [2]not available; [3]removal efficiency; [4]Class A river water quality criteria established by TEPA (TEPA 2016); BOD, biochemical oxygen demand; SS, suspended solids; NH₃-N, ammonia nitrogen; TN, total nitrogen; TP, total phosphorus, C_0, initial concentration of each indicator variable, C_e, effluent concentration of each indicator variable, RE, removal efficiency of each indicator variable

Table 2. Influent and effluent concentrations and RE for major water quality indicators of the wetland System B.

items	unit	BOD	SS	NH₃-N	TN	TP	chlorophyll a
C_0	mg/L	33 ± 18	24 ± 13[1]	3.2 ± 1.2	11.5 ± 6.1	0.7 ± 0.3	38 ± 31
C_e	mg/L	12 ± 9	41 ± 24	0.7 ± 0.3	2.9 ± 1.6	0.3 ± 0.2	43 ± 22
RE[3]	%	64	–[2]	78	75	57	–
Standard[4]	mg/L	2	25	0.3	–	0.05	–

[1]mean ± standard deviation; [2]not available; [3]removal efficiency; [4]Class A river water quality criteria established by TEPA (TEPA 2016); BOD, biochemical oxygen demand; SS, suspended solids; NH₃-N, ammonia nitrogen; TN, total nitrogen; TP, total phosphorus, C_0, initial concentration of each indicator variable, C_e, effluent concentration of each indicator variable, RE, removal efficiency of each indicator variable

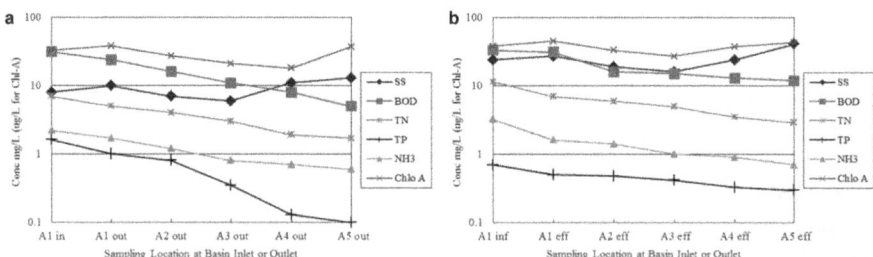

Figs. 6a and 6b. Distributions of the influent (C_0) and effluent (C_e) concentrations for SS, BOD, TN, TP, NH₃, and chlorophyll a for Systems A and B.

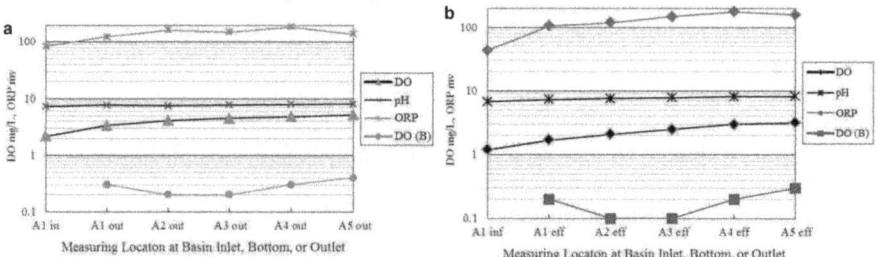

Figs. 7a and 7b. Distributions of the influent and effluent DO, pH, and ORP measurements for Systems A and B [DO (B) is the DO measurement taken at the bottom of the water column].

In System A, the investigation results show that removal efficiencies for BOD, NH_3-N, TN, and TP were 69, 73, 76, and 94%, respectively. In System B, the investigation results show that removal efficiencies for BOD, NH_3-N, TN, and TP were 64, 78, 75, and 57%, respectively.

This indicates that moderate to good removal efficiencies for the major water pollutants (organic chemicals and nutrients) are being obtained in the wetland systems. Thus, the wetland systems can purify the polluted water from Ju-Liao Stream and polish the effluent from the wastewater treatment plant of the paper mill. The water quality can be improved before it flows into the downstream Kaoping River. The major mechanisms causing the pollutants removal include biological, chemical, and physical processes. Moreover, decrease in nitrate concentrations and low nitrite concentrations (less than 0.05 mg/L) were observed in both wetland systems. This indicates the occurrence of both nitrification and denitrification processes in the wetland basins.

Increased SS and chlorophyll a concentrations were detected in both systems. This was due to the fact that the influents contained nutrients, which enhanced the algal growth. The growth of algae resulted in the increase in SS and chlorophyll a concentrations. This also indicates the wetland systems had eutrophication problems.

Floating-leaves wetland plants can be planted in the last wetland basins to provide the shades for algal growth control. This can also minimize the increase in SS concentrations in wetland effluents. This is also a green and eco-technology for algal control.

Because carbonates (CO_3^{2-}) and OH^- were produced during the CO_2 consumption by algae through photosynthesis processes, increased pH values were observed in wetland basins (Wallace et al. 2016). Due to the occurrence of natural turbulence and algal induced photosynthesis in wetland basins, increase in dissolved oxygen DO and oxidation-reduction potentials ORP values was detected. Therefore, photosynthesis and natural reaeration processes are the main mechanisms resulting in the increased DO and ORP values.

Relatively low DO values were detected at the wetland bottom. This was due to the high BOD concentrations in the influent and some organic matter accumulated in wetland sediments. Photosynthesis and reaeration mechanisms had lower influence on sediments and in the bottom zone of the wetland basins. Thus, decreased DO was observed at the bottom zone of the water columns. However, decrease in DO could enhance the occurrence of the denitrification mechanism, and this caused the decrease in nitrate concentrations in the wetland environments. Denitrification and nitrification can occur simultaneously in wetlands environments. Both NH_3-N and NO_3-N dropped significantly in the wetland system. As high BOD was detected in the wetland influents, carbon was available for the occurrence of denitrification process. Results also indicate that the produced nitrite (NO_2-N) was degraded in the wetland system, and the accumulation of nitrite was not observed. Removed nutrients may return to the wetland environment if sediments and plants remain in the system. Therefore, increase in nutrient and organic concentrations might be observed during the latter part of the wetland operation.

Although reductions of NH_3-N, BOD, and TP concentrations were observed after the CW treatment, investigation results indicate that the NH_3-N, BOD, and TP concentrations in CW effluents still could not meet the Class A river water quality criteria established by TEPA. To meet the Class A river water quality criteria, further wetland management is a necessity to improve the pollutant treatment efficiency. The proposed CW management strategies include frequent plant harvesting and sediment excavation from the wetland basins. Results also indicate that the wetland environment favored the algal growth, which resulted in the increased chlorophyll a concentrations. Planting floating-leaves in the last wetland basins may be a cost-effective and effective measure for the control of algal growth and chlorophyll a.

Site and environmental conditions (e.g., wildlife and recreation activities, plant muturity and coverage, sediment quality, ambient temperature) can also affect the biodegradation efficiencies and decay rates of the pollutants.

Wetland Sediment

Sediment samples were collected from the six wetland basins (A1, A3, and A5 from System A and B1, B4, and B7 from System B) for sediment property analyses. Table 3 shows the average analytical results of the sediments collected from six wetland basins. The wetland sediments were mainly sandy silt, and thus, the fine particle property caused significant sorption effects. Analytical results show that high sediments total phosphorus (STP), sediments oxygen demand (SOD), total organic matter (TOM), and sediments total nitrogen (STN) values were observed in sediments. This implies that high concentrations of organic chemicals and nutrient (N and P) were accumulated in sediments causing the variations of sediment properties. The organic-abundant sediments also resulted in the drop of ORP (oxidation-reduction potential) values.

Results of metal analyses for sediment samples are shown in Table 4. Results show that higher concentrations of Fe, Mn, Cu, Zn, and Cr were detected because the wetland influents included industrial wastewaters, which contained different kinds of metals. Thus, significant metal accumulation in sediments occurred after a long-term operation, and this caused the increase in metal concentrations in sediments.

Compared to the metal concentrations in sediments from A3 and B3 basins, higher metal concentrations were detected in sediments collected from A1 and B1 basins. When water flowed into the wetland basins, decreased flow velocity resulted in the sedimentation of SS. Thus, the metals in influents were adsorbed onto sediments in

Table 3. Analytical results of sediments collected from the six wetland basins.

Basin Location	pH	ORP (mV)	WC (%)	STP (g kg^{-1})	SOD (g O$_2$ m^{-2}d^{-1})	TOM (%)	STN (g kg^{-1})
A1[a]	7.3 ± 0.37[c]	-95 ± 39	65 ± 8	1.4 ± 0.8	3.9 ± 1.2	6.5 ± 1.7	17.2 ± 4.4
A3	7.4 ± 0.4	-125 ± 52	53 ± 9	0.89 ± 0.4	4.1 ± 1.1	4.2 ± 1.8	15.7 ± 6.2
A6	7.4 ± 0.5	-174 ± 47	50 ± 8	1.61 ± 0.3	2.8 ± 0.9	2.4 ± 0.9	$15.7 \pm .3$
B1[b]	7.2 ± 0.4	-138 ± 41	57 ± 9	1.83 ± 0.4	6.5 ± 1.3	2.9 ± 1.1	19.1 ± 3.5
B4	7.5 ± 0.3	-248 ± 63	61 ± 8	1.21 ± 0.5	5.1 ± 1.9	4.1 ± 1.5	17.1 ± 4.2
B7	7.4 ± 0.4	-352 ± 85	52 ± 8	1.44 ± 0.3	3.2 ± 2.1	1.7 ± 0.8	19.6 ± 5.8

ORP: oxidation-reduction potentials, WC: water content, STP: sediments total phosphorus, SOD: sediments oxygen demand, TOM: total organic matter, STN: sediments total nitrogen
[a]A1, A3, A6 are the first, third, and sixth basins in System A, respectively
[b]B1, B4, B7 are the first, fourth, and seventh basins in System B, respectively
[c]average ± standard deviation

Table 4. Results of metal analyses in sediment samples.

Basin Location	Cr (mg kg⁻¹)	Cu (mg kg⁻¹)	Fe (mg kg⁻¹)	Mn (mg kg⁻¹)	Ni (mg kg⁻¹)	Pb (mg kg⁻¹)	Zn (mg kg⁻¹)
A1[a]	125 ± 33[c]	278 ± 46	18,439 ± 4,437	213 ± 22	41 ± 7	28 ± 7	261 ± 24
A3	27 ± 12	53 ± 15	17,558 ± 3,522	158 ± 26	31 ± 5	12 ± 8	173 ± 26
B1[b]	52 ± 13	72 ± 17	20,441 ± 6,995	216 ± 37	29 ± 8	18 ± 9	144 ± 21
B3	38 ± 15	59 ± 14	23,465 ± 5,337	236 ± 25	38 ± 9	17 ± 6	165 ± 19

[a]A1 and A3 are the first and third basins in System A, respectively
[b]B1 and B3 are the first and third basins in System B, respectively
[c]average ± standard deviation

the wetland system (Gati et al. 2016). This caused the increase in metal concentrations in sediments, especially in the first wetland basin. Metal accumulation in sediments can be minimized via the pretreatment measures. For example, a sand or an activated carbon filter, a densely and fully vegetated zone, or a chemical coagulation/flocculation system for heavy metal removal can be installed. However, increased operational and maintenance cost would be expected if a pretreatment system is installed. To prevent the release of the pollutants into the wetland system, periodical sediment excavation is required. Frequent plant harvesting is also required to prevent the release of metals from the plant litters and dead plants into the wetland basins. A higher plant uptake rate can be achieved with frequent plant harvesting activity.

Operation and Maintenance

To evaluate the effectiveness of CW construction in biodiversity establishment, an ecological survey was conducted in the CW system. Land animals (e.g., reptiles, birds, mollusks, insects, mammals), aquatic animals (e.g., fish, aquatic insects, mollusks, crustaceans, and amphibians), vegetation (e.g., terrestrial and aquatic plants), zooplankton (e.g., molluscs, wheels animal, cladocera, protozoa, copepod, aquatic insects), and phytoplankton (e.g., green algae, euglenoids, diatoms, and blue-green algae) were monitored and recorded following the methods described in Ishiyama et al. (2014), Mazumdar et al. (2015), and Marazzi et al. (2017).

Except for stream water quality improvement and ecosystem protection, the KRRBCW has also been used for eco-education purpose.

The KRRBCW has been operated and maintained by volunteers from the Dashu Region Old Bridge Association, which is a local environmental protection organization. The KRRBCW was adopted by this association in 2001. The main goal of the association is to assure that the KRRBCW can play an important role in water purification, water quality improvement, wastewater polishing, environmental education, ecosystem rehabilitation, and recreation.

There are more than 150 unpaid volunteers for daily wetland operation and maintenance. The income of the association mainly comes from government support, public donation, and revenues from the souvenir and coffee shop at KRRBCW.

The association holds and organizes different social activities (e.g., workshops, athletic games, training courses, fishing games, singing and painting competitions, summer camps) with government authorities, universities, and other non-government organizations to draw more people's attention so that the latter can value this environmentally friendly ecosystem.

Conclusions

Results from this performance evalaution study indicate the following activities can be undertaken to enhance the water quality improvement function of the CW: (1) Wetland plant harvesting needs to be performed frequently to maintain a high pollutant uptake rate by wetland plants; (2) Appropriate pretreatment unit might be a necessity to lower the organic and nutrient concentrations in influents; (3) Wetland sediments monitoring is required to evaluate the impacts of sediment contamination on the water quality and ecosystem. Sediment management strategies (e.g., excavation) need to be taken if deterioration of sediment quality is observed; and (4) To establish a sustainable operational program, a local volunteer association needs to be organized for wetland operation and maintenance service. The KRRBCW accomplishes the main goal for wastewater polishing and stream water purification for downstream water body (Kaoping River) protection. It also creates a pleasing and high-grade environment for public and wildlife. The KRRBCW has become one of the most important and successful multi-functional CWs in Taiwan.

Acknowledgements

This project was funded by Taiwan Environmental Protection Administration and Environmental Protection Agency of Kaohsiung City Government. The authors would like to thank Prof. Andy Hong at University of Utah, US, as well as Prof. W.P. Sung at National Chin-Yi University, Taiwan for their review and valuable comments.

References Cited

Audet, J., C.C. Hoffmann, P.M. Andersen, A. Baattrup-Pedersen, J.R. Johansen, S.E. Larsen et al. 2014. Nitrous oxide fluxes in undisturbed riparian wetlands located in agricultural catchments: Emission, uptake and controlling factors. Soil Biol. Biochem. 68: 291–299.

Ávila, C., J.B. Bayona, I. Martin, J.J. Salas and J. Garcia. 2015. Emerging organic contaminant removal in a full-scale hybrid constructed wetland system for wastewater treatment and reuse. Ecol. Eng. 80: 108–116.

Białowiec, A., A. Albuquerque and P.F. Randerson. 2014. The influence of evapotranspiration on vertical flow subsurface constructed wetland performance. Ecol. Eng. 67: 89–94.

Chao, A., N.J. Gotelli, T.C. Hsieh, E.L. Sander, K.H. Ma, R.K. Colwell et al. 2014. Rarefaction and extrapolation with hill numbers: a framework for sampling and estimation in species diversity studies. Ecol. Monogr. 84(1): 45–67.

Collins, A.R. and N. Gillies. 2014. Constructed wetland treatment of nitrates: removal effectiveness and cost efficiency. J. Am. Water Resour. As. 50(4): 898–908.

Corbella, C., J. Puigagut and M. Garfi. 2017. Life cycle assessment of constructed wetland systems for wastewater treatment coupled with microbial fuel cells. Sci. Total Environ. 584-585: 355–362.

Construction and Planning Agency Ministry of the Interior (CPAMI). 2017. Construction and Planning Agency Ministry of the Interior, National Sewage Users to take over the Penetration Rate and the Overall Sewage Treatment Rate Statistics. Taipei, Taiwan.

de Klein, J.J.M. and A.K. van der Werf. 2014. Balancing carbon sequestration and GHG emissions in a constructed wetland. Ecol. Eng. 66: 36–42.

Duflot, R., S. Aviron, A. Ernoult, L. Fagrig and F. Burel. 2015. Reconsidering the role of semi-natural habitatin agricultural landscape biodiversity: a case study. Ecol. Res. 30(1): 75–83.

Dunne, E.J., M.F. Coveney, V.R. Hoge, R. Conrow, R. Naleway, E.F. Lowe et al. 2015. Phosphorus removal performance of a large-scale constructed treatment wetland receiving eutrophic lake water. Ecol. Eng. 79: 132–142.

Filoso, S., S.M.C. Smith, M.R. Williams and M.A. Palmer. 2015. The efficacy of constructed stream–wetland complexes at reducing the flux of suspended solids to Chesapeake Bay. Envir. Sci. and Tech. 49(15): 8986–8994.

Calheiros, S.S.C., A.O.S.S. Rangel and P.M.L. Castro. 2014. Constructed wetlands for tannery wastewater treatment in Portugal: ten years of experience. Int. J. Phytoremediat. 16(9): 859–870.

Gati, G., C. Pop, F. Brudaşcă, A.E. Gurzău and M. Spînu. 2016. The ecological risk of heavy metals in sediment from the Danube Delta. Ecotoxicology. 25(4): 688–696.

He, J., F. Moffette, R. Fournier, J.P. Revéret, J. Théau, J. Dupras, J.P. Boyer et al. 2015. Meta-analysis for the transfer of economic benefits of ecosystem services provided by wetlands within two watersheds in Quebec, Canada. Wetl. Ecol. Manag. 23(4): 707–725.

Hsieh, C.Y., E.T. Liaw and K.M. Fan. 2015. Removal of veterinary antibiotics, alkylphenolic compounds, and estrogens from the Wuluo constructed wetland in southern Taiwan. J. Environ. Sci. Heal, Part A. 50(2): 151–160.

Hsueh, M.L., L. Yang, L.Y. Hsieh and H.J. Lin. 2014. Nitrogen removal along the treatment cells of a free-water surface constructed wetland in subtropical Taiwan. Ecol. Eng. 73: 579–587.

Ishiyama, N., T. Akasaka and F. Nakamura. 2014. Mobility-dependent response of aquatic animal species richness to a wetland network in an agricultural landscape. Aquat. Sci. 76(3): 437–449.

Kaplan, D.I., S.W. Buettner, D. Li, S. Huang, P.G.K. van Groos, P.R. Jaffé et al. 2016. *In situ* porewater uranium concentrations in a contaminated wetland: Effect of seasons and sediment depth. Appl. Geochem. 85: 128–136.

Labella, A., D. Caniani, T. Hughes-Riley, R.H. Morris, M.I. Newton, P. Hawes et al. 2015. Assessing the economic suitability of aeration and the influence of bed heating on constructed wetlands treatment efficiency and life-span. Ecol. Eng. 83: 184–190.

Lai, D.Y. 2014. Phosphorus fractions and fluxes in the soils of a free surface flow constructed wetland in Hong Kong. Ecol. Eng. 73: 73–79.

Leung, H.M., N.S. Duzgoren-Aydin, C.K. Au, S. Krupanidhi, K.Y. Fung, K.C. Cheung et al. 2017. Monitoring and assessment of heavy metal contamination in a constructed wetland in Shaoguan (Guangdong Province, China): bioaccumulation of Pb, Zn, Cu and Cd in aquatic and terrestrial components. Environ. Sci. Pollut. R. 24(10): 9079–9088.

Lin, J.L., W.C. Kuo, Y.M. Chang, R.Y. Surampalli and C.M. Kao. 2015. Development of a Natural Treatment System for Stream Water Purification: Mechanisms and Environmental Impacts Evaluation. J. Environ. Eng. 141(11): 04015029.

Li, M., W. Yang, T. Sun and Y. Jin. 2016. Potential ecological risk of heavy metal contamination in sediments and macrobenthos in coastal wetlands induced by freshwater releases: A case study in the Yellow River Delta, China. Mar. Pollut. Bull. 103(1): 227–239.

Li, Y., G. Zhu, W.J. Ng and S.K. Tan. 2014. A review on removing pharmaceutical contaminants from wastewater by constructed wetlands: design, performance and mechanism. Sci. Total Environ. 468: 908–932.

Ligi, T., K. Oopkaup, M. Truu, J.K. Preem, H. Nõlvak, W.J. Mitsch et al. 2014. Characterization of bacterial communities in soil and sediment of a created riverine wetland complex using high-throughput 16S rRNA amplicon sequencing. Ecol. Eng. 72: 56–66.

Lu, S., X. Zhang, J. Wang and L. Pei. 2016. Impacts of different media on constructed wetlands for rural household sewage treatment. J. Cleaner Prod. 127(20): 325–330.

Lv, T., Y. Zhang, L. Zhang, P.N. Carvalho, C.A. Arias and H. Brix. 2016. Removal of the pesticides imazalil and tebuconazole in saturated constructed wetland mesocosms. Water Res. 91: 126–136.

Mander, Ü., G. Dotro, Y. Ebie, S. Towprayoon, C. Chiemchaisri, S.F. Nogueira et al. 2014. Greenhouse gas emission in constructed wetlands for wastewater treatment: a review. Ecol. Eng: 66: 19–35.

Marazzi, L., E.E. Gaiser, V.J. Jones, F.A.C. Tobias and A.W. Mackay. 2017. Algal richness and life-history strategies are influenced by hydrology and phosphorus in two major subtropical wetlands. Freshwater Biol. 62(2): 274–290.

Mazumdar, K. and S. Das. 2015. Phytoremediation of Pb, Zn, Fe, and Mg with 25 wetland plant species from a paper mill contaminated site in North East India. Environ Sci Pollut R. 22(1): 701–710.

McGenity, T.J. 2014. Hydrocarbon biodegradation in intertidal wetland sediments. Curr. Opin. in Biotech. 27: 46–54.

Menon, R. and M.M. Holland. 2014. Phosphorus release due to decomposition of wetland plants. Wetlands. 34(6): 1191–1196.

Ministry of the Interior (MOI). 2017. Ministry of the Interior. Static Statistics of Population, Taipei, Taiwan.

Morató, J., F. Codony, O. Sánchez, L.M. Pérez, J. García and J. Mas. 2014. Key design factors affecting microbial community composition and pathogenic organism removal in horizontal subsurface flow constructed wetlands. Sci. Total Environ. 481: 81–89.

Paing, J., V. Serdobbel, M. Welschbillig, M. Calvez, V. Gagnon and F. Chazarenc. 2015. Treatment of high organic content wastewater from food-processing industry with the French vertical flow constructed wetland system. Water Sci. Technol. 72(1): 70–76.

Papaevangelou, V.A., G.D. Gikas, Z. Vryzas and V.A. Tsihrintzis. 2017. Treatment of agricultural equipment rinsing water containing a fungicide in pilot-scale horizontal subsurface flow constructed wetlands. Ecol. Eng. 101: 193–200.

Pelissari, C., C. Ávila, C.M. Trein, J. García, R.D. de Armas and P.H. Sezerino. 2017. Nitrogen transforming bacteria within a full-scale partially saturated vertical subsurface flow constructed wetland treating urban wastewater. Sci. Total Environ. 574: 390–399.

Prasse, C.E., A.H. Baldwin and S.A. Yarwood. 2015. Site history and edaphic features override the influence of plant species on microbial communities in restored tidal freshwater wetlands. Appl. Environ. Microb. 81(10): 3482–3491.

Rossmann, M., A.T. Matos, E.C. Abreu, F.F. Silva and A.C. Borges. 2013. Effect of influent aeration on removal of organic matter from coffee processing wastewater in constructed wetlands. J. Environ. Manage. 128: 912–919.

Saeedi, M. and A. Jamshidi-Zanjani. 2015. Development of a new aggregative index to assess potential effect of metals pollution in aquatic sediments. Ecol. Indic. 58: 235–243.

Semeraro, T., C. Giannuzzi, L. Beccarisi, R. Aretano, A.D. Marco, M.R. Pasimeni and I. Petrosillo. 2015. A constructed treatment wetland as an opportunity to enhance biodiversity and ecosystem services. Ecol. Eng. 82: 517–526.

Sultana, M.Y., C. Mourti, T. Tatoulis, C.S. Akratos, A.G. Tekerlekopoulou and D.V. Vayenas. 2016. Effect of hydraulic retention time, temperature, and organic load on a horizontal subsurface flow constructed wetland treating cheese whey wastewater. J. Chem. Technol. Biot. 91(3): 726–732.

Syrbe, R.U., E. Michel and U. Walz. 2013. Structural indicators for the assessment of biodiversity and their connection to the richness of avifauna. Ecol. Indic. 31: 89–98.

Taiwan Environmental Protection Administration (TEPA). 2016. Taiwan Environmental Protection Administration. Environmental Regulations and Standards. Taipei, Taiwan.

Taiwan Environmental Protection Administration (TEPA). 2017a. Taiwan Environmental Protection Administration. Environmental Resource Database. Taipei, Taiwan.

Taiwan Environmental Protection Administration (TEPA). 2017b. Taiwan Environmental Protection Administration. Nationally Important Wetland Conservation Program. Taipei, Taiwan.

Tian, W., K. Qiao, H. Yu, J. Bai, X. Jin, Q. Liu et al. 2016. Remediation of aquaculture water in the estuarine wetlands using coal cinder-zeolite balls/reed wetland combination strategy. J. Environ. Manage. 181: 261–268.

Torrijos, V., O.G. Gonzalo, A. Trueba-Santiso, I. Ruiz and M. Soto. 2016. Effect of by-pass and effluent recirculation on nitrogen removal in hybrid constructed wetlands for domestic and industrial wastewater treatment. Water Res. 103: 92–100.

Tu, Y.T., P.C. Chiang, J. Yang, S.H. Chen and C.M. Kao. 2014. Application of a constructed wetland system for polluted stream remediation. J. Hydrol. 510: 70–78.

Valenzuela, E.I., A. Prieto-Davó, N.E. López-Lozano, A. Hernández-Eligio, L. Vega-Alvarado, K. Juárez et al. 2017. Anaerobic methane oxidation driven by microbial reduction of natural organic matter in a tropical wetland. Appl. Environ. Microb. 83(11) doi: 10.1128/AEM.00645-17. Print 2017 Jun 1.

Vymazal, J. 2013a. Emergent plants used in free water surface constructed wetlands: A review. Ecol. Eng. 61: 582–592.

Vymazal, J. 2013b. The use of hybrid constructed wetlands for wastewater treatment with special attention to nitrogen removal: A review of a recent development. Water Res. 47(14): 4795–4811.

Vymazal, J. and T. Březinová. 2015. The use of constructed wetlands for removal of pesticides from agricultural runoff and drainage: A review. Environ. Int. 75: 11–20.

Wang, C.Y., D.J. Sample, S.D. Day and T.J. Grizzard. 2015. Floating treatment wetland nutrient removal through vegetation harvest and observations from a field study. Ecol. Eng. 78: 15–26.

Wangchuk, K., A. Darabant, P.B. Rai, M. Wurzinger, W. Zollitsch and G. Gratzer. 2014. Species richness, diversity and density of understory vegetation along disturbance gradients in the Himalayan conifer forest. J. Mt. Sci. 11(5): 1182–1191.

Wallace, J., P. Champagne and G. Hall. 2016. Multivariate statistical analysis of water chemistry conditions in three wastewater stabilization ponds with algae blooms and pH fluctuations. Water Res. 96: 155–165.

Wu, S., S. Wallace, H. Brix, P. Kuschk, W.K. Kirui, F. Masi et al. 2015. Treatment of industrial effluents in constructed wetlands: challenges, operational strategies and overall performance. Environ Pollut. 201: 107–120.

Xiong, Y., S. Peng, Y. Luo, J. Xu and S. Yang. 2015. A paddy eco-ditch and wetland system to reduce non-point source pollution from rice-based production system while maintaining water use efficiency. Environ. Sci. Pollut. R 22(6): 4406–4417.

Yan, Y. and J. Xu. 2014. Improving winter performance of constructed wetlands for wastewater treatment in northern China: A review. Wetlands 34(2): 243–253.

Yi, X.H., D.D. Jing, J. Wan, Y. Ma and Y. Wang. 2016. Temporal and spatial variations of contaminant removal, enzyme activities, and microbial community structure in a pilot horizontal subsurface flow constructed wetland purifying industrial runoff. Environ. Sci. Pollut. R 23(9): 8565–8576.

Zalewski, M. 2014. Ecohydrology, biotechnology and engineering for cost efficiency in reaching the sustainability of biogeosphere. Ecohydrol. and Hydrobiol. 14(1): 14–20.

Zhai, J., J. Xiao, M.H. Rahaman, Y. John and J. Xiao. 2016. Seasonal variation of nutrient removal in a full-scale artificial aerated hybrid constructed wetland. Water 8(12): 551–561.

8

Constructed Wetlands Technology in Cuba

Research Experiences

Irina Salgado-Bernal,[1,*] *Maira M. Pérez-Villar,*[2] *Lizandra Pérez-Bou,*[1]
Mario Cruz-Arias,[1] *Margie Zorrilla-Velazco*[2] and
María E. Carballo-Valdés[1]

INTRODUCTION

The current situation of the shortage of water is a global problem, and Cuba does not escape from this situation. The impact in the vulnerability of this resource in Cuba is due to the concentration of industrial installations in urban zones, the uses of obsolete technologies, a low introduction level of cleaner production practices in industries and services (Suárez et al. 2012), and the dumping, in some communities, of domestic wastewaters without the appropriate treatment to the water receptor bodies.

The Cuban archipelago has 38.1 km^3 of fresh water. However, not all this water constitutes the offer, because just one part is usable water. About 35% of water is available from the total (Triana 2017). In addition, there are insufficient results to cover residential, agricultural, and industrial needs, due to different phenomena such as prolonged dry periods, water overexploitation, pollution, saline intrusion, deficit of the forest cover, and lack of reuse and recycling of water (Suárez et al. 2012).

---

[1] Departamento de Microbiología y Virología, Facultad de Biología, Universidad de La Habana, La Habana, Cuba.
[2] Centro de Estudio de Química Aplicada, Universidad Marta Abreu de Las Villas, Cuba.
* Corresponding author: irina@fbio.uh.cu

Water treatment plants with the application of different technologies exist in the country, but many of them present maintenance deficit, justified partly by the high costs. This has originated their bad operation and the inadequate purification of the residual waters. For this reason, more consideration should be given to natural systems such as artificial or constructed wetlands (CWs), as an alternative to conventionally engineered wastewater treatment plants. Studies have shown that these natural systems work well in tropical and developing countries. These environmentally friendly systems show good efficiency in urban and per urban areas, while having generally lower capital and maintenance costs than a conventional wastewater treatment plants (García-Armisen et al. 2008; Wei et al. 2014). Another advantage is the possibility of reusing the treated water (Lyu et al. 2016). Different experiences in the research with these treatment systems have been achieved in Cuba, some of which show up in this chapter. However the knowledge about complex processes that occur in the plants, microorganisms and soil matrix in the wetlands is still incomplete, which is a challenge in the studies of constructed wetlands to world level. This is a need for a better appraisal on the purifying processes that occur in wetlands under different climatic and operational conditions.

Current Situation of Water Pollution in Cuba. Generalities

Cuba is the most populated country in the Caribbean Sea, with 11,242,628 people and the population growth rate was 0.3% in 2009, which was lower than the world average of 1.2% (Peláez 2010).

Average rainfall of Cuba is 1,059 mm in the rainy season (May–October) and 316 mm in the dry season (November–April) with relative annual humidity that averages 78%. Significant elements in the climate of Cuba are the hurricanes that affect the country on an average of once every two years. The tropical hurricanes are areas of drop pressures of among 300–500 km of diameter that cause winds, rains, and extremely strong sea surf that usually have catastrophic effects in the regions where they cross (Hernández 2001). The very long and narrow configuration of the island of Cuba gives place to the existence of short course rivers with a reduced flow in their majority. Cauto is the longest river with a length of 370 km and Toa is the biggest. The fluvial basins in a same way have relatively little extension and there are a total of 632 bigger than 5 km² with a fluvial glide of 31,682 million cubic meters (SYC 2010).

Environmental Policy

The environment protection and the rational use of natural resources have been the common heritage of Cuban society. The first National Environmental Strategy (NES) was adopted in 1997, and represented the effort results that were spear-headed by the Ministry of Science, Technology and Environment (*CITMA* in Spanish), which was created in 1994 as the lead agency for the activity, together with a number of other Cuban institutions and bodies that are involved in the economic and social development of the country.

At present, the environmental policy and management are based on the National Environmental Strategy approved in 2007, this is supplemented by the National

Biodiversity Strategy; the National Strategy for Environmental Education; the National Action Program against Desertification and Drought; and the National Strategy for Biological Safety. Other instruments include the 1997 Environment Act; the Sectorial strategies of the Organizations of the State Central Administration (OSCA); and the Territorial Strategies (NES 2007–2010).

The framework of the integrated management of natural resources are special regions for sustainable development: the mountain ecosystems, the hydrographic basins; the National Council of Hydrographic Basins, constituted in 1997 and contains Provincial Councils; National and Provincial Groups for Bays and Harbors, directed towards integrated management; and the beaches, swamps and protected areas which fall within a National System, resulting from a Global Environment Facility program of the United Nations Development Program (GEF/UNDP) (Suárez et al. 2012).

Environmental Problems

The main environmental problems identified in the National Environmental Strategy are soil degradation, deforestation, pollution, loss of biological diversity, and lack of water (Suárez et al. 2012).

Pollution

In Cuba, the pollution of the water, land, and atmosphere is due to different causes such as the concentration of industrial installations in urban zones; the uses of obsolete technologies; a low introduction level of cleaner production practices in industries and services.

Actually, the National Environmental Strategy incorporates actions in order to prevent, reduce, and control the pollution such as, the introduction of cleaner production in the industries, increasing the level of recycling solid and liquid wastes, and monitoring discharges, emission, and pollution charges among others.

The recycling of wastes, from industry and from the population at large, is among the elements of the waste-management process that require additional organization and financial resources (Suárez et al. 2012).

Lack of Water

As a result of a large program for the construction of dams, including micro dams, the water storage capacity of the country increased from 48 million cubic meters in 1958, to more than 9,600 million cubic meters at the beginning of the present decade. This has allowed a substantial increase in the safe water coverage for the population reaching up to 95.6% (Suárez et al. 2012) and a per capita volume of 3,361 m^3 per person per year, according to a study published four years ago by the Latin American Bank of Development.

Considering this fact, Cuba was the fourth country with less hydric availability in Latin America, just overcome by Haiti, Puerto Rico, and Dominican Republic. However, dates published by Cuban Statistical Annual located the country in a worse situation that this previous study, because at present Cuba has only 1,215.7 m^3 per person per year (Triana 2017). There are insufficient results to cover residential, agricultural and industrial needs, due to different phenomena such as a prolonged dry periods, water overexploitation, pollution, saline intrusion, deficit of the forest cover, and lack of reuse and recycling of water. The solution of these problems requires significant levels of environmental sanitation investments in human settlements. The majority of these activities receive the support of the Environmental National Program, the Global Environmental Facility (GEF) through the United Nations Development Program (UNDP) and the United Nations Environment Program (UNEP) (Suárez et al. 2012).

Water Sources

Cuban archipelago has 38.1 km^3 of fresh water. But not all this water constitutes the offer, because just one part is usable water. About 13.5 km^3 of water is available from the total, this is 35%.

Contrary to the world, 75% of the water is superficial, due to the construction of hundreds of dams in the whole country (Triana 2017).

Water supply in Cuba consists of two general categories, surface water and groundwater. These sources are managed at a national level by the "*Instituto Nacional de Recursos Hidráulicos*" in Spanish (National Institute of Hydraulic Resources, INRH).

Surface Water

Surface water is originated from watersheds containing a wide array of rivers and streams. These watersheds provide more than 65% of Cuba's available water and make up more than 80% of the potential water supply. Watersheds in the island consist of an estimated 31.7 billion cubic meters of water, and are generally found in the north and south of a dividing topographic ridge running through the center of the island (García 2006).

In total, these watersheds encompass slightly more than 15% of Cuba's territory, supplying water to approximately 40% of the population and contributing to 60% of all economic activity. Eight high priority Cuba watersheds are the followings: Cuyaguateje, Ariguanabo, Almendares-Vento, Hanabanilla, Zaza, Cauto, Guantanamo-Guasa, and Toa (García 2006).

Ground Water

The nation of Cuba benefits from a sizable amount of groundwater totaling an estimated 6.4 billion cubic meters. This groundwater is mostly found within carbonate rocks in the form of calcium-bicarbonate type. There are several regions in the country in which groundwater is predominantly utilized. These regions include Ciudad de La Habana, Matanzas, Ciego de Ávila, and Camagüey. In total, groundwater accounts for approximately 35% of all available water used in the country (Cubagua 2017). Accessible water is defined as water that can be readily obtained through the country's existing infrastructure.

Water Demand and Treatment Capacity

Cuban *INRH* estimated that during the same year the total water demand in Cuba was 7,260 million m^3. This 7260 million m^3 yr^{-1} is equivalent to 19.9 million m^3 yr^{-1} or 1.77 m^3 person^{-1} d^{-1}. This daily per capita number accounts for all water use classifications, and includes approximately 20% domestic water use, 5.5% industrial, and 57.9% agricultural; 17% goes to other destinations (Triana 2017).

The total demand for domestic water use is 3.78 million m^3 d^{-1} or 1,380 million m^3 yr^{-1}. Thus, the total capacity of Cuba's 59 surface water plants, when fully operational, is approximately 1.15 million m^3 d^{-1}, or 420 million m^3 yr^{-1}. The amount of groundwater treated at chlorination stations is reported to be roughly 2.5 times the amount of the potable or drinking water produced by the surface water plants, namely 1,050 million m^3 yr^{-1} (Cubagua 2017). This brings the total potable water production capacity, neglecting losses in distribution, to 1,470 million m^3 yr^{-1} or 0.36 m^3 d^{-1} capita^{-1}. Comparing this capacity to the aforementioned domestic demand (0.34 m^3 d^{-1} capita^{-1}) indicates that Cuba has enough nominal water treatment capacity (e.g., assuming that the plants are actually working properly) to satisfy domestic water demand. Yet, due to a variety of issues primarily relating to the condition of the water distribution system, in particular the excessive amount of leaks, the demand is often left unmet not because of lack of nominal treatment capacity but because of inefficiencies of the distribution system.

Wastewater Treatment Systems

Most of the current wastewater infrastructure in Cuba was built over 50 years ago and has not been well maintained or rehabilitated throughout its service life. Furthermore, the wastewater treatment system is not as extensive as the water distribution systems and generally lacks complete integration of a wastewater collection network with a wastewater treatment plant. The sanitary sewer coverage for the entire island is considered to be at 94% for the current population of 11.20 million. The 94% coverage has two major components: 38% consists of connections to wastewater collection systems, and 56% consists of *in situ* wastewater systems; 6% are without service (Pan American Health Organization 2000). Rural areas have a relatively lower percentage of sanitation coverage, with 84% of the population with access to wastewater collection facilities; in urban areas 97% of the population has access. However, 56.3% of Cuba's population under coverage is being served with the use of septic tanks and latrines: 49.6% in the urban areas and 76.9% in the rural areas (Cubagua 2017). In the year 2000, an estimated 19% of the wastewater collected underwent some sort of treatment (PAHO 2000) and this was reduced to 4% as of 2007. The rest of the wastewater is discharged into nearby water-ways with minimal or no treatment. In 2006, *INRH* reported a total of 121 wastewater pump stations, 470 septic tanks, and 862,121 seepage pits throughout the entire island.

The most common form of treatment is through the use of stabilization lagoons that range from anaerobic, to aerobic and facultative. As of 2006, the *INRH* was directly responsible for 302 lagoon systems. Of these lagoon systems, 241 (79.5%) have been re-habilitated and only 53 (17.5%) perform an efficient treatment (Cubagua 2017). *INRH* also plays a role in the assessment of approximately 1200 oxidation lagoons that belong to other entities. Facultative stabilization lagoons are the most common form of industrial wastewater treatment on the island. In addition to stabilization lagoons, wetland systems or trickling filters are also utilized in some areas.

According to the *INRH*, there are eight wastewater treatment plants in Cuba. However, through further investigation, a total of 11 existing plants have been identified, two of the plants are located in The Havana City "Maria del Carmen" and "Sistema Central" and two of them are located outside the city. Wastewater treatment plants of low capacities can be found in Santiago "La Cuba" and Villa Clara "Ensenachos", while the other known plants can be found in Cayo Coco, "EDAR Unidad Cojimar", and Varadero, "Siguapa and Taino I" to serve the tourism industry. García-Armisen et al. (2008) highlights a constructed pilot solar aquatic system facility built in the Havana City for alternative wastewater treatment.

In 2005, only five of the aforementioned wastewater treatment plants were considered *efficient* and only four were efficient in 2007. However, a more recent issue of *Options*, an economic newspaper in Cuba, dated September 2008, includes an interview of Nobel Rovirosa of the National Center of Hydrology and Water Quality in Cuba (*CENHICA* in Spanish) which reveals that in the entire country, only two treatment plants are operational: Quibu and Maria del Carmen. There are also several wastewater treatment plants that are being planned and constructed throughout Cuba, mostly in The Havana City, with the use of funding for specific sanitary infrastructure projects.

Within Havana City there are two conventional wastewater treatment plants (Maria del Carmen and Sistema Central) and one small unconventional plant (Solar Aquatic System). The conventional wastewater treatment plants serve two of the six existing collection networks. These six networks consist of more than 1,570 km of sewer mains and laterals, 23 pump stations, and 15 stabilization lagoons. Originally built for almost 1.2 million inhabitants many years ago, these six major collection networks (Sistema Central, Maria del Carmen, Almendares Sur, Cotorro, Alamar, and Puentes Grandes) serve approximately 55% of the city population of over 2.2 million people. The sanitation coverage for the rest of the population is managed by septic tank coverage (26%) and smaller wastewater collection networks (12%). The remaining 7% of the population in the nation's capital does not receive any type of sanitation coverage (Alonso and Mon 1996).

The Maria del Carmen wastewater treatment plant is situated in the middle region of the Almendares watershed within the Maria del Carmen network which serves a population of about 20,500, and consists of approximately 89 km of sewer lines built in the 1960s (Alonso-Hernandez and Mon 1996). This wastewater treatment plant consists of primary settling, a trickling filter process, secondary sedimentation, and finally, sludge digestion. Even though the plant was out of service for several years and would discharge its effluent directly into the Almendares River in Havana (Olivares-Rieumont et al. 2005); now, the treatment plant seems to have been rehabilitated and is currently in operation, although at very low capacity and with treatment at sub-optimal levels.

The last of the operational treatment plants in Havana City is a small operational pilot wastewater treatment facility which was built in the lower region of the Almendares watershed. The solar aquatic system plant was built to treat the effluent of 5,000 inhabitants. This plant was found to be working efficiently in a study comparing three types of wastewater treatment facilities in Cuba including a conventional wastewater treatment plant (Maria del Carmen) and a constructed wetland.

Quibu is another wastewater treatment plant in Havana City actually operating. Built several kilometers west of The Havana City on the Quibu River, the Quibu Wastewater Treatment Plant had an approximate construction cost of $2.0 million (GEF-UNDP 2002). It has a design capacity of 12,960 m^3 d^{-1}, but is running at approximately 6,910 m^3 d^{-1}. The general treatment configuration of the Quibu wastewater treatment plants is roughly the same as for the Maria del Carmen plant. The treatment process involves screening and then primary sedimentation using settling tanks. Then the wastewater flows through a biological trickling filter and undergoes secondary sedimentation before it is discharged as effluent. The sludge is processed through an anaerobic digester. The Quibu Wastewater Treatment Plant is expected to double its capacity to 25,920 m^3 d^{-1} in the near future (Gutiérrez et al. 2007).

Along with the expansion of the Maria del Carmen wastewater treatment plant, there are two other wastewater treatment plants proposed in order to improve the sanitary infrastructure in Havana City along the Almendares River. The first treatment plant, El Pitirre, lies in the upper region of the Almendares watershed within the wastewater collection system of Cotorro which was built in the late 1960s and early 1970s. The municipality of Cotorro consists of approximately 72,000 inhabitants of which almost one-third is connected to the 23.5 km sewer network. This situation is of high concern as it may be very detrimental to the primary drinking water supply for the Havana region: the Vento Aquifer. This aquifer is hydraulically connected to the Almendares River, especially in upstream reaches of the river which is where the wastewater is discharged from the municipality of Cotorro. The second proposed wastewater treatment plant to be funded through OPEC (Organization of the Petroleum Exporting Countries) was to be located in the lower region of the Almendares watershed in the Puentes Grandes network designed to serve up to 200,000 people (Artiles and Gutiérrez 1997). Even though the system has 80% of its population connected to lateral sewer lines, the system lacks a main trunk line or a wastewater treatment plant, and most of the wastewater is discharged directly into the Almendares River (Alonso and Mon 1996; Artiles and Gutiérrez 1997). Due to its elevated cost, the construction of the Puentes Grandes Wastewater Treatment Plant has been postponed and a more cost-efficient alternative using a subaqueous outfall to the ocean is being evaluated.

Besides the two aforementioned plants that are waiting to be completed as part of the comprehensive plan to improve the sanitary infrastructure along the Almendares River, there is also a comprehensive sanitation plan to improve the sanitary infrastructure surrounding the Havana Bay. The Cuban government, with the support of the United Nations Development Program, has plans to construct four new treatment plants (referred to here as Upper Luyano, Lower Luyano, Luyano 3, and Martín Pérez) to treat all the municipal wastewater from the tributary rivers flowing towards Havana Bay. These plants will collectively treat wastewater generated by over 100,000 persons. Three of the plants will be constructed along the Luyano River, the main source of organic and industrial wastes to the bay, and the fourth plant will be constructed on the Martín Pérez River (GEF-UNDP 2002).

Approximately 1 km upstream of Lower Luyano Wastewater Treatment Plant is the site of a second treatment plant on the Luyano River (referred to here as the Upper Luyano Wastewater Treatment Plant).

The project includes the construction of an advanced treatment plant that will include primary and secondary treatment including clarifiers, and tertiary treatment with nutrient removal of nitrogen and phosphorus.

The wastewater treatment plants will also include a demonstration project focusing on utilizing sludge for fertilizer and/or energy. When completely finalized, the Upper Luyano Wastewater Treatment Plant will serve a population of about 70,000 people in the Luyano River area with a total capacity of 51,840 m^3 d^{-1} divided into three equal phases.

Some of the funding will also go into constructing 7.5 km of new sewer lines in the Luyano River area (GEF-UNDP 2002).

Wastewater Treatment Demand and Capacity

Assuming the domestic wastewater flow to be 80% of the potable water demand, the domestic wastewater flow for the island of Cuba is estimated to be 1103 million m^3 yr^{-1} or approximately 3.02 million m^3 d^{-1}, which gives us an approximate wastewater demand per person to be 0.27 m^3 d^{-1} in Cuba. Taking into account that approximately 8.5 million of Cuba's population lives in urban areas, the average urban wastewater treatment demand would be about 2.30 million m^3 d^{-1}.

If all of Cuba's existing treatment plants were fully operational, the country could count on a wastewater treatment capacity of 107,900 m^3 d^{-1}. However, since we know that probably only the Maria del Carmen and Quibu treatment plants are properly functioning, and at approximately a third of their full capacity, the current operational wastewater treatment capacity for Cuba can be estimated to be 21,600 m^3 d^{-1}. Once completed, the two plants on the Luyano River in Havana are expected to add 138,240 m^3 d^{-1} to the nation's wastewater treatment capacity. If all the wastewater treatment plants were rehabilitated to full capacity and new ones currently in construction were finished, the wastewater treatment capacity for the island would only be enough to treat approximately 400,000 m^3 d^{-1}, less than one fifth of the country's urban daily wastewater flow.

Havana City's priorities in wastewater infrastructure are aligned with the protection of the water supply, with an emphasis on the upper and middle regions of the Almendares watershed which have a strong connection with the Vento Aquifer, the primary drinking water source for Havana's population. Regarding the choice of type of treatment system, due consideration should be given to natural systems such as constructed wetlands and solar aquatic systems, as alternatives to conventionally engineered wastewater treatment plants. Studies have shown that these natural systems work well in tropical and developing countries such as Cuba. These environmentally friendly systems show good efficiency in urban and per urban areas, while having generally lower capital and maintenance costs than a conventional wastewater treatment plant. A study performed in Havana City comparing the performance of the Maria del Carmen wastewater treatment plant, a solar aquatic system and a small scale constructed wetlands concluded that natural systems effluent quality was much better than that of conventional treatment (García-Armisen et al. 2008). As part of the comprehensive sanitation plan for the Havana Bay, the performance of small scale constructed wetlands in Havana should be further studied in order to determine the feasibility of larger scale implementation throughout the island (GEF-UNDP 2002).

The present condition of Cuba's water and wastewater infrastructure warrants extensive improvements to both systems. The evaluation conducted has highlighted the major challenges the country faces with respect to water sanitation. The existing surface water treatment plants and ground water chlorination stations appear to have adequate capacity to meet the domestic water demand for the entire island, yet due to the heavy deterioration of the distribution piping; the system is not capable of meeting the needs of the population.

For wastewater, the existing infrastructure does not adequately meet the demands of the growing populations within urban centers. There are not enough wastewater treatment plants on the island, and a large majority of the existing plants are nonoperational or running at low capacities. Most cities do not have extensive sewer networks and existing lines are not complete and do not cover their proposed population due to lack of integration with a treatment plant. Even though construction of new wastewater infrastructure is taking place, the rate at which these projects are advancing have been unusually slow and this can lead to cost overruns and depletion of project financing.

The wastewater treatment system is in need of integrated management and a much stronger emphasis needs to be placed in the rehabilitation of the existing wastewater infrastructure as well the construction of new wastewater infrastructure in locations of high importance such as densely populated urban centers and locations where the integrity of the drinking water aquifer is a concern. Due to the nature of wastewater disposal with regards to population density in urban centers as well as lack of rural cost-effectiveness, prioritization led to only the top 15 cities with population of over 100,000 being taken into account in the wastewater improvement cost estimates. These improvements include restoration and construction of wastewater treatment plants, restoration and construction of wastewater collection pump stations, as well as the construction of the needed wastewater collection pipe networks.

Use of Constructed Wetlands Technology in Cuba

The Cuban economy and technological condition situation need the development of sustainable wastewater treatment technologies to be more environmental friendly, easy to operate, less energy-intensive and cost-effective.

Review of the literature shows evidence of limited research of the constructed wetland technology in the tropical regions. Many of the developing countries without the wastewater treatment infrastructure are located within this belt and could therefore, benefit more from the application of this type of systems.

The constructed wetlands offer effective and reliable treatment of wastewater in a simple and inexpensive way, requiring minimal maintenance (Reed 1995; Kadlec and Knight 1996).

In Cuba the prohibitive operating and maintenance costs of conventional treatment systems limited the satisfactory use of these technologies, so that the potential of constructed wetlands as a low-cost intervention deserves to be explored.

However, experience with the use of these treatment systems is still limited within the island despite the benefits, but an asset is that most of the current existing treatment plants are in a highly degraded state.

Case Studies in the Central Region of Cuba

Case studies are presented for the treatment of wastewaters with different pollutants; the studies were based on either pilot or laboratory scale experimental systems.

1. Pilot scale vertical flow constructed wetlands for domestic wastewater treatment
2. Pilot scale horizontal flow constructed wetlands for domestic wastewater treatment
3. Laboratory scale horizontal flow constructed wetlands for oily wastewater treatment
4. Laboratory scale horizontal flow constructed wetlands for heavy metals wastewater treatment

Constructed Wetland for Domestic Wastewater Treatment

The constructed wetlands were designed to create environments that promote the removal of total suspended solids (TSS), pathogenic microorganisms, organic matter: biochemical oxygen demand (BOD) and chemical oxygen demand (COD), and excess nutrients. However, there are other factors that affect the efficiency of contaminant removal mechanisms such as wetland designs, spatial constraints, climate, hydraulic loading rates (HLR), hydraulic residence times (HRT), and the presence of macrophytes.

Pilot Scale Vertical Flow Constructed Wetlands. The vertical flow subsurface wetland is situated next to a weld rails industry in the central region of Cuba (Pérez et al. 2014). This industry does not use water in the production process. Their wastewaters are only domestic. There is a primary treatment with a septic moat and a buffer tank for the wastewater. The primary treatment does not satisfy the Cuban regulatory limits for discharge (NC27 2012). The surface area of the experimental wetland is about 20 m².

Young *Cyperus alternifolius* (Yan et al. 2016; Zhai et al. 2016) were collected and planted by hand (four plants per m²). Iron-rich soil was used for support medium. A layer of 0.60 m depth was placed over a gravel (10–15 mm) layer of 0.15 m of depth, which aimed to prevent clogging of the perforations in the discharge pipe. The support media contained a high content of iron, which favors phosphorus removal and a sorption capacity of 0.96 g kg⁻¹ at 25°C (Pérez et al. 2014). Also perforated pipes, as inlet and discharges structures, and vertical pipes were placed to oxygenate the deep layers. Clay was used as sealing material to line the treatment bed. The flow of wastewater, pumped at rate of 0.375 m³ hr⁻¹, was fed intermittently in two batches per week lasting 4 hours on each occasion. The pilot vertical subsurface wetland operated with a superficial loading rate of 19 g BOD m⁻² d⁻¹. High removal efficiencies were achieved for most pollutants (Table 1). The efficiency of phosphorus removal was high compared to that commonly reported for gravel support media 20–30% (Prochaska and Zouboulis 2006; Vohlaa et al. 2011). This may result from the use of iron-rich soil as support media. Iron oxides constitute an efficient sorption medium for phosphorus.

The effluent concentrations for all parameters complied with the Cuban regulation standards (NC27 2012). There was an increase of dissolved oxygen in the wastewater effluent (5.5 mg L⁻¹) compared to the influent wastewater (1.1 mg L⁻¹). An increase

Table 1. Summary of influent and effluent characteristics of wetland wastewater.

Parameter	Regulatory level Receiver body A	Influent	Effluent	η (%)
pH	6.5–8.5	7.3 ± 0.4	7.04 ± 0.02	-
Electrical conductivity (μS cm^{-1})	1,400	690 ± 106	677 ± 18	-
COD (mg L^{-1})	70	98 ± 17	10 ± 4	90
BOD$_5$ (mg L^{-1})	30	42 ± 6	6 ± 4	85
Dissolved oxygen (mg L^{-1})	-	1.1 ± 0.4	5.4 ± 0.2	-
Total phosphorus (mg L^{-1})	2	7.1 ± 0.8	1.7 ± 0.2	76
Kjeldahl nitrogen (mg L^{-1})	5	14.0 ± 3.4	2,4 ± 0.2	83
NH$_4$-N (mg L^{-1})	-	2.6 ± 1.8	1.5 ± 0,2	82
NO$_2$ Nitrogen (mg L^{-1})	-	0.03 ± 0.02	0.03 ± 0.001	-
NO$_3$ Nitrogen (mg L^{-1})	-	5.0 ± 2.7	8.0 ± 0.3	-
Total suspended solids (mg L^{-1})	-	20 ± 4	0.4 ± 0.1	98

Means, standard deviations n = 4; η: removal efficiency

in the dissolved oxygen in the effluent is an advantage of vertical wetland, which has previously been reported by several authors (Molle et al. 2006; Saeed and Sun 2011; Kim et al. 2016).

Pilot Scale Horizontal Flow Constructed Wetland. The horizontal flow subsurface wetland is located at the Santa Clara city in the central region of Cuba. It was designed to treat the wastewater of the service unit of the electric company of Villa Clara. Their wastewaters are domestic because they originate from the dining kitchen. There is a primary treatment with a grease interceptor for the wastewater. The primary treatment does not satisfy the Cuban regulatory limits for discharge (NC27 2012). The superficial area of the experimental wetland is about 3 m^2. The beds contained 0.15 m of 10–15 mm gravel media and 0.40 m of iron-rich soil, and are planted with *Cyperus alternifolius* at 4 seedlings per m^2 (Yan et al. 2016; Zhai et al. 2016). A continuous flow of wastewater of 0.075 m^3 h^{-1} was applied.

The results showed that the horizontal flow constructed wetland were significantly efficient in the removal of organic matter (Table 2) and effluent concentrations for all parameters complied with the Cuban regulation standards (NC27 2012).

Table 2. Summary of characteristics of influent and effluent wetland wastewater.

Parameter	Regulatory level Receiver body B	Influent	Effluent	η (%)
pH	6–9	5.47 ± 0.34	7.12 ± 0.04	-
Electrical conductivity (μS cm^{-1})	2.000	1,365 ± 1.208	723 ± 24	-
COD (mg L^{-1})	90	1,600 ± 35	78 ± 14	95
BOD$_5$ (mg L^{-1})	40	695 ± 135	35 ± 3	95
Total phosphorus (mg L^{-1})	4	14.2 ± 8.4	3.4 ± 0.5	76
Kjeldahl nitrogen (mg L^{-1})	10	20.6 ± 1.4	8.6 ± 0.9	58

Means, standard deviations n = 3; η: removal efficiency

Low removal efficiency of Kjeldahl nitrogen was achieved compared to that vertical flow subsurface wetland (82%). It is hypothesized that low oxygen availability resulted in low removal of nitrogen due to limited nitrogen oxidation for this type of horizontal subsurface wetland (Molle et al. 2006; Saeed and Sun 2011; Kim et al. 2016).

Laboratory Scale Horizontal Flow Constructed Wetlands for Different Wastewater Pollutants

The experimental systems for all pollutants evaluated is composed of a CW of horizontal subsurface flow of 0.45 m long, 0.33 m wide and 0.2 m deep, planted with *Cyperus alternifolius*. This emergent plant has shown high resistance and ability to remove different pollutants (Pérez et al. 2014; Yan et al. 2016; Zhai et al. 2016).

Oily Wastewater Treatment

Cuba has increased the number of decentralized plants for generating electric power. In these facilities a fuel cleaning stage fuel (diesel or fuel) is required before it is used in internal combustion engines. This stage produces wastewater with high fat and oil content. There is a primary treatment with a buffer tank and grease interceptor for the wastewater. The effluent wastewater for the primary treatment was supplied to the experimental wetland. The substrate used in this wetland was gravel 10–15 mm in size. A significant reduction in effluent COD concentration was observed (Table 3).

A positive effect on greases, oil, and hydrocarbons removal efficiency was also recorded. The effluent concentrations met the discharge limits. The study showed that horizontal wetland system is an effective method to treat oily wastewaters.

Heavy Metals Wastewater Treatment. The wastewaters from the galvanic industry have high contents of heavy metals, mainly chromium and nickel. Horizontal subsurface wetland laboratory scale to removal these heavy metal was evaluated. Red ferralitic soil that has a high mineral content was used as support media, favoring the elimination of pollutants. Standards solutions of 10 and 20 mg L^{-1} of these metals were applied to the experimental wetlands. Removal efficiencies greater than 99% for both metals were obtained after 2 hours of residence time of the wastewater in the wetland.

The concentrations of chromium and nickel in the plants were generally high, especially in root and leaves tissues (Fig. 1), with a higher concentration of nickel

Table 3. Summary of characteristics of influent and effluent wetland wastewater.

Parameter	Influent	Effluent	η (%)
pH	7.6	7.3	-
Total Solids (mg L^{-1})	102	34	67
COD (mg L^{-1})	165.1	37.2	77.5
Greases, oils, and hydrocarbons (mg L^{-1})	80	< 10	> 87

Means, standard deviations n = 3; η: removal efficiency

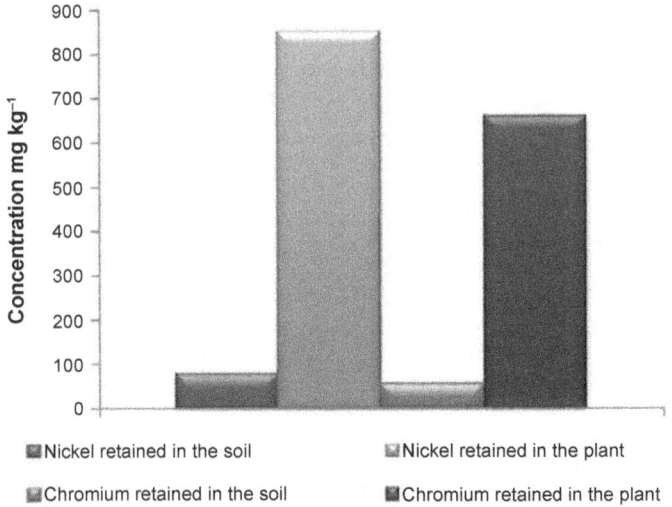

Fig. 1. Heavy Metals Wastewater Treatment in Laboratory Scale Horizontal Flow Constructed Wetlands. Metals retained in the soil and plant.

in the leaves and a higher concentration of chromium in the roots. From the current study, *Cyperus alterniufolius* can be used for the removal of nickel and chromium in constructed wetlands.

Perspectives for Constructed Wetlands Systems Optimization

Limitations of Constructed Wetlands

Constructed wetlands are considered in general as an example of clean, green or environmental technology, because in them, environmental science is applied to conserve the environment and resources, supported by the concept of sustainable development: to satisfy the necessities of the present generations without committing the possibilities of those of the future to assist their own necessities (Brundtland Commission, ONU 1987). Nevertheless, the work with this type of systems still presents numerous problems, which means CWs can be considered a cleaner technology when comparing it with other methods currently used for the water treatment, but it still does not clean up to 100%.

Constructed or artificial wetlands present strengths like the simplicity of their operation, the minimum or null energy consumption, keeping in mind that the energy used by the system fundamentally support the biological agents that participate in the removal of pollutants processes and that the energy is recycled inside the same system, among the different organisms. Other aspects are the low residues production during the system operation, low installation, and maintenance costs, compared with physical, chemical, and biological conventional systems, the reliability of the operation, the low environmental sound impact, and the good integration within the environment (Chen 2011; Curia et al. 2011; Norton-Brandao et al. 2013; Wei et al. 2014). Additionally, they can be built in small communities of rural areas, where traditional treatments of wastewater are not feasible (Babatunde et al. 2008; Chung et al. 2008). They can settle down in the same place where the water is generated and the maintenance can be carried out for personal relatively untrained (Chung et al. 2008). Another advantage is the possibility of reusing the treated water, according to the physical, chemical, and biological characteristics of the output water, because CWs are considered one of the current technologies applied in wastewater reuse (Lyu et al. 2016).

They also present disadvantages, like the most part of all technologies, in this case associated with the requirement of very big areas for their installation, which some authors like Wu et al. (2015) identify it as the main limiting factor for their application, approximately from 2 to 5 m^2 per person. Specifically superficial flow wetlands can generate mosquitoes and smell problems, the design approaches and current operation are still imprecise and they require a long period for starting up and stabilizing, from some months or a year in systems with sub-superficial flow, up to several years in systems with superficial flow (Kadlec et al. 2000), due to the time requiring for plants adaptation. In the case of artificial horizontal sub-superficial flow wetlands (HSSF) can possibly exist emissions of greenhouse gases, due to the anaerobic processes. Nevertheless these emissions are insignificant when comparing with other sources.

Additional to these aspects, the work with wetlands can generate other problems. Examples of derived consequences of the lack of maintenance or inadequate design for a vertical sub-superficial flow wetland (VSSF) exist: For example, when previous pre-treatment to CW does not exist, the system can be overloaded; because the solids accumulate on the surface. When this problem does not exist suspended solids accumulate to smaller rates and overall life span may be reduced to ten or twenty years. This excessive accumulation of solids can block the distribution of the influent in the system and the porous spaces. Also, the pruning of the plants is important and recommended every two to three years (when it is applicable in winter). The pruning has advantages like the export of nutrients from the system and the prevention of the formation of dead vegetation thick layers with stagnated water which is ideal for plagues. If it is not done, it may also present advantages like the formation of an isolated layer of dead vegetable material, the provision of a detritus layer for the adsorption of trace metals, the provision of a carbon source for denitrification, and the continuity or no change of wetlands ecological conditions.

Enhancing Strategies for Constructed Wetlands

CWs have indeed been proven to be efficient for the treatment of conventional (organic matter, nutrients, microorganisms, etc.), nonconventional (such as heavy metals and hydrocarbons), and even emerging pollutants (pharmaceuticals, steroid hormones, biocides, etc.) in a variety of wastewaters such as domestic sewage, agricultural wastewater, industrial effluent, mine drainage, leachates, contaminated groundwater, urban runoff, and raw drinking water worldwide (Vymazal 2011; Verlicchi and Zambello 2014). In spite of this in the past decades, numerous studies have focused on the design, operation, and performance of these systems. In addition, in order to expand the applications of CWs for wastewater treatment, great research efforts have currently been made for improving the treatment performance of the wetlands and minimizing the negative effects and enhancing the positive ones of different influencing factors (Meng et al. 2014; Yan and Xu 2014). Considering these limitations, different future challenges have been identified for the improvement of CWs. Some of them are:

- To uncover the CW black box.
- To identify the biochemical processes, through the use of microbial labeling techniques, among others.
- To quantify the processes (e.g., oxygen releasing from the roots and the reduction of sulfates).
- To model the processes involved: Mathematical models that represent the complex net of interacting processes. They can be used to optimize the design and operation.

Other authors like Wu et al. (2014) have identified other alternatives as:

- Thermal insulation
- Tidal flow operation
- Step-feeding
- Effluent recirculation
- Supply of external carbon sources (addition of organic carbon and application of organic substrates)
- Harvest of biomass
- Bioaugmentation
- Addition of earthworms

Regarding the importance of microorganisms in the removal of pollutants in CWs through complex biochemical reactions, the bioaugmentation strategy will be approached in this chapter. The importance of the metagenomics in the exploration of wetlands black box will also be explained.

Bioaugmentation

Bioaugmentation is included among the new strategies for the improvement of bioremediation processes efficiency in CWs. Bioaugmentation or "sowing" consist on the addition of microbial highly concentrated and specialized populations, of simple strains or consortia, to a given polluted place (Tyagi et al. 2011; Cerqueira et al. 2012; Meng et al. 2014).

This technique is considered appropriate for places without enough microbial cells or in which the native population does not present the metabolic capacities and necessary characteristic for the interaction with the polluting compounds (Tyagi et al. 2011). The options more commonly employed to attack this strategy are: the addition of bacterial pure strains or pre-adapted consortia, the introduction of genetically modified bacteria, and the introduction of gene groups outstanding in the interaction with the pollutant, by means of a vector, to be transferred to the indigenous microorganisms (Fantroussi and Agathos 2005).

Microbial inocula are cell homogeneous suspensions, which, obtained under optimum conditions, may often suffer stress with the contact of the natural habitats complexity. In the real cases, the introduced population begins to fall after being added, due to multiple abiotic and biotic factors (Hosokawa et al. 2009). Among these factors highlight the competition for nutrients, with the autochthonous microorganisms (Fantroussi and Agathos 2005), the pH that can inhibit the degradation and removal processes and the temperature that influences the microbial growth and the degradation and removal potential. For this reason, the bioaugmentation by itself is not enough, it should generally be accompanied by physical and environmental alterations of the treated area (Tyagi et al. 2011). To increase the probability of initial establishment and effectiveness of a long term inoculum, additional conditions should be provided to guarantee the protection in the event of the adverse circumstances of the atmosphere. Some techniques such as encapsulation and immobilization are used, which facilitate a more efficient and quicker interaction with the pollutants regarding the free cells (Moslemy et al. 2003; Tyagi et al. 2011), and advantages like biofilms formation (Fantroussi and Agathos 2005).

In spite of the great number of elements to consider and some contradictory results of different investigation groups, the bioaugmentation represents one of the most promissory alternatives to improve bioremediation efficiency (Tyagi et al. 2011), because the selectivity and specialization of the inoculating microorganisms defines the effectiveness of a remediation process with biological agents (Hamdi et al. 2007). For this reason, the search of competent microorganisms for the application in each place or polluted component is probably at the present time one of the best solutions (Tyagi et al. 2011). Bioaugmentation using compound microbial inocula can fine tune the bacterial population and enhance the pollutants removal efficiency of a constructed wetland system. This has been demonstrated in studies like Zhao et al. (2016), where compound microbial inocula were enriched and applied to a pilot-scale constructed wetland system to investigate their bioaugmentation effect on nitrogen removal under cold temperature (10°C). The results showed a 10% higher removal efficiency of ammonia and total nitrogen compared to a control (unbioaugmented) group. Nutrient removal in CWs added with a consortium of six denitrifying bacteria for treating polluted river water was evaluated by Shao et al. (2014). The removal efficiencies were found to be 75% for COD, 96% for TN, 96% for NH_4^+-N, and 90% for TP.

Within the research group on Microbial Biotechnology of the Faculty for Biology, University of Havana, investigations related with this topic have been carried out. One of the objectives of the study "Removal of pollutants of residual waters for autochthonous rizobacteria with application in artificial wetlands" (Salgado 2012), was to bioaugment CWs for the treatment of synthetic domestic wastewater, with and without the presence of heavy metals. The bioaugmentation was done with a consortium named CAD/1S, designed for organic matter, ammonium, and phosphate removal, for the case of domestic wastewater and with the consortium CAM/G+, designed for lead, chromium, and mercury removal (Fig. 2).

In general, in this study was observed that in the bioaugmented CWs, for the treatment of domestic wastewater, the removal of the three indicators was reached, in a stable way within the expected time and with removal percentages of 100% for organic matter, superior to 70% for ammonium nitrogen, and higher than 55% for phosphates.

Fig. 2. Constructed wetlands bioaugmentation (microcosm scale) operating in the Laboratories 301-302-303 in Mexico, Faculty of Chemistry, UNAM, using microbial consortia designed by Salgado et al. (2012).

The behavior of bioaugmented CWs for metals removal showed removal percentages superior to 95% in the lead, above 50% for the chromium and superior to 80% for the mercury.

The results of the bioaugmented wetlands presented significant differences when comparing with wetlands without the presence of the consortia.

Also pollutant concentrations in the effluents presented significant differences with the initial concentrations of these compounds in the influent.

Metagenomics for Discovering the "Black Box"

Future research into the elucidating structure and distribution of the related microbial communities and identifying the related microbes which assist in removal processes will further help to optimize the treatment performance of CW systems.

Metagenomics can be a useful tool for this purpose, because it allows us to know how many and what species are present in a certain atmosphere and what ecological function play each species in an individual way. The term metagenomics was coined by Handelsman et al. in 1998. Among the concepts defined in the literature, one of the most synthetic belongs to Chen and Pachter (2005), that defines it as the application of modern genomic techniques for the study of microbial communities directly in their natural atmospheres. Other authors like Riesenfeld et al. (2004) define it as the investigation of collective microbial genomes retrieved directly from environmental samples and does not rely on cultivation or prior knowledge of the microbial communities. Traditional cultivation methods and traditional genomics can at best access 1%. However, metagenomics can in principle access 100% of the genetic resources in an environment. DNA is directly extracted from the environment and cloned into cosmid, fosmid, or bacterial artificial chromosomes (BAC) vectors producing large insert libraries.

Theoretically, a metagenomic library contains clones representing the entire genetic complement of a single habitat. The information held within a metagenomic library can be used to determine community diversity and activity, the presence of specific microorganisms or biosynthetic pathways as well as simply searching for the presence of individual genes (Steele and Streit 2005).

In a CW system, the metagenomics might contribute to the pre-valuation of the biodegradative capacities of the present microorganisms, through the evaluation of the microbial diversity *in situ* and the estimation of the functional capacity. It could also be useful in the monitoring *in situ* of biodegradation behavior and community stability. This is particularly interesting in the case of wetlands bioaugmentation with specialized strains, because once the inoculation of the exogenous microorganisms in the system happens, it is important to check their establishment in the time, as well as their implication in the removal of the pollutants.

All of the molecular approaches available for community structure and function analysis have advantages and limitations associated with them, and none provides complete access to the genetic and functional diversity of complex microbial communities. A combination of several techniques should be applied to interrogate the diversity, function, and ecology of microorganisms. Culture-based and culture-independent molecular techniques are neither contradictory nor excluding and should be considered as complementary (Rastogi and Sani 2011).

Conclusions

Although constructed wetlands have been mostly exploited in Europe and United States, even in different countries of Latin America, the studies in the Caribbean region are still incipient. Some regions exist in Cuba where they have begun to be seen as an option for the management of wastewater, but presently deeper studies are needed for the optimization of their operation and the diversification of their functions. These aspects become extensive to the studies with these systems at world level.

The application of alternatives as the bioaugmentation of wetlands with well characterized microbial inocula could constitute a tool to develop the behavior of the system in the removal of pollutants.

Also, the setting in practice of molecular studies as the metagenomics would help to the best understanding in the biochemical processes and the discovery of new processes that involves the complexity of the constructed wetlands.

Acknowledgements

The authors gratefully acknowledge the International Foundation for Science (IFS), Stockholm, Sweden, for the Grants awarded to Irina Salgado (IFS Projects: W/4860-1-2) and the logistic support of Laboratories 301-302-303 of the Faculty of Chemistry, UNAM, Mexico. All of them have contributed to the research in the topic of constructed wetlands in the Faculty for Biology, University of Havana. Acknowledgement is given to Armando Martínez-Sardiñas, MSc., from the Facultad de Biología, Universidad de La Habana, a specialist in Environmental Microbiology and Biotechnology, for his valuable peer review of this contribution.

References Cited

Alonso, J. and E.A. Mon. 1996. Caracterización del Abastecimiento de Agua Potable y Saneamiento de la Ciudad de La Habana. Document Number 50443 - 1011/A81/036473.
Artiles, R. and J. Gutiérrez. 1997. Saneamiento de la Cuenca Almendares. Document Number 90469 - CD/2300/A81/034796. AIDIS (Interamerican Sanitary and Environmental Engineering Association) Sao Paulo, Brazil and CEPIS (Panamerican Health Organization, Centro Panamericano de Ingeniería Sanitaria y Ciencias del Ambiente) Lima, Perú. http://www.bvsde. paho.org/bvsaidis/puertorico/xxxii.pdf. In Spanish.
Babatunde, A.O., Y.Q. Zhao, M. O'Neill and B. O'Sullivan. 2008. Constructed wetlands for environmental pollution control: A review of developments, research and practice in Ireland. Environment International 34(1): 116–126.
Brundtland Commission United Nations. 1987. Report of the World Commission on Environment and Development: Our Common Future.
Cerqueira, V.S., E.B. Hollenbach, F. Maboni, F.A.O. Camargo, M.C.R. Peralba and F.M. Bento. 2012. Bioprospection and selection of bacteria isolated from environments contaminated with petrochemical residues for application in bioremediation. World Journal of Microbiology and Biotechnology 28(3): 1203–1222.
Chen, H. 2011. Surface-flow constructed treatment wetlands for pollutant removal: Applications and perspectives. Wetlands 31: 805–814.
Chung, A.K.C., Y. Wub, N.F.Y. Tam and M.H. Wong. 2008. Nitrogen and phosphate mass balance in a sub-surface flow constructed wetland for treating municipal wastewater. Ecological Engineering 32: 81–89.
Cubagua. 2017. http://www.hidro.cu/In Spanish.
Curia, A.C., J.C. Koppe, J. Costa, L.A. Féris and W.D. Gerber. 2011. Application of pilot-scale-constructed wetland as tertiary treatment system of wastewater for phosphorus and nitrogen removal. Water Air Soil Pollut. 218: 131–143.
El Fantroussi, S. and S.N. Agathos. 2005. Is bioaugmentation a feasible strategy for pollutant removal and site remediation? Current Opinion in Microbiology 8: 268–275.
García, J.M. 2006. Experiencias cubanas en la institucionalización del manejo integrado de cuencas. Voluntad Hidráulica 98: 15–28. In Spanish.
García-Armisen, T., J. Prate, Y. Marrero and P. Servais. 2008. Faecal bacterial indicators removal in various wastewater plants located in Almendares River watershed (Cuba). Water Sci. Technol. 58(4): 773–779.
Global Environment Facility—United Nations Development Programme (GEF-UNDP). 2002. Regional (Cuba, Jamaica): Demonstrations of Innovative Approaches to the Rehabilitation of Heavily Contaminated Bays in the Wider Caribbean. http://www.undp.org.cu/proyectos/bahia. html).
Gutiérrez, E.J., M. Bollo, R. Marsán, Z. Castaño and Y. Pérez. 2007. Contaminación de las aguas en la cuenca urbanizada del Río "Quibú", en Ciudad de La Habana. V FRIEND World Conference, Havana, Cuba. 2006. Hydrological Impacts of Climate Variability and Change: Selected Presentations on Latin America and the Caribbean. Technical Document IHP–LAC, N°11 published in 2007 by the International Hydrological Programme (IHP), Regional Office for Science for Latin America and the Caribbean (Montevideo, Uruguay), Regional Office for Culture for Latin America and the Caribbean (Havana, Cuba) of the United Nations Education, Scientific and Cultural Organizations.
Hamdi, H., S. Benzarti, L. Manusadzianas, I. Aoyama and N. Jedidi. 2007. Bioaugmentation and biostimulation effects on PAH dissipation and soil ecotoxicity under controlled conditions. Soil Biol. Biochem. 39: 1926–1935.

Handelsman, J., M.R. Rondon, S.F. Brady, J. Clardy and R.M. Goodman. 1998. Molecular biological access to the chemistry of unknown soil microbes: a new frontier for natural products. Chemistry & Biology 5(10): R245–R249.

Hernández, P. 2001. Geografía de Cuba. Tabloide, Semanario Juventud Rebelde, Havana, Cuba 1: 13. In Spanish.

Hosokawa, R., N. Motonori, M. Morikawa and H. Okuyama. 2009. Autochthonous bioaugmentation and its possible application to oil spills. World J. Microbiol. Biotechnol. 25: 1519–1528.

Kadlec, R.H. and R. Knight. 1996. Treatment wetlands. Lewis Publishers, Florida.

Kadlec, R.H., R.L. Knight, J. Vymazal, H. Brix, P. Cooper and R. Haberl. 2000. Constructed wetlands for pollution control. Scientific and Technical Report No. 8. IWA. London, UK.

Kim, S.H., J.S. Cho, J.H. Park, J.S. Heo, Y.S. Ok, R.D. Delaune et al. 2016. Long-term performance of vertical-flow and horizontal-flow constructed wetlands as affected by season, N load, and operating stage for treating nitrogen from domestic sewage. Environ. Sci. Pollut. R. 23: 1108–1119.

Lyu, S., W. Chen, W. Zhang, Y. Fan and W. Jiao. 2016. Wastewater reclamation and reuse in China: Opportunities and challenges. Journal of Environmental Sciences 39: 86–96.

Meng, P., H. Pei, W. Hu, Y. Shao and Z. Li. 2014. How to increase microbial degradation in constructed wetlands: influencing factors and improvement measures. Bioresour. Technol. 157: 316–326.

Molle, P., A. Lienard, A. Grasmick and A. Iwema. 2006. Effect of reeds and feeding operations on hydraulic behavior of vertical flow constructed wetlands under hydraulic overloads. Water Res. 4: 606–612.

Moslemy, P., R.J. Neufeld, D. Millette and S.R. Guiot. 2003. Transport of gellan gum microbeads through sand: an experimental evaluation for encapsulated cell bioaugmentation. J. Environ. Manage. 69: 249–259.

NC27/2012.2012. Vertimiento de aguas residuales a las aguas terrestres y al alcantarillado. Oficina Nacional de Normalización. La Habana, Cuba. In Spanish.

NES. National Environment Strategy 2007–2010. 1st ed. Editorial Academia. La Habana, Cuba. In Spanish.

Norton-Brandao, D., S.M. Scherrenberg and J.B. van Lier. 2013. Reclamation of used urban waters for irrigation purposes—A review of treatment technologies. J. Environ. Manag. 122: 85–98.

Olivares-Rieumont, S., D. de la Rosa, L. Lima, D.W. Graham, K. D'Alessandro, J. Borroto et al. 2005. Assessment of heavy metals in the Almendares River sediments—Havana City, Cuba. Water Research 39: 3945–3653.

Pan American Health Organization (PAHO). 2000. Evaluación de los Servicios de Agua Potable y Saneamiento 2000 en las Américas, Informe Analítico. Washington D.C. http://www.bvsde. paho. In Spanish.

Peláez, O. 2010. Creció la población cubana en el. 2009. Diario Granma, Habana, Cuba, January 5, 2010. 1: 3. In Spanish.

Pérez, M., J.M. Hernández, J. Bossens, E. Rosa and F. Tack. 2014. Vertical flow constructed wetlands: Kinetics of nutrient and organic matter removal. Water Sci. Technol. 70: 76–81.

Prochaska, C.A. and A. Zouboulis. 2006. Removal of phosphates by pilot vertical-flow constructed wetlands using a mixture of sand and dolomite as substrate. Ecol. Eng. 26: 293–303.

Rastogi, G. and R.K. Sani. 2011. Molecular techniques to assess microbial community structure, function, and dynamics in the environment. Chapter 2. pp. 29–57. In: Ahmad, I.F. and J. Pitchel [eds.]. Microbes and Microbial Technology. Agricultural and Environmental Applications. DOI 10.1007/978-1-4419-7931-5_2. Springer Science+Business Media, LLC.

Reed, S.C. 1995. Natural systems for waste management and treatment. 2nd edition. McGraw Hill. New York. US.

Riesenfeld, C.S., P.D. Schloss and J. Handelsman. 2004. Metagenomics: genomic analysis of microbial communities. Annu. Rev. Genet. 38: 525–552.

Saeed, T. and G.Z. Sun. 2011. A comparative study on the removal of nutrients and organic matter in wetland reactors employing organic media. Chem. Eng. J. 171: 439–47.

Salgado, I. 2012. Remoción de contaminantes de aguas residuales por rizobacterias autóctonas con aplicación en humedales artificiales. Ph.D. Thesis, University of Havana, Havana, Cuba. In Spanish.

Salgado-Bernal, I., M.E. Carballo-Valdés, A. Martínez-Sardiñas, M. Cruz-Arias and C. Durán-Domínguez. 2012. Interacción de aislados bacterianos rizosféricos con metales de importancia ambiental. Tecnol. Ciencias Agua 3(3): 83–95. In Spanish.

Shao, Y.Y., H.Y. Pei, W.R. Hu, C.P. Chanway, P.P. Meng, Y. Ji et al. 2014. Bioaugmentation in lab scale constructed wetland microcosms for treating polluted river water and domestic wastewater in northern China. Int. Biodeter. Biodeg. 95: 151–159.

Steele, H.L. and R.W. Streit. 2005. Metagenomics: Advances in ecology and biotechnology. FEMS Microbiology Letters 247: 105–111.

Suárez, J.A., P.A. Beatón, R.F. Escalona and O. Pérez. 2012: Energy, environment and development in Cuba. Renewable and Sustainable Energy Reviews 16: 2724–2731.

SYC. Statistical Yearbook of Cuba. 2010. 1st ed. Havana, Cuba: Edited by the National Office of Statistics. ISBN: 959-7119-19-6.

Triana, J. 2017. http://oncubamagazine.com/columnas/cuando-el-agua-regrese-a-la-tierra/.

Tyagi, M., M. Da Fonseca and C. De Carvalho. 2011. Bioaugmentation and biostimulation strategies to improve the effectiveness of bioremediation processes. Biodegradation 22: 231–241.

Verlicchi, P. and E. Zambello. 2014. How efficient are constructed wetlands in removing pharmaceuticals from untreated and treated urban wastewaters? A review. Sci. Total Environ. 470-471: 1281–1306.

Vohlaa, C., M. Kõiva, J.H. Bavorb, F. Chazarencc and Ü. Mandera. 2011. Filter materials for phosphorus removal from wastewater in treatment wetlands—A review. Ecol. Eng. 37: 70–89.

Vymazal, J. 2011. Constructed wetlands for wastewater treatment: Five decades of experience. Environ. Sci. Technol. 45: 61–69.

Wei, X.Z., P. Guand and G.H. Zhang. 2014. Research progress in the new model of pollutants in water and their removing methods. Ind. Water Treat. 34(5): 8–12.

Wu, H., J. Zhang, H.H. Ngo, W. Guo, Z. Hu, S. Liang et al. 2015. A review on the sustainability of constructed wetlands for wastewater treatment: Design and operation. Bioresour. Technol. 175: 594–601.

Wu, S., P. Kuschk, H. Brix, J. Vymazal and R. Dong. 2014. Development of constructed wetlands in performance intensifications for wastewater treatment: A nitrogen and organic matter targeted review. Water. Res. 57C: 40–55.

Yan, Q., G. Feng, X. Gao, C. Sun, J. Guo and Z. Zhu. 2016. Removal of pharmaceutically active compounds (PhACs) and toxicological response of *Cyperus alternifolius* exposed to PhACs in microcosm constructed wetlands. J. Hazard. Mater. 301: 566–575.

Yan, Y. and J. Xu. 2014. Improving winter performance of constructed wetlands for wastewater treatment in northern China: A review. Wetlands 34: 243–253.

Ye, F. and Y. Li. 2009. Enhancement of nitrogen removal in towery hybrid constructed wetland to treat domestic wastewater for small rural communities. Ecol. Eng. 35: 1043–1050.

Zhai, J., H. Rahaman, J. Ji, Z. Luo, Q. Wang, H. Xiao et al. 2016. Plant uptake of diclofenac in a mesocosm-scale free water surface constructed wetland by *Cyperus alternifolius*. Water Sci. Technol. 73: 3008–3016.

Zhao, X., J. Yang, S. Bai, F. Ma and L. Wang. 2016. Microbial population dynamics in response to bioaugmentation in a constructed wetland system under 10°C. Bioresour. Technol. 205: 166–173.

9

Greenhouse Gas Emissions and Treatment Performance in Constructed Wetlands with Ornamental Plants

Case Studies in Veracruz, Mexico

María Elizabeth Hernández-Alarcón[1],* and
José Luis Marín-Muñiz[2]

INTRODUCTION

Clean water is a necessity around the world. However, population growth, excessive use of water, deterioration of freshwater due to pollution, and lack of treatment plants are situations that make it impossible to have sufficient water in good condition (Kivaisi 2001). The deficit of conventional treatments such as activated sludge systems is related to the high cost of construction, operation, and maintenance. This occurs mainly in rural communities where the scarcity of resources is common. In recent years, the use of Eco-technologies such as constructed wetlands (CWs) for treatment of wastewater has gained popularity as an economic and ecological alternative to conventional treatments (Mitsch and Gosselink 2015; Vymazal 2015). CWs are systems that mimic the environmental service of natural wetlands as a wastewater filter, but within a more controlled environment. Such systems, according to the water flow, can be divided into surface (SCW) or subsurface (SSCW) constructed wetlands.

[1] Biotechnological Management of Natural Resources Network, Institute of Ecology, Veracruz, Mexico.
[2] Department of Sustainability and Regional Development, Veracruz College, Xalapa, Veracruz, Mexico.
* Corresponding author: elizabeth.hernandez@inecol.mx

In the first type of wetlands, the water is in direct contact with the atmosphere, while in the second type, the water flows through a filter media (soil, gravel, zeolite, rocks, sand, etc.). In the SSCW, the flow direction would be in horizontal or vertical form. The three elemental components of CWs are microorganisms, filter media, and vegetation (Kadlec and Wallace 2009; Mitsch and Gosselink 2015).

In CWs, the microorganisms (bacteria, fungi, and algae) play a key role in the degradation and mineralization of nutrients and biodegradable organic compounds (Valipour and Ahn 2016). The media materials affect the hydraulic conductivity, the macrophyte growth, and provide conditions for microorganisms to attach and act as a filtration/adsorption medium for pollutants (Valipour et al. 2014; Haynes 2015). In the case of vegetation, it reduces wind speed and thus supports sedimentation and prevents resuspension, provides substrate for periphyton and bacteria, helps in the prevention of medium clogging, stores and uptakes nutrients, and in carbon-limited systems provides carbon for denitrification during biomass decomposition. Vegetation also provides a surface for microbial attachment and the roots release gas and exudates (Shelef et al. 2013; Vymazal 2013). Despite of all advantages of CW, several reactions that occur under anoxic and anaerobic conditions in this type of system produce greenhouse gases (GHG) such as methane (CH_4), carbon dioxide (CO_2), and nitrous oxide (N_2O), which are released into the atmosphere.

In Mexico, the use of CWs is recent (15 years). According to García-García et al. (2016) there are only 71 experimental studies reported, and only 5 in a consolidated phase, operated in order to solve water quality problems. It is important to highlight that some studies carried out in Mexico have used ornamental plants as vegetation, which is a feature less common in other countries (Belmont and Metcalfe 2003; Zurita et al. 2009; Zurita and Carreón-Álvarez 2015; Hernández 2016). Using ornamental plants in wetlands makes the system more aesthetic and is an option for the production of commercial flowers. Greenhouse gases contribute to the radiative forces in the atmosphere and consequently affect climate change. There is no information about the effect of ornamental plants on GHG emissions. The aim of this study is to analyze the water problems in the State of Veracruz and present three experiences of treatment performance and one experience of greenhouse gas emissions in CW with ornamental plants to mitigate water pollution, under tropical and subtropical conditions in the State of Veracruz, Mexico. CWs for one single family residence wastewater treatment, CWs for student dormitories wastewater treatment, and CWs for communities' wastewater treatment were evaluated.

The Wastewater Problems in Veracruz, Mexico

One of the objectives of the National Water Program, derived from the National Development Plan 2013–2018 for Mexico is to strengthen the water supply and access to potable water, sewerage, and sanitation services (CONAGUA 2014a). Unfortunately, the efforts have been made parsimoniously. The number of municipal wastewater treatment plants has been increasing slowly, 653 treatment plants cleaning 91 723 L s^{-1} of water in 2011 and 779 plants in 2014 treating 96 275 L s^{-1} of water were reported by the National Water Commission (CONAGUA 2011, 2014b). These treatment plants have been mostly installed in urban zones. In most states, municipal wastewater treatment plants in rural communities with less than 2,500 inhabitants have not been taken into consideration, even though more than a quarter of the Mexican population, which is over 119 million inhabitants, lives in rural communities (INEGI 2015). In these areas, the direct or indirect use of untreated wastewater for crop irrigation or the discharges in local rivers are still a common practice. It is well known that the use of untreated municipal wastewater in an agricultural setting puts human health at risk mainly due to the potential presence of excreta-related pathogens (viruses, bacteria, protozoan, and multicellular parasites) (Zurita and Carreón-Álvarez 2015).

Specifically, for the State of Veracruz, where the population is over 8 million, there are only 15 treatment plant systems in operation (CONAGUA 2014a) basically in urban areas, while, in more than 20,000 rural communities, where almost 39% of the state's inhabitants live (INEGI 2015), the installation of systems is scarce. Such places are suitable areas for the implementation of ecological and economical wastewater treatments such as CWs.

Methods

Study Sites

The three CWs sites involved in this study are located in the State of Veracruz and are described in more detail below (Fig. 1).

These ecological systems solve three distinct water problematics (domiciliary, community, and student dormitories wastewaters).

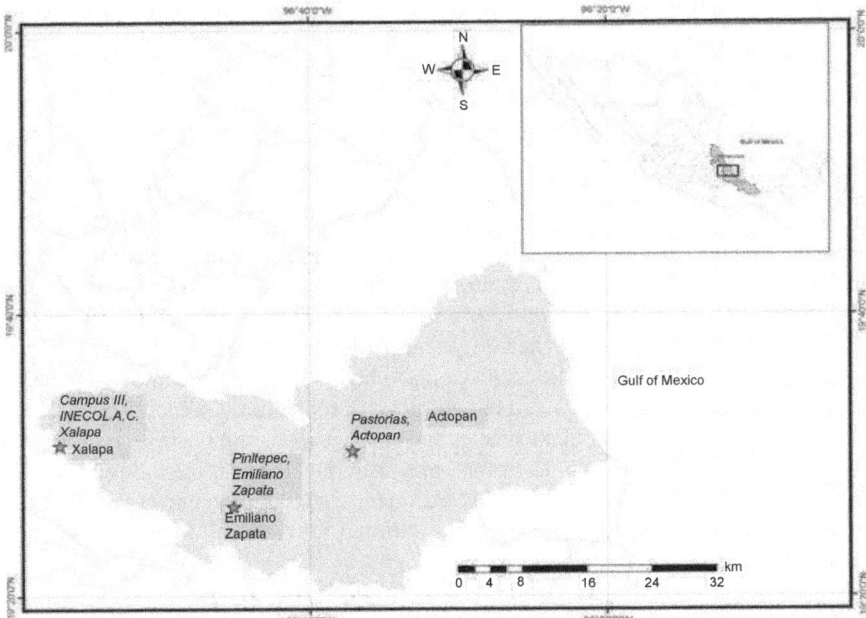

Fig. 1. Study sites locations in Veracruz, Mexico.

Actopan

Four wetland cells were constructed in the backyard of one single family residence in San José Pastorías (Municipality of Actopan), Veracruz, Mexico (−96° 57′08″N and 19°55′83″S) at 260 m AMSL (above mean sea level). Weather in the region is tropical with an annual precipitation of 947.1 mm and annual average temperature of 24.3°C (25.2°C during the study; from August to February). The dimensions of each cell were 1.5 m long, 0.23 m wide, and 0.6 m deep, and were filled with porous river rock (average porosity of 50%, water flow 10 cm below surface) collected from the local river Topiltepec in the Pastorías community. A 40 cm upper layer of rock with particle size of about 4 cm was settled over a second layer (10 cm) with particle size of about 12 cm in order to prevent clogging. The four cells were planted in August 2015 with 2 plants of *Alpinia purpurata* + 2 plants of *Hedychium coronarium* + 2 plants of *Canna hybrids* (polyculture). Two cells were kept flooded for a week with tap water, and thereafter with dilutions of wastewater for vegetation adaptation until 01 October 2015. Then they were continuously fed with gray wastewater from one single family residence without dilutions until the end of the study (25 February 2016). Water was stored in a 1100 L^{-1} tank and its flow rate was adjusted for each wetland cell to have a hydraulic residence time (HRT) of 3 days.

Emiliano Zapata

A pilot scale treatment wetland was constructed in the community of Pinoltepec Municipality, of Emiliano Zapata, Veracruz (96° 45′18″W 19° 26′45″N) at 780 m AMSL. Weather in the region is humid tropical with an annual precipitation of 2779.1 mm and annual average temperature of 25.2°C (27.2°C during the study; from May to October). The pilot treatment wetland was supplied with settled municipal wastewater. The population in Pinoltepec is over 649 inhabitants and has 168 residences. The treatment wetland consisted of two parallel concrete cells (20 m length, 1 m width, and 0.6 m depth) with subsurface water flow (0.10 m diameter volcanic gravel, 0.5 m depth, water flow 10 cm below surface). Both cells were planted with the same array of vegetation and received the same flow of wastewater. Vegetation in the cells were distributed, from the inflow to the outflow, as follows: 0–5 m *Typha* sp., 5–12 m *Cyperus papyrus*, 12–13 m *Lillium* sp., 13–17 m *Zantedeschia aethiopica*, 17–18 m *Anthurium andraeanum*, 18–20 m *Hedychium coronarium*. This arrangement was established to enhance flower production of *Zantedeschia aethiopica* since previous studies showed that a high nutrient load could stimulate the growth of *Zantedeschia aethiopica* but not flower production (Galindo-Zetina 2012). On the other hand, wetland plant species such as *Typha* sp. and *Cyperus papyrus*, are tolerant to high nutrient loads. Therefore, they were planted near the inflow where the nutrient concentration was higher. This treatment wetland was supplied with settled municipal wastewater and flow rates were adjusted for each cell to have a hydraulic residence time, HRT, of 40 hours.

Xalapa

The pilot scale CW was constructed on Campus III of the Institute of Ecology in Xalapa, the capital of the State of Veracruz (97° 01″ W 19° 33.5″N) at 1560 m AMSL. The weather in the region is humid subtropical with an annual precipitation of 1509.1 mm and annual average temperature of 18°C (19°C during the study; from June to November). The treatment wetland consisted of two parallel concrete cells (12 m length, 1.2 m width, and 0.6 m depth) with subsurface water flow. They used volcanic gravel (0.15 m diameter) as substrate, with 0.5 m depth, water flow 10 cm below surface. Both cells were planted with the same array of vegetation and the wastewater flow was adjusted for each cell to have the same HRT of 7 days. Vegetation in the cells was distributed from the inflow to the outflow as follows: 0–5 m *Cyperus alternifolius*, 5–7 m *Lillium* sp., 7–8 m *Hedychium coronarium*, 8–11 m *Zantedeschia aethiopica*, and 11–12 m *Anthurium andraeanum*. This treatment wetland was supplied with settled domestic wastewater from the visitors' dormitories of Campus III facilities at the Institute of Ecology. Wastewater flow rates were adjusted for each cell to have a HRT of 6 days.

Water Sampling

Water samples (200 mL) were taken from influent and effluent of each cell every other week, from June to October 2013 in Emiliano Zapata, from June to November 2016 in Xalapa, and from October 2015 to February 2016 in Actopan.

Analytical Methods

The samples were analyzed for chemical oxygen demand (COD), ammonia nitrogen ($N-NH_3$), nitrate ($N-NO_3^{-1}$), orthophosphate ($P-PO_4^{3-}$), total phosphorus (TP), total nitrogen (TN), total organic carbon (TOC), total solids (TS), and total suspended solids (TSS). COD was measured using the oxidation of $K_2Cr_2O_7$ micro-method (APHA 1998), $N-NH_3$ was analyzed by the Nessler method, and total nitrogen was analyzed by the Kjeldahl method (APHA 1998). Phosphate phosphorus, $P-PO_4$, was quantified using the ascorbic acid method by Sandell and Onsh (1978). Total phosphorus was analyzed by persulfate digestion followed by the ascorbic acid method. All these colorimetric analyses were performed using a UV-Vis Jenway-Genova spectrophotometer (Jenway-Essex, UK). A Total Organic Carbon Analyzer (Torch, Teledyne Tekmar) was used to analyze TOC in water. TSS and TS were analyzed by gravimetric methods (APHA 1998).

For the Xalapa and Actopan water samples, only $P-PO_4$, $S-SO_4$, and $N-NO_3$ were quantified using ion chromatography equipment Dionex ICS-1100 (Dionex, Sunnyvale CA, US), consisting of an isocratic pump, an anion pre-column (Dionex AG11, 2 mm), an anion separator column (Dionex Ion Pac AS23, 4 mm) coupled with an anion self-regenerating suppressor (ASRS-300, 2 mm), and a conductivity detector. High-purity reagents were used throughout together with Milli-Q purified water, which was deionized and then filtered through a 0.2 mm Whatman membrane. A mixture solution of 4.5 mM Na_2CO_3 and 0.8 mM $NaHCO_3$ (Dionex) was used as eluent with a flow rate of 0.25 ml min^{-1}. Nutrient standard solutions were prepared using Dionex Seven Anion Standard Solution II. Wastewater samples were filtered with a 0.2 mm Whatman membrane, and then placed in vials with filter caps to be injected (25 μL) by Dionex AS-DV into the Ion Chromatograph.

Removal efficiencies (RE) were calculated with the following formula: RE = $[(C_i - C_e)/C_i]$ x 100%, where C_i is the influent pollutant concentration and C_e is the pollutant concentration in the effluent.

Plant Growth

Individual plant height was measured every month using a measuring tape. The number of flowers was also registered during the study. Only the data for *H. coronarium* is reported in the three sites. In particular, the *C. hybrids* growth is reported for Actopan, and the growth of *Z. aethiopica* plants reported in Emiliano Zapata and Xalapa.

Greenhouse Gases Emissions

Methane and nitrous oxide emissions were measured in the pilot scale treatment wetland in Pinoltepec, Emiliano Zapata, Veracruz, using the closed chamber method (Altor and Mitsch 2006; Hernandez and Mitsch 2006). The chambers consisted of permanent bases made with PVC pipe (0.25 m height × 0.10 m ID) and removable covers (0.10 m height × 0.30 m ID). Chambers were placed in both cells, at 3, 9 and 15 m from inflow to outflow in *Typha* sp., *Cyperus papyrus*, and *Zantedeschia aethiopica* vegetation zones, respectively.

When the gas fluxes were measured, the cover was placed on top and sealed to the base using water in the collar. Internal gas samples were collected from the chamber every 5 minutes, during a 40-minute period by using a syringe fitted with a Luer stopcock. Gas samples were stored in 10 ml vials previously vacuum evacuated and kept at 4°C until their analysis (which was within three days after they were collected). The gas samples were analyzed for their CH_4 and N_2O concentrations simultaneously using a gas chromatograph (Perkin Elmer Clarus 500) equipped with a 2 m Porapak Q 80/100 column, a flame ionization detector (FID) and Electron Capture Detector (ECD). Operation conditions were: 13 mL min^{-1} of nitrogen as the carrier gas, temperatures of the FID, ECD, injector, and oven were set at 150, 360, 95, and 40°C, respectively.

Results

Plant Growth

For the pilot scale CWs, two of the most abundant ornamental plants in the polycultures were chosen to show their growth (Fig. 2). In Emiliano Zapata, in six months (May to October), *Z. aethiopica* plants increased their height from 0.20 to 0.57 m and *H. coronarium* from 0.25 to 0.78 m. In Xalapa, in six months (from June to November), *Z. aethiopica* plants increased from 0.13 to 0.30 m and *H. coronarium* from 0.18 to 0.68 m. In Actopan, the *Hedychium coronarium* growth was very low, increasing its height from 0.3 to 0.45 m in seven months (from August to February), while *C. hybrid* grew faster, increasing from 0.15 to 0.6 m in five months (from August to December).

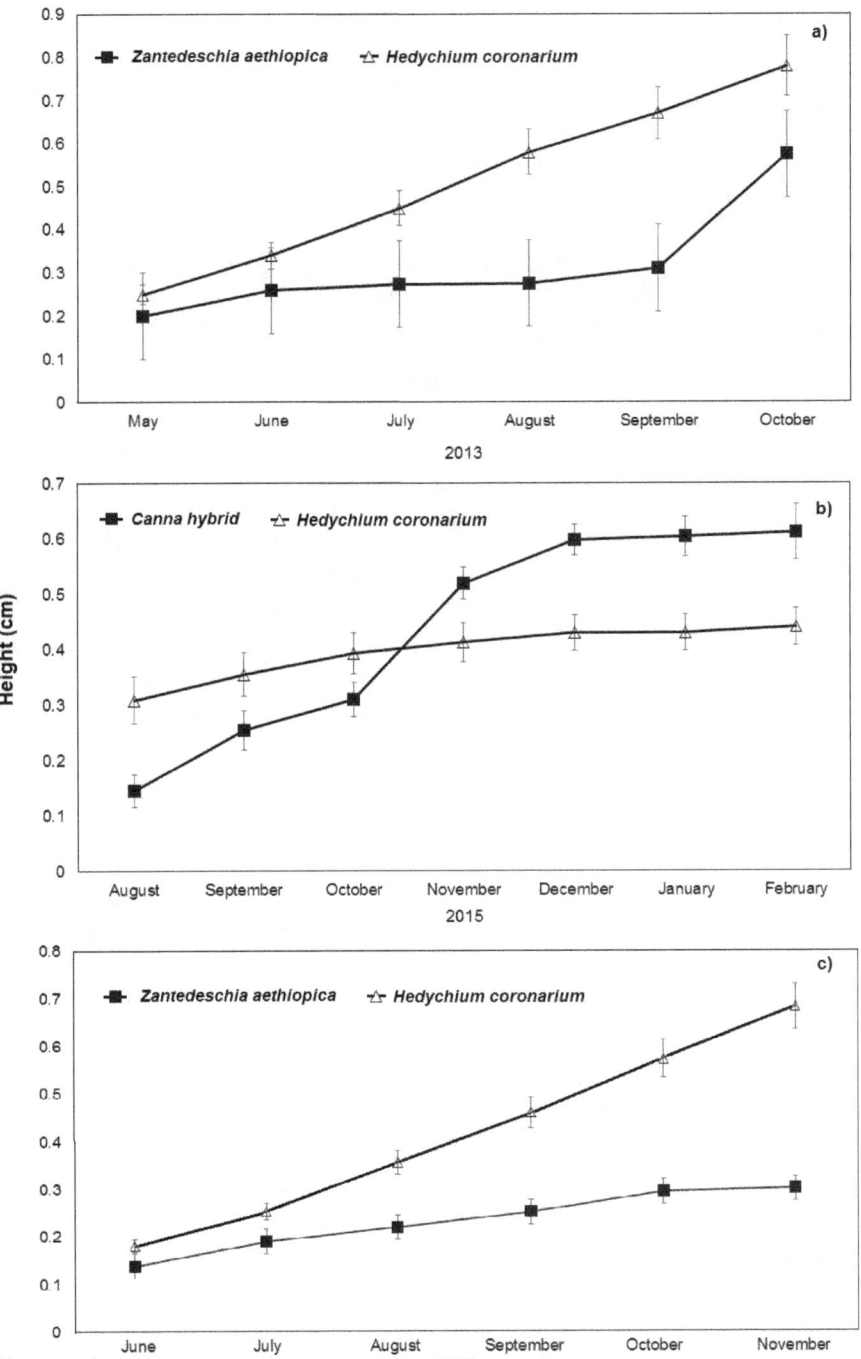

Fig. 2. Height increase of three ornamental plants in CWs used to treat domestic wastewater in (a) Emiliano Zapata, Ver. (b) Actopan, Ver., and (c) Xalapa, Ver., Mexico.

Flower Production

In Emiliano Zapata wetland, *Z. aethiopica* produced few flowers, while in Xalapa, flowers of this species were abundant (Fig. 3 and Table 1). *Canna hybrid* in Actopan also produced abundant flowers. In both Xalapa and Emiliano Zapata wetlands, *H. coronarium* produced a good number of flowers while in Actopan, this plant did not produce any.

Fig. 3. *Hedychium coronarium* (a, l), *Zantedeschia aethiopica* (b, c, h) and *Canna hybrid* flowers (d, i, j, k) produced at the constructed wetlands of Emiliano Zapata (e), Actopan (f), and Xalapa (g).

Table 1. Flower production by different species of ornamental plants in CWs treating wastewater in three different locations of Veracruz, Mexico.

Plant	Flowering months	Location	Numbers of flowers observed	Total area of the wetlands planted with the species (m²)
Zantedeschia aethiopica	August-October	Emiliano Zapata	4	10
	October–November	Xalapa	27	6
Hedychium coronarium	September–October	Emiliano Zapata	15	4
	June–November	Xalapa Actopan	0	2
Canna hybrid	September–November	Actopan	9	0.34

Treatment Performance

Actopan

The concentrations of N-NO$_3$ were very small (0 to 0.7 mg L^{-1}) in the CWs with polycultures and similar between the influent and the effluent. Meanwhile, removal efficiency for P-PO$_4$, N-NH$_3$, and S-SO$_4$ were 80.6, 65.2, and 49.5%, respectively. In addition, TS was removed by 31.6% (Table 2).

Table 2. Average pollutant concentrations ± standard error (SE) at the inflow of a domiciliary CW planted with polyculture of ornamental plants, and the removal percentage achieved during the first five months of operation (October 2015–February 2016) with a HRT of three days in Actopan, Veracruz, Mexico.

Parameter	Inflow Concentration (mg L^{-1})	Removal (%)
Ammonia nitrogen N-NH$_3$	23 ± 11	65 ± 11
Phosphates P-PO$_4$	5 ± 1	81 ± 17
Nitrates N-NO$_3$	0.11 ± 0.07	NO
Sulfates S-SO$_4$	39 ± 10	50 ± 10
Total solids TS	1197 ± 167	32 ± 5

NO: Not observed

Xalapa

Nutrients such as N-NH$_3$ and P-PO$_4$, were removed with efficiencies of 72 and 61%, respectively, in the CWs with polycultures of macrophytes and ornamental plants (Table 3). Sulfates were removed with high efficiency (81%), while organic carbon was less removed (50%).

Table 3. Average concentrations of pollutants ± SE at the inflow of a pilot scale CW planted with polyculture of macrophytes and ornamental plants and the removal percentage achieved during the first five months of operation (June–November 2016) with a HRT of six days in Xalapa, Veracruz.

Parameter	Inflow concentration (mg L^{-1})	Removal (%)
Ammonia-Nitrogen N-NH$_4$	31 ± 5	72 ± 5
Phosphates P-PO$_4$	6 ± 1	61 ± 7
Sulfates S-SO$_4$	19 ± 4	81 ± 4
Total Organic Carbon TOC	55 ± 14	51 ± 8

Emiliano Zapata

The best removal efficiencies (higher than 50%) observed in Emiliano Zapata CW planted with a polyculture of macrophytes and ornamental plants were obtained for COD, TSS, and TN. However, for other parameters such as TP, TS, P-PO$_4$, and N-NH$_4$ removal efficiencies were less than 50% (Table 4).

Greenhouse Gases Emissions

No nitrous oxide emissions were observed during the study period in this pilot scale CW. However, CH$_4$ emissions were detected in the CW at Emiliano Zapata, the highest emissions being observed in August and September and, in particular, more in the zones with *Zantedeschia* than in the zones with native plants *Typha* sp. and *Cyperus* sp. (Fig. 4).

Table 4. Average concentrations of pollutants ± SE at the inflow and outflow of a pilot scale CW planted with polyculture of macrophytes and ornamental plants and the removal percentage achieved during the first five months of operation (June–October 2013) with a HRT of 40 hr in Pinoltepec, Emiliano Zapata, Veracruz, Mexico.

Parameter	Inflow concentration (mg L^{-1})	% Removal
Total Nitrogen TN	119 ± 20	47 ± 12
Total Phosphorus TP	12 ± 3	33 ± 6
O-phosphates O-P	8 ± 1	25 ± 2
Ammonia Nitrogen N-NH$_3$	33 ± 4	27 ± 3
Chemical Oxygen Demand COD	378 ± 31	67 ± 14
Total Solids TS	720 ± 52	34 ± 9

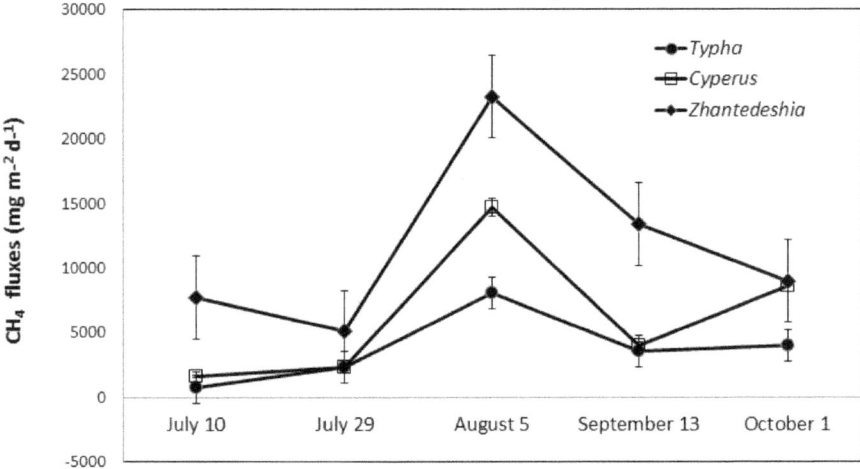

Fig. 4. Methane emissions in zones planted with native and ornamental plants in a CW treating municipal wastewater in Emiliano Zapata, Veracruz, Mexico.

Discussion

The use of ornamental plants in CWs has been described in the last decade in tropical countries (Belmont and Metcalfe 2003; Zhang et al. 2007; Konnerup et al. 2009; Calheiros et al. 2015; Zurita et al. 2006, 2009). However, few of them report flower production. In this study, the feasibility of flower production in three independent CWs was documented. The CWs operated with different nutrient loading rates and climatic conditions.

Z. *aethiopica* grew well in Emiliano Zapata and Xalapa. However, flower production was better in Xalapa. This may be explained by the fact that the Emiliano Zapata CW had higher nutrient loading than the Xalapa CW. Galindo-Zetina (2012) reported that high nutrient loading stimulates Z. *aethiopica* growth but without flower production whereas with low nutrient loading, flower production was observed. Other important factors related with the growth of plants are light and temperature. Casierra-Posada et al. (2012) reported that the flower quality in Z. aethiopica was affected by the solar radiation with better results when the plants grew under shadow. In this study, in all sites the plants have this requirement. On the other hand, Zhang et al. (2011) and Casierra-Posada et al. (2012) mentioned that flower production of *Z. aethiopica* is optimum in temperatures ranges between 15 and 26°C, temperatures observed in Xalapa during the study. Temperatures near to 30°C causes stress in the plant and lower flower production (Zhang et al. 2011), in Emiliano Zapata the temperature were higher than 26°C, and this might explain the lower flower production in this site.

   H. coronarium produced more flowers per m² in Emiliano Zapata than Xalapa, and the CW in Actopan did not produced any flowers. This may suggest that this species prefers high nutrient loading. The suitable range of temperature for the flowering production of this specie is between 20 and 25°C, as observed by the temperatures registered in Emiliano Zapata and Actopan; however in Actopan, the nutrient loading was low. Some authors (Argollo and Rodrigues 2007; Chandra and Anju 2015) have described that the production of flowers of *H. coronarium* is common in the spring and the early summer, without being demanding in light requirements, these periods of flowering agreed with the observed in this study in Emiliano Zapata.

   On the other hand, *C. hybrid* grew well in Actopan and produced big and healthy flowers. This can be related to the low nutrient loading in the gray water, in light of what Espinosa et al. (2009) has described that the *Cannaceae* family is undemanding in requirements of nutrients and grows regardless of whether it is shaded or exposed to the sun.

   Among the three CWs evaluated, the one in Emiliano Zapata resulted in lower removal efficiencies, which may be explained by the short HRT (less than 2 days) compared with 3 and 6 days in Actopan and Xalapa, respectively. In Emiliano Zapata, the HRT was short because the CW needed to treat large volumes of municipal wastewater in a relatively small area. The outcome of this strategy was a trade off with pollutant removal efficiencies. After the evaluation, we recommended increasing HRT to improve nitrogen and phosphorus removal efficiencies. On the other hand, the CW in Xalapa and Actopan showed similar removal efficiencies for phosphates but in Xalapa, better ammonia nitrogen removals were achieved. This may be also explained by the longer HRT used in the CW located in Xalapa.

   The survival conditions and the removal of pollutants observed for ornamental plants is an important feature when selecting constructed wetland plants. In this case, *H. coronarium*, *Z. aethiopica* and *C. hybrid* used in these three independent CWs resulted beneficial, because they add better aesthetic appearance to the system by using flowers than typical plants used in CWs such as *Typha* or *Scirpus* spp. (Vymazal 2011, 2013). Furthermore, CWs were efficient in the removal of pollutants, attaining removals that were mostly greater than 50%. Some recent studies in experimental CWs using *H. coronarium* have also reported the adaptability and evaluation of removal contaminants in an application to treat swine wastewater (Panazzolo et al. 2013), observing a reduction of nitrogen and phosphorous compounds that was lower than 40%. Similarly, Garzón et al. (2016) evaluated CWs with *Z. aethiopica* and *H. coronarium* for domestic wastewater treatment. They observed removals of only 15 and 25% for P-PO$_4$ and COD when they used 0.2 m³ d⁻¹ of flow. Moreover, Zurita and White (2014) studied a CW system treating gray water with *Z. aethiopica* and *Canna indica* and detected lower that 23% reduction in total nitrogen and phosphorus from influent to effluent.

The better pollutant removals found in this work in comparison with the studies described above, show the relevance of using ornamental plants for removal of pollutants in CWs in the local climate conditions of State of Veracruz. Other studies using the same species have been conducted with synthetic wastewater (Li et al. 2014; Hu et al. 2016). However, in this chapter, the treatment performance of three CWs with ornamental plants solving a real water contamination problem were described. With the generated knowledge, the replication of this sustainable alternative in other households and communities is suggested.

Regarding GHG emissions, the fact that N_2O emission were not observed might be related to the low ammonia removal efficiencies found, which might indicate that poor nitrification and denitrification activity occurred during the study period.

The higher CH_4 emissions in the *Zantedeschia aethiopica* zones compared with the zones near the inflow planted with *Typha* sp. and *Cyperus gigantus* was contrary to what was expected, because in the inflow zones (planted with *Typha* and *Cyperus*), COD concentrations were higher, and near the outflow, the COD concentrations in water were lower due to the treatment performed on the wetland.

High methane emissions in the zone with *Zantedeschia* might be related to differences in the radial oxygen loss, ROL, between the plant species (Mei et al. 2014) that causes differences in the methane production and consumption within the CW. Thus, more studies on this subject are needed.

Comparing the methane emissions observed in the Emiliano Zapata CW wetland with data from literature (Table 5), the values observed in zones with *Typha* sp. were higher that those reported for this species in temperate countries like Sweden and

Table 5. Methane emissions from CW with different plants in several countries.

Country	Plant	Type of wastewater	CH_4 emissions mg m^{-2} d^{-1}	Reference
Sweden	*Typha latifolia*	Secondary effluent	37 to 1739	Johansson et al. (2004)
Japan	*Zizania latifolia*	Synthetic wastewater	0 to 1680	Inamori et al. (2007)
	Phragmites australis	Synthetic wastewater	0 to 1080	Inamori et al. (2007)
Estonia	*Phragmites australis*	Raw municipal Wastewater	−0.20 to 644	Mander et al. (2008)
Estonia	*Typha latifolia*	Raw municipal Wastewater	−0.16 to 996	Mander et al. (2008)
Estonia	*Phragmites australis and Scirpus sylvaticus*	Wastewater from a hospital	−0.22 to 8924	Mander et al. (2008)
Check Republic	*Phragmites australis*	Sewage and stormwater runoff	0 to 2232	Picek et al. (2007)
Spain	*Phragmites australis*	Sludge from wastewater plant treatment	700 to 14092	Uggetti et al. (2012)
Colombia	*Phragmites australis*	Pre-treated domestic wastewater	−5000 to 11000	Silva-Vinasco and Velarde Solis (2011)
	Heliconia sp.		−4000 to 15000	
Mexico	*Typha* sp.	Settled domestic waste water	800 to 5000	This study
	Zantedeschia aethiopica		5000 to 23000	This study
	Cyperus payrus		1000 to 12000	This study

Estonia, but they are in the range observed for other species in countries like Spain and Colombia.

Conclusions

Flower production and pollutant removal was possible in these CWs treating domestic wastewater at three different scales in Veracruz, Mexico. Flower production for *Z. aethiopica* and *C. hybrids* was favored by low nutrient loading rates, while for *H. coronarium* the opposite was observed: this species produced flowers with high nutrient loading rates. The pollutant removal performance was dependent on HRT, with higher removal efficiencies with longer HRT. Considering these findings, we argued for the viability of the use of CWs for mitigating water pollution, with the advantage of flower production that can be commercialized when ornamental plant production is high, especially in the case of CWs treating wastewater volume for communities with more than 500 people. Here it is important to have caution with the hazardous pollutants in wastewater such as heavy metals or toxic organic compounds that can be bioaccumulated in plants and flowers. However, in rural communities, municipal wastewater is mainly sewage, consisting of biodegradable organic matter that will produce safer flowers. Having domiciliary CWs with flowers in the backyard can make the system more attractive and the space more comfortable. These features may allow this type of system to be used not just as a treatment system, but instead as a large outdoor planter in the gardens of the houses or as leisure areas in rural communities. It would be important to study more species that can produce flowers with different colors, in order to make the systems more colorful. Considering the results, we recommend increasing the construction of CWs to mitigate water pollution problems and to assist with public health in rural areas of Mexico, where conventional wastewater treatment plants are scarce. It is also important to monitor greenhouse gas emissions in CW with ornamental flowers to establish if there is an effect of this type of plant on methane and nitrous oxide emissions.

Acknowledgements

The authors acknowledge Dr. Zhang Li for her valuable comments to improve this manuscript.

References Cited

Altor, A.E. and W.J. Mitsch. 2008. Pulsing hydrology, methane emissions, and carbon dioxide fluxes in created marshes: A 2-year ecosystem study. Wetlands 28: 423–438.

APHA. 1998. Standard Methods for the Examination of Water and Wastewater. APHA-AWWA-WPCF, American Public Health Association, American Water Works Association, Water Pollution Control Federation. Washington, D.C. US.

Argollo, J. and M. Rodrigues. 2007. Biología floral de *Hedychium coronarium* Koen. (Zingiberaceae). Rev. Brasileira Horticultura Ornamental 13(1): 21–30.

Belmont, M.A. and C.D. Metcalfe. 2003. Feasibility of using ornamental plants (*Zantedeschia aethiopica*) in subsurface flow treatment wetlands to remove nitrogen, chemical oxygen demand and nonylphenol ethoxylate surfactants—a laboratory-scale study. Ecol. Eng. 21: 233–247.

Calheiros, C., V. Bessa., R. Mesquita, H. Brix, A. Rangel and P. Castro. 2015. Constructed wetlands with a polyculture of ornamental plants for wastewater treatment at a rural tourism facility. Ecol. Eng. 79: 1–7.

Casierra-Posada, F., P.J. Nieto and C. Ulrichs. 2012. Growth, production and flower quality in calla lily (*Zantedeschia aethiopica* L.K. Spreng) exposed to different light quality. Agronomía Colombiana 15(1): 97–105.

Chandra, T. and G. Anju. 2015. A comprehensive review on *Hedychium coronarium* J. Koenig. (Dolanchampa/Kapurkachri). Int. J. Res. Ayurveda Pharm. 6(1): 98–100.

CONAGUA. 2011. Comisión Nacional del Agua. Situación del Subsector Agua Potable, Drenaje y Saneamiento. Secretaría de Medio Ambiente y Recursos Naturales. México D. F. 82 p. In Spanish.

CONAGUA. 2014a. Comisión Nacional del Agua. Programa Nacional Hídrico 2014–2018. Comisión Nacional del Agua. México D. F. 139 p. In Spanish.

CONAGUA. 2014b. Comisión Nacional del Agua. Inventario Nacional de plantas municipales de potabilización y de tratamiento de aguas residuales en operación. Diciembre 2014. Secretaría de Medio Ambiente y Recursos Naturales. México D. F. 308 p. In Spanish.

Espinosa, A., M. Beltrán, M. Colinas, J. Mejía, M. Rodríguez and A. Ubarkzyk. 2009. Catálogo nacional de especies y variedades comerciales de plantas y flores producidas en México y Universidad Autónoma de Chapingo. México. In Spanish.

Galindo-Zetina, M. 2012. Emisión de gases invernadero, remoción de contaminantes y crecimiento de plantas ornamentales en humedales construidos para el tratamiento de aguas residuales. Ph.D. Thesis. Facultad de Biología. Instituto Tecnológico de Zacpoaxtla, Puebla. Mexico. In Spanish.

García-García, P., L. Ruelas-Monjardín and J.L. Marín-Muñiz. 2016. Constructed wetlands: A solution to water quality issues in Mexico? Water Policy 18(3): 654–669.

Garzón, M., J. González and R. García. 2016. Evaluación de un sistema de tratamiento doméstico para reúso de agua residual. Rev. Int. Contam. Amb. 32(2): 199–211. In Spanish.

Haynes, R.J. 2015. Use of industrial wastes as media in constructed wetlands and filter beds—prospects for removal of phosphate and metals from wastewater streams. Crit. Rev. Environ. Sci. Technol. 45(10): 1041–1103.

Hernández, M.E. and W.J. Mitsch. 2006. Influence of hydrologic pulses and vegetation on nitrous oxide emissions from created riparian marshes in Midwestern US. Wetlands 26(3): 862–877.

Hernández, M.E. 2016. Humedales ornamentales con participación comunitaria para el saneamiento de aguas municipales en México. Revista Internacional de Desarrollo Regional Sustentable 1(2): 01–12.

Hu, Y., F. He, L. Ma, Y. Zhang and Z. Wu. 2016. Microbial nitrogen removal pathways in integrated vertical-flow constructed wetland systems. Bioresour. Technol. 207: 339–345.

Inamori, R., P. Gui, P. Dass, M. Matsumura, K.Q. Xu, T. Kondo et al. 2007. Investigating CH_4 and N_2O emissions from eco-engineering wastewater treatment processes using constructed wetland microcosms. Process Biochemistry 42: 363–373.

INEGI. 2015. Instituto Nacional de Estadística y Geografía. Crecimiento poblacional México. www.inegi.org.mx. In Spanish.

Johansson, A.E., A.M. Gustavsson, M.G. Öquist and B.H. Svensson. 2004. Methane emissions from a constructed wetland treating wastewater, seasonal and spatial distribution and dependence on edaphic factors. Water Research 38: 3960–3970.

Kadlec, R.H. and S.D. Wallace. 2009. Treatment Wetlands. 2nd ed. Taylor & Francis Group. Boca Raton, Florida.

Kivaisi, A. 2001. The potential for constructed wetlands for wastewater treatment and reuse in developing countries: a review. Ecol. Eng. 16: 545–560.

Konnerup, D., T. Koottatep and H. Brix. 2009. Treatment of domestic wastewater in tropical subsurface flow constructed wetlands planted with Canna and Heliconia. Ecol. Eng. 35(2): 248–257.

Li, J., X. Liu, Z. Yu, X. Yi, Y. Ju, J. Huang et al. 2014. Removal of fluoride and arsenic by pilot vertical-flow constructed wetlands using soil and coal cinder as substrate. Water Sci. Technol. 70(4): 620–626.

Mander, U., K. Lõhmus, S. Teiter, T. Mauring, K. Nurk and J. Augustin. 2008. Gaseous fluxes in the nitrogen and carbon budgets of subsurface flow constructed wetlands. Science of the Total Environment 404: 343–353.

Mei, X., Y. Yang, N.F. Tam, Y.W. Wang and L. Li. 2014. Roles of root porosity, radial oxygen loss, Fe plaque formation on nutrient removal and tolerance of wetland plants to domestic wastewater. Water Research 50: 147–159.

Mitsch, W.J. and J.G. Gosselink. 2015. Wetlands, 4th edn. John Wiley & Sons, New York.

Panazzolo, A., A. Carraro and A. Teixeira. 2013. Effect of cultivated species and retention time on the performance of constructed wetlands. Environ. Technol. 35(8): 961–965.

Picek, T., H. Cızkova and J. Duseka. 2007. Greenhouse gas emissions from a constructed wetland— Plants as important sources of carbon. Ecological Engineering 31: 98–106.

Sandell, E. and H. Onish. 1978. Photometric determination of traces of metals. John Wiley and Sons Inc. US.

Shelef, O., A. Gross and S. Rachmilevitch. 2013. Role of plants in a constructed wetland: Current and new perspectives. Water 5: 405–419.

Silva-Vinasco, J.P. and A. Valverde-Solís. 2011. Assessment of greenhouse effect gases in sub-superficial flow constructed wetlands. Ing. Univ. Bogotá (Colombia) 15(2): 519–553.

Uggetti, E., J. Garcia, S.E. Lind, P. Martikainen and I. Ferrer. 2012. Quantification of greenhouse gas emissions from sludge treatment wetland. Water Research 46: 1744–1762.

Valipour, A., N. Hamnabard, K.S. Woo and Y.H. Ahn. 2014. Performance of high-rate constructed phytoremediation process with attached growth for domestic wastewater treatment: effect of high TDS and Cu. Environ. Manag. 145: 1–8.

Valipour, A. and Y. Ahn. 2016. Constructed wetlands as sustainable ecotechnologies in descentralization practices: A review. Environ. Sci. Pollut. Res. 23: 180–197.

Vymazal, J. 2011. Plants used in constructed wetlands with horizontal subsurface flow: A review. Hydrobiologia 20: 133–156.

Vymazal, J. 2013. Emergent plant used in free water surface constructed wetlands: A review. Ecol. Eng. 61P: 582–592.

Vymazal, J. 2015. The role of natural and constructed wetlands in nutrient cycling and retention on the landscape. Springer Cham Heidelberg New York Dordrecht London.

Zhang, X.B., P. Liu, Y.S. Yang and W.R. Chen. 2007. Phytoremediation of urban wastewater by model wetlands with ornamental hydrophytes. J. Environ. Sci. (China) 19(8): 902–909.

Zhang, X., Q. Wu, X. Li, S. Zheng, S. Wang, L. Guo et al. 2011. Haploid plant production in *Zantedeschia aethiopica* 'Hong Gan' using anther culture. Scientia Horticulturae 129: 3335–342.

Zurita, F., J. De Anda and M. Belmont. 2006. Performance of laboratory-scale wetlands planted with tropical ornamental plants to treat domestic wastewater. Water Qual. Res. J. Canada 41(4): 410–417.

Zurita, F., J. De Anda and M.A. Belmont. 2009. Treatment of domestic wastewater and production of commercial flowers in vertical and horizontal subsurface-flow constructed wetlands. Ecol. Eng. 35(5): 861–869.

Zurita, F. and J. White. 2014. Comparative study of three two-stage hybrid ecological wastewater treatment systems for producing high nutrient, reclaimed water for irrigation reuse in developing countries. Water 6: 213–228.

Zurita, F. and A. Carreón-Álvarez. 2015. Performance of three pilot-scale hybrid constructed wetlands for total coliforms and *Escherichia coli* removal from primary effluent—a 2-year study in subtropical climate. J. Water Health 13.2: 446–458.

10

Treatment of Wastewater from Livestock Activities with Artificial or Constructed Wetland

José Marrugo-Negrete,* Juan Figueroa-Sánchez,
Iván Urango-Cárdenas and *Germán Enamorado-Montes*

INTRODUCTION

In Colombia, small-scale livestock farming occupies an important area for the production of milk and meat. By 2012, about 400,000 small farmers depended solely on dairy activity. However, there exists a predominance of informality in around 44% of these farmers for the execution of the different field tasks and even in the productive stages (Jaramillo and Areiza 2013). It is precisely this informality that poses a greater risk to the deterioration of the environment, in the sense that it can lead to the indiscriminate use of agrochemicals, the use of which is given great attention by the fact that they may pose a potential risk to human health, due to its possible presence in the finished wastewater (Márquez 2008).

The most recent National Water Study in Colombia established that the livestock sector alone uses 8.5% of the total water used in this country. However, it is timely to indicate the lack of information on the impact to this resource, and special emphasis is placed on the fact that there isn't a methodology to collect all the data related to the use and disposal of agrochemicals (IDEAM 2015). The state of disinformation about the degree of contamination caused by the livestock sector is of critical concern.

University of Cordoba, Faculty of Basic Sciences, Water, Applied and Environmental Chemistry Group, Montería, Colombia.
Emails: jgfplayer@hotmail.com; ivaild@hotmail.com; genamoradomontes@correo.unicodoba.edu.co
* Corresponding author: joseluismarrugo@gmail.com

Steinfeld et al. (2006) summarized some of the main impacts of the livestock sector on water resources. They divided the use of water according to their destination within the supply chain in: (1) consumption water and (2) water for maintenance services. However, the water from both these activities is returned to the environment with less quality than the original water quality. The water destined for consumption is what the animal needs to survive, and it is out of the scope of the present work. Similarly, this consumption water negatively impacts the environment through the different pathways of the animal's excretion. Water for maintenance service refers to the water used for cattle bath and the cleaning of stables. This is the type of water we will focus on this chapter, and it generally returns to the environment directly in important proportions depending on the type of livestock: intensive, extensive, or small scale. Also, two mechanisms of contamination are usually distinguished: from a point source (e.g., after washing the stables where manure and urine are present) and from a non-point source (that occurs from runoff from fields grazing where cattle usually defecate). Regardless of the pollution mechanism, livestock wastewater contains high nutrient loads from livestock excreta, which can lead to eutrophication of aquatic ecosystems.

Different authors have evaluated the quality of wastewater generated in productive farms. Li et al. (2012) reported total organic carbon levels between 279 and 1,505 mg L^{-1} and total phosphorus (TP) between 2.8 and 13.4 mg L^{-1}. Zhang et al. (2012) found levels of chemical organic demands (COD) around 1,103 mg L^{-1} and TP values of 14.4 mg L^{-1}. Meanwhile, Carvalho et al. (2013) reported COD values of 1,042 mg L^{-1} with pH between 7.3 and 8.1. Lee and Shoda (2008) observed COD levels between 5,000 and 5,700 mg L^{-1} and pH between 8.4 and 8.7. Finally, Othman et al. (2013) analyzed concentrations of 3,600 mg L^{-1} for COD and 380 mg L^{-1} for TP.

The artificial or constructed wetlands have been widely used since their first large-scale application in the late 1960s (Liu et al. 2015). Taking into account the low costs, the easy operation and maintenance, and their environment friendly methods, the constructed wetlands could be considered as a green treatment technology; when simulating the natural wetland processes these systems have a great versatility to treat various types of wastewater (Wu et al. 2015).

In the present work, horizontal subsurface flow wetlands were planted with different macrophytes commonly grown in areas close to livestock farms, operated at different flow regimes, in order to improve the quality of livestock wastewater and to prevent eutrophication of ecosystems where this type of wastewater is commonly dumped without any treatment.

Materials and Methods

Description of the Livestock Farm

The livestock farm was located in the rural area of Cordoba Department (Colombia), near the facilities of the University of Cordoba, where pilot systems of artificial or constructed wetland were installed. The amount of cattle at the time of the study was less than fifty. Thus, the type of livestock farming could be considered small-scale. According to information given by the foreman, the livestock of this farm has a dual purpose: farm production is focused on milk or meat according to market prices (Niño

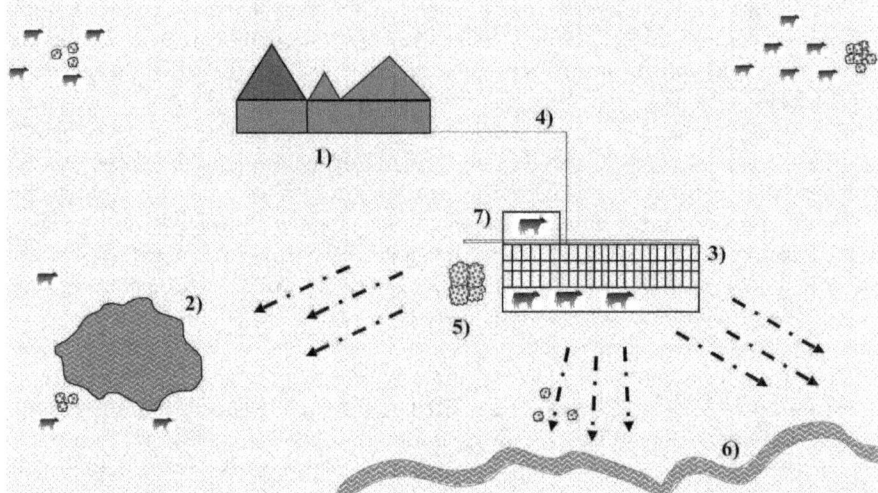

Fig. 1. Components present on a small-scale livestock farm in the department of Córdoba, Colombia: 1. Foreman family house; 2. Pool; 3. Cow stables; 4. Tap water source; 5. Manure waste; 6. Surface water source; 7. channel waste.

and Alarcón 2015). It has one or two cow stables where milking, bathing, vaccination, and weighing activities are carried out. It has disrupted the supply of drinking water from a nearby house where the foreman's family lives. It also has a pool to store rainwater or surface water from a source that crosses the property, which can be commonly used as a drinking water supply for the family and livestock. The wastewater from cattle baths and manure waste can easily reach this pool. The components described above are characteristic of many subsistence and small-scale livestock production systems in the Cordoba Región. An outline of these components is illustrated in Fig. 1.

Sampling Collection

The wastewater to be treated in the wetlands system was obtained from a cattle farm, whose strategy for the removal of ecto-parasites consisted in the bath of cattle using a knapsack sprayer, this method also called hand spaying use a veterinary active ingredient such as cypermethrin. A trial sample was collected after the washing of the stables and was stored in plastic containers of 250 L previously washed. Different subsamples of 1 L were taken for characterization. To evaluate the efficiency of both, pretreatment systems and artificial or constructed wetlands, samples were taken at the inflow and outflow of each system, which were stored in 1 L amber containers and refrigerated at 4°C until analysis. For the monitoring of wetlands, *in situ* parameters of pH, temperature, dissolved oxygen, electrical conductivity, and turbidity were measured using the equipments that are shown in Table 1.

Chemical Analysis

All conventional water quality analyses of biological organic demand (BOD_5), phosphorus ($P-PO_4^{3-}$), total suspended solids (TSS), settlable solids, ammonia (N-

Table 1. Equipments employed for *in situ* monitoring of constructed wetlands and for the cypermethrin determination.

Measured variable	Model	Units
pH and temperature	Hanna HI9126	-, °C
Dissolved oxygen	YSI 550A	mg L^{-1}
Electrical conductivity	Hanna 8733	µS cm^{-1}
Turbidity	Hanna HI93703	NTU
Cypermethrin	Dionex Ultimate 3000 Series	mg L^{-1}

NH$_4^+$), and *in situ* parameters were performed following the protocols described by the Standard Methods (APHA-AWWA-WEF 2005). pH, temperature, electrical conductivity (EC), and dissolved oxygen (DO) measurement were carried out by electrode methods; turbidity was measured by the nephelometric method; TSS by the gravimetric method (dried at 103–105°C, SM 2540D); P-PO$_4^{3-}$ by ascorbic acid method (SM 4500-P.B.4); the chemical oxygen demand (COD) by the closed reflux—titrimetric method (SM 5220C); the BOD$_5$ by incubation at five days in Winkler bottles with electrometric determination (SM 4500-OG); the N-NH$_4^+$ was determined by the Nessler method (4500-NH$_3$ C). The analysis of cypermethrin was performed by liquid-liquid extraction using acetonitrile as solvent and saturation with sodium chloride. The extract was vacuum evaporated upto approximately 1 mL and subsequently, dried in a gentle current of nitrogen, and finally, reconstituted up to 1 mL with acetonitrile and a known amount of fluoranthene-d$_{10}$ internal standard (Loper and Anderson 2003). The determination was done by using a high resolution liquid chromatograph equipped with a UV diode array detector (Table 1) at a wavelength of 220 nm. An isocratic method was used (85% methanol, 15% water) at a flow rate of 0.8 mL min^{-1} for 10 min in a C18 reverse phase column of 100 mm × 3.0 mm × 10 µm, which stood at 30°C. The injection volume was 10 µL. Quantification was carried out using a calibration curve method. Analytical quality control of cypermethrin analysis included instrument calibration using high purity analytical standards from Chem Service, analysis of matrices spiked, and procedural blank. The correlation coefficients (r) for the calibration curves were all > 0.999. A standard of cypermethrin was analyzed every 10 samples to check instrument calibration. The method detection limit (MDL), calculated as three standard deviations of the mean from ten blank solutions, was 0.25 mg/L. The spiked recoveries ranged from 85.7–96.4%, and the relative standard deviations (RSD) ranged from 0.8–5.3%.

Pretreatment Systems

Descending-Ascending Sand Filter. It consisted of two vertical units connected in series: the first unit was a descending dynamic coarse filter (FGDi) while the second unit was an ascending gross filter (FGAC). The bed of the two units was composed of gravel quarry of various sizes and coarse and fine river sand. A detailed design was published by Durango et al. (2014). The operation of the system consisted of passing 1000 L of livestock wastewater from the cattle bath, through pulses of 20 L. This process was usually carried out in less than a day. A FGDi unit held larger solids, drained towards the FGAC unit, which led the filtered water into a tank that was interconnected with the wetlands system.

Sedimentation Tank. A plastic container of 1000 L was employed. The operation of the system consisted of adding 1000 L livestock wastewater from the cattle bath in the sedimentation tank, with a residence time of five days, after which the supernatant water spreading carefully through an outlet valve connected to a tank that carried water to the wetlands system.

Artificial or Constructed Wetland Systems

Wetlands experiment was carried out in a greenhouse located at the Faculty of Basic Sciences, Central Campus of the University of Cordoba, Colombia (8° 47′ 32.0″ N, 75° 51′ 41.9″ W), with dimensions: 7.0 × 4.0 × 4.5 meters (long, wide, and high), surrounded by a metal mesh and covered by greenhouse plastic. The horizontal subsurface flow constructed wetlands consisted of rectangular container of fiberglass, with dimensions of 80 × 80 × 40 cm (long, wide, and high). Each was filled with 39 cm of gravel quarry, previously washed and sieved to a size between 1–3 cm, with initial porosity of 0.42, D_{60} = 10 mm and D_{10} = 7.5 mm. The height of the water column was 34 cm, regulated by an adjustable PVC arrangement in the exit pipe. Each system remained at an angle of 1° to promote the flow of water. For continuous feeding mode, the system worked by gravity and operated at flows of 9.0 and 18.0 mL min^{-1} approximately, corresponding to 2 and 4 days of hydraulic residence time (HRT), respectively, which were reviewed every 8 hours and reset in cases where required. For intermittent mode, 4 pulses per day of 3.25 and 5.50 L were added, for 2 and 4 days of HRT respectively. The plant species *Typha latifolia* was obtained from rainwater channels near the University of Cordoba. Meanwhile *Cyperus alternifolius* was bought in a nursery in the zone, where it is commonly sold for ornamental use. All plants were subjected to an adaptation period of a month in ten liter plastic containers with gravel based support media of similar size to what was used in the constructed wetlands. The plants that presented a better development were chosen to be planted in the respective wetlands (Marrugo-Negrete et al. 2016).

Calculations of Removal Efficiency for the Treatment Systems

In the case of the pretreatment systems, the removal efficiency was calculated according to equation 1:

$$Removal\ Percentage = \frac{C_i - C_f}{C_i} \times 100\% \qquad (1)$$

where C_i is the inlet concentration of the response variable (cypermethrin, COD, TSS) to the inlet of constructed wetland, while C_f is the concentration in the wetland outlet.

The removal efficiency of artificial or constructed wetlands systems was calculated in accordance with input and output loads of each response variable (cypermethrin, P-PO$_4^{3-}$, N-NH$_4^+$, and BOD$_5$) expressed in mg m^{-2} d^{-1}, according to equation 2:

$$Removal\ Percentage = \frac{Load_e - Load_s}{Load_e} \times 100\% \qquad (2)$$

where Load$_e$ and Load$_s$ expressed in mg m^{-2} d^{-1}, are the inlet and outlet loads of the response variables (cypermethrin, P-PO$_4$$^{3-}$, N-NH$_4$$^+$ and BOD$_5$) in each constructed wetland, calculated according to equation 3:

$$Load_{e,s} = \frac{C_{i,f} \times F_{i,f} \times 10}{Wetland\ Area} \tag{3}$$

where C$_{i,f}$ are cypermethrin, P-PO$_4$$^{3-}$, N-NH$_4$$^+$, and BOD$_5$ concentrations in mg/L at, respectively, the inlet and outlet of wetlands, while F$_{i,f}$ are the flow at the inlet and outlet of wetlands and Wetland Area is the area in cm^2.

Experimental Design and Data Treatment

The pretreatment systems experiment was designed as one factor with two levels. The factor was the type of pre-treatment and levels were sedimentation and filtration. The response variables were cypermethrin, TSS, and COD removal. The experiment design to evaluate constructed wetlands system was a 2 × 2 × 2 factorial design, with type of plant (2 levels: *T. latifolia* and *C. alternifolius*), hydraulic residence time (2 levels: 2 and 4 days), and feeding mode (continuous and intermittent flow) as factors. The response variables used were cypermethrin, P-PO$_4$$^{3-}$, N-NH$_4$$^+$, and BOD$_5$ removals.

The results are shown as mean ± standard deviation. Analyses were performed with the statistical package Statistica version v.10.0 from Statsoft. Homogeneity and homoscedasticity were tested (Levene test and Shapiro-Wilk test). In some cases, logarithmic and root square data transform was performed to satisfy the assumptions. The influence of factors on the response variable was evaluated by one-way ANOVA and factorial ANOVA for the pretreatment system and constructed wetland experiment, respectively. Multiple comparison Tukey tests were performed in the cases where it was required. All analyses were carried out at a confidence level of 95%.

Results and Discussion

Characterization of Livestock Wastewater

Table 2 presents a characterization of the wastewater from the cattle bath, including conventional parameters such as chemical organic demand (COD), phosphate (P-PO$_4$$^{3-}$), ammonia (N-NH$_4$$^+$), settlable and suspended solids, some *in situ* physicochemical parameters like pH, temperature, turbidity, and dissolved oxygen (DO), and the insecticide cypermethrin.

Levels were below those reported by Reaves et al. (1994) where they reported 1120, 135.6, and 402.1 mg L^{-1} for total suspended solids (TSS), P-PO$_4$^{3} and N-NH$_4$$^+$, respectively, for the dairy waste in Lagrange country, Indiana. These concentrations may vary from one farm to another, depending on factors such as the number of cattle, the type of livestock, and even how the waste collection is performed. In this study, a significant amount of tap water was used for washing the cows after each day of bathing with the insecticide cypermethrin solution. The amount of water added for cleaning the cow could vary greatly between bathings and between farms, even for the same number of animals. The cypermethrin levels detected in this study could be considered highly harmful to fish and mammals. Different studies have shown that

Table 2. Characteristics of wastewater from cattle baths from a small-scale livestock production system of the Cordoba Department, Colombia.

Parameter	Units	Mean	Standard deviation	Minimum	Maximum
Temperature	°C	27.9	1.6	26.1	29.2
pH	-	7.45	± 0.35	7.11	7.80
COD	mg L^{-1}	1,876.78	± 205.48	1,543.85	2,103.10
P-PO$_4$$^{3-}$	mg L^{-1}	32.9	± 11.4	19.9	40.9
N-NH$_4$$^+$	mg L^{-1}	48.67	± 23.08	14.21	68.13
TSS	mg L^{-1}	119.9	± 24.8	92.8	141.5
Cypermethrin	mg L^{-1}	18.63	± 6.58	11.91	29.73
Dissolved oxygen	mg L^{-1}	1.65	0.20	1.45	1.85
Turbidity	NTU	874	-	-	-
Settlable solids	mg L^{-1}	24	± 1	23	25

this molecule can be up to a thousand times more harmful to fish than to mammals and birds (Bradbury and Coats 1989; Edwards et al. 1986; Loteste et al. 2013). This is why it is very important to prevent it from reaching the surface water sources. Manure waste from livestock activities can be incorporated into the crop soil as composting or through a liquid preparation that is sprayed on crop fields. However, in both cases a detailed mass balance should be carried out to prevent nonpoint source pollution. There may be situations with loss of air quality due to the ease with which these wastes release odors (Blanes-Vidal et al. 2009; Won et al. 2017).

Pretreatment Systems

Inlet (C$_i$) and outlet (C$_f$) concentrations and removal of COD, TSS, and Cypermethrin for both sedimentation and filtration pretreatment systems are shown in Table 3. The highest removal values were achieved with the sedimentation tank for all response variables studied. COD removal by sedimentation was 5% higher than by filtration, while on the other hand, 23 and 25% increase of the removal in TSS and cypermethrin, respectively, was obtained for the sedimentation tank with respect to the ascending-descending filtration system.

The one-way ANOVA showed that type of pretreatment was not statistically significant (p > 0.05) for cypermethrin and TSS removal. However, statistically significant differences were found for COD (p < 0.05). This could be associated with the longer residence time in the sedimentation tanks which, due to the operating characteristics, is assimilated to an anaerobic batch reactor, although the predominant separation principle would be of a physical type (particle sedimentation) where some degradation of organic matter could occur (Gutiérrez et al. 2014). Both physical pretreatment processes provide a tool that is able to fulfill the proposed target, that is, decreasing the amount of solids to prevent wetland clogging. This is achieved by means of different mechanisms: while filtration is a substrate-dependent process where the media determines the surface area available for the interaction with the pollutants to be treated, in sedimentation, on the other hand, it depends on the weight of the particles, added to various interaction forces between them, leading to a decantation.

Tabla 3. Removal percentage of cypermethrin, COD, and TSS. Italic numbers denote standard deviations.

Treatment	Parameter	C_i, mg/L	C_f, mg/L	Removal, %
Filtration	Cypermethrin	17 ± 5	9.5 ± 4.5	44 ± 16
	COD	1996 ± 86	936 ± 99	53 ± 6
	TSS	120 ± 25	70 ± 4	40 ± 11
Sedimentation	Cypermethrin	28 ± 24	14 ± 0.7	49 ± 2
	COD	1638 ± 133	360 ± 50	78 ± 1
	TSS	17 ± 77	66 ± 36	63 ± 4

Other authors have used highly efficient complex pretreatment units like anoxic type of reactors coupled with artificial or constructed wetland (Gonzalo et al. 2017). Meanwhile, Bosak et al. (2016) employed a two-unit pretreatment, sedimentation and filtration, connected in series coupled to wetlands. In all the cases, pretreatment was selected for a particular purpose, ranging from economic to geographic aspects. In the case of this study, potential sites for implementation of treatment systems are rural areas of a developing country like Colombia, where there is limited skilled labor for the system operation. Therefore, a simple low cost technology is mandatory. Other types of pretreatment units such as septic tanks (Cui et al. 2003, 2006; Abdel-Shafy et al. 2012) and stabilization ponds (Hammer 1992; Shappell et al. 2007) have been habitually employed. However, both require a qualified and continuous maintenance, which may be less suitable given the site conditions described above.

Changes for the "in situ" Variables

Environmental conditions within each constructed wetland were monitored. The average water temperature that entered the wetland systems was 27.5°C (continuous flow) and 30.2°C (intermittent). Artificial or constructed wetlands slightly decreased the temperature of the water. These changes were always higher in intermittent mode feeding. However, it was observed that in both feeding modes studied, the temperature changes were on average less than 2°C. In addition, only the system planted with *T. latifolia* operated in continuous flow showed an increase in the temperature (Fig. 2a). Seasonality is often the most influential factor on temperature in constructed wetland systems, and it is widely accepted that in cold climates efficiency can be reduced compared to warm climates (Kivaisi 2001; Zhang et al. 2015). However, other aspects such as population density of plants and the type of macrophyte (emergent or submerged) are usually relevant to generate changes in temperature (Reeder 2011). Nevertheless, more commonly significant variations in temperature are observed in surface flow wetlands compared to subsurface flow wetlands. Therefore, systems with temperatures above 15°C are preferable to avoid significant changes in this variable, especially when the removal of nitrogen species is needed (Caselles-Osorio and García 2007; Ng and Gunaratne 2011), as is the case of cattle bath water.

On the other hand, the average pH value was 7.83 for the input of the systems operated in continuous flow and 7.22 for the operation in intermittent mode. Artificial wetlands showed limited capacity to modify the pH of the water to be treated,

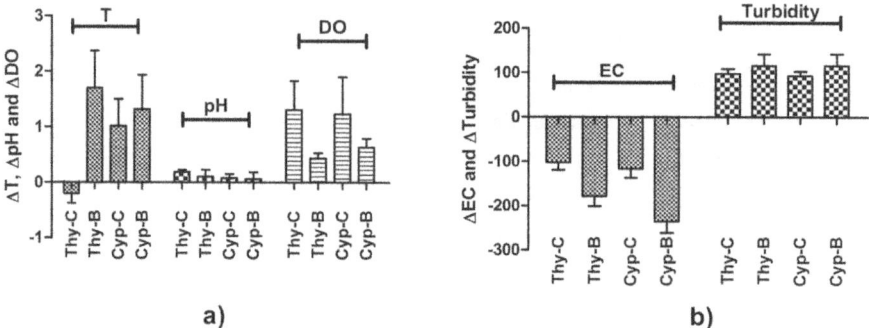

a)   **b)**

Fig. 2. Changes for the "*in situ*" variables in the artificial of constructed wetland. Thy-C: continuous flow with *T. latifolia*; Thy-B: intermittent mode with *T. latifolia*; Cyp-C: continuous flow with *C. alternifolius*; Cyp-B: intermittent mode with *C. alternifolius*; (a) Changes of Temperature (T, °C), Dissolved Oxygen (DO, mg L⁻¹), and pH; (b) Changes in electrical conductivity (EC, µS cm⁻¹) and turbidity (NTU).

regardless of the feeding mode and the type of planted macrophyte. The treatments studied showed a small decrease in pH in the range of 0.06 to 0.19 (Fig. 2a). The pH values for both input and output systems are within the reference range (6 to 9) considered by the main national regulations. In addition, pH monitoring is usually very important because of its close relationship with the removal of phosphorus and some forms of nitrogen (Kadlec and Wallace 2009). Despite the dependence of important processes on a specific pH range, constructed or artificial wetlands have been successfully used for the treatment of effluents with pH values lower than 6 (Nyquist and Greger 2009; Närhi et al. 2012) and higher than 9 (Mayes et al. 2009a; Buckley et al. 2016), in which the potential buffer effect mediated by the substrate is exploited. On the one hand, alkaline effluents can be neutralized by CO_2 production, $CaCO_3$ precipitation, cation exchange, and the presence of organic acids present in organic matter; on the other hand, acidic effluents can be regulated by the dissolution of carbonates present in the gravel, the reduction of the iron plates, and indirectly, by the presence of sulfate-reducing bacteria (Mayes et al. 2009b).

The DO levels decreased in all systems, compared to input levels; this is a characteristic behavior of subsurface flow wetland systems because in these systems there is no possibility of water aeration by external processes. The incorporation of oxygen is only associated with the transport of atmospheric air towards the roots of the macrophytes. Continuous flow operation decreased DO levels by 2 to 3 times compared to intermittent operated systems (Fig. 2a). These differences between the two modes of operation can be attributed to the fact that during the operation of wetlands under continuous flow, they remain flooded all the time, which would lead to a marked decrease of DO in the water (Saeed and Sun 2012; García and Corzo 2008), while during the intermittent operation, there could be a slight incorporation of oxygen promoted by three pulses per day at high flow. Knight et al. (2000) observed similar changes in DO levels, with decreases of 2.5 to 1.6 mg L⁻¹ for systems operated under continuous flow.

The average EC of the influent was between 209 and 307 µS cm⁻¹, with a significant increase in all the systems reflected in the negative values plotted in Fig. 2b. During the continuous flow operation, this increase was around 100 µS/cm, doubling for intermittent systems. Wetlands planted with *C. alternifolius* reached EC levels slightly higher than those planted with *T. latifolia*. This difference can be attributed to the type of emergent macrophyte in artificial or constructed wetlands, since increases in EC are mainly due to two reasons: (1) the salts released by plants and (2) the loss of water within the system, expressed as evapotranspiration (Kadlec and Wallace 2009; Stefanakis et al. 2011), which is affected by the production of vegetal biomass and is usually elevated in tropical environments such as the present study.

As for the turbidity behavior, a significant reduction in this parameter was observed in all systems. During the continuous flow operation with an initial average value of 123.7 NTU, a minimum of 25.9 NTU was reached with the macrophyte *T. latifolia*. Meanwhile, during intermittent operation, the turbidity of the system input was 134.2 NTU, decreasing to 17.7 and 17.8 NTU for *C. alternifolius* and *T. latifolia*, respectively. In general, the turbidity decreased around 100 NTU, with little appreciable differences between macrophytes and between modes of operation (Fig. 2b). Subsurface flow wetlands typically reduce turbidity levels, taking advantage of the combined action of two of their main components, plants and substrate, which promote filtration, sedimentation, and precipitation of compounds that cause turbidity (Mtavangu et al. 2017).

Performance of the Artificial or Constructed Wetland Systems

Nutrients removal (N, P). The average input loads of P-PO$_4^{3-}$ were 81 and 113 mg m^{-2} d^{-1}, for residence times of 4 and 2 days, respectively. The lowest output loads were observed for the HRT of 4 days, which was reflected in higher removals compared with the 2 days HRT (Fig. 3a). This meant that this variable was highly significant (p < 0.01) for the removal of orthophosphates. The feeding mode slightly affected P-PO$_4^{3-}$ removals, although statistically non-significantly (p = 0.08). A minor effect was presented with the plant type (p = 0.12). The P-PO$_4^{3-}$ removals for planted wetlands with *T. latifolia* are significantly higher at 4 days HRT compared to 2 days HTR for both feeding modes, whereas for *C. alternifolius*, only the intermittent mode showed statistically significant differences (p < 0.05). The results showed a behavior similar to that reported for wastewater from dairy waste and pig farms with efficiencies between 35–96% (Newman et al. 2000; Schaafsma et al. 2000; Hunt and Poach 2001). The mechanisms of phosphate removal in constructed wetlands include the sedimentation, adsorption onto the substrate, and plant absorption, the latter being the most important (Ansola et al. 2003; Healy et al. 2007).

The mean input loads for N-NH$_4^+$ were 255 mg m^{-2} d^{-1} for 4 days HRT and 635 mg m^{-2} d^{-1} for 2 days HRT. The output loads fluctuated between 5.95 and 156.12 mg m^{-2} d^{-1} with the 4 days HRT, whereas with the lower HRT they were between 37.58 and 550.58 mg m^{-2} d^{-1}. The two species of macrophytes showed a similarity for the removal of ammonia in the wastewater of the cattle bath (p = 0.78), obtaining the highest removals above 75% for the operations in continuous flow (Fig. 3a). Both the feeding mode and its interaction with HRT significantly affected (p < 0.01) the removal of N-NH$_4^+$. During the operation of the wetlands under continuous flow, the removals were increased with increasing residence time in the wetlands planted with both macrophytes. Unexpectedly, an opposite trend was observed for intermittent mode operation. In attempting to explain this situation, we

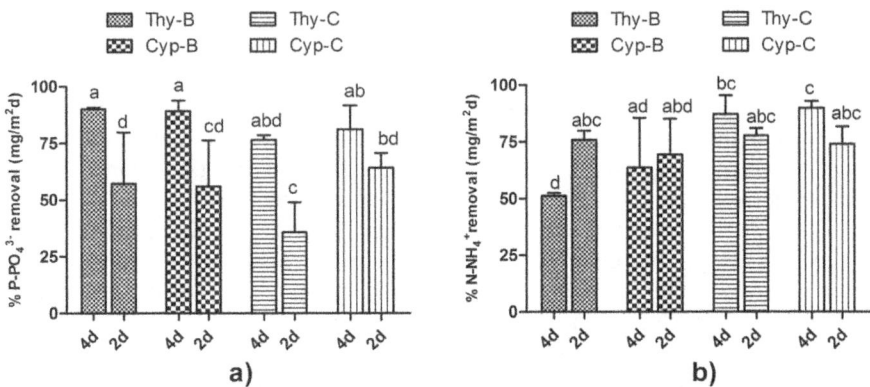

Fig. 3. Removal of the nutrient loads in the Artificial of Constructed Wetland. Error bars denote standard deviations. Bars that do not share a letter showed significant differences (Tukey test, $p < 0.05$) between treatments with the two plants (Thy-C: continuous flow with *T. latifolia*; Thy-B: intermittent mode with *T. latifolia*; Cyp-C: continuous flow with *C. alternifolius*; Cyp-B: intermittent mode with *C. alternifolius*) (a). Removal percentage of P-PO$_4^{3-}$ loads with intermittent or batch (B) and continuous (C) flow mode at two hydraulic residence times (HRT); (b) Removal percentage of N-NH$_4^+$ loads.

could rely on the levels of dissolved oxygen, since for this particular treatment, the lowest decreases were observed. It is possible that this was promoted by the effect of rapid mixing between the water that was in the system and incoming water. This daily and sharper exchange for 2-day HRT favors the incorporation of oxygen into the system, since one of the main routes of ammonia removal in artificial or constructed wetlands is ammonification, which is strongly dependent on the presence of oxygen (Lee et al. 2009). The removals achieved in the present study are lower than those reported by Vymazal and Kröpfelová (2015), who obtained efficiencies up to 98% in multistage constructed wetland systems that are specifically designed to improve the efficiency of removal of nitrogen compounds. In other studies with constructed wetlands for the removal of NH$_4^+$ contained in livestock wastewater, averages were obtained between 50 and 82% (Knight et al. 2000; Gottschall et al. 2007), which are consistent with those of the present study.

Organic matter removal (BOD$_5$, Cypermethrin). The input loads of BOD$_5$ were 4206.2 and 6240.4 mg m^{-2} d^{-1} for 4 and 2 days of HRT. The mean removals reached in the experiment were between 81 and 96% (Fig. 4a). The HRT and its interaction with the feeding mode were highly significant ($p < 0.01$). In all cases, the removals increased as HRT increased. For the intermittent mode, the differences between residence times of each planted system were statistically significant ($p < 0.05$), whereas in continuous flow this same factor was only significant for the system planted with *C. alternifolius* ($p < 0.05$). In this sense, 2 days HRT is more favorable to continuous flow BOD$_5$ removals (around 89% for *C. alternifolius* and 91% for *T. latifolia*) than to intermittent mode BOD$_5$ removals (*C. alternifolius* 81% and *T. latifolia* 85%). Since BOD$_5$ represents the biodegradable carbon, it is expected that the high removals achieved can be an indicator of a significant microbial activity within the artificial wetlands. However, systems with greater possibility of incorporation of oxygen as the vertical wetlands are more efficient than horizontal wetlands (Stefanakis et al. 2014). According to a review made by Vymazal and Kröpfelová (2009), the removal of BOD$_5$

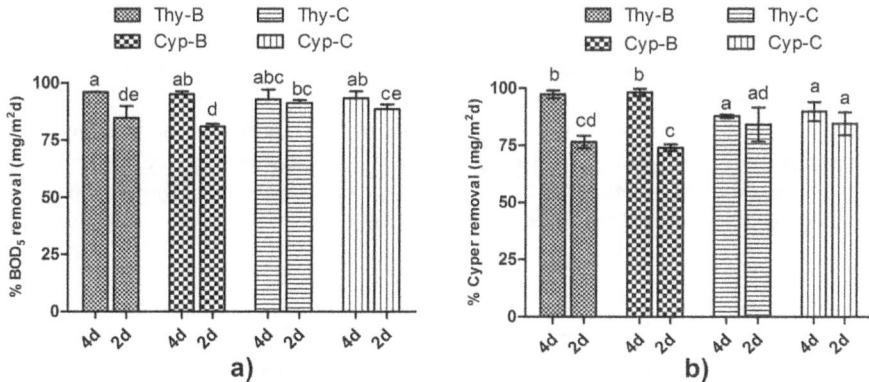

Fig. 4. Removal of the organic loads in the Artificial of Constructed Wetland. Error bars denote standard deviations. Bars that do not share a letter showed significant differences (Tukey test, p < 0.05) between treatments with the two plants (Thy-C: continuous flow with *T. latifolia*; Thy-B: intermittent mode with *T. latifolia*; Cyp-C: continuous flow with *C. alternifolius*; Cyp-B: intermittent mode with *C. alternifolius*) (a). Removal percentage of BOD_5 loads with intermittent (B) and continous (C) flow mode at two hydraulic residence times (HRT); (b) Removal percentage of cypermethrin loads.

is closely related to the type of wastewater. In the case of landfill leachates, the amount of removals can be very low, corresponding to only 32.8%, while for municipal wastewater from secondary treatment it may be up to 80%. These differences can be explained, in part, by the type of organic matter that predominates in each of these residues. It is clear that there is a significant fraction of biodegradable organic matter in the livestock wastewater, this fact increase the potential of constructed or artificial wetlands use for the treatment of these effluents. This consideration is reaffirmed by high removals reported by Reaves et al. (1994), Knight et al. (2000), and Zhang et al. (2017) with means between 61 and 95%, for wastewaters similar to the present study.

The highest removals of all parameters studied were obtained for cypermethrin (98%, Fig. 4b). These were achieved for loads between 362.21 and 633.14 mg/m² day for 4 and 2 days of HRT, respectively. The factorial ANOVA showed that only HRT and interaction between HRT and feeding mode factors were significant (p < 0.01). In general, a similar behavior to that observed for BOD_5 removals was presented. The intermittent mode accentuated the differences between residence times, since only under this flow regime statistical differences (p < 0.01) were observed between HRT for systems planted with *T. latifolia* and *C. alternifolius*. Pesticides can be removed in wetlands constructed by the combined action of several decomposition processes by microorganisms (Lakshmi et al. 2009), which are linked to both the substrate and the roots of plants. This pesticide-microorganism interaction generally leads to pesticide mineralization (Xu et al. 2007). Removals similar to those of the present study were reported by Budd et al. (2009), who observed the behavior and evaluated the removal of pyrethroids present in agricultural irrigation return waters, using surface flow wetlands planted with *Paspalum distichum*, *Polygonum lapathifolium*, and *Echinochloa crus-galli*, reporting removals between 95 and 100%. However, the input concentrations were much lower than those in the present study. Marrugo et al. (2016) also evaluated the removal of cypermethrin in livestock wastewaters, using subsurface flow wetlands planted with *Cyperus papyrus*, *Limocharias flavas*, *Alpinia purpurata*, and an unplanted bed, where removals of cypermethrin ranged from 83 to 98% using horizontal subsurface flow wetlands. Vymazal and Březinová (2015) emphasized the nature of the pesticide as a key factor in the removal mechanisms within constructed wetland systems. The physicochemical properties such as solubility and partition coefficients soil/water and octanol/water become relevant at the time of studying removals; pesticides with a high octanol/water partition coefficient can be expected to be removed more efficiently. However, there is no conclusive evidence corroborating this hypothesis. Other authors assign greater importance to the macrophyte population density (Stehle et al. 2011; Elsaesser et al. 2013).

Conclusions

Liquid waste generated in livestock activities developed by small scale farms constitute a source of pollution that receives very little attention, even though high amounts of nutrients and other contaminants such as pesticides are present in them. The use of artificial or constructed wetlands allowed the removal of $P-PO_4^{3-}$, $N-NH_4^+$, BOD_5, and cypermethrin, regardless of the feeding mode. These results become an important tool to minimize operating costs, since horizontal subsurface flow wetlands are usually operated in continuous flow, requiring additional equipment to maintain the water course. This could be of little use in livestock farms located in rural areas.

References Cited

Abdel-Shafy, H.I. and M.A. El-Khateeb. 2012. Integration of septic tank and constructed wetland for the treatment of wastewater in Egypt. Desalin. Water Treat 51(16-18): 3539–3546.

Ansola, G., J.M. González, R. Cortijo and E. deLuis. 2003. Experimental and full-scale pilot plant constructed wetlands for municipal wastewater treatment. Ecol. Eng. 21(1): 43–52.

APHA, AWWA, WEF. 2005. Standard methods for the examination of water and wastewater. American Public Health Association—American Water Works Association—Water Environment Federation. 21th edition. Port City Press, Maryland, US.

Blanes-Vidal, V., M.N. Hansen, A.P.S. Adamsen, A. Feilberg, S.O. Petersen and B.B. Jensen. 2009. Characterization of odor released during handling of swine slurry: Part II. Effect of production type, storage, and physicochemical characteristics of the slurry. Atmos. Environ. 43(18): 3006–3014.

Bosak, V., A. VanderZaag, A. Crolla, C. Kinsley and R. Gordon. 2016. Performance of a constructed wetland and pretreatment system receiving potato farm wash water. Water 8(183): 1–14.

Bradbury, S.P. and J.R. Coats. 1989. Comparative toxicology of the pyrethroid insecticides. Rev. Environ. Contam. Toxicol. 108: 133–177.

Buckley, R., T. Curtin and R. Courtney. 2016. The potential for constructed wetlands to treat alkaline bauxite residue leachate: Laboratory investigations. Environ. Sci. Pollut. Res. Int. 23(14): 14115–14122.

Budd, R., A. O'Geen, K.S. Goh, S. Bondarenko and J. Gan. 2011. Removal mechanisms and fate of insecticides in constructed wetlands. Chemosphere 83: 1581–1587.

Carvalho, P.N., J.L. Araújo, A.P. Mucha, M.C. Basto and C.M. Almeida. 2013. Potential of constructed wetlands microcosms for the removal of veterinary pharmaceuticals from livestock wastewater. Bioresour. Technol. 134: 412–416.

Caselles-Osorio, A. and J. García. 2007. Impact of different feeding strategies and plant presence on the performance of shallow horizontal subsurface-flow constructed wetlands. Sci. Total Environ. 378(3): 253–262.

Cui, L.H., S.M. Luo, X.Z. Zhu and Y.H. Liu. 2003. Treatment and utilization of septic tank effluent using vertical-flow constructed wetlands and vegetable hydroponics. J. Environ. Sci. (China) 15(1): 75–82.

Cui, L.H., W. Liu, X.Z. Zhu, M. Ma, X.H. Huang and Y.Y. Xia. 2006. Performance of hybrid constructed wetland systems for treating septic tank effluent. J. Environ. Sci. (China) 18(4): 665–669.

Durango, J.D, I.D. Urango, J.J. Pinedo, S.M. Burgos, A.J. Estrada, J.G. Ortega et al. 2014. Evaluación de un filtro lento de arena, de tipo descendente-ascendente, para el tratamiento de efluentes ganaderos contaminados con cipermetrina. *In*: 189–193. Memorias del VII Seminario Internacional de Gestión Ambiental y II Seminario de Ciencias Ambientales Sue-Caribe. Santa Marta, Colombia. In Spanish.

Edwards, R., P. Millburn and D.H. Hutson. 1986. Comparative toxicity of cis-cypermethrine in rainbow trout, frog, mouse, and quail. Toxicol. Appl. Pharmacol. 84: 512–522.

Elsaesser, D., A.G. Buseth, A. Geist, T. Mæhlum and R. Schulz. 2011. Assessing the influence of vegetation on reduction of pesticide concentration in experimental surface flow constructed wetlands: Application of the toxic units. Ecol. Eng. 37: 955–962.

García, J. and A. Corzo. 2008. Depuración con humedales construidos. Guía Práctica de Diseño, Construcción y Explotación de Sistemas de Humedales de Flujo Subsuperficial. Universidad Politécnica de Catalunya. Spain. In Spanish.

Gonzalo, O.G., I. Ruiz and M. Soto. 2017. Integrating pretreatment and denitrification in constructed wetland systems. Sci. Total Environ. 584-585: 1300–1309.

Gottschall, N., C. Boutin, A. Crolla, C. Kinsley and P. Champagne. 2007. The role of plants in the removal of nutrients at a constructed wetland treating agricultural (dairy) wastewater. Ecol. Eng. 29(2): 154–163.

Gutiérrez, N., E. Valencia and A. Aragón. 2014. Eficiencia de remoción de DBO$_5$ y SS en un sedimentador y lecho filtrante para el tratamiento de aguas residuales del beneficio de café (*Coffea arabica*). Colombia Forestal 17(2): 151–159. In Spanish.

Hammer, D.A. 1992. Designing constructed wetlands systems to treat agricultural nonpoint source pollution. Ecol. Eng. 1(1-2): 49–82.

Healy, M., M. Rodgers and J. Mulqueen. 2007. Treatment of dairy wastewater using constructed wetlands and intermittent sand filters. Bioresour. Technol. 98: 2268–2281.

Hunt, P. and M. Poach. 2001. State of the art for animal wastewater treatment in constructed wetlands. Water Sci. Technol. 44(11-12): 19–25.

IDEAM. 2015. Estudio Nacional del Agua 2014. Instituto de Hidrología, Meteorología y Estudios Ambientales, Bogotá, Colombia. In Spanish.

Jaramillo, A.R. and A.M. Areiza. 2013. Análisis del Mercado de Leche y Derivados Lácteos en Colombia (2008–2012). Superintendencia de Industria y Comercio, Bogotá, Colombia. In Spanish.

Kadlec, R. and S. Wallace. 2009. Treatment Wetlands. 2nd edition. CRC Press, Boca Raton, US.

Kivaisi, A. 2001. The potential for constructed wetlands for wastewater treatment and reuse in developing countries: A review. Ecol. Eng. 16(4): 545–560.

Knight, R.L., V.W.E. Payne, R.E. Borer, R.A. Clarke and J.H. Pries. 2000. Constructed wetlands for livestock wastewater management. Ecol. Eng. 15(1-2): 41–55.

Lakshmi, C.V., M. Kumar and S. Khanna. 2009. Biodegradation of chlorpyrifos in soil by enriched cultures. Curr. Microbiol. 58: 35–38.

Lee, H. and M. Shoda. 2008. Removal of COD and color from livestock wastewater by the Fenton method. J. Hazard. Mater. 153(3): 1314–1319.

Lee, C., T. Fletcher and G. Sun. 2009. Nitrogen removal in constructed wetland systems. Engineering Life Science 9(1): 11–22.

Li, C.Y., S.B. Wu, F. Sun, T. Lv, R.J. Dong and C.L. Pang. 2012. Performance of lab-scale tidal flow constructed wetlands treating livestock wastewater. Adv. Mat. Res. 518: 2631–2639.

Liu, R., Y. Zhao, L. Doherty, Y. Hu and X. Hao. 2015. A review of incorporation of constructed wetland with other treatment processes. Chem. Eng. J. 279: 220–230.

Loper, B. and K. Anderson. 2003. Determination of pyrethrin and pyrethroid pesticides in urine and water matrixes by liquid chromatography with diode array detection. Journal of AOAC International 86(6): 1236–1240.

Loteste, A., J. Scagnetti, M.F. Simoniello, M. Campana and M.J. Parma. 2013. Hepatic enzymes activity in the fish *Prochilodus lineatus* (Valenciennes, 1836) after sublethal cypermethrin exposure. Bull. Environ. Contam. Toxicol. 90(5): 601–604.

Márquez, D. 2008. Residuos químicos en alimentos de origen animal: Problemas y desafíos para la inocuidad alimentaria en Colombia. Revista Corpoica–Ciencia y Tecnología Agropecuaria 9(1): 124–135. In Spanish.

Marrugo-Negrete, J.L., J.G. Ortega-Ruíz, A.E. Navarro-Frómeta, G.H. Enamorado-Montes, I.D. Urango-Cárdenas, J.J. Pinedo-Hernández et al. 2016. Remoción de cipermetrina presente en el baño de ganado utilizando humedales construidos. Revista Corpoica–Ciencia y Tecnología Agropecuaria. 17(2): 203–216.

Mayes, W., J. Aumônier and A. Jarvis. 2009a. Preliminary evaluation of a constructed wetland for treating extremely alkaline (pH 12) steel slag drainage. Water Sci. Technol. 59(11): 2253–2263.

Mayes, W, L. Batty, P. Younger, A. Jarvis, M. Kõiv, C. Vohla et al. 2009b. Wetland treatment at extremes of pH: A review. Sci. Total Environ. 407(13): 3944–3957.

Mtavangu, S., A. Rugaika, A. Hilonga and K. Njau. 2017. Performance of constructed wetland integrated with sand filters for treating high turbid water for drinking. Water Practice & Technology 12(1): 25–42.

Närhi, P., M. Räisänen, M. Sutinen and R. Sutinen. 2012. Effect of tailings on wetland vegetation in Rautuvaara, a former iron–copper mining area in northern Finland J. Geochem. Explor. 116-117: 60–65.

Newman, J., J. Clausen and J. Neafsey. 2000. Seasonal performance of a wetland constructed to process dairy milkhouse wastewater in Connecticut. Ecol. Eng. 14(1-2): 181–198.

Ng, W. and G. Gunaratne. 2011. Design of tropical constructed wetlands. pp. 69–94. In: Tanaka, N., W. Ng and K. Jinadasa [eds.]. Wetlands for Tropical Application: Wastewater Treatment by Constructed Wetland. Imperial College Press, London, UK.

Niño, J. and D. Alarcón. 2015. Moooi Dairy Opportunities for Colombia-Duch Colaboration: Fact-Finding Study Dairy Sector in Colombia. Netherlands Enterprise Agency, Netherlands.

Nyquist, J. and M. Greger. 2009. A field study of constructed wetlands for preventing and treating acid mine drainage. Ecol. Eng. 35(5): 630–642.

Othman, I., A.N. Anuar, Z. Ujang, N.H. Rosman, H. Harun and S. Chelliapan. 2013. Livestock wastewater treatment using aerobic granular sludge. Bioresour. Technol. 133: 630–634.

Reaves, R.P., P.J. DuBowy and B.K. Miller. 1994. Performance of a constructed wetland for dairy waste treatment in Lagrange Country, Indiana. pp. 43–52. In: Proceedings of a Workshop on Constructed Wetlands for Animal Waste Management. Lafayette, US.

Reeder, B. 2011. Assessing constructed wetland functional success using diel changes in dissolved oxygen, pH, and temperature in submerged, emergent, and open-water habitats in the Beaver Creek Wetlands Complex, Kentucky (USA). Ecol. Eng. 37(11): 1772–1778.

Saeed, T. and G. Sun. 2012. A review on nitrogen and organics removal mechanisms in subsurface flow constructed wetlands: Dependency on environmental parameters, operating conditions, and supporting media. J. Env. Manage. 112(0): 429–448.

Schaafsma, J.A., A.H. Baldwin and C.A. Streb. 2000. An evaluation of a constructed wetland to treat wastewater from a dairy farm in Maryland, US. Ecol. Eng. 14(1-2): 199–206.

Shappell, N.W., L.O. Billey, D. Forbes, T.A. Matheny, M.E. Poach, G.B. Reddy et al. 2007. Estrogenic activity and steroid hormones in swine wastewater through a lagoon constructed-wetland system. Environ. Sci. Technol. 41(2): 444–450.

Stefanakis, A., P. Komilis and A. Tsihrintzis. 2011. Stability and maturity of thickened wastewater sludge treated in pilot-scale sludge treatment wetlands. Water Res. 45(19): 6441–6452.

Stefanakis, A., C.S. Akratos and V.A. Tsihrintzis. 2014. Vertical Flow Constructed Wetlands: Eco-engineering Systems for Wastewater and Sludge Treatment. Elsevier, Amsterdam.

Steinfeld, H., P. Gerber, T.D. Wassenaar, V. Castel and C. Haan. 2006. Livestock's long shadow: environmental issues and options. Food and Agriculture Organization of the United Nations–FAO, Roma.

Stehle, S., D. Elsaesser, C. Gregoire, G. Imfeld, E. Niehaus, E. Passeport et al. 2011. Pesticide risk mitigation by vegetated treatment systems: a meta-analysis. J. Environ. Qual. 40: 1068–1080.

Vymazal, J. and L. Kröpfelová. 2009. Removal of organics in constructed wetlands with horizontal sub-surface flow: A review of the field experience. Sci. Total Environ. 407(13): 3911–3922.

Vymazal, J. and T. Březinová. 2015. The use of constructed wetlands for removal of pesticides from agricultural runoff and drainage: A review. Environ. Int. 75: 11–20.

Vymazal, J. and L. Kröpfelová. 2015. Multistage hybrid constructed wetland for enhanced removal of nitrogen. Ecol. Eng. 84: 202–208.

Won, S., S.-M. Shim, B.-G. You, Y.-S. Choi and C. Ra. 2017. Nutrient production from dairy cattle manure and loading on arable land. Asian-Australas. J. Anim. Sci. 30(1): 125–132.

Wu, H., J. Zhang, H.H. Ngo, W. Guo, Z. Hu, S. Liang et al. 2015. A review on the sustainability of constructed wetlands for wastewater treatment: Design and operation. Bioresour. Technol. 175: 594–601.

Xu, G., Y. Li, W. Zheng, X. Peng, W. Li and Y. Yan. 2007. Mineralization of chlorpyrifos by co-culture of Serratia and *Trichosporon* spp. Biotechnol. Lett. 29: 1469–1473.

Zhang, Z., Y. Li, S. Chen, S. Wang and X. Bao. 2012. Simultaneous nitrogen and carbon removal from swine digester liquor by the Canon process and denitrification. Bioresour. Technol. 114: 84–89.

Zhang, D., K. Jinadasa, R. Gersberg, Y. Liu, S. Tan and W. Ng. 2015. Application of constructed wetlands for wastewater treatment in tropical and subtropical regions (2000–2013). J. Environ. Sci. (China) 30: 30–46.

Zhang, X., T. Inoue, K. Kato, H. Izumoto, J. Harada, D. Wu et al. 2017. Multi-stage hybrid subsurface flow constructed wetlands for treating piggery and dairy wastewater in cold climate. Environ. Technol. 38(2): 183–191.

11

REAGRITECH Project

Regeneration and Reuse of Runoff and Drainage Water from Agriculture by Treatment Wetlands

Lorena Aguilar,[1] *Ángel Gallegos,*[1]
Carlos A. Arias,[2] and *Jordi Morató*[1,*]

INTRODUCTION

In recent decades, agricultural practices have been modified to obtain greater production with the main aim of increasing food production. This increase has been achieved through intensification of practices such as the use of irrigation, the addition of nutrients, and the application of fungicides and pesticides. These new trends in agricultural practices have generated direct impacts on soil, air, water, biodiversity, and landscape, as well as indirect impacts.

Increased farming and agricultural activity has put natural resources under pressure, generating a strong impact on the environment. The extension of irrigated crops area increases the demand for water. This situation requires a constant research that could help to minimize the anthropic impacts. In particular, the water used for crop irrigation can drag nutrients and pesticides and therefore, generate a polluted runoff that can negatively affect rivers and groundwater. The presence of these pollutants in soil, groundwater, and water runoff is a major problem for agricultural activities and for its negative impact on the environment.

[1] UNESCO Chair on Sustainability, Universitat Politècnica de Catalunya-BarcelonaTech, Carrer Colom 1, TR1, ESEIAAT, Terrassa, 08222 Spain.
[2] Department of Bioscience, University of Aarhus, Ole Worms Allé 1, Building 1135, Arhus C., 8000 Denmark.
Email: carlos.arias@bios.au.dk.
* Corresponding author: jordi.morato@upc.edu

The runoff from water used to irrigate crops can cause pollution of surface waters and aquifers since this water can transport compounds such as nutrients, fungicides, and pesticides, as well as polluting ground water once they infiltrate. The consumption of water resources as well as the pollution of natural waters are the most significant impacts of this type of activity.

In the EU, 50 to 80% of the nitrogen and 30 to 80% of the phosphorus present in water is due to agriculture (Isermann 1990). In Spain, 80% of the groundwater has nitrate concentration above 25 mg/L (European Environment Agency 2007) and 12.7% of the national territory has been declared vulnerable to nitrate water pollution (Fernández-Ruiz 2007).

The type of nitrate transport in agricultural works is called diffuse transport. Diffuse transport or diffuse pollution does not require a physical movement to occur; it only needs a differential gradient that permits the chemical combination between different zones in the subsoil. Several soil and pollutant characteristics are necessary to better improve the knowledge about pollution and specially to know pollution source: plot scale (to know how water bodies are affected), water pathways (throughout quantitative and qualitative dynamic analysis), and chemical pollutants and their hydrochemical relationships and retention time (Tournebize et al. 2008, 2017). Moreover, the transport of pesticides applied to soil and water bodies is mainly due to runoff (Runes et al. 2003).

There are many available technologies designed to mitigate the impact generated by agricultural runoff. Constructed wetlands (CWs) and especially subsurface CWs are treatment systems that use natural processes and no external energy input to improve water quality through a combination of physical, chemical, and biological processes. They can be successfully used for domestic wastewater treatment and for effluents of different types, including agricultural runoff (Babatunde et al. 2008).

CWs have been widely used to control and prevent non-point source pollution of water bodies acting as a barrier between agricultural areas and receiving water bodies, so they can mitigate the impact of pesticides (and specifically fungicides) and nutrients carried by runoff (Ghadiri and Rose 1991). Once the potential capacity to pollute is reduced by the CWs, water can be reused for irrigation or other productive activities.

Constructed wetlands and strips of vegetation (buffer strips) are natural and appropriate technologies that provide an effective mean of purifying water polluted by nitrates, and therefore, treating diffuse pollution produced by farming practices.

In order to treat wastewater effectively, several factors have to be taken into account, for example, the system's capacity, the plant species used, colonization characteristics of certain microbial groups, and the interactions of biogenic compounds and particular contaminants (wastewater components) with the filter bed material (Stottmeister et al. 2003). Although filtration is considered an important process in these removal mechanisms, additional interactions occur among media, plants, and water. Many processes and relations between them take place: microbial-mediated processes, chemical networks, volatilization, sedimentation, sorption, photodegradation, plant uptake, and transpiration flux and accretion (Kadlec and Wallace 2009). The importance of microbial processes has been further studied as many reactions are microbiologically mediated (Stottmeister et al. 2003; Kadlec and Wallace 2009).

Increased removal efficiency of nitrogen from wastewater is one of the key issues for further development of constructed wetlands and other decentralised technologies. The diversity of microorganisms involved in the N-cycle is expected to be high in these systems. In fact, previous studies have suggested that archaeal nitrifiers, denitrifying fungi, aerobic denitrifying bacteria, and heterotrophic nitrifying microorganisms may play an important role in nitrogen transformations in treatment wetlands (Truu et al. 2005). Most importantly, the effects of biofilms on nitrogen transformation and removal have not been adequately studied and modelled. As microorganisms affect processes like nitrification, denitrification, uptake, and sedimentation, they, that is, microorganisms, have to be taken into consideration when modelling the transformation and removal of nitrogen from wastewater (Mayo and Bigambo 2005). Previous studies have shown that shifts in the structure of bacterial communities can be associated with changes in a number of soil properties, including soil texture and soil nitrogen availability (Dong and Reddy 2010).

The support medium is an important component since it supports plant growth (in case of planted wetland systems), as well as the establishment of a microbial biofilm, and it influences the hydraulic processes (Stottmeister et al. 2003). Additionally, recent studies concluded that the type of support medium is one of the main factors influencing bacterial communities (Vacca et al. 2005; Calheiros et al. 2009). However, none of these studies took place in real constructed wetlands; both of them consisted in different pilot systems, with the same influent water.

Different authors (Ibekwe et al. 2003; Calheiros et al. 2009) indicated that the diversity of the bacterial community in the constructed wetlands systems might influence the final effluent quality, and so the engineering should be directed to develop a higher diversity in order to enhance processes such as nitrification and denitrification (Ibekwe et al. 2003). The removal of nitrogen in constructed wetlands is usually limited by the nitrification process, and in order to reach high total nitrification rates, it is important that biological nitrification takes place. Additionally, in order to increase denitrification rates in the unsaturated systems, the establishment of recycling or an additional step is a must.

Vymazal (2007) reports that removal efficiencies of total nitrogen in CWs varied between 40 and 55%, with removing load ranging between 250 and 630 g N m^{-2} yr^{-1} depending on CWs type and influent loading. The treatment allows for direct reuse of treated effluent for irrigation.

First order degradation models for CW design are recognized as the best approach to date. The PkC* first order model includes the effect of hydraulics, pollutant kinetics degradation, as well as the background pollutant concentration inherent to CWs (Kadlec and Wallace 2009).

Based on preliminary characterization of the selected site, the discharge standards and the use of the PkC* model (where P is the hydraulic flow efficiency, k is the pollutant specific rate of biological degradation, and C* is the desired outflow concentration (Co) in relation to inflow pollutant levels, and system production of pollutant from degrading biological material in the system), three pilot constructed wetland plants were designed and installed by the REAGRITECH project on three plots in the Lleida region, Catalonia, Spain (Sudanell, Vilanova de la Barca, and Bellvís), with a maximum treated flow of 750 L/d for all the prototypes. Each pilot was designed as a hybrid CW including: first, a vertical flow cell (VFCW) followed by a horizontal flow cell (HFCW), filled with different media, namely 32 mm Ø gravel and 12 mm Ø cork granulates, in order to test both media performance.

All systems were completely automated, increasing its efficiency and productivity by the extensive monitoring of all components (Pérez et al. 2006). Information and Communication Technologies (ICTs) play an important role in agricultural value chains (Calera and Campos 2010) and allow researchers to obtain useful information in real time for improving decision making. On the other hand, agricultural production areas present needs and requirements for ICTs such as the monitoring of local weather (temperature, rain, and other climatic variables) or the automatic control of irrigation, that can support technical processes with greater efficiency, reliability, and security (Morillo et al. 2005; Chaparro and Pérez 2013; Gallegos et al. 2016). The automated control is described in detail by Gallegos et al. (2016).

REAGRITECH has been a demonstration project with the main objective to reduce the consumption of water resources and improve its quality, integrating natural systems for wastewater treatment controlling nutrient and pesticides in the generation source (agricultural activity). The main aim of REAGRITECH Project was to demonstrate a sustainable method for water reuse of agricultural runoff, in order to optimize it, at parcel scale and minimize diffuse pollution. At the same time, other specific objectives were:

- Reuse of runoff and infiltration groundwater from irrigated crops.
- Increase the efficiency of water treatment from agricultural activities using natural and appropriate treatment systems.
- Improve the quality of surface water and groundwater and reducing water consumption and related costs.
- Restore and improve riparian ecosystems.
- Reduce carbon footprint and human anthropic impact.

The expected longer-term results for the REAGRITECH Project were the implementation of the innovative systems to treat agricultural runoff on a large scale, improving water quality and the riparian ecosystems at the same time. The application of these natural water treatment systems is supported by the European Union (EU) environmental policies and legislation, and may play a basic role for sustainable water management, especially for river basins affected by agricultural activities.

Material and Methods

Study of Potential Alternative Sites

A multicriteria assessment has been performed in order to determine the most appropriate location and to select the parcels of land where the prototypes would be installed, near the river stretches of the irrigation areas at the Urgell and Segarra-Garrigues channels, in Lleida, Catalonia (Spain). Some of the used indicators were linked to the hydrogeological requirements and the nitrate vulnerable areas, such as *available area*, *slope*, *accessibility,* and *location*. Other important variables to be known were the type of irrigation systems, new irrigation implantation systems, and real availability of plots, among others.

In order to define the most suitable plots, all locations were analyzed taking into account geotechnical analysis with several site investigations: trial pits, soil sample extractions, drilling of boreholes, and performance of some pumping tests. Assessment of water quality had also to be included in the comparison among all possible locations.

Once the implementation areas were identified and the necessary permits by the landowners were obtained, the corresponding authorizations by all public administration entities were obtained. After the final plots were chosen, hydrological studies for Segarra–Garrigues and Urgell irrigation plots were accomplished. The final selected locations were the following ones:

- One planted plot within the transformation from upland to irrigation area, from the *Segarra-Garrigues waterway* irrigation system, in the Sudanell municipality.

- Two planted plots within the irrigation area of *Urgell waterway*, with two pilot plants in Vilanova de la Barca and Bellvís municipalities.

Prototype's Design and Build up

Three hybrid CW were designed and constructed including a vertical flow cell followed by a horizontal flow cell, and placed at three irrigation areas in Lleida (Catalonia). Each pilot plant was built as a compact, modular, and mobile system, using 6 m (20 ft) shipping containers that can be transported and installed at research sites. Water was taken from infiltrated water by means of well pumps.

The innovative modular system was designed to produce an effluent with concentrations below 5 mgL^{-1} of NH$_4$-N and 10 mgL^{-1} of NO$_3$-N. The system helped to determine operational parameters, to compare different media for the removal of agricultural pollutants, and to allow the reduction of operational and maintenance costs.

The treatment wetland pilot plants were designed with agricultural use of water in mind. A mobile design, energy-efficient, autonomous, and able to remotely collect information was clearly needed. Therefore, a new innovative concept of mobile treatment wetland was established (Figs. 1 and 2), reducing at the same time the on-site construction phase. Besides, the mobile pilot plants could be tested before in the University facilities, prior to its set up in the field. Each prototype system had three main different elements (Fig. 3): (1) a source point where the groundwater was captured from a well (750 Ld^{-1} of equivalent daily flow) using piezometers, (2) a treatment zone composed by a Subsurface Treatment Wetland, obtaining an effluent with a concentration below 10 mg NO$_3$ L^{-1}, and, (3) an infiltration zone, using a buffer strip area (0.5 ha) or for agricultural reuse. The piezometers design was made according to the *Guide of Setting-Up Piezometers recommendations*, developed by the Catalan Water Agency (ACA). The initial depth of well probes was scheduled to completely drill the coating of the quaternary soils above the bedrock, according to the depth determined by the stratigraphy obtained from the geological study in the parcel.

The first pilot plant was constructed and installed in a field at Sudanell in 2014, to check all components and to test the reliability of the system, during 2014–2015. This hybrid CW was designed and constructed including first a two-chambered

Fig. 1. Reagritech Mobile Pilot Plant (front view).

Fig. 2. Reagritech Mobile Pilot Plant (rear view).

sedimentation tank as the primary system, and in second place a 5.5 m² vertical flow (VF) cell, filled with 1 m depth of 4 mm diameter (Ø) coarse sand, followed by two 4.1 m parallel horizontal flow (HF) cells, filled with 80 cm depth of 32 mm Ø gravel and 12 mm Ø cork granulates respectively (Fig. 2). The hydraulic controls allowed the remote operation, monitoring, and recirculation of desired flows, resulting in higher removal performance.

A second version of the pilot plant was designed one year later (2016), integrating different improvements in the system and its components, and with only one horizontal subsurface flow compartment. This second unit was installed in Bellvís using cork as medium and Vilanova de la Barca with gravel as medium.

As mentioned before, all the treatment plants were designed following the PkC* model, a first order degradation model to include the hydraulic effects, pollutant kinetics degradation, as well as the pollutant concentration background. The maximum daily treated water flow for each single system was 750 Liters of treated flow per day

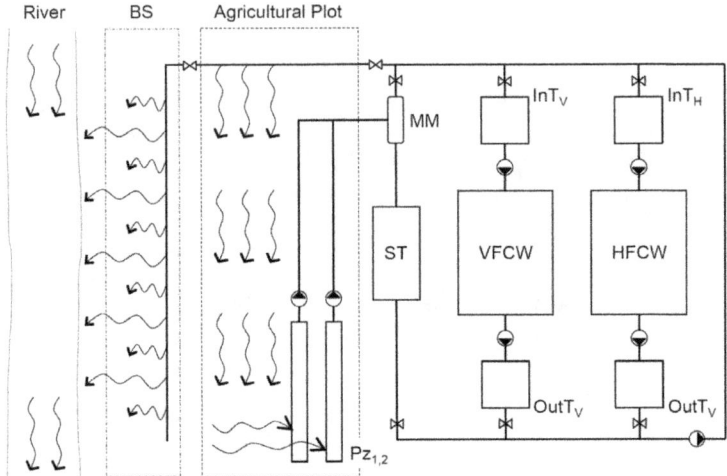

Fig. 3. Functional diagram of the Sudanell prototype (Pz, Piezometer; MM, Multiparameter Water Quality Meter; ST, Sedimentation Tank; InT, Inlet Tank; OutT, Outlet Tank; VFCW, Subsurface Vertical Flow Constructed Wetland; HFCW, Subsurface Horizontal Flow Constructed Wetland; BS, Buffer Strip).

per plot (L_{tf} d^{-1}·pl^{-1}). All units were designed to produce an effluent at the outlet with concentrations of NH_4–N and NO_3–N below the limit required for irrigation in Spain.

Buffer Strips were the third element of the project, carried out for controlling diffuse pollution control. The REAGRITECH Project was directed towards the environmental and ecological improvement of the studied area through a series of actions based on the creation of buffer strips to increase biodiversity and improve the ecological status of the riparian ecosystems, following all guidelines from the Water Directive Framework. The ecological status of the river basins in the agricultural areas was studied using the Riparian Forest Quality Index (QBR) (Munné et al. 1998), considering the percentage of the riparian coverage, the coverage and its quality, and the river channel's naturalness degree.

The most important activity for buffer strip creation was to select native vegetation and at the same time, choose the most suitable area according to its diversity and heterogeneity.

The final selected vegetation was the typical and representative for the natural ecosystems in the area, where main species on tree strata were *Salix alba*, *Fraxinus* sp., and *Populus nigra*, whereas for shrub strata, *Tamarix* sp. and *Sambucus* sp. were selected.

Buffer strips reforestation with native species was carried out in all three locations during the spring of 2016 and a second plantation during winter of the same year, using an irregular pattern, and with a distance of 4 m between each plant. An excavator was used to dig the holes for planting trees, and some shrubs were manually planted. Knowing the extreme weather conditions during the summer period, an irrigation system to water the buffer strip vegetation was installed at each location. These irrigation systems were intended to guarantee the plantation survival, and fed using treated water effluent from the constructed wetland as a new source for buffer strip maintenance.

Monitoring of Treatment Wetland

Multi-parametric probes were used for on-line measuring of nitrates, electrical conductivity, pH, dissolved oxygen, and temperature, monitoring both, inlet and outlet water. Other control parameters were analyzed at monthly intervals with three consecutive day campaigns, following standard methods (APHA 2002), such as BOD_5, COD, TSS, ammonia nitrogen, total Kjeldahl nitrogen, TKN, nitrite and nitrate nitrogen, and total phosphorus among others.

Results and Discussion

Riparian Forest Quality

The diagnosis showed an irregular vegetation cover, poor vegetal structure, significant presence of invasive species, and presence of artificial obstacles. Based on internal assessments, all three areas under study were to be placed in the POOR category (yellow and orange categories), according to the ecological ranges established by QBR index.

After the buffer strip reforestation with native species, carried out in all three locations during the spring of 2016 and a second plantation during winter of the same year, the riparian forest quality evolved to a better category.

However, and even with the irrigation system working, public efforts should be maintained in order to guarantee the plantation survival, and therefore, the improvement of the ecological status.

Validation of the Pilot Plant Constructed Wetland

In Sudanell, no significant levels of nitrates were detected at the inlet of the constructed wetland. However, this plot was selected because a long-term study can be developed during next years, in order to test the evolution of nitrate pollution from a newly established irrigation area.

Table 1 shows the inlet average values during the entire period of sampling (12 months).

Figure 4 shows the nitrate removal efficiency for both prototypes at Bellvís (with cork) and Vilanova de la Barca (with gravel) during the year 2016. The variability from inlet concentration had not affected the nitrate removal. Data obtained for nitrate ($mg\ NO_3^-\ L^{-1}$) removal from the cork Bellvís pilot plant showed the best removal rates rather than the gravel plant at Vilanova de la Barca. Both systems, HF and VF (data not shown) showed a comparative good operation, with removal rates ranging from 80 to 99% for the HFCW.

Data obtained (not shown) for nitrate ($mg\ NO_3^-\ L^{-1}$) removal in Vilanova de la Barca using gravel showed a lower efficiency for the Vertical Subsurface Flow Constructed Wetland (VFCW) compared to the Horizontal Subsurface Flow Constructed Wetland (HFCW); there was a slight efficiency decrease during the summer months and a better operating during the spring months. The HF gravel wetland showed the best performance, with removal rates ranging from 4 to 46%.

Table 1. Water quality at the inlet in Vilanova de la Barca and Bellvís.

Inlet values	Vilanova	Bellvis
COD (mg L^{-1})	11.19	23.67
BOD (mg L^{-1})	5.17	18.50
NO$_3^-$ (gd^{-1})	29.20	23.41
NO$_2^-$ (mg L^{-1})	< 0.2	< 0.2
NH$_4^+$ (mg/l)	< 2.5	< 2.5
TKN (mg L^{-1})	13.32	6.29
P (mg L^{-1})	< 1	< 1
pH	7.72	7.65
Electric Conductivity (mScm^{-1})	2.40	3.16
TSS (mg L^{-1})	< 5	4.51
Turbididy (NTU)	4.26	9.07

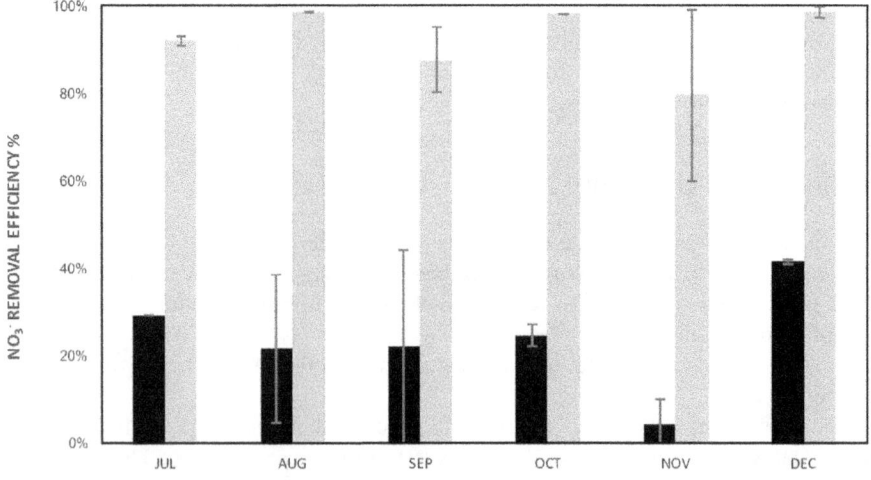

■ VILANOVA-GRAVEL   ▨ BELLVIS-CORK

Fig. 4. Comparison of the removal efficiency for nitrate (NO$_3^-$) in the Vilanova de la Barca prototype with gravel (dark) and Bellvís with cork (light).

According to the results obtained during the first half of 2016, it was decided to by-pass the VFCW and to pass all the water through the HFCW because it gave better results. The better operation of HFCW was clearly linked to the low oxygen concentrations, which are a must to improve the performance of denitrifying bacteria, compared with VFCW that operated under oxygenated conditions.

During the summer season, a high nitrate concentration (more than double the normal concentrations) entered the system in Bellvís. The operation of the VFCW system was also bypassed to the HFCW during the summer months, just as it was done in Vilanova de la Barca. This operation had been carried out in order to check the treatment effectiveness in the HFCW without the influence of the VFCW system. Despite the greater influent nitrate concentration (reaching a maximum value of 52.8 mg/L), the cork pilot plant showed significant higher removal rates ($p < 0025$) compared with the gravel systems during the entire year, and with no significant differences ($p > 0.41$) between the VFCW and the HFCW during most of the time. In fact, the performance of the cork pilot plant proved to be excellent, with removal rates ranging above 90% during 6 of the 10 months analyzed.

Some very important conclusions can be drawn from the results obtained during the operation for both pilot plants, using gravel and cork and for vertical (VFCW) and horizontal flow (HFCW) constructed wetlands.

In summary, although significant removal rates of nitrates could be achieved with conventional gravel artificial wetlands (4–46% removal of NO_3^-), higher significant removal rates were obtained by cork constructed wetlands (62–98% removal of NO_3^-).

The removal rates attained with cork system convert this pilot plant into a promising system to be used for the specific removal of nitrate from agriculture polluted waters. Thus, cork can be an excellent alternative material to mitigate the environmental impact generated by nitrogen pollution from agriculture. With the gravel pilot plants tested, a removal of 0.98 $gd^{-1}m^{-2}$ had been achieved, whereas the result increased to a higher 1.47 $gd^{-1}m^{-2}$ using cork. Naturally this material should be further studied since it will tend to react with the microbial communities present in the artificial wetlands.

Hydrogeological Transport Modeling

A hydrogeological model for the REAGRITECH Project implementation area was developed in order to show how the groundwater's quality could be improved according to different scenarios. The developed model could allow researchers to increase the knowledge of the natural environment of the area, and at the same time to analyse the impact and the possibility to improve the "polluted conditions" in the adjacent areas of the treatment plants and its surroundings.

In order to evaluate the evolution of the area after the implementation of the project, several scenarios with different concentrations and different flows were considered, all of them in two different situations: with or without buffer strip plantation. The purpose of these scenarios was to evaluate:

– The importance of the existence of the buffer strip.

– The relevance of nitrate concentration in irrigation/infiltration water.

– The spatial scope of the quality improvement.

In order to improve the understanding of the system and its operation, a model was developed for each studied area, one for the plot in Sudanell and two for the plots of Vilanova de la Barca and Bellvís in the Urgell area.

Table 2. Evolution of nitrate concentration related with time.

Effluent validation (3.2 mgL⁻¹)		
Time (days)	Concentration (mg L⁻¹)	Removal rate
0	40	-
366	34.7	13.22%
1096	25.8	35.59%
1826	21.5	46.33%
3650	19.5	51.27%

The construction and development of a mathematical model requires several phases of calibration, validation, and analysis of sensitivity and uncertainties, in order to achieve the best possible adjustment between experimental data and model predictions. For each pilot plot, the groundwater's quality was simulated and validated within the current project situation and using the different pre-determined scenarios, and to show the potential quality overtime considering buffer strips development, different effluent nitrate concentrations, and treatment flows.

A model with groundwater flow was used, taking into account the withdrawal from the most productive well, the recharge by irrigation in the reforested area (Buffer Strip), and the rain recharge, extracted from the historical rainfall by subtracting evapotranspiration, ET.

In Table 2 a validation start scenario is shown in order to test the evolution of nitrate concentration in relationship with time. The final nitrate concentration was directly related to elapsed time after the treatment process.

In this situation, the final nitrate concentration in the buffer strip area is less than 5 mgL⁻¹ and the effect to decrease nitrates concentration reaches to a distance greater than 50 m. Therefore, the two treatment systems combination (wetland and buffer strips) played a key role for improving groundwater quality.

Conclusions

REAGRITECH project demonstrated the feasibility of an innovative combined system for the removal of diffuse agricultural pollutants such as nitrates. These natural appropriate treatments, like sub-surface flow constructed wetlands and buffer strips (vegetation of native plant species), remove pollutants with low maintenance costs, low or zero energy requirements, no technical skills requirements for maintenance, and no production of sludge.

REAGRITECH managed to build and test an innovative prototype that is compact, modular, and mobile, allowing it to be moved to different areas and to be installed easily and low cost. As an innovation, some of the CW pilot plants used cork in the granular bed, as an alternative to standard gravel material. Although substantial removal of nitrates could be achieved with conventional gravel CWs (4–46% removal of NO_3^-), higher removals (62–98% removal of NO_3^-) were obtained by cork constructed wetlands.

Thus, cork can be an excellent alternative material to mitigate the environmental impact generated by nitrogen pollution from agriculture. With the gravel pilot plants tested, a removal of 0.98 g/d/m^2 had been achieved, whereas the result increased to a higher 1.5 gd^{-1}m^{-2} using cork.

A hydrogeological model for the implementation area was developed in order to show how the groundwater's quality could be increased according to different scenarios. After the validation was finished, it was concluded that:

- Buffer Strip presence is essential, as a complementary system to improve the groundwater quality.

- If greater flow is added to the medium with low nitrate concentration, groundwater quality is improved and the affected area with positive effects increased.

- The dilution effect is the primary improving quality factor. Depending on the hydrogeological conditions, the affected area is variable.

Further research will be directed to analyze and quantify some microbial genes as biological markers of denitrification, such as nirS and nosz, using quantitative PCR real time and Illumina MiSeq high-throughput sequencing.

Acknowledgments

To LIFE Programme, the EU's financial instrument supporting environmental, nature conservation and climate action projects throughout the EU, that supports REAGRITECH LIFE 11ENV/ES/579. Authors gratefully acknowledge Dr. José Carlos Mendoza for its manuscript revision.

References Cited

APHA. 2002. Standard Methods for Examination of Water and Wastewater. American Public Health Association. AWWA, American Water Works Association. Washington, D.C. US.

Babatunde, A.O., Y.Q. Zhao, M. O'Neill and B. O'Sullivan. 2008. Constructed wetlands for environmental pollution control: A review of developments, research and practice in Ireland. Environment International 34: 116–126.

Calera, A. and I. Campos. 2010. El uso de la teledetección en la gestión del agua en la agricultura de regadío. pp. 13–15. In: Incorporación de la teledetección a la gestión del agua en la agricultura. Monográfico Riegos del Alto Aragón. Enero. http://www.infoterra.es/asset/cms/file/monografico_riegos_alto_aragon.pdf. In Spanish.

Calheiros, C.S.C., A.F. Duque, A. Moura, I.S. Henriques, A. Correia, A.O. Rangel et al. 2009. Substrate effect on bacterial communities from constructed wetlands planted with Typhalatifolia treating industrial wastewater. Ecol. Eng. 35: 744–753.

Chaparro, J. and B. Pérez. 2013. Agro-Tic's una visión tecnológica integrada e interactiva de información para el desarrollo agrícola en Venezuela. In 11th Latin American and Caribbean Conference for Engineering and Technology. August 14-1. Cancún, Mexico. In Spanish.

Dong, X. and G. Reddy. 2010. Soil bacterial communities in constructed wetlands treated with swine wastewater using PCR-DGGE technique. Bioresource Technology 101: 1175–1182.

European Environment Agency. 2007. Present concentration of nitrate in groundwater bodies in European countries.

Fernández-Ruiz, L. 2007. Los nitratos y las aguas subterráneas en España. Enseñanza de las Ciencias de la Tierra 15(3): 257–265. In Spanish.

Gallegos, A., L. Aguilar, I. Campos, P. Caro, S. Sahuquillo, C. Pérez et al. 2016. TIC para la determinación de los parámetros operacionales de humedales construidos diseñados para el tratamiento de aguas contaminadas por nitratos. Revista Ingenierías Universidad de Medellín 15(28): 53–70. In Spanish.

Ghadiri, H. and C.W. Rose. 1991. Overland flow transport of organochlorine pesticides and their enrichment in the eroded sediment. *In*: Organochlorine Residues Strategies for Management and Research in Australian Agriculture, Volume 2. Primary Tasks Pty. Ltd. Victoria, Australia.

Ibekwe, A.M., C.M. Grieve and S.R. Lyon. 2003. Characterization of microbial communities and composition in constructed dairy wetland wastewater effluent. Appl. Environ. Microbiol. 69: 5060–5069.

Isermann, K. 1990. Share of agriculture in nitrogen and phosphorus emissions into the surface waters of Western Europe against the background of their eutrophication. Fertil. Res. 26: 253–269.

Kadlec, R. and S. Wallace. 2009. Treatment Wetlands. 2nd edition, Taylor & Francis Group.

Mayo, A.W. and T. Bigambo. 2005. Nitrogen transformation inhorizontal subsurface flow constructed wetlands. I: Model development. Phys. Chem. Earth 30: 658–667.

Morillo, J., M. Bellido and E. Gordillo. 2005. La agricultura y las nuevas tecnologías. In La agricultura y las ganaderías extremeñas. Ch. 19: 383–401. In Spanish.

Munné, A., C. Solà and N. Prat. 1998. QBR: Un índice rápido para la evaluación de la calidad de los ecosistemas de ribera. Tecnología del Agua 175: 20–37. In Spanish.

Pérez, A., M. Milla and M. Mesa. 2006. Impacto de las tecnologías de la información y la comunicación en la agricultura. Cultivos Tropicales 27(1): 11–17. In Spanish.

Runes, H.B., J.J. Jenkins, J.A., Moore, P.J. Bottomley and B.D. Wilson. 2003. Treatment of atrazine in nursery irrigation runoff by a constructed wetland. Water Res. 37(3): 539–550.

Stottmeister, U., A. Wiessner, P. Kuschk, M.K. Kappelmeyer, R.A. Bederski, H. Muller et al. 2003. Effects of plants and microorganisms in constructed wetlands for wastewater treatment. Biotechnol. Adv. 22: 93–117.

Tournebize, J., M.-P. Arlot, C. Billy, F. Birgand, J.-P. Gillet and A. Dutertre. 2008. Quantification et maîtrise des flux de nitrate: de la parcelledrainéeaubassinversant. Ingénieries (Special Number), pp. 5–25. http://www.set-revue.fr/sites/default/files/articles-eat/pdf/DG2008-PUB00024193. pdf. In French.

Tournebize, J., C. Chaumont and U. Mander. 2017. Implications for constructed wetlands to mitigate nitrate and pesticide pollution in agricultural drained watersheds. Ecological Engineering 103: 415–425.

Truu, M., K. Nurk, J. Juhanson and U. Mander. 2005. Variation of microbiological parameters within planted soil filter fordomestic wastewater treatment. J. Environ. Sci. Health. Part Toxic/Hazard. Subst. Environ. Eng. 40: 1191–1200.

Vacca, G., H. Wand, M. Nikolausz, P. Kuschk and M. Kastner. 2005. Effect of plants and filter materials on bacteria removal in pilot-scale constructed wetlands. Water Res. 39: 1361–1373.

Vymazal, J. 2007. Removal of nutrients in various types of constructed wetlands. Sci. Total Environ. 380: 48–65.

12

Bioremediation of Shrimp Aquaculture Effluents

The Convenience of Artificial Wetlands

*Otoniel Carranza-Díaz** and *José Guillermo Galindo-Reyes*

INTRODUCTION

On the Pacific Ocean coasts of Mexico, and particularly in Sinaloa, the development of shrimp farming has been carried out since its inception in a disordered way, using shrimp monoculture biotechnology. This is presently an obsolete method since it has resulted in water pollution and environmental impact in areas where farms are located. Today, this has become a serious problem that requires prompt solutions, because the farm effluents contain a high amount of organic matter, pesticides (in the range of ng/L to µg/L), coliform bacteria, and other contaminants that are negatively affecting the ecosystems adjacent to the farms. Faced with this problem, one of the most viable alternatives is the use of constructed wetlands (Lin et al. 2002, 2003, 2005; Idris et al. 2012; Jiabo et al. 2014).

Artificial wetlands using filtering bivalve mollusks, such as oysters (*Crassostrea* spp.), mussels (*Mytilus* spp.), or other species, in effluents from shrimp ponds represent a possible solution. In this chapter, the characterization of a shrimp farm effluent in Sinaloa, Mexico, comparing the water quality before and after treatment with an integrated oyster bank, exploring the potential of this method and other wetland remediation techniques were examined. Total suspended solids (TSS), heavy metals

Facultad de Ciencias del Mar, Universidad Autónoma de Sinaloa. Paseo Claussen S/N, Col. Los Pinos. Mazatlán, Sinaloa, México.
Email: guillermo_galindo_reyes@hotmail.com
* Corresponding author: otoniel.carranza@uas.edu.mx

(Hg, Pb), selected pesticides, and total coliform bacteria amounts were determined in the water discharged by a selected shrimp farm. The average values recorded were: TSS (53.5 mg/L), Hg (5.5 µg/L), Pb (289 µg/L), and total coliforms (2210 MPN/100 mL). Also, different pesticides were found, which ranged from 0.1 to 1.75 µg/L. Subsequently, a pilot test was conducted by placing thousand oysters on the discharge area (natural wetland) of farm effluent during thirty days. At the end of this time, the same pollutants were quantified, finding a reduction of pollutant concentrations by around 45 percent for pesticides and 55 percent for TSS and heavy metals. The total coliform bacteria decreased from 3 to 5 times MPN/100 mL. These results demonstrated that using bivalves mollusks as a bioremediation technique in shrimp farms is very promising from an economic and environmental point of view. In addition, perspectives regarding the use of constructed wetlands for treatment of marine aquaculture effluents are presented and discussed.

Aquaculture Production Systems

In many countries around the world, the aquaculture activities have grown vertiginously during the last decades. In fact, aquatic organisms farming represent an important part of the food industry worldwide. In 2014, aquaculture production registered 73.8 million tons worldwide, equivalent to 44.1 percent of the total world production by fishing and aquaculture (FAO 2016). Among the most sought after aquatic organisms are marine crustaceans such as the Pacific white shrimp (*Litopenaeus vannamei*). Shrimp has reached a production level that has placed it as the second most demanded fishery product worldwide (FAO 2016). Consequently, shrimp production systems technification has advanced in recent years (Ray and Lotz 2017). The shrimp cultivation in ponds under controlled conditions is one of the most commonly used techniques. However, this biotechnology has been developed with many failures, particularly because it is based on non-ecological principles. Thus, it has resulted in economic losses in the shrimp farms, and also an increasingly negative environmental impact due to deforestation of mangroves and salt marsh as well as wastewater discharges into the environment. The shrimp farms wastewater contain an excess of organic matter, nutrients, solids and other pollutants that drain into coastal ecosystem where a lot of these farms are located (Paez-Osuna et al. 2003). In Mexico's northwest region, there is a large aquaculture production which includes larval production laboratories and shrimp fattening farms (CESASIN 2012). Traditionally, fattening units consist of semi-extensive cropping systems characterized by requiring large areas generally in the order of hectares. In addition, modern techniques of white shrimp farming have been implemented, among which those intensive and hyper-intensive farming systems, new shrimp lines and aquaculture feeding strategies development (Krummenauer et al. 2014). The semi-extensive practice in aquaculture involves frequent replacements of water which are taken from the environment, used for farming, and then dumped into the sea, in some cases without any treatment (Liang et al. 2017).Water exchange in aquaculture may carry different contaminants such as nutrients, heavy metals, pesticides, and microorganisms such as bacteria, viruses, parasites, and fungi (Páez-Osuna et al. 2003). While bacterial impact in farming systems can be controlled with the use of antimicrobials; viruses can cause total losses (Escobedo-Bonilla et al. 2008), and organic pollutants and heavy metals residues in penaeid shrimp may represent a risk of being transferred to the trophic chain (Galindo-Reyes et al. 1999; Ruelas-Inzunza et al. 2004). This situation has led many producers worldwide to opt for bio-safe aquaculture production systems, that is, systems with greater control over cultures that are generally closed, with low rates of water exchange and farming high densities. Currently, intensive and hyper-intensive shrimp farming systems have gained popularity because it is possible to grow organisms at high densities with limited or even zero water exchanges (De Souza et al. 2012). Other variants are farming systems with bio-flocs or BFT (*BioFloc Technology*) (Schveitzer et al. 2013) as well as recirculating aquaculture systems (Badiola et al. 2012). Therefore, using these techniques types, aquaculture production may increase which require additional energy for their derived implementation of extra feed intake, prebiotics or probiotics addition, aeration systems operation, and pumps to move tons of daily water use. Therefore, increasing production costs and impacts the environment.

Bioremediation of Aquaculture Effluents

Aquaculture effluents contain nitrogen and phosphorus concentrations which in high amounts are harmful to the marine environment (Páez-Osuna 2001). Eutrophication of coastal waters is one of the most serious environmental problems arising from aquaculture effluents discharge into the sea (Páez-Osuna et al. 2003). Originally, shrimp farms were settled as monoculture systems in semi-intensive to intensive modalities, with an increasing use of fertilizers, formulated food, biocides, disinfectants based on chlorine, etc., to guarantee good harvests, and principally, economic gains. Furthermore, the shrimp farming has been developed following the technology used in Ecuador which is now obsolete, where since the 1980s a number of problems of mismanagement and a misconception of aquaculture had already occurred, resulting in a series of losses in productivity and considerable damage to the environment (FAO 2017). Aquaculture effluents bioremediation consists of the remediation of these waters through biological strategies generally based on natural processes occurring in the environment (Martins et al. 2017). Bioremediation is a practice that has been implemented worldwide for the treatment of aquaculture effluents (Gutierrez-Wing et al. 2006). There are examples of aquaculture bioremediation systems using mangroves (Su et al. 2011), microalgae systems (Marinho-Soriano et al. 2011; Yu et al. 2014), constructed wetlands (Lin et al. 2002), and filter feeders such as bivalve mollusks (e.g., *Anadara tuberculosa*) to decontaminate aquaculture effluents (Nieves-Soto et al. 2011). The use of filtering bivalve mollusks such as oyster (*Crassostrea* spp.) and clam (*Anadara* spp.) on the effluents from shrimp culture ponds would reduce the particulate organic matter, biocides, coliform bacteria, and some heavy metals, which are contaminants of concern in farm effluents (Galindo-Reyes et al. 1999). This biotechnology is actually economical and environmentally friendly, and could additionally generate an extra value to the productivity of the farms, since the mollusks could be used as raw material, for instance to glycogen production, which has diverse uses in alimentary industry (Anacleto et al. 2015). Furthermore, stabilization ponds (Von Sperling 2007) are also used for aquaculture effluents treatment, representing an interesting option for producers due to their low construction and operation costs. In practice, stabilization ponds are perhaps the most commonly used treatment systems for shrimp aquaculture effluent remediation. Stabilization ponds can store wastewater for a certain time, retain particles, and biodegrade some contaminants, hence, reducing the impact on the environment (Páez-Osuna 2001). However, stabilization ponds in the aquaculture industry are complex systems, of which there is little documented information to identify their ecological benefit on receiving water bodies. If the stabilization pond's design is inappropriate or there is no training of the personnel operating them, poor water treatment performance may result. On the other hand, stabilization ponds normally lack of maintenance, resulting in proliferation of aquatic plants and organisms such as amphibians, crustaceans, mollusks, fish, which can reduce hydraulic residence time and consequently water treatment efficiency. In this sense, artificial or constructed wetland systems are a possible option to replace stabilization ponds, thereby mitigating the effect on the adjacent ecosystems.

A Case Study in Sinaloa, Mexico

For the purposes of this work, a case study on bioremediation using bivalve mollusks to reduce pollution caused by aquaculture activities will be presented. A series of measurements and determinations were carried out in a shrimp farm located in the south of Sinaloa. Figure 1 shows a scheme of the selected shrimp farm for this study.

Figure 1 shows the shape, volume, and the pond arrangement of the study farm facility. The farm cultivates white shrimp (*Litopenaeus vannamei*) in five ponds in monoculture. The bivalve mollusks of the genus *Crassostrea* spp. were collected in the Piaxtla bay, located in the municipality of San Ignacio, Sinaloa (Mexico), which is a site without reports of contamination. Then the mollusks were moved to the Faculty for Marine Sciences of the Sinaloa Autonomous University (*UAS*, in Spanish), where they were kept for three weeks under controlled conditions of clean seawater, supplied with aeration and fed with microalgae. In this way, the organisms became self-depurated. Once clean, the mollusks were settled in the discharge area (natural wetland) of the shrimp farm (Fig. 2). The number of these organisms per unit area was determined based on the amount of particulate organic matter, biocides, and heavy metals previously quantified in the effluent. The effluent flow rates discharged of the shrimp farm were provided by the manager of the farm. The total volume discharged per day was estimated considering the percentage of water replacement of the ponds per day, the volume of the ponds, and the water replacement frequency. To assess this, a flow meter was used measuring the time of discharge and obtaining the total volume of water discharged per unit of time (hour, day, or week). Single samples for quantifying TSS, metals, and pesticides were taken at the input channel, output channel, and E1, E2, E3 sampling sites of the farm ponds (Figs. 1 and 2), before and after 30 days of utilizing the bivalve mollusks as bioremediation organisms. Then, the samples were analyzed in the laboratory. The physicochemical parameters pH, temperature, salinity, and dissolved oxygen were analyzed *in situ* at the mentioned sampling locations. The salinity of water was measured with an Atago S/Mill refractometer. Temperature was recorded using a mercury thermometer and dissolved oxygen was determined by the method of Winkler, following the protocol presented in Strickland and Parson (1972). The pH values were measured with a Hanna HI 9811 potentiometer, and total suspended solids (TSS) were determined according to APHA-AWWA-WPCF (1992). Furthermore, selected heavy metals and pesticides were measured in water

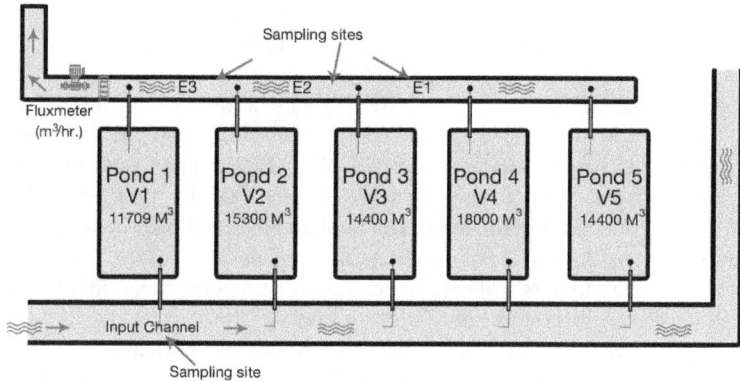

Fig. 1. Schematic diagram of the shrimp farm facility.

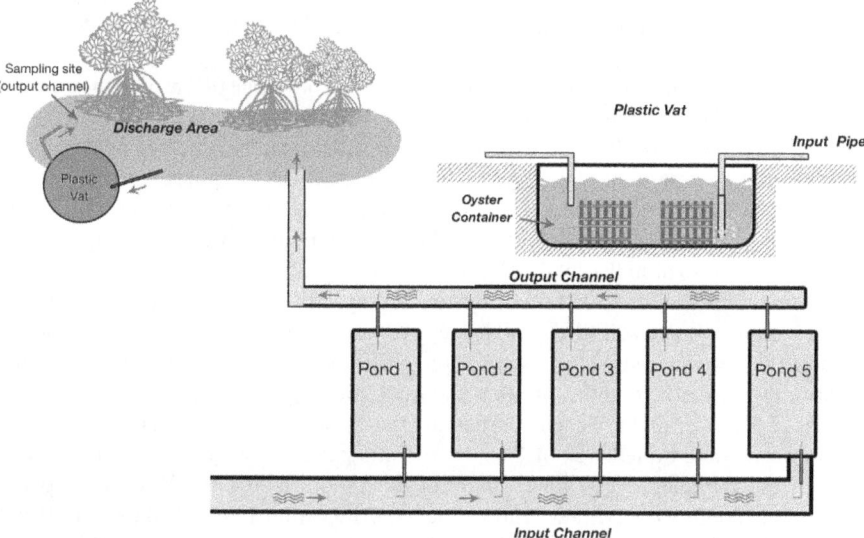

Fig. 2. Bioremediation system and growth of oysters in the shrimp farm discharge area (natural wetland).

samples at the five sampling locations (Figs. 1 and 2). The heavy metals content (lead and mercury) in water was quantified by the dithizone method presented in APHA-AWWA-WPCF (1992) whereas the total coliform bacteria were determined by the MPN method (Escartín-Fernández 1981; APHA-AWWA-WPCF 1992). In addition, the EPA surrogate pesticides (α-HCH, Lindane, β-HCH, Heptachlor, δ-HCH, Endosulfan 1, DDE, Dieldrin, Endrin, DDD, Endosulfan 2, DDT, Endrin Aldehyde, Metoxichlor) were selected as target compounds due to their environmental relevance (Woodruff et al. 1994). The pesticides were first extracted from the water matrix by a liquid-liquid system according to APHA-AWWA-WPCF (1992). Subsequently, the extracts were concentrated on a vacuum rotating evaporator and final analysis was performed using a Shimadzu GC-17A gas chromatograph equipped with flame ionization and electron capture detectors following the protocols of AOAC (1990) and Capelli et al. (1992). The pesticide analyses were only conducted at the output channel, before and after the oyster placement (Fig. 2). In order to evaluate the mollusk growth rate and removal of pollutants from the effluent water of the farm (bio-remediation), several experiments were carried out using lots of oysters between 37 to 61 mm, previously treated by depuration process as above indicated. The oysters were then transferred to the farm and placed in a 1000-liter plastic pond, placed a few meters from the discharge channel of the farm (Fig. 2). The pond was connected by a 0.07 m (3-inches) PVC pipe to the discharge channel, and the flow was adjusted to 1224 L/h. The oysters used in this work were collected from a place nearby the city of Mazatlan (Piaxtla bay) without pollution reports as mentioned before. Then, they were acclimated according to personal experience in a plastic pond using filtered sea water and fed with microalgae cultures during one month.

Figure 2 shows a scheme of the shrimp farm and the discharge area. The input water was pumped from the estuarine area to the ponds during the flood tide periods. The output water was discharged to an adjacent zone (natural wetland) not directly connected to the estuarine area. During the experimental period the shrimps were fed with commercial food (Purina®). The hydraulic residence time of the water in the ponds was 0.84 d (Fig. 1).

After one month, some mollusks were collected from a plastic vat, and the amount of pesticides, heavy metals (mercury and lead), TSS, and total coliform bacteria remaining in the water phase were determined by the methods indicated above.

In order to determine the amount of bivalve mollusks required to improve the water quality discharged by the farm effluent, the filtration rate (L/hr) and ingestion rate (mg/hr) of oysters were quantified. Thus, the amount of organisms required for contaminant removal in the total volume water of effluent was estimated.

To evaluate the filtration rate of oysters, 10 oysters were placed in three aquaria with 8 L of clean sea water filtered it through 10 μm pore size filters. 2 L of microalgae cultures (*Tetraselmis* spp., *Chaetoceros* spp., and *Isochrysis* spp.) were added to each aquarium. At time zero and 2 hr after, the microalgae quantity was counted using a Neubauer chamber and a Wesco composite microscope. With these data, the average filtration rate determined was of 7.69×10^6 cells per L-hr-oyster. Since the average mass of the microalgae cell reported by other authors is 301×10^6 mg cells per L (Rodríguez 2000; López-Prado 2003), then the ingestion rate calculated is 39 mg $(\text{hr-oyster})^{-1}$. These values are similar to those reported by other authors (Madrigal-Castro et al. 1985).

Oyster-based System Treatment Performance

The following Tables 1 and 2 and Fig. 3, show the physicochemical parameters and pollutants values registered in the shrimp farm before and after oysters were introduced into the discharge area. The oyster-based system treatment performance was conducted comparing the concentrations between the sampling dates at each sampling point before and after the oyster placement. The principal source of metals is the former mining industry whereas coliform bacteria are from urban source and cattle paddock. As can be seen, after the placement of the oyster bank into the farm's discharge area (natural wetland), the amounts of TSS, heavy metals, and total coliform bacteria were reduced to around 55 percent of the initial TSS and heavy metal concentration and total coliform bacteria, from 3 to 5 times the MPN/100 mL.

Table 1. Values of the physicochemical variables in the shrimp farm performed before (May 29, 2015) and after (June 29, 2015) introducing oysters.

Place	Temperature (°C)		Salinity (‰)		pH Value		Dissolved O_2 (mg/L)	
	Before	After	Before	After	Before	After	Before	After
E 1	28.2	32.2	45	31	8.8	7.6	4.33	5.83
E 2	28.6	33.6	45	31	8.8	7.6	4.42	5.92
E 3	28.1	32.1	44	30	8.9	7.7	4.11	5.81
Input channel	26.8	32.8	45	31	8.3	7.1	4.23	5.83
Output channel	28.5	32.5	46	32	8.8	7.6	4.13	5.63

Measurements were done *in situ* measuring one time per sampling point during each sampling campaign

Table 2. Total suspended solids (TSS), mercury, lead, and total coliform bacteria measured as the most probable number (MPN) in the study farm performed before (May 29, 2015) and after (June 29, 2015) introducing oysters.

Place	TSS (mg/L)		Mercury (µg/L)		Lead (µg/L)		Total coliform bacteria (MPN/100 mL)	
	Before	After	Before	After	Before	After	Before	After
E 1	40.7 ± 1.9	20.7 ± 3.1	2.2 ± 0.2	1.0 ± 0.2	180 ± 14.5	100 ± 16.5	980	180
E 2	41.2 ± 2.4	21.2 ± 3.3	2.4 ± 0.3	1.4 ± 0.2	190 ± 33.8	130 ± 11.0	990	190
E 3	43.5 ± 2.0	21.5 ± 3.2	2.3 ± 0.2	1.6 ± 0.1	190 ± 30.0	130 ± 9.6	900	200
Input channel	35.4 ± 3.0	17.4 ± 2.1	1.8 ± 0.4	0.9 ± 0.1	140 ± 1.8	90 ± 5.0	600	200
Output channel	53.5 ± 2.1	23.5 ± 2.6	5.5 ± 0.3	1.7 ± 0.2	289 ± 18.5	160 ± 8.1	2210	310

Analyses of TSS, Mercury and Lead were done by triplicate. The analysis for total coliform bacteria were done according to MPN considering 5 dilutions (APHA-AWWA-WPCF 1992)

Additionally, for pesticides except Endrin Aldehyde, the concentrations were reduced below 40 percent of the initial contaminant concentration (Fig. 3). Considering the total average load value (the average of the sum before and after the oysters per sampling point) of total suspended solids (TSS) measured in E1, E2, E3 and output channel for this investigation ($\bar{x}$ = 66.45 mg L^{-1}) in round numbers of 70 mg L^{-1}, and the volume discharged by it (732.8 m^3 hr^{-1}), then the amount of TSS discharged per hour, becomes 51,296,000 mg hr^{-1}. Moreover, the average filtration rate of the oysters was calculated to 7.69 × 10^6 cells per L-hr-oysters, and taking the average mass of microalgae reported by other authors of around 301 × 10^6 mg cells per L (Rodríguez 2000; López-Prado 2003), the ingestion rate turns out to be 39 mg per hr-oyster. These values are similar to those reported by other authors (Madrigal-Castro et al. 1985).

Therefore, the number of oysters required to retain (biofiltrate) this amount of TSS would be approximately to 1,315,282 oysters. In round numbers, it would be required 1.3 million of oysters, which at a density of 160–170 oysters per m^2 (the oyster cages have standard measures of 0.49 m^2 that holds approximately 80 oysters), will require an area of 7,879 m^2, with a 0.9 m depth in the discharge area

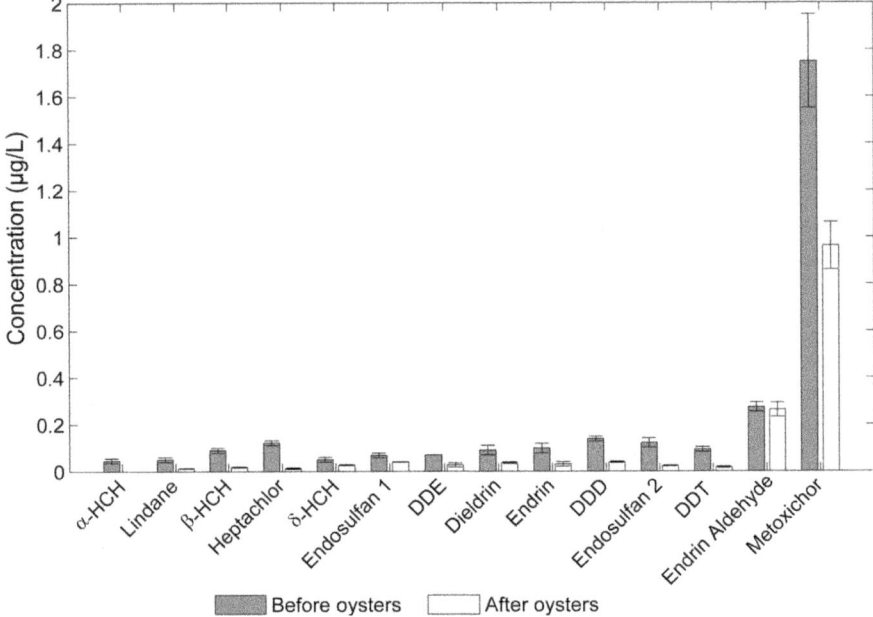

Fig. 3. Pesticides analyses at the output channel before (gray) and after (white) the oyster placement (Analysis of pesticides were done by triplicate).

(less than one hectare). These results demonstrate that the use of bivalve mollusks as a bioremediation technique in shrimp farms is very promising from the technical, economic, and environmental point of view.

However, care has to be taken regarding the human consumption of the contaminated mollusks. Previously the mollusks must be treated for eliminating contaminants such as metals and pesticides accumulated in their tissues. If the treatment is adequate, they might become potentially edible even for human consumption (Anacleto et al. 2015).

Artificial Wetlands Technology for Water Treatment in the Aquaculture Industry

Artificial wetland systems for wastewater treatment have been successfully studied worldwide and represent a technological option to be used in aquaculture effluents bioremediation (Liang et al. 2017). Although artificial wetlands have been predominantly studied as water treatment systems in low salinity aquaculture cultures (Lin et al. 2002; Schulz et al. 2003; Sindilariu et al. 2009), there are studies of artificial wetlands employed for marine aquaculture water treatment (Neori et al. 2000; Shpigel et al. 2013). For example, Webb et al. (2013) found nutrient removal efficiencies (total dissolved inorganic nitrogen) greater than 81.8 percent using artificial wetlands for marine aquaculture wastewater treatment.

The study of aquatic plants in artificial wetlands for aquaculture effluents treatment is essential for efficient water treatment (Kadlec and Wallace 2009).

Generally, halophytes have been used in saline waters which have the capacity to adapt to salinity mainly growing in coastal areas throughout the world (Buhmann and Papenbrock 2013). Some of the aquatic plants used in artificial wetland systems for the treatment of wastewater from the aquaculture industry are shown in Table 3. As it can be seen, the use of artificial wetlands has been used in aquaculture farming covering a wide spectrum of salinities, from freshwater to mesohaline, oligohaline, and polyhaline seawater. Although Table 3 does not seek to be a definitive comparative baseline, it shows artificial wetland systems' versatility from the point of view of plant selection, water type, wetland type, farming type, and its potential to be used for aquaculture effluents bioremediation. In general, aquaculture effluents treatment using artificial wetlands has not been much studied in comparison to the use of this systems other types of wastewater (Kadlec and Wallace 2009). In the context of sustainable aquaculture, it is necessary to improve farming systems development integrating efficient water use and decontamination strategies (Martins et al. 2010).

Artificial wetlands are extremely versatile water treatment systems which may significantly reduce water treatment costs and pollution in coastal areas (Brown et al. 1999). Additionally, artificial wetlands are a low energy requirements strategy so its implementation in aquaculture production units can be attractive (Liang et al. 2017).

Aquatic plants can be easily integrated into artificial wetland systems for aquaculture effluents treatment. However, in marine aquaculture, the salt content in effluents requires careful selection of plants (De Lange et al. 2013). The plants' rhizodeposits in the wetland, oxygen transport to the rhizosphere, radicular filtration, and nutrient uptake from water may vary among selected plant species (Buhmann and Papenbrock 2013; Stottmeister et al. 2003). Furthermore, some coastal plants are creeping plants so artificial wetlands design for marine aquaculture effluent treatment must be adapted to such conditions. Artificial wetland elements such as depth, porous media selection (gravel or sands), and hydraulic regime (saturation *vs.* percolation) play an important role in promoting vegetation rooting, solute retention as well as redox conditions in the wetland. Besides, there is a broad spectrum of plants that can be used in artificial wetland systems for marine aquaculture effluents treatment that have not yet been studied. For example, *Sedumpachy phyllum* and *S. clavatum* are succulent plants native to Mexico that have been used in green roofs in hot and dry climates (Farrell et al. 2012). These plants have the capacity to tolerate salinity and high temperatures so their use in wetland systems for marine aquaculture effluents bioremediation is not to be ruled out. Other plants such as *Salicornia bigelovii* and *Suaeda esteroa* have the potential to be implemented in artificial wetlands systems (Páez-Osuna 2001). In particular, *Salicornia bigelovii* is of interest given its oleaginous characteristics that can be used for biofuels or animal feed supplement production (previously testing its innocuity) (Glenn et al. 1991).

Aquaculture effluents generally contain concentrations of dissolved oxygen in the order of 4 mg/L or higher which allows the survival of culture organisms (Lin et al. 2005; Ray et al. 2017). In order to promote the biological transformation of nutrients such as nitrogen in aquaculture effluents in artificial wetlands system, nitrification and denitrification of the effluent should be considered. Thus, the study of nitrogen transformations in artificial wetlands should be oriented to a wide range of redox gradients (Stottmeister et al. 2003). On the other hand, toxic hydrogen sulfide is a compound that can be present in aquaculture effluents (Herbeck et al. 2014).

Table 3. Selection of wetlands plants used for aquaculture wastewater treatment.

Plant	Culture	Salinity ‰/ conductivity	Wetland type/ Scale	Reference
Acrostichum aureum	Shrimp	36–42	HSSF/pilot scale	Sansanayuth et al. (1996)
Salicornia europaea	*Litopenaeus vanammei*	23.9–31.7	SSF/small scale	Webb et al. (2013)
Salicornia persica	*Sparus aurata*	20	Hybrid/pilot scale	Shpigel et al. (2013)
Ruppia maritima *Chara* spp. *Pithophora* spp. *Nymphaea odorata* *Hydrochola carolinensis* *Typha latifolia* *Juncus effusus* *Sesbania drummondii* *Borrichia frutescens* *Avicennia germinans*	*Litopenaeus vanammei*	3–15	SF/full scale	Tilley et al. (2002)
Phragmites australis *Spartina alterniflora* Loisel *Scirpus mariqueter*	*Litopenaeus vanammei*	8.25–8.26‰	Hybrid/full scale	Shi et al. (2011)
Phragmites australis *Spartina alterniflora* *Scirpus mariqueter*	*Takifugu obscurus*	7.4–7.6‰	Hybrid/full scale	Jiabo et al. (2014)
Ipomoea aquatica *Paspalum vaginatum* *Phragmites australis*	*Chanos chanos*	0.5‰	Hybrid/pilot scale	Lin et al. (2002)
Typha angustifolia *Phragmites australis*	*Litopenaeus vannamei*	0.3‰	Hybrid/full scale	Lin et al. (2005)
Arundo donax *Phragmites australis*	Finfish	3,600 µS cm⁻¹	HSSF/pilot scale	Idris et al. (2012)
Phragmites communis *Phalaris arundinacea*	Trout	723 µS cm⁻¹	SSF/pilot scale	Sindilariu et al. (2009)
Phragmites australis	*Oncorhynchus mykiss*	677 µS cm⁻¹	HSSF/small scale	Schulz et al. (2003)
Canna sp. *Scirpus* sp.	Red Tilapia (*O. mossambicus* x *O. aureus*)	n.r.	HSSF/full scale	Zachritz II et al. (2008)
Canna indica *Typha orientalis* *Acorus calamus*	catfish	n.r.	VF/full scale	Zhong et al. (2011)

SF: Surface Flow, HSSF: Horizontal Sub-Surface Flow, SSF: Sub-Surface Flow, VF: Vertical Flow, n.r. not reported

Nevertheless, the sulfur transformations in artificial wetlands for marine aquaculture effluents treatment have not been widely studied yet.

The versatility of artificial wetlands is again highlighted as being able to be used in their variants as intensified and hybrid systems (Nivala et al. 2013; Ávila et al. 2014). For example, elements such as bivalve mollusks can be incorporated into artificial wetland systems. Bivalve mollusks as filtering organisms have the advantage of absorbing contaminants within their organisms (Nieves-Soto et al. 2011). The hybrid artificial wetland-bivalve mollusks system can be used for aquaculture effluents bioremediation but has not yet been researched. The project of a hybrid artificial wetland with bivalve mollusks must consider the mollusks' fate as they should not be directly used for human consumption. The contaminat load by pathogenic organisms and heavy metals in bivalve mollusks could lead to public health problems if they are consumed without a previous depuration treatment.

Aquaculture industry wastewater treatment is the first step in establishing water reuse strategies. Aquaculture effluents contain nutrients that, although in excess can cause ecological problems, can also be reused. Among the nutrients that can be recovered from aquaculture effluents are unconsumed food and trace oligoelements. The use of artificial wetland systems in recirculating aquaculture systems is a promising activity since they can function as water treatment modules that reduce the pollutant load of effluents but recover some nutrients from water (Lin et al. 2003). In general, the use of artificial wetlands in recirculating aquaculture systems (RAS) has been studied in low salinity farming, while there are few studies of artificial wetlands in marine aquaculture RAS systems (Tilley et al. 2002; Shi et al. 2011). Although water treatment techniques for recirculating aquaculture systems have been extensively developed and technified worldwide (Van Rijn 2013), treatment processes are frequently mechanized and require external energy sources which increase operating costs (Martins et al. 2010). The use of artificial wetlands may represent a low-cost strategy for marine aquaculture in RAS systems. Additionally, aquaculture effluents' disinfection strategies could include metallic aggregates that form part of the support medium in artificial wetlands (Magaña-López et al. 2016) and that may eventually replace conventional systems based on UV light in RAS systems, but have not yet been researched.

The use of artificial wetland systems for the treatment of intensive farming effluents with bio-flocs has not been studied to date. Intensive and hyper-intensive farming systems with bio-flocs are technological innovations that generate large volumes of solids contained in wastewater (Schveitzer et al. 2013). Bio-flocs are microorganism conglomerates which contain nutrients that can be recovered from effluents. The design of artificial wetlands for the treatment of effluents from intensive farming (e.g., Biofloc technology) should consider the fate of the solids, either to recover some of them or to retain them in the artificial wetland without promoting clogging. Furthermore, the design and construction of artificial wetlands for the aquaculture industry should be based on area availability. In many cases, the area is limited so that stabilization ponds can be conditioned to function as an artificial wetland.

At the global level, the use of antimicrobial substances in the aquaculture industry is increasing (Santaeufemia et al. 2016). Antimicrobials are biocidal substances designed to have a biological effect on bacteria or parasites (Seoane et al. 2014).

Aquaculture effluents may contain traces of many of these substances, which appear normally in concentrations of the order of ng L^{-1} to μg L^{-1} (Hussain et al. 2012). Antimicrobials are generally used as prophylactic treatment in the aquatic organisms culture. In Mexico, the antimicrobials use is not regulated and, in general, chemical substances in the aquaculture industry can lead to drug resistance in pathogenic bacteria (Páez-Osuna 2001), as well as accumulation of chemicals in their bioactive form or as metabolites in different environmental compartments. When entering coastal ecosystems, antimicrobials can have a negative effect on the environment even though they are in low concentrations (Schwarzenbach et al. 2006). Among the substances used in the aquaculture industry, the most common ones are chloramphenicol, iodine, formalin, malachite green (Páez-Osuna 2001), oxytetracycline, enrofloxacin (Bôto et al. 2016), florfenicol (Seoane et al. 2014), monensin and salinomycin (Hussain et al. 2012). Bôto et al. (2016) investigated the reduction of oxytetracycline and enrofloxacin concentrations in artificial wetland systems, showing satisfactory results (> 99 percent removal). However, it is still necessary to advance in the research towards reducing the use of antimicrobials or investigate their fate in artificial wetlands systems for marine aquaculture effluents treatment in order to elucidate aspects such as the role of plants, salinity, type and size of the wetland (pilot scale or full scale), and organic load. Additionally, it is necessary to overcome the analytical challenge involved in the processing of samples and access to high precision instruments such as liquid chromatography coupled to mass spectrometry for the detection and quantification of drug residues as well and pesticides in the aquaculture industry (Nakata et al. 2005).

Final Remarks

Feeding strategies and culture densities are the main factors leading to water pollution derived from shrimp production. In addition, food additives and molasses used for intensive cultures lead to wastewater containing high concentrations of carbon and nitrogen. According to this review on the application of an artificial wetlands system for shrimp farm wastewater treatment, the following experiments should contemplate a variety of issues. Model experiments are needed in order to understand the response of constructed wetlands for treating high loads of organics and nutrients contained in shrimp effluents. While most studies have focused on effluents from shrimp fattening, artificial wetlands can also be used for the treatment of the shrimp nursery's wastewater, effluents derived from shrimp-larvae production as well as wastewater coming from live food cultures such as rotifers, artemia, and other crustaceans.

Farmers are particularly concerned about shrimp diseases which have caused economical loses. Unfortunately, the use of pharmacological products in aquaculture is expected to increase in future. Therefore, further experiments should focus not only on the removal of antimicrobials by constructed wetlands but also on the reduction of pathogenic bacteria such as *Vibrio parahaemolyticus* as well as viruses like WSSV (White Spot Syndrome Virus) which can be investigated in model experiments employing molecular biology approaches (Escobedo-Bonilla et al. 2008). Investigation on integrated multi-trophic aquaculture coupled with constructed wetlands could also be an interesting option to reduce the use of antimicrobial in shrimp production.

In several countries, the legislation for wastewater discharges to coastal areas is becoming stricter due to a growing understanding of the environmental impacts and concerns about climate change and ocean life protection. Artificial or constructed wetlands are a promising technology which can be implemented with oyster-based treatment modules in order to mitigate the environmental impact of shrimp farming. Further experiments should focus on the role halophyte plants in oyster based constructed wetlands systems as well as in recirculating aquaculture systems towards promoting water reuse at pilot and full scale.

Acknowledgments

Otoniel Carranza-Diaz acknowledges the support of the Project PROFAPI 2015/060 (Autonomous University of Sinaloa). Authors are thankful to Dipl.-Ing. Miriam S. Müller for revising this chapter.

References Cited

Anacleto, P., A.L. Maulvault, M.L. Nunes, M.L. Carvalho, R. Rosa and A. Marques. 2015. Effects of depuration on metal levels and health status of bivalve mollusks. Food Control 47: 493–501.
AOAC. 1990. Official Methods of Analysis. 15th Edition. Association of Official Analytical Chemists, Inc. Virginia, US.
APHA-AWWA-WPCF. 1992. Standard Methods for the Examination of Water and Wastewater, 17 ed. Spanish translation, Ediciones Diaz de Santos SA. Madrid, Spain.
Ávila, C., V. Matamoros, C. Reyes-Contreras, B. Piña, M. Casado, L. Mita et al. 2014. Attenuation of emerging organic contaminants in a hybrid constructed wetland system under different hydraulic loading rates and their associated toxicological effects in wastewater. Sci. Tot. Environ. 470-471: 1272–1280.
Badiola, M., D. Mendiola and J. Bostock. 2012. Recirculating Aquaculture Systems (RAS) analysis: Main issues on management and future challenges. Aquacul. Eng. 51: 26–35.
Bôto, M., C.M.R. Almeida and A.P. Mucha. 2016. Potential of constructed wetlands for removal of antibiotics from saline aquaculture effluents. Water 8: 465.
Brown, J.J., E.P. Glenn, K.M. Fitzsimmons and S.E. Smith. 1999. Halophytes for the treatment of saline aquaculture effluent. Aquaculture 175: 255–268.
Buhmann, A. and J. Papenbrock. 2013. Biofiltering of aquaculture effluents by halophytic plants: Basic principles, current uses and future perspectives. Environ. Exp. Bot. 92: 122–133.
Capelli, R., V. Contardi, V. Fossato and R. Frache. 1992. Metodologie Analitiche per lo Studio della Qualita dell'Ambiente Marino. Consiglio Nazionale delle Ricerche. Genova, Italia. In Italian.
CESASIN. 2012. Sinaloa líder en producción de camarón de cultivo en México. Comité de Sanidad Acuícola de Sinaloa. Industria Acuícola. 8(5): 16–22. In Spanish.
De Lange, H.J., M.P.C.P. Paulissen and P.A. Slim. 2013. 'Halophyte filters': The potential of constructed wetlands for application in saline aquaculture. Int. J. Phytoremediation 15: 352–364.
De Souza, D.M., S.M. Suita, F.P.L. Leite, L.A. Romano, W. Wasielesky and E.L.C. Ballester. 2012. The use of probiotics during the nursery rearing of the pink shrimp *Farfantepenaeus brasiliensis* (Latreille, 1817) in a zero exchange system. Aquacult. Res. 43: 1828–1837.
Escartín-Fernández, E. 1981. Microbiología Sanitaria. Agua y Alimentos. Vol. 1. Ediciones de la Universidad de Guadalajara. Guadalajara, México. In Spanish.
Escobedo-Bonilla, C.M., V. Alday-Sanz, M. Wille, P. Sorgeloos, M.B. Pensaert and H.J. Nauwynck. 2008. A review on the morphology, molecular characterization, morphogenesis and pathogenesis of white spot syndrome virus. J. Fish Dis. 31: 1–18.
FAO. 2016. El estado mundial de la pesca y la acuicultura 2016. Contribución a la seguridad alimentaria y la nutrición para todos. Roma. http://www.fao.org/3/a-i5798s.pdf. In Spanish.
FAO. 2017. National Aquaculture Sector Overview. Ecuador. National Aquaculture Sector Overview Fact Sheets. In: FAO Fisheries and Aquaculture Department. Rome. http://www.fao.org/fishery/countrysector/naso_ecuador/en.

Farrell, C., R.E. Mitchell, C. Szota, J.P. Rayner and N.S.G. Williams. 2012. Green roofs for hot and dry climates: Interacting effects of plant water use, succulence and substrate. Ecol. Eng. 49: 270–276.

Galindo-Reyes, J.G., V.U. Fossato, C. Villagrana-Lizarraga and F. Dolci. 1999. Pesticides in water, sediments, and shrimp from a coastal lagoon off the Gulf of California. Marine Poll. Bull. 38(9): 837–841.

Glenn, E.P., J.W. O´Leary, M.C. Watson, T.L. Thompson and R.O. Kuehl. 1991. *Salicornia bigelovii* Torr.: An oilseed halophyte for seawater irrigation. Science 251(4997): 1065.

Gutierrez-Wing, M.T. and R.F. Malone. 2006. Biological filters in aquaculture: Trends and research directions for freshwater and marine applications. Aquacul. Eng. 34: 163–171.

Herbeck, L.S., M. Sollich, D. Unger, M. Holmer and T.C. Jennerjahn. 2014. Impact of pond aquaculture effluents on seagrass performance in NE Hainan, tropical China. Marine Poll. Bull. 85: 190–192.

Hussain, S.A., S.O. Prasher and R.M. Patel. 2012. Removal of ionophoric antibiotics in free water surface constructed wetlands. Ecol. Eng. 41: 13–21.

Idris, S.M., P.L. Jones, S.A. Salzman, G. Croatto and G. Allinson. 2012. Evaluation of the giant reed (*Arundo donax*) in horizontal subsurface flow wetlands for the treatment of recirculating aquaculture system effluent. Environ. Sci. Pollut. Res. 19: 1159–1170.

Jiabo, X., S. Yonghai, Z. Genyu, L. Jianzhong and Z. Yazhu. 2014. Effect of hydraulic loading rate on the efficiency of effluent treatment in a recirculating puffer aquaculture system coupled with constructed wetlands. J. Ocean Univ. China. 13(1): 146–152.

Kadlec, R.H. and S. Wallace. 2009. Treatment Wetlands, Second ed. CRC Press, Boca Raton, Florida. US.

Krummenauer, D., T. Samocha, L. Poersch, G. Lara and W. Wasielesky, Jr. 2014. The reuse of water on the culture of Pacific white shrimp, *Litopenaeus vannàmei*, in BFT system. J. World Aquacult. Soc. 45(1): 3–13.

Liang, Y., H. Zhu, G. Bañuelos, B. Yan, Q. Zhou, X. Yu et al. 2017. Constructed wetlands for saline wastewater treatment: A review. Ecol. Eng. 98: 275–285.

Lin, Y.-F., S.-R. Jing, D.-Y. Lee and T.-W. Wang. 2002. Nutrient removal from aquaculture wastewater using a constructed wetlands system. Aquaculture 209: 169–184.

Lin, Y.-F., S.-R. Jing and D.-Y. Lee. 2003. The potential use of constructed wetlands in a recirculating aquaculture system for shrimp culture. Environ. Pollut. 123(1): 107–113.

Lin, Y.-F., S.-R. Jing, D.-Y. Lee, Y.-F. Chang, Y.-M. Chen and K.-C. Shi. 2005. Performance of a constructed wetland treating intensive shrimp aquaculture wastewater under high hydraulic loading rate. Environ. Pollut. 134: 411–421.

López-Prado, I.C. 2003. Alimentación del rotífero *Braquionus plicatilis* con dos microalgas marinas y su efecto nutricio para larvas mysis de *Litopenaeus vannamei*. B.S. Thesis (Tesis de licenciatura). Universidad Autónoma de Sinaloa. Mazatlán. México. In Spanish.

Madrigal-Castro, E., O. Pacheco-Urpí, E. Zamora-Madriz, R. Quesada-Quesada and J. Alfaro-Montoya. 1985. Tasa de filtración del ostión de manglar (*Crassostrea rhizophorae* Guilding 1828), a diferentes salinidades y temperaturas. Rev. Biol. Trop. 33(1): 77–79. In Spanish.

Magaña-López, R., V.M. Luna-Pabello, J.A. Barrera-Godínez, M.T. Orta de Velásquez and G. Fernández-Villagómez. 2016. Effect of mineral aggregates on the morphology and viability of *Toxocaracanis* eggs. Ecol. Eng. 90: 125–134.

Marinho-Soriano, E., C.A.A. Azevedo, T.G. Trigueiro, D.C. Pereira, M.A.A. Carneiro and M.R. Camara. 2011. Bioremediation of aquaculture wastewater using macroalgae and Artemia. Int. Biodeter. Biodegr. 65: 253–257.

Martins, C.I.M., E.H. Eding, M.C.J. Verdegem, L.T.N. Heinsbroek, O. Schneider, J.P. Blancheton et al. 2010. New developments in recirculating aquaculture systems in Europe: A perspective on environmental sustainability. Aquacul. Eng. 43: 83–93.

Martins, M.C., E.B.H. Santos and C.R. Marques. 2017. First study on oyster-shell-based phosphorous removal in saltwater—Aproxy to effluent bioremediation of marine aquaculture. Sci. Tot. Environ. 574: 605–615.

Nakata, H., K. Kannan, P.D. Jones and J.P. Giesy. 2005. Determination of fluoroquinolone antibiotics in wastewater effluents by liquid chromatography–mass spectrometry and fluorescence detection. Chemosphere 58: 759–766.

Neori, A., M. Shpigel and D. Ben-Ezra. 2000. A sustainable integrated system for culture of fish, seaweed and abalone. Aquaculture 186: 279–291.

Nieves-Soto, M., F. Enriquez-Ocaña, P. Piña-Valdez, A.N. Maeda-Martínez, J.R. Almodóvar-Cebreros and H. Acosta-Salmón. 2011. Is the mangrove cockle *Anadara tuberculosa* a candidate for effluent bioremediation? Energy budgets under combined conditions of temperature and salinity. Aquaculture 318: 434–438.

Nivala, J., T. Headley, S. Wallace, K. Bernhard, H. Brix, M. van Afferden et al. 2013. Comparative analysis of constructed wetlands: The design and construction of the ecotechnology research facility in Langenreichenbach, Germany. Ecol. Eng. 61: 527–543.

Páez-Osuna, F. 2001. The environmental impact of shrimp aquaculture: Causes, effects, and mitigating alternatives. Environ. Manag. 28: 131–140.

Páez-Osuna, F., A. Gracia, F. Flores-Verdugo, L.P. Lyle-Fritch, R. Alonso-Rodríguez, A. Roque et al. 2003. Shrimp aquaculture development and the environment in the Gulf of California ecoregion. Marine Poll. Bull. 46(7): 806–815.

Ray, A.J. and J.M. Lotz. 2017. Shrimp (*Litopenaeus vannamei*) production and stable isotope dynamics in clear-water recirculating aquaculture systems versus biofloc systems. Aquacult. Res. 1–9.

Rodríguez, B.B. 2000. Evaluación de las técnicas de producción de microalgas en el laboratorio comercial Generación Cincuenta, S.A. de C.V. B.S. Thesis (Tesis de Licenciatura). Universidad Autónoma de Sinaloa. Mazatlán. México. In Spanish.

Ruelas-Inzunza, J., S.B. Garcia-Rosales and F. Páez-Osuna. 2004. Distribution of mercury in adult penaeid shrimps from Altata-Ensenada del Pabellon lagoon (SE Gulf of California). Chemosphere 57: 1657–1661.

Sansanayuth, P., A. Phadungchep, S. Ngammontha, S. Ngdngam, P. Sukasem, H. Hoshino et al. 1996. Shrimp pond effluent: Pollution problems and treatment by constructed wetlands. Water Sci. Technol. 34(11): 93–98.

Santaeufemia, S., E. Torres, R. Mera and J. Abalde. 2016. Bioremediation of oxytetracycline in seawater by living and dead biomass of the microalga *Phaeodactylum tricornutum*. J. Hazard. Mater. 320: 315–325.

Schulz, C., J. Gelbrecht and B. Rennert. 2003. Treatment of rainbow trout farm effluents in constructed wetland with emergent plants and subsurface horizontal water flow. Aquaculture 217: 207–221.

Schveitzer, R., R. Arantes, P.F.S. Costódio, C.M. do Espírito Santo, L.V. Arana, W.Q. Seiffert et al. 2013. Effect of different biofloc levels on microbial activity, water quality and performance of *Litopenaeus vannamei* in a tank system operated with no water exchange. Aquacult. Eng. 56: 59–70.

Schwarzenbach, R.P., B.I. Escher, K. Fenner, T.B. Hofstetter, C.A. Johnson, U. von Gunten et al. 2006. The challenge of micropollutants in the aquatic environment. Science 313: 1072–1077.

Seoane, M., C. Rioboo, C. Herrero and A. Cid. 2014. Toxicity induced by three antibiotics commonly used in aquaculture on the marine microalga *Tetraselmis suecica* (Kylin) Butch. Mar. Environ. Res. 101: 1–7.

Shi, Y., G. Zhang, J. Liu, Y. Zhu and J. Xu. 2011. Performance of a constructed wetland in treating brackish wastewater from commercial recirculating and super-intensive shrimp grow out systems. Bioresour. Technol. 102: 9416–9424.

Shpigel, M., D. Ben-Ezra, L. Shauli, M. Sagi, Y. Ventura, T. Samocha et al. 2013. Constructed wetland with Salicornia as a biofilter for mariculture effluents. Aquaculture 412-413: 52–63.

Sindilariu, P.-D., A. Brinker and R. Reiter. 2009. Factors influencing the efficiency of constructed wetlands used for the treatment of intensive trout farm effluent. Ecol. Eng. 35: 711–722.

Stottmeister, U., A. Wießner, P. Kuschk, U. Kappelmeyer, M. Kaestner, O. Bederski et al. 2003. Effects of plants and microorganisms in constructed wetlands for wastewater treatment. Biotechnol. Adv. 22: 93–117.

Strickland, J.D.H. and T.R. Parson. 1972. A Practical Handbook of Sea Water Analysis. Bull. 167. Fisheries Res. Board of Canada. Ottawa, Canada.

Su, Y.-M., Y.-F. Lin, S.-R. Jing and P.-C.L. Hou. 2011. Plant growth and the performance of mangrove wetland microcosms for mariculture effluent depuration. Marine Poll. Bull. 62: 1455–1463.

Tilley, D.R., H. Badrinarayanan, R. Rosati and J. Son. 2002. Constructed wetlands as recirculation filters inlarge-scale shrimp aquaculture. Aquacul. Eng. 26: 81–109.

Van Rijn, J. 2013. Waste treatment in recirculating aquaculture systems. Aquacul. Eng. 53: 49–56.

Von Sperling, M. 2007. Waste Stabilization Ponds Biological Wastewater Treatment Series. Vol. 3. IWA Publishing. London, UK.

Woodruff, T.J., A.D. Kyle and F.Y. Bois. 1994. Evaluating health risks from occupational exposure to pesticides and the regulatory response. Environ. Health Perspect 102(12): 1088–1096.

Yu, Z., X. Zhu, Y. Jiang, P. Luo and C. Hu. 2014. Bioremediation and fodder potentials of two *Sargassum* spp. in coastal waters of Shenzhen, South China. Marine Poll. Bull. 85: 797–802.

Webb, J.M., R. Quintã, S. Papadimitriou, L. Norman, M. Rigby, D.N. Thomas et al. 2013. The effect of halophyte planting density on the efficiency of constructed wetlands for the treatment of wastewater from marine aquaculture. Ecol. Eng. 61: 145–153.

Zachritz II, W.H., A.T. Hanson, J.A. Sauceda and K.M. Fitzsimmons. 2008. Evaluation of submerged surface flow (SSF) constructed wetlands for recirculating tilapia production systems. Aquacul. Eng. 39: 16–23.

Zhong, F., W. Liang, T. Yu, S.P. Cheng, F. He and Z.B. Wu. 2011. Removal efficiency and balance of nitrogen in recirculating aquaculture system integrated with constructed wetlands. J. Environ. Sci. Health A 46: 789–794.

13

Fate of Contaminants of Emerging Concern in Constructed Wetlands

Víctor Matamoros[1,*] and *María Hijosa-Valsero*[2]

INTRODUCTION

Climate change and population increase have raised the pressure on water sources worldwide (Garnier et al. 2015). One of the most relevant aspects affecting water resources is water pollution, which is mainly due to the discharge of wastewater treatment plants (WWTPs) effluents into surface water bodies. In fact, conventional WWTPs have not been designed to attenuate a new generation of pollutants, such as the so called contaminants of emerging concern (CECs) (Petrie et al. 2015). CECs are defined as chemicals of synthetic origin or deriving from a natural source that have been recently discovered to have possible harmful effects on environmental or public health, with the extent of such risk is yet to be established (Naidu et al. 2016).

In some cases, the release of emerging chemical or microbial contaminants into the environment has likely occurred for a long time, but may not have been recognized until new detection methods were developed. They include organic compounds such as pharmaceuticals, personal care products, pesticides, nanoparticles, and microplastics, but also antibiotic resistance genes and prions, among others (Halden 2015). The occurrence of such compounds in water bodies may cause different adverse effects on aquatic biota. Some of the potential environmental effects of the presence of CECs in surface waters are the reduction of macroinvertebrate diversity in rivers (Muñoz et al. 2009), declining fish populations in lakes (Kidd et al. 2007), and physiological

---

[1] Department of Environmental Chemistry, IDAEA-CSIC, c/Jordi Girona, 18-26, E-08034, Barcelona, Spain.

[2] Centro de Biocombustibles y Bioproductos, Instituto Tecnológico Agrario de Castilla y León (ITACyL), Villarejo de Órbigo, Parcelas 2-6, E-24358, León, Spain.
Email: hijvalma@itacyl.es; mhijv@unileon.es

* Corresponding author: victor.matamoros@cid.csic.es

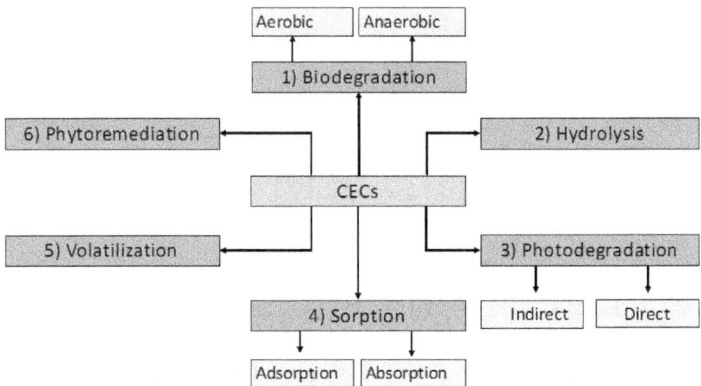

Fig. 1. Summary of the different mechanisms involved in the attenuation of contaminants of emerging concern in constructed wetlands.

stress in freshwater mussels (Gillis et al. 2014). Furthermore, there is major public concern regarding agricultural applications of treated wastewater (TWW) due to the introduction of CECs from irrigation waters to crops via plant uptake and the risk that accumulated residues pose to humans via consumption of edible parts (Malchi et al. 2014; Marsoni et al. 2014).

Although conventional WWTPs cannot completely attenuate all CECs, this issue can be solved by using advanced wastewater treatments such as oxidation or membrane-based technologies (Luo et al. 2014). However, advanced treatment systems require a substantial outlay and high operating and maintenance costs, and they are not integrated into the landscape (Matamoros and Bayona 2013).

In contrast, constructed wetlands (CWs) have emerged in recent decades as an ecologically efficient solution for solving all these issues (Kadlec and Wallace 2009). CWs are designed to simulate the processes that occur in natural wetlands but in a more controlled environment. The main attenuation processes for CECs occurring in CWs are biodegradation (aerobic and anaerobic), photodegradation (direct or indirect), sorption (adsorption or absorption), phytoremediation (plant uptake or biodegradation enhancement), hydrolysis, and volatilization (Fig. 1).

In this chapter, the capability of using CWs for removing CECs will be discussed for various compound groups (pharmaceuticals, personal care products, endocrine disrupting compounds, pesticides, antibiotic resistance genes, nanoparticles, and other contaminants of emerging concern), taking into consideration their removal processes.

Pharmaceutical Compounds

About 3,000 different compounds are used as ingredients in pharmaceuticals, including painkillers, antibiotics, steroids, antidiabetics, betablockers, contraceptives, lipid regulators, antidepressants, and illicit, impotence, and cytostatic drugs (Richardson and Kimura 2016).

Several reviews and book chapters devoted to studying the attenuation of pharmaceutical compounds by CWs have been recently published (Matamoros and Bayona 2013; Carvalho et al. 2014; Li et al. 2014; Verlicchi and Zambello 2014;

Table 1. Design and management parameters affecting the attenuation of CECs in CWs.

Factors	Relevance of the attenuation of CECs
(1) CW configuration	SFCW (aerobic biodegradation and photodegradation) and VFCW configurations remove pharmaceutical compounds more efficiently than HFCW configuration (anaerobic conditions).
(2) HLR/HRT	The higher the HRT and longer the HLR are, greater removal of CECs.
(3) Water depth	In HFCWs, lower water depth result in a greater removal of CECs due to lower redox potential.
(4) Clogging	The accumulation of different types of solids and biofilm lead to the reduction of the infiltration capacity of the gravel bed and CECs.
(5) Sorption material	The use of new support matrix materials such as cork increases the attenuation of CECs.
(6) Plant effect	The presence of vegetation increases the attenuation of CECs due to the release of oxygen and chemical exudates that increase biodegradation and plant uptake.
(7) Batch *versus* continuous feeding	Drain and fill cycles in batch mode operation can introduce air into the pore spaces of the soil matrix and thereby enhance the microbial oxidation of organic matter and CECs.
(8) Seasonality	The attenuation of CECs in the warm season is higher than in the cold season. Higher temperature and light radiation in warm season result in a greater biodegradation and photodegradation rates, respectively.
(9) Recirculation	It is used to increase the attenuation of nutrients and it may also increase the attenuation of CECs.

Zhang et al. 2014a; Ávila and García 2015). Table 1 shows some of the different factors affecting their removal in CWs.

The first of these factors (1) is the CW configuration. The three main configurations include horizontal flow (HF), vertical flow (VF) and surface flow (SF) CWs. VF systems remove more efficiently biodegradable compounds than HF wetlands (e.g., ibuprofen, naproxen or caffeine). This is can be explained by the fact that VFCWs work under predominantly aerobic conditions which are more effective for removing pharmaceuticals than the anaerobic conditions under which HFCWs operate. Finally, the use of SFCWs results in higher removal efficiencies due to the concurrence of photodegradation in addition to aerobic biodegradation (Matamoros et al., 2008). Although biodegradation kinetics in SFCWs are generally slower due to the lower organic matter content and consequently, lower biomass, their greater hydraulic residence time (HRT) results in a higher removal efficiency of pharmaceuticals in comparison with subsurface flow (SSF) CWs (HF and VF). Photodegradable compounds such as diclofenac, triclosan, or ketoprofen are easily removed by SFCWs. Nevertheless, SFCWs normally work under low mass loading rates and they are used as tertiary treatment or polishing systems, whereas HF and VFCWs operate under high mass loading rates and can treat raw (less frequent) or primary treated wastewater effluents. In this sense, there is a concern about the feasibility of using wetlands as a cost-effective method because wetlands typically require a low hydraulic loading rate (HLR) and a long HRT to achieve efficient CEC removal (2). This means that wetland treatment methods may need a large surface area. Different studies have shown that there is an exponential relationship between the surface area required by a CW and the removal efficiency achieved for a specific pharmaceutical compound. For instance, Ranieri et al. (2011) found that paracetamol removal in a pilot plant HFCW varied from 52% at HLR of 240 mm d^{-1} to 87% at HLR of 120 mm d^{-1} and 99% at HLR of 30 mm d^{-1}. The next factor is the water depth in HFCWs (3). It has been observed that the shallower the CW is, the less negative the redox potential is and therefore, greater removal efficiencies of pharmaceuticals can be achieved. Clogging (4) is maybe one of the most important issues of using CWs (mainly in SSF systems) since it results in the gradual blocking of the porous medium and therefore, in hydraulic malfunction and/or reduced treatment performance of the CWs for removing pharmaceutical compounds (Matamoros and Bayona 2006; Aiello et al. 2016). The selection of the sorption material (5) for HF and VFCWs is an important design parameter. Coarse granite gravel or sand media are the regular support matrix for CWs, but in recent years, the use of alternative sorbents has increased. Dordio et al. (Dordio et al. 2010; Dordio et al. 2011) carried out different laboratory studies in which they found that the use of light expanded clay aggregates (LECA) or biosorbents such as cork can improve the removal efficiency of some pharmaceutical compounds in CWs. The presence of plants (6) is one of the main differences between sand filters and CWs. The presence of plants in CWs is generally considered beneficial as they can take up and assimilate nutrients, stabilize the bed surface, act as an anchoring surface for biofilm, release exudates that can aid the biodegradation processes, pump and release oxygen to the bottom of the systems, provide good conditions for physical filtration, and insulate against low temperatures. Another important operational factor is the wastewater feeding strategy used (7). Ávila et al. (2013) observed that the mode of operation in batch feeding resulted in a prevailing higher redox status, when compared to functioning under saturated conditions, which in turn significantly enhanced the elimination of pharmaceutical compounds. An additional factor affecting the attenuation of pharmaceuticals in CWs is seasonality (8); both temperature and sun-light intensity play an important role on the biodegradation and photodegradation of pharmaceutical compounds. For example, Dordio et al. (2010) studied winter (12°C) and summer (26°C) ibuprofen, carbamazepine, and clofibric acid removal efficiencies in planted microcosm CWs and observed better efficiencies in summer (96% for ibuprofen, 97% for carbamazepine, and 75% for clofibric acid) than in winter (82% for ibuprofen, 88% for carbamazepine, and 48% for clofibric acid). Reyes-Contreras et al. (2012) studied several small-scale CWs fed with urban wastewater during 39 months and observed that winter removal efficiencies for salicylic acid and caffeine (53.3–90.3% and 21.9–83.8%, respectively) were lower than those recorded in summer (84.3–99.6% and 81.5–99.5%, respectively). Finally, recirculation (9) seems to play an important role to enhance the attenuation of both nutrients and pharmaceutical compounds, but further studies are necessary to find clear evidences on that (Ávila et al. 2017).

To improve the effectiveness of CWs for removing pollutants, and to exploit the different degradation pathways between systems, the use of hybrid systems has emerged in the last years. Hybrid systems consist of various types of CWs staged in series (Vymazal 2013). For instance, the use of a VFCW located as a first step would be able to attenuate pharmaceutical compounds by aerobic pathways, whereas a HFCW afterwards is able to attenuate compounds by anaerobic processes; finally, the inclusion of a SFCW can attenuate photodegradable compounds (Hijosa-Valsero et al. 2010; Ávila et al. 2014a). Ávila et al. (2014a) studied the capacity of a hybrid CW system consisting of two VFCWs working sequentially in parallel, one HFCW, and one SFCW in series to eliminate 8 CECs under three different hydraulic loading rates (HLRs) (0.06, 0.13, and 0.18 m d^{-1} considering the area of the two VF beds) and found that the hybrid system was capable of removing pharmaceutical compounds up to 80% at all three HLRs. This study demonstrated for the first time that the use of hybrid CW systems permits to increase HLRs without decreasing the removal of CECs, which makes this technology highly competitive in comparison with conventional activated sludge WWTPs.

Finally, novel CW developments such as aerated HFCWs have shown a high capacity to treat high mass-loaded wastewaters such as those from landfill leachate (Nivala et al. 2007), but also urban wastewater influents. For instance, Auvinen et al. (2017a) showed that the removal of pharmaceuticals from urban wastewater significantly increased when continuous aeration is applied (99 ± 1% *vs.* 68 ± 32% for metformin and 99 ± 1% *versus* 17 ± 19% for valsartan). Nevertheless, Ávila et al. (2014b) observed that the use of active aeration in a saturated gravel bed wetland generally improved the treatment performance compared to the free-draining gravel bed, but achieved a similar performance to the free-draining sand-based VF wetlands. Further research at full-scale is needed to find out to what extent this technology can be feasible for removing pharmaceuticals from high mass loaded wastewaters such as hospital effluents.

Personal Care Products

Personal care products (PCPs) are a diverse group of CECs used in all kind of cosmetic formulas such as soaps, lotions, toothpastes, fragrances, or sunscreens. Unlike pharmaceuticals, which are intended for internal use, PCPs are products intended for external use on the human body and thus are not subjected to metabolic alterations; therefore, large quantities of PCPs enter the environment unaltered through regular usage (Brausch and Rand 2011). Musk fragrances such as galaxoldie and tonalide have been extensively studied since they belong to the list of High Production Volume (HPV) chemicals. They are highly hydrophobic compounds (log K_{ow} > 5) that have a high interaction with the organic matter retained in CWs. Matamoros et al. (2006) observed that these compounds are removed in the first meters of HFCWs and therefore their removal is related to their high sorption onto the organic matter and biofilm that grows on gravel surface. On the other hand, in a 20 year-old CW for treating sludge from a WWTP, Matamoros et al. (2012) observed the attenuation of hydrophobic micropollutants (galaxolide, cashmeran, celestolide, and DEHP) from the top to the bottom layer of the CW ranging from 40 to 98%, except for tonalide which increased significantly with sludge depth. Methyl dihydrojasmonate, a hydrophilic fragrance (log K_{ow} = 2.98) used in detergents and cosmetics, has been observed to be better removed in summer (52.6–98.9%) than in winter (38.5–76.1%) in different pilot CW configurations treating wastewater (Hijosa-Valsero et al. 2010). In those works, methyl dihydrojasmonate removal was more efficient and stable in a *Typha*-SF-CW (81.9–96.1% in summer; 71.3–73.3% in winter) than in other types of CWs (SF/HF; *Typha/Phragmites*/unplanted). This compound has also been removed at an efficiency rate of 99% in both HF and VFCWs in warm conditions treating urban wastewater (Matamoros and Bayona 2006; Matamoros et al. 2007). The attenuation

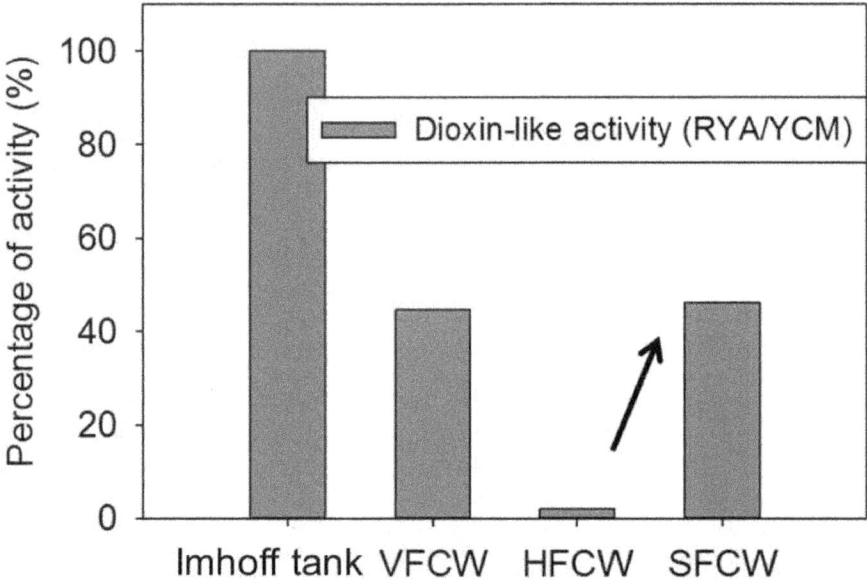

Fig. 2. Dioxin-like activity (RYA/YCM) of wastewater samples according with the treatment stage (adapted from Ávila et al. 2014).

of some sunscreen/UV filter compounds such as oxybenzone or benzophenone-3 has also been assessed in CWs. This compound has been found to be easily removed by HFCWs as well as VFCWs, with removal efficiencies above 90% (Matamoros and Bayona 2006; Matamoros et al. 2007). Nevertheless, the removal efficiency of both oxybenzone and methyl dihydrojasmonate in SFCWs has been observed to be much lower (64–77%). This is due to the fact that SFCWs treat secondary effluents, in which the concentration of these compounds is usually lower (100–200 ng L^{-1}). In this sense, there may be a minimum concentration below which a compound is not further degraded. The minimal threshold concentration depends mainly on the kinetic parameters of growth and metabolism, but also on the thermodynamics of the overall transformation reaction. Actually, the support media affinity constant is the most important parameter with respect to the biodegradation of contaminants to very low concentrations (Fetzner 2002). Triclosan, an antiseptic compound found in toothpaste, is generally removed at high percentages in CWs due to its high interaction with organic matter (log K_{ow} = 4.8). Ávila et al. (2014a) found that this compound can be removed up to 90% after a hybrid CW consisting of VFCW-HFCW-SFCW. Nevertheless, Fig. 2 shows that although dioxin-like activity decreased as triclosan was attenuated, an increase on dioxin-like activity was observed after the SFCW, indicating the photodegradation of triclosan and a probable production of dioxin-like compounds (Mezcua et al. 2004).

Endocrine Disrupting Compounds

Endocrine disrupting compounds (EDCs) are an exogenous substance or a mixture that alters function(s) of the endocrine system and consequently causes adverse health effects in an intact organism, or its progeny, or (sub)populations (WHO/UNEP 2013). Although there are pesticide and pharmaceutical compounds that are EDCs, this section will be focused on the application of CWs for removing bisphenol A, octylphenol, nonylphenol, and steroidal sex hormones.

Apparently, CWs are an efficient technology for the attenuation of EDCs from wastewater. For instance, Vymazal et al. (2015) reported that the average removal of estrogens (estrone, 17β-estradiol, estriol, and 17α-ethinylestradiol, also known as E1, E2, E3, and EE2), testosterone and progesterone in three full-scale HFCWs treating urban wastewater in summer exceeded 95%. Qiang et al. (2013) studied seven SF-CWs treating wastewater in rural areas and reported a clear effect of seasonality on the removal of some compounds, like E1 (85% summer, 60% winter), EE2 (75% summer, 40% winter), bisphenol A (50% summer, 30% winter), and 4-nonylphenol (75% summer, 30% winter). However, other substances were unaffected by seasonality, like E2 (70% summer, 75% winter) and E3 (75% summer, 80% winter). Conversely, hybrid systems consisting of a series of ponds and CWs treating dairy wastewater or manure from a swine livestock showed constant high removal efficiencies throughout the year for estrogens and androgens (92–95% removal) (Cai et al. 2012) and E1, E2, and E3 (98% removal) (Shappell et al. 2007), respectively.

The presence of plants in CWs seems to enhance the removal of EDCs. For instance, A et al. (2017) assessed the effectiveness of VFCWs planted with *Phragmites australis* and unplanted beds for treating a synthetic leachate, and observed that the removal of bisphenol A (9–99%) and 4-tert-butylphenol (18–100%) was affected by leachate composition, HRT, and the presence of plants. They suggested that adsorption and subsequent biodegradation were the main elimination processes. In addition, the type of vegetal species chosen could also be important for EDCs elimination, as proposed by Toro-Vélez et al. (2016), who compared HFCWs with *Heliconia psitacorum*, *P. australis* or unplanted, and stated that the best removal performance was obtained with *Heliconia psitacorum* (73% for bisphenol A and 63% for nonylpehnols), whereas the worst results were recorded in the unplanted system.

The effect of bed depth has been evaluated by Song et al. (2009) with small-scale VFCWs planted with *P. australis* and fed with tertiary effluent from a WWTP. The shallowest wetland bed showed the highest removal efficiency for E1 (68%), E2 (84%), and EE2 (75%), which suggests that unsaturated and aerobic conditions favor estrogen removal. Adsorption accounted for less than 12% of the removed estrogens irrespective of depth, indicating that biotic processes play a major role in the removal of these compounds. In any case, the removal of EDCs (4-tert-octylphenol, E1 or bisphenol A) in CWs has been suggested to be compound-dependent with positive trends associated with log K_{ow} (Dai et al. 2017).

Similarly, as it occurred for other CECs, the use of hybrid systems has shown to improve the overall removal efficiency of EDCs. For instance, removal efficiencies of 45–99% for bisphenol A (Ávila et al. 2015; Dai et al. 2017), 95% for estrogens (Cai et al. 2012), 92% for androgens (Cai et al. 2012), 48% for E1 (Dai et al. 2017), 90–95% for E2 (Masi et al. 2004), 90–95% for EE2 (Masi et al. 2004), 58% for 4-tert-octylphenol (Dai et al. 2017), 32% for 4-nonylphenol (Dai et al. 2017) or 55–91% for nonylphenol ethoxylates (Hsieh et al. 2013) have been reported in hybrid systems.

Pesticides

Due to the need to increase crop yields, both the use of pesticides and the development of new chemical products have risen considerably in recent years. In this regard, modern agrochemical technology has produced hundreds of new synthetic organic chemicals for use as pesticides, but only some of them have actually been banned by Environmental Protection Agencies (Water Framework Directive in EU and Clean Water Act in the Unites States). Agricultural and urban run-off of highly-polar pesticides has been extensively detected for mobile compounds such as phenoxyacid herbicides, fungicides, or triazinic compounds. For instance, a maximum peak concentration of 100 µg L^{-1} was recorded for mecoprop after sugarcane land irrigation (Davis et al. 2013), a peak concentration of 190 µg L^{-1} was recorded for atrazine after a rainfall event in a corn field (Boyd et al. 2003), and a peak concentration of 500 µg L^{-1} for mecoprop was recorded in roof run-off waters after a rainfall episode in Switzerland (Bucheli et al. 1998). Two different wetland systems have been studied: CWs and vegetated filter strips (VF).

CWs have been used to treat pesticides from agricultural run-off or urban wastewater. The attenuation of these pesticides in raw wastewater is normally low, since the HRT of these systems is also low (2–4 days). Nevertheless, in the case of agricultural run-off, retention wetlands are designed to have higher HRTs, and therefore higher removal efficiencies have been reported (Vymazal and Březinová 2015). Similar to pharmaceutical compounds, different design parameters affect the attenuation of pesticides in CWs. For instance, Elsaesser et al. (2011) reported that wetland cells with vegetation reduced peak pesticide concentration more effectively (up to 91%) than cells without vegetation (72%) in a field study carried out in Norway. Rose et al. (2006) concluded that CWs designed for pesticide retention should include both open water and vegetated zones, to increase the potential for complementary chemical, photolytic, microbial, and plant-mediated pesticide breakdown. Lv et al. (2016) observed the effect of plant presence on the attenuation of imazalil and tebuconazole in HFCWs fed with synthetic wastewater and found that mesocosms planted with *Typha latifolia*, *P. australis*, *Iris pseudacorus*, *Juncus effusus*, and *Berula erecta* showed significantly higher removal efficiencies than unplanted control mesocosms. For a 2-day HRT, the average removal in planted systems for imazalil was 90–100% in summer and 75–90% in winter, whereas tebuconazole removal attained 80–100% in summer and 45–75% in winter. The main mechanisms suggested for the removal of these two compounds were plant uptake (5–6% in summer), plant-stimulated microbial degradation in the bed support media, and bacterial nitrification. A review carried out by Gregoire et al. (2009) pointed out that the removal of pesticides in CWs could be improved considerably by increasing HRT and using adequate adsorbing materials to favor the contact between pesticides and biocatalysts.

In this sense, Dordio et al. (2007) assessed the efficiencies of three different materials, LECA (in two different particle sizes), expanded perlite, and sand to remove the herbicide MCPA from water. LECA with smaller particle sizes exhibited a high sorption capacity for MCPA (near 100%) for initial concentrations of 1 mg L^{-1}, which encourages testing this support medium as support matrix in CWs. Even so, there are some pesticides which show low attenuation in CWs. For instance, a new group of pesticides, namely neocotinoids, have shown to persist during conventional and wetland treatment, and to pose potential risk in effluent-dominated, receiving surface waters (Sadaria et al. 2016).

In addition to the use of retention wetlands to treat agricultural run-off, buffer zones between 2 and 10 m wide have been established along rivers and around lakes in order to minimize the risk of off-site pollution caused by spray drift, drain-flow, and run-off. In some EU countries, such as Denmark and Belgium, these systems are already mandatory. VFs have been shown to attenuate the sediment, nutrient, and pesticide losses to receiving water bodies (Reichenberger et al. 2007; Vymazal and Březinová 2015). Nevertheless, data from different studies have shown that pesticide mass retention depends on their sorption capacity. Hence, the respective averages (and ranges) for weakly (Koc < 100), moderately (100 < Koc < 1,000) and strongly sorbed pesticides (Koc > 1,000) were 61% (0–100%), 63% (0–100%), and 76% (53–100%), respectively (Arora et al. 2010). The high variability in mass retention can be explained by both different experimental conditions and the specific pesticide compounds. The factors reported to affect the herbicide retention in VFs are as follows (Krutz et al. 2005): width of VF (positive correlation between VF width and the retention of herbicides), the presence of vegetation increases the attenuation of pesticides, source area to VF area ratio (not many differences have been reported), inflow pesticide concentration (greater retention at higher concentration), herbicide properties (log Koc), and antecedent moisture (pesticide retention is negatively correlated with VFs' antecedent moisture content).

Antibiotics and Antibiotic Resistance Genes

Due to the growing interest on the occurrence of antibiotics in the aquatic environment and the consequent increase of antibiotic resistance genes (ARGs) in the aquatic environment, an entire section has been devoted to explore the capability of using CWs for their attenuation. After administration, antibiotic residues can be released into the receiving environments through discharge of the feces or urine, thus posing potential risks to human health and ecosystems (Grenni et al. 2017; Tao et al. 2010; Underwood et al. 2011). The use of antibiotics in humans and animals could also lead to development and spread of antibiotic-resistant bacteria (ARB) and ARGs in the environment (Xu et al. 2015). In this regard, the escalating of ARGs has become a serious threat to human health because of the reduced susceptibility of disease-causing microorganisms to antibiotics in medical treatment (Yang et al. 2014). Different studies have assessed the attenuation of both antibiotics and ARGs in CWs.

Hijosa-Valsero et al. (2011) assessed the attenuation of antibiotics in different CW configurations. They observed that CWs were efficient for the removal of sulfamethoxazole (59 ± 30–87 ± 41%), a compound which was also removed by a conventional WWTP, and, in addition, CWs removed trimethoprim (65 ± 21–96 ± 29%). The elimination of other antibiotics in CWs was limited by the specific system-configuration: amoxicillin (45 ± 15%) was only eliminated by a SFCW planted with *Typha angustifolia*; doxycycline was removed in a SFCW planted with *T. angustifolia* (65 ± 34–75 ± 40%), in a *Phragmites australis*-floating macrophytes system (62 ± 31%) and in conventional HFCWs (71 ± 39%); clarithromycin was partially eliminated by an unplanted SFCW system (50 ± 18%); erythromycin could only be removed by a *P. australis*-HFCW system (64 ± 30%); and ampicillin was eliminated by a *T. angustifolia*-floating macrophytes system (29 ± 4%). Lincomycin was not removed by any of the studied systems (WWTP or CWs). The presence or absence of plants, the vegetal species (*T. angustifolia* or *P. australis*), the flow type, and the CW design characteristics regulated the specific removal mechanisms.

Chen et al. (2016a) observed that in various mesocosm-scale HFCWs planted with *Cyperus alternifolius* L. the removal efficiency of total antibiotics (erythromycin-H_2O, lincomycin, monensin, ofloxacin, sulfamerazine, sulfamethazine, and novobiocin) ranged from 18 to 99%, while those for the total ARGs varied between 50 and 86%. The same authors concluded that both, microbial degradation and physical sorption processes, were responsible for the removal of antibiotics and ARGs in CWs. In another study, Chen et al. (2016b) observed that the removal efficiencies of total antibiotics in different mecoscosm-scale CWs ranged from 76 to 99%, while those of total ARGs varied between 64 and 84%. Since these compounds are stated to be removed by sorption, CWs with different sorption materials were tested. Under the same HLR, the removals of antibiotics in those CWs with different support media had the following order: zeolite > medical stone > ceramic > oyster shells. The presence of plants was beneficial to the removal of pollutants, and HF and VFCWs had higher pollutant removal than SFCWs, especially for antibiotics. This is in keeping with the fact that antibiotics and ARGs are mainly removed by biodegradation and sorption processes. Nevertheless, it is worth noticing that antibiotics like tetracyclines are easily removed by photodegradation, but they were not considered in this study (Garcia-Rodríguez et al. 2013). In summary, antibiotics and ARGs can be removed by biodegradation, photodegradation, or sorption processes.

The removal efficiencies of antibiotics and ARGs achieved by CWs were relatively similar or even higher than those found in conventional WWTPs (Xu et al. 2015). Nevertheless, the use of advanced oxidation processes such as Fenton oxidation ($Fe_2 +/H_2O_2$) and UV/H_2O_2 has shown to be capable of achieving a much higher reduction of ARGs, ranging from 1.6 to 3.5 logs (Zhang et al. 2016).

Nanoparticles

Nano-enabled materials are produced at growing volumes which increases the likelihood of nanoparticles being released into the environment. One of the most well-known groups of nanoparticles is silver nanoparticles (Ag-NPs). They are used in common household products, such as textiles, biocidal sprays, food packaging material, and toys, because of the antimicrobial activity of the ionic Ag^+ (Kaegi et al. 2013). Button et al. (2016) observed that wastewater contaminated with AgNP did not affect the CW microbial community function or structure following a 28-day exposure (100 μg L^{-1}). In fact it was suggested that wetland microbial communities can tolerate and develop resistance to lower levels of *in situ* AgNP exposure in a relatively short time. In this sense, Auvinen et al. (2017b) observed that HFCWs are able to remove most of the Ag (80–90%) and the largest fraction of Ag is found in/on the biofilm. Detailed electron microscopy analyses suggested that Ag-NPs are transformed into Ag_2S in all microcosm experiments due to the sulfidation of NPs. This sulfidation decreases available Ag^+ due to the low aqueous solubility of $Ag_2S(s)$, and may limit the bioavailability or toxicity of AgNPs to aquatic organisms (Lowry et al. 2012). The results suggested that Ag-NPs were retained with the suspended solids in CWs and, therefore, that the removal efficiency of TSS is an important factor determining the discharge of Ag-NPs from CWs. The removal of Ag from the water phase can be partly explained by aggregation-sedimentation. The addition of an aeration system to CWs has been reported to enhance the removal of nutrients and pharmaceutical compounds (Fan et al. 2013; Auvinen et al. 2017a), but Liu et al. (2011) suggested that aeration may result in the release of Ag^+ due to the oxidative dissolution of Ag-NPs. Ag^+ may be of even greater environmental concern than Ag in its particulate form (Behra et al. 2013). Nevertheless, Auvinen et al. (2017b) did not observe differences in the distribution of Ag in comparison to non-aerated bed system.

In another study, the same authors (Auvinen et al. 2016) assessed the use of different support media materials in microcosms HFCWs to attenuate Ag-NPs. The results showed that sand, zeolite, and biofilm-coated gravel induce efficient removal (85, 55, and 67%, respectively) of Ag from the water phase indicating that citrate-coated Ag-NPs are efficiently retained in CWs. Plants are a minor factor in retaining Ag, as a large fraction of the recovered Ag remains in the water phase (0.42–0.58). Most Ag associated with the plant tissues is attached to or taken up by the roots, and only negligible amounts (maximum 3%) of Ag are translocated to the leaves. Biofilm on the gravel increased the Ag removal efficiency by 67% in comparison to gravel alone, suggesting an important mechanism of Ag sequestering in CWs. The authors stated that support media covered by biofilm will increase with operation time, indicating that more biomass will be able to accumulate Ag-NPs in full-scale CWs than in the microcosms studied. Also, plant roots, occupying a large volume in full-scale CWs, may offer an important attachment site for biofilm and hence for Ag-NPs.

Other NPs such as copper oxide (CuO) NPs used intensively in electronics, ceramics, films, polymers inks, and coatings, and cadmium sulphide/zinc sulphide (CdS/ZnS) quantum dots (QDs) have been found to be uptaken by aquatic plants (*Schoenoplectus tabernaemontani*). The root uptake percentage (of the total initial NP mass) for CuO NP treatment ranged from 40.6 to 68.4%, while the values were 8.7 to 21.3% for CdS QDs (Zhang et al. 2014b). The authors pointed out the high potential of the studied aquatic plants for the bioaccumulation of these NPs from industrial and urban wastewater effluents.

Other Contaminants of Emerging Concern

In this section, four groups of organic compounds included in the list of CECs will be discussed in accordance with their removal in CWs: Phthalates, anticorrosive agents, disinfection by-products, and iodinated contrast media.

Phthalic acid esters (PAEs) are used to produce plastic for numerous consumer products, commodities, and building materials. Worldwide PAE production is estimated at six million tons per year. Xiaoyan et al. (2015) showed that PAE removal efficiencies differed based on chemical properties (45–83%). Water flow type in the wetland was the most important factor in eliminating PAEs and the VFCW was the most effective configuration. Another important group of CECs are the anticorrosive agents used in aircraft deicing and antifreeze fluids but also used in dishwasher detergents. For instance, benzotriazoles have been studied in different CW configurations, again VFCWs were found to provide the best conditions for their removal. Their removal efficiencies ranged from 60 to 65% and from 89 to 93% in HFCW and

VFCW, respectively (Matamoros et al. 2010). Chlorination of treated wastewater has been reported to produce disinfection-by-products (DBPs), some of which are carcinogenic or toxic. Results showed that most of the 11 DBPs (trihalomethanes and haloketones) and nitrogenous DBPs (haloacetonitriles and trihalonitromethanes) were efficiently removed (> 90%) in six HFCWs with HRT of 5 d and there were no significant differences among the systems (Chen et al. 2014). The last group of organic compounds is iodinated X-ray contrast media. They are metabolically-stable substances which are used in human medicine for imaging of organs or blood vessels during diagnostic tests and are intravascularly administered (Pérez and Barceló 2007). Drewes et al. (2001) pointed out that organic iodine compounds are not removed from wastewater by advanced oxidation techniques like ozonation, but can be efficiently removed by reverse osmosis and partially removed by biodegradation under anoxic conditions. In fact, they reported that a CW working as a tertiary treatment for urban wastewater under fully anaerobic conditions was able to remove 37% of iodinated media. In a different study, Zhang et al. (2017) assessed several small-scale VFCWs working under saturated conditions and planted with different vegetal species (*Typha latifolia*, *Phragmites australis*, *Iris pseudacorus*, *Juncus effusus*, *Berula erecta*, or unplanted) for the removal of iohexol from artificial wastewater. The highest removal efficiency (above 80%) was achieved by *B. erecta*-systems under low HLR and high initial iohexol concentrations. Removal by sorption onto the support media and phytoaccumulation were estimated to be below 0.75%, and it was suggested that biodegradation was the main pathway for the elimination of iohexol in these saturated CWs.

Future Trends

Although the attenuation of CECs by using CWs has been studied for more than 20 years, there are still a lot of gaps or needs that can be identified.

- Future research should include the study of other CECs, as well as the toxicity of both influent and effluent waters. For example, only some of the pollutants included in the EU Watch list have been studied until now, whereas most of the studied CECs have only been assessed in one CW configuration.

- The attenuation of CECs may result in an increase of toxicity due to the production of the so-called transformation products (TPs). Currently, most of the studies performed in CWs only take into consideration the attenuation of the parental CECs, but they miss to study the formation of TPs or at least to assess how that affects toxicity.

- There are a lot of scientific evidences that prove that the presence of plants enhances the attenuation of CECs. In this regard, future research should include studies to understand removal mechanisms, interactions soil-plant (rhizosphere), and explore the capacity of plant exudates and uptake for removing pollutants.

- New developments such as the use of aerated CWs or the combination of CWs with microbial fuel cells are attracting a lot of attention due to its potential capacity to attenuate pollutants combined with energy production.

- The combined use of CWs with advanced oxidation technologies. For instance, if a water effluent is discharged into a highly sensitive surface water body and there is not enough surface area to set a CW, an advanced oxidation technology can be built instead, but including a small SFCW afterwards to polish the water and increase its biodiversity.
- The use of modeling for the prediction of CEC attenuation in CWs. Future studies of modeling of CECs will be of high interest to predict their behavior and design more efficient CW systems.

Concluding Remarks

CWs are capable of removing a great number of CECs, from pharmaceuticals to nanomaterials. Some of the most important design and management parameters for improving the attenuation of pharmaceutical compounds by using CWs are: CW configuration, being VF and SFCWs the most efficient, HRT or HLR, water depth in HFCWs, clogging, the selection of the sorption material, the presence of plants, seasonality, and recirculation. Hybrid systems and aerated CWs have been observed to increase the attenuation of pharmaceutical compounds. Musk fragrances such as galaxolide and tonalide have been observed to be removed by hydrophobic interaction, whereas other personal care products are attenuated by biodegradation (oxybenzone and methyl dihydrojasmoante). EDCs such as bisphenol A, octylphenol, or ethinylestradiol are efficiently removed by CWs, mainly due to hydrophobic interactions with organic matter and consequent biodegradation. The use of CWs in agriculture, such as retention wetlands or buffer strips, has been observed to be relevant for reducing the discharge of pesticides into surface water bodies. The possibility of using CWs for removing some of the most recent concerning CECs such as nanoparticles, antibiotics, and ARGs is making this technology attractive for society, but future research is needed to explore other CECs and their associated toxicity. Overall, a lot of different CW configurations have been studied, and in most of the cases, CWs have been observed to be at least as efficient for removing CECs as conventional WWTPs, and in some cases even as advanced wastewater technologies.

Acknowledgements

Dr. V.M. would like to acknowledge a Ramon y Cajal contract from the Spanish Ministry of Economy and Competitiveness (RYC-2013-12522). MH-V thanks a postdoctoral contract (DOC-INIA, grant number DOC 2013-010) funded by the Spanish Agricultural and Agrifood Research Institute (INIA) and the European Social Fund. The authors kindly thank Prof. Josep M. Bayona for the revision of the chapter.

References Cited

A, D., D. Fujii, S. Soda, T. Machimura and M. Ike. 2017. Removal of phenol, bisphenol A, and 4-tert-butylphenol from synthetic landfill leachate by vertical flow constructed wetlands. Sci. Total Environ. 578: 566–576.

Aiello, R., V. Bagarello, S. Barbagallo, M. Iovino, A. Marzo and A. Toscano. 2016. Evaluation of clogging in full-scale subsurface flow constructed wetlands. Ecol. Eng. 95: 505–513.

Arora, K., S.K. Mickelson, M.J. Helmers and J.L. Baker. 2010. Review of pesticide retention processes occurring in buffer strips receiving agricultural runoff. J. Am. Water Resour. As. 46: 618–647.

Auvinen, H., V.V. Sepúlveda, D.P.L. Rousseau and G. Du Laing. 2016. Substrate- and plant-mediated removal of citrate-coated silver nanoparticles in constructed wetlands. Environ. Sci. Pollut. Res. 23: 21920–21926.

Auvinen, H., I. Havran, L. Hubau, L. Vanseveren, W. Gebhardt, V. Linnemann et al. 2017a. Removal of pharmaceuticals by a pilot aerated sub-surface flow constructed wetland treating municipal and hospital wastewater. Ecol. Eng. 100: 157–164.

Auvinen, H., R. Kaegi, D.P.L. Rousseau and G. Du Laing. 2017b. Fate of silver nanoparticles in constructed wetlands—a microcosm study. Water Air Soil Poll. 228: 97.

Ávila, C., C. Reyes, J.M. Bayona and J. García. 2013. Emerging organic contaminant removal depending on primary treatment and operational strategy in horizontal subsurface flow constructed wetlands: Influence of redox. Water Res. 47: 315–325.

Ávila, C., V. Matamoros, C. Reyes-Contreras, B. Piña, M. Casado, L. Mita et al. 2014a. Attenuation of emerging organic contaminants in a hybrid constructed wetland system under different hydraulic loading rates and their associated toxicological effects in wastewater. Sci. Total Environ. 470-471: 1272–1280.

Ávila, C., J. Nivala, L. Olsson, K. Kassa, T. Headley, R.A. Mueller et al. 2014b. Emerging organic contaminants in vertical subsurface flow constructed wetlands: Influence of media size, loading frequency and use of active aeration. Sci. Total Environ. 494–495: 211–217.

Ávila, C., J.M. Bayona, I. Martín, J.J. Salas and J. García. 2015. Emerging organic contaminant removal in a full-scale hybrid constructed wetland system for wastewater treatment and reuse. Ecol. Eng. 80: 108–116.

Ávila, C. and J. García. 2015. Chapter 6—Pharmaceuticals and personal care products (PPCPs) in the environment and their removal from wastewater through constructed wetlands. pp. 195–244. *In*: Eddy, Y.Z. [ed.]. Comprehensive Analytical Chemistry 67. Elsevier.

Ávila, C., C. Pelissari, P.H. Sezerino, M. Sgroi, P. Roccaro and J. García. 2017. Enhancement of total nitrogen removal through effluent recirculation and fate of PPCPs in a hybrid constructed wetland system treating urban wastewater. Sci. Total Environ. 584–585: 414–425.

Behra, R., L. Sigg, M.J.D. Clift, F. Herzog, M. Minghetti, B. Johnston et al. 2013. Bioavailability of silver nanoparticles and ions: from a chemical and biochemical perspective. J.R. Soc. Interface 10: 20130396.

Boyd, P.M., J.L. Baker, S.K. Mickelson and S.I. Ahmed. 2003. Pesticide transport with surface runoff and subsurface drainage through a vegetative filter strip. Agricultural and Biosystems Engineering Publications 46: 9.

Brausch, J.M. and G.M. Rand. 2011. A review of personal care products in the aquatic environment: Environmental concentrations and toxicity. Chemosphere 82: 1518–1532.

Bucheli, T.D., S.R. Müller, A. Voegelin and R.P. Schwarzenbach. 1998. Bituminous roof sealing membranes as major sources of the herbicide (R,S)-mecoprop in roof runoff waters: Potential contamination of groundwater and surface waters. Environ. Sci. Technol. 32: 3465–3471.

Button, M., H. Auvinen, F. Van Koetsem, B. Hosseinkhani, D. Rousseau, K.P. Weber et al. 2016. Susceptibility of constructed wetland microbial communities to silver nanoparticles: A microcosm study. Ecol. Eng. 97: 476–485.

Cai, K., C.T. Elliott, D.H. Phillips, M.-L. Scippo, M. Muller and L. Connolly. 2012. Treatment of estrogens and androgens in dairy wastewater by a constructed wetland system. Water Res. 46: 2333–2343.

Carvalho, P.N., M.C.P. Basto, C.M.R. Almeida and H. Brix. 2014. A review of plant–pharmaceutical interactions: from uptake and effects in crop plants to phytoremediation in constructed wetlands. Environ. Sci. Pollut. R. 21: 11729–11763.

Chen, J., X.-D. Wei, Y.-S. Liu, G.-G. Ying, S.-S. Liu, L.-Y. He et al. 2016a. Removal of antibiotics and antibiotic resistance genes from domestic sewage by constructed wetlands: Optimization of wetland substrates and hydraulic loading. Sci. Total Environ. 565: 240–248.

Chen, J., G.-G. Ying, X.-D. Wei, Y.-S. Liu, S.-S. Liu, L.-X. Hu et al. 2016b. Removal of antibiotics and antibiotic resistance genes from domestic sewage by constructed wetlands: Effect of flow configuration and plant species. Sci. Total Environ. 571: 974–982.

Chen, Y., Y. Wen, Z. Tang, L. Li, Y. Cai and Q. Zhou. 2014. Removal processes of disinfection byproducts in subsurface-flow constructed wetlands treating secondary effluent. Water Res. 51: 163–171.

Dai, Y.-N.V., R. Tao, Y.-P. Tai, N.F.-Y. Tam, Dan A. and Y. Yang. 2017. Application of a full-scale newly developed stacked constructed wetland and an assembled bio-filter for reducing phenolic endocrine disrupting chemicals from secondary effluent. Ecol. Eng. 99: 496–503.

Davis, A.M., P.J. Thorburn, S.E. Lewis, Z.T. Bainbridge, S.J. Attard, R. Milla et al. 2013. Environmental impacts of irrigated sugarcane production: Herbicide run-off dynamics from farms and associated drainage systems. Agriculture, Ecosystems & Environment 180: 123–135.

Dordio, A.V., J. Teimão, I. Ramalho, A.J.P. Carvalho and A.J.E. Candeias. 2007. Selection of a support matrix for the removal of some phenoxyacetic compounds in constructed wetlands systems. Sci. Total Environ. 380: 237–246.

Dordio, A., A.J.P. Carvalho, D.M. Teixeira, C.B. Dias and A.P. Pinto. 2010. Removal of pharmaceuticals in microcosm constructed wetlands using *Typha* spp. and LECA. Bioresour. Technol. 101: 886–892.

Dordio, A.V., P. Gonçalves, D. Texeira, A.J. Candeias, J.E. Castanheiro, A.P. Pinto et al. 2011. Pharmaceuticals sorption behaviour in granulated cork for the selection of a support matrix for a constructed wetlands system. Int. J. Environ. An. Ch. 91, 615–631.

Drewes, J.E., P. Fox and M. Jekel. 2001. Occurrence of iodinated X-ray contrast media in domestic effluents and their fate during indirect potable reuse. J. Environ. Sci. Heal. A 36: 1633–1645.

Elsaesser, D., A.G.B. Blankenberg, A. Geist, T. Mæhlum and R. Schulz. 2011. Assessing the influence of vegetation on reduction of pesticide concentration in experimental surface flow constructed wetlands: Application of the toxic units approach. Ecol. Eng. 37: 955–962.

Fan, J., B. Zhang, J. Zhang, H.H. Ngo, W. Guo, F. Liu et al. 2013. Intermittent aeration strategy to enhance organics and nitrogen removal in subsurface flow constructed wetlands. Bioresour. Technol. 141: 117–122.

Fetzner, S. 2002. Biodegradation of xenobiotics. *In*: Silva, D.D. [ed.]. Encyclopedia of Life Support Systems (EOLSS), Biotechnology. Eolss Publishers, developed under the Auspices of the UNESCO, Oxford, UK.

Garcia-Rodríguez, A., V. Matamoros, C. Fontàs and V. Salvadó. 2013. The influence of light exposure, water quality and vegetation on the removal of sulfonamides and tetracyclines: A laboratory-scale study. Chemosphere 90: 2297–2302.

Garnier, M., D.M. Harper, L. Blaskovicova, G. Hancz, G.A. Janauer, Z. Jolánkai et al. 2015. Climate change and European water bodies, a review of existing gaps and future research needs: Findings of the ClimateWater project. Environ. Manage. 56: 271–285.

Gillis, P.L., F. Gagné, R. McInnis, T.M. Hooey, E.S. Choy, C. André et al. 2014. The impact of municipal wastewater effluent on field-deployed freshwater mussels in the Grand River (Ontario, Canada). Environ. Toxicol. Chem. 33: 134–143.

Gregoire, C., D. Elsaesser, D. Huguenot, J. Lange, T. Lebeau, A. Merli et al. 2009. Mitigation of agricultural nonpoint-source pesticide pollution in artificial wetland ecosystems. Environ. Chem. Lett. 7: 205–231.

Grenni, P., V. Ancona and A. Barra Caracciolo. 2017. Ecological effects of antibiotics on natural ecosystems: A review. Microchem. J. (in press).

Halden, R.U. 2015. Epistemology of contaminants of emerging concern and literature meta-analysis. J. Hazard. Mater. 282: 2–9.

Hijosa-Valsero, M., V. Matamoros, R. Sidrach-Cardona, J. Martín-Villacorta, E. Bécares and J.M. Bayona. 2010. Comprehensive assessment of the design configuration of constructed wetlands for the removal of pharmaceuticals and personal care products from urban wastewaters. Water Res. 44: 3669–3678.

Hijosa-Valsero, M., G. Fink, M.P. Schlüsener, R. Sidrach-Cardona, J. Martín-Villacorta, T. Ternes et al. 2011. Removal of antibiotics from urban wastewater by constructed wetland optimization. Chemosphere 83: 713–719.

Hsieh, C.-Y., L. Yang, W.-C. Kuo and Y.-P. Zen. 2013. Efficiencies of freshwater and estuarine constructed wetlands for phenolic endocrine disruptor removal in Taiwan. Sci. Total Environ. 463-464: 182–191.

Kadlec, R.H. and S.D. Wallace. 2009. Introduction to treatment wetlands. pp. 3–20. Treatment Wetlands, Second Edition. CRC Press, Boca Raton.

Kaegi, R., A. Voegelin, C. Ort, B. Sinnet, B. Thalmann, J. Krismer et al. 2013. Fate and transformation of silver nanoparticles in urban wastewater systems. Water Res. 47: 3866–3877.

Kidd, K.A., P.J. Blanchfield, K.H. Mills, V.P. Palace, R.E. Evans, J.M. Lazorchak et al. 2007. Collapse of a fish population after exposure to a synthetic estrogen. P. Natl. Acad. Sci. US 104: 8897–8901.

Krutz, L.J., S.A. Senseman, R.M. Zablotowicz and M.A. Matocha. 2005. Reducing herbicide runoff from agricultural fields with vegetative filter strips: A review. Weed Sci. 53: 353–367.

Li, Y., G. Zhu, W.J. Ng and S.K. Tan. 2014. A review on removing pharmaceutical contaminants from wastewater by constructed wetlands: Design, performance and mechanism. Sci. Total Environ. 468-469: 908–932.

Liu, J., K.G. Pennell and R.H. Hurt. 2011. Kinetics and mechanisms of nanosilver oxysulfidation. Environ. Sci. Technol. 45: 7345–7353.

Lowry, G.V., B.P. Espinasse, A.R. Badireddy, C.J. Richardson, B.C. Reinsch, L.D. Bryant et al. 2012. Long-term transformation and fate of manufactured Ag nanoparticles in a simulated large scale freshwater emergent wetland. Environ. Sci. Technol. 46: 7027–7036.

Luo, Y., W. Guo, H.H. Ngo, L.D. Nghiem, F.I. Hai, J. Zhang et al. 2014. A review on the occurrence of micropollutants in the aquatic environment and their fate and removal during wastewater treatment. Sci. Total Environ. 473-474: 619–641.

Lv, T., Y. Zhang, L. Zhang, P.N. Carvalho, C.A. Arias and H. Brix. 2016. Removal of the pesticides imazalil and tebuconazole in saturated constructed wetland mesocosms. Water Res. 91: 126–136.

Malchi, T., Y. Maor, G. Tadmor, M. Shenker and B. Chefetz. 2014. Irrigation of root vegetables with treated wastewater: Evaluating uptake of pharmaceuticals and the associated human health risks. Environ. Sci. Technol. 48: 9325–9333.

Marsoni, M., F. De Mattia, M. Labra, A. Bruno, M. Bracale and C. Vannini. 2014. Uptake and effects of a mixture of widely used therapeutic drugs in *Eruca sativa* L. and *Zea mays* L. plants. Ecotox. Environ. Safe. 108: 52–57.

Masi, F., G. Conte, L. Lepri, T. Martellini and M. Del Bubba. 2004. Endocrine disrupting chemicals (EDCs) and pathogens removal in an hybrid CW system for a tourist facility wastewater treatment and reuse. *In*: Lienard, A. [ed.]. Proceedings of the 9th IWA Specialized Group Conference on 'Wetland systems for Water Pollution Control', Avignon, France.

Matamoros, V. and J.M. Bayona. 2006. Elimination of pharmaceuticals and personal care products in subsurface flow constructed wetlands. Environ. Sci. Technol. 40: 5811–5816.

Matamoros, V., C. Arias, H. Brix and J.M. Bayona. 2007. Removal of pharmaceuticals and personal care products (PPCPs) from urban wastewater in a pilot vertical flow constructed wetland and a sand filter. Environ. Sci. Technol. 41: 8171–8177.

Matamoros, V., J. García and J.M. Bayona. 2008. Organic micropollutant removal in a full-scale surface flow constructed wetland fed with secondary effluent. Water Res. 42: 653–660.

Matamoros, V., E. Jover and J.M. Bayona. 2010. Occurrence and fate of benzothiazoles and benzotriazoles in constructed wetlands. Water Sci. Technol. 61: 191–198.

Matamoros, V., L.X. Nguyen, C.A. Arias, S. Nielsen, M.M. Laugen and H. Brix. 2012. Musk fragrances, DEHP and heavy metals in a 20 years old sludge treatment reed bed system. Water Res. 46: 3889–3896.

Matamoros, V. and J.M. Bayona. 2013. Chapter 12—Removal of pharmaceutical compounds from wastewater and surface water by natural treatments. Comprehensive Analytical Chemistry 62: 409–433.

Mezcua, M., M.J. Gómez, I. Ferrer, A. Aguera, M.D. Hernando and A.R. Fernández-Alba. 2004. Evidence of 2,7/2,8-dibenzodichloro-p-dioxin as a photodegradation product of triclosan in water and wastewater samples. Anal. Chim. Acta 524: 241–247.

Muñoz, I., J.C. López-Doval, M. Ricart, M. Villagrasa, R. Brix, A. Geiszinger et al. 2009. Bridging levels of pharmaceuticals in river water with biological community structure in the Llobregat River basin (Northeast Spain). Environ. Toxicol. Chem. 28: 2706–2714.

Naidu, R., V.A. Arias Espana, Y. Liu and J. Jit. 2016. Emerging contaminants in the environment: Risk-based analysis for better management. Chemosphere 154: 350–357.

Nivala, J., M.B. Hoos, C. Cross, S. Wallace and G. Parkin. 2007. Treatment of landfill leachate using an aerated, horizontal subsurface-flow constructed wetland. Sci. Total Environ. 380: 19–27.

Pérez, S. and D. Barceló. 2007. Fate and occurrence of X-ray contrast media in the environment. Anal. Bioanal. Chem. 387: 1235–1246.

Petrie, B., R. Barden and B. Kasprzyk-Hordern. 2015. A review on emerging contaminants in wastewaters and the environment: Current knowledge, understudied areas and recommendations for future monitoring. Water Res. 72: 3–27.

Qiang, Z., H. Dong, B. Zhu, J. Qu and Y. Nie. 2013. A comparison of various rural wastewater treatment processes for the removal of endocrine-disrupting chemicals (EDCs). Chemosphere 92: 986–992.

Ranieri, E., P. Verlicchi and T.M. Young. 2011. Paracetamol removal in subsurface flow constructed wetlands. J. Hydrol. 404: 130–135.

Reichenberger, S., M. Bach, A. Skitschak and H.-G. Frede. 2007. Mitigation strategies to reduce pesticide inputs into ground- and surface water and their effectiveness: A review. Sci. Total Environ. 384: 1–35.

Reyes-Contreras, C., M. Hijosa-Valsero, R. Sidrach-Cardona, J.M. Bayona and E. Bécares. 2012. Temporal evolution in PPCP removal from urban wastewater by constructed wetlands of different configuration: A medium-term study. Chemosphere 88: 161–167.

Richardson, S.D. and S.Y. Kimura. 2016. Water analysis: Emerging contaminants and current issues. Anal. Chem. 88: 546–582.

Rose, M.T., F. Sanchez-Bayo, A.N. Crossan and I.R. Kennedy. 2006. Pesticide removal from cotton farm tailwater by a pilot-scale ponded wetland. Chemosphere 63: 1849–1858.

Sadaria, A.M., S.D. Supowit and R.U. Halden. 2016. Fate of Neonicotinoid Pesticides During Wastewater and Wetland Treatment. Assessing Transformation Products of Chemicals by Non-Target and Suspect Screening—Strategies and Workflows Volume 1. American Chemical Society pp. 121–131.

Shappell, N.W., L.O. Billey, D. Forbes, T.A. Matheny, M.E. Poach, G.B. Reddy et al. 2007. Estrogenic activity and steroid hormones in swine wastewater through a lagoon constructed-wetland system. Environ. Sci. Technol. 41: 444–450.

Song, H.-L., K. Nakano, T. Taniguchi, M. Nomura and O. Nishimura. 2009. Estrogen removal from treated municipal effluent in small-scale constructed wetland with different depth. Bioresour. Technol. 100: 2945–2951.

Tao, R., G.-G. Ying, H.-C. Su, H.-W. Zhou and J.P.S. Sidhu. 2010. Detection of antibiotic resistance and tetracycline resistance genes in Enterobacteriaceae isolated from the Pearl rivers in South China. Environ. Pollut. 158: 2101–2109.

Toro-Vélez, A.F., C.A. Madera-Parra, M.R. Peña-Varón, W.Y. Lee, J.C. Bezares-Cruz, W.S. Walker et al. 2016. BPA and NP removal from municipal wastewater by tropical horizontal subsurface constructed wetlands. Sci. Total Environ. 542: 93–101.

Underwood, J.C., R.W. Harvey, D.W. Metge, D.A. Repert, L.K. Baumgartner, R.L. Smith et al. 2011. Effects of the antimicrobial sulfamethoxazole on groundwater bacterial enrichment. Environ. Sci. Technol. 45: 3096–3101.

Verlicchi, P. and E. Zambello. 2014. How efficient are constructed wetlands in removing pharmaceuticals from untreated and treated urban wastewaters? A review. Sci. Total Environ. 470-471: 1281–1306.

Vymazal, J. 2013. The use of hybrid constructed wetlands for wastewater treatment with special attention to nitrogen removal: A review of a recent development. Water Res. 47: 4795–4811.

Vymazal, J. and T. Březinová. 2015. The use of constructed wetlands for removal of pesticides from agricultural runoff and drainage: A review. Environ. Int. 75: 11–20.

Vymazal, J., T. Březinová and M. Koželuh. 2015. Occurrence and removal of estrogens, progesterone and testosterone in three constructed wetlands treating municipal sewage in the Czech Republic. Sci. Total Environ. 536: 625–631.

WHO/UNEP. 2013. State of the science of endocrine disrupting chemicals—2012. United Nations Environment Programme and the World Health Organization. *In*: Bergman, Å., J. Heindel, S. Jobling, K. Kidd, R. Zoeller [eds.]. WHO Press, Geneva.

Xiaoyan, T., W. Suyu, Y. Yang, T. Ran, D. Yunv, A. Dan et al. 2015. Removal of six phthalic acid esters (PAEs) from domestic sewage by constructed wetlands. Chem. Eng. J. 275: 198–205.

Xu, J., Y. Xu, H. Wang, C. Guo, H. Qiu, Y. He et al. 2015. Occurrence of antibiotics and antibiotic resistance genes in a sewage treatment plant and its effluent-receiving river. Chemosphere 119: 1379–1385.

Yang, Y., B. Li, S. Zou, H.H.P. Fang and T. Zhang. 2014. Fate of antibiotic resistance genes in sewage treatment plant revealed by metagenomic approach. Water Res. 62: 97–106.

Zhang, D., R.M. Gersberg, W.J. Ng and S.K. Tan. 2014a. Removal of pharmaceuticals and personal care products in aquatic plant-based systems: A review. Environ. Pollut. 184: 620–639.

Zhang, D., T. Hua, F. Xiao, C. Chen, R.M. Gersberg, Y. Liu et al. 2014b. Uptake and accumulation of CuO nanoparticles and CdS/ZnS quantum dot nanoparticles by Schoenoplectus tabernaemontani in hydroponic mesocosms. Ecol. Eng. 70: 114–123.

Zhang, Y., Y. Zhuang, J. Geng, H. Ren, K. Xu and L. Ding. 2016. Reduction of antibiotic resistance genes in municipal wastewater effluent by advanced oxidation processes. Sci. Total Environ. 550: 184–191.

Zhang, Y., T. Lv, P.N. Carvalho, L. Zhang, C.A. Arias, Z. Chen et al. 2017. Ibuprofen and iohexol removal in saturated constructed wetland mesocosms. Ecol. Eng. 98: 394–402.

14

Electrochemically Assisted Artificial Wetlands

Generating Electricity from Wastewater Treatment

María Guadalupe Salinas-Juárez

INTRODUCTION

The prevailing need of satisfying the power demand by sustainable ways urges the pursuit of new and efficient technologies that generate electricity with reduced environmental impacts. Under this motivation, during the last decade a bio-electrochemical device has been developed, which can produce power *in situ* in a sustainable form with reduced CO_2 emissions. This device incorporates a microbial fuel cell with a vegetal species as continuous fuel source; the bio-electrochemical system takes advantage of the chemical energy capture coming from the sun by the photosynthesis process in the plant and converts the chemical energy into electricity in the microbial fuel cell. Fuel consists of the organic compounds released by the plant (rhizodeposition); consequently these devices are called plant-microbial fuel cells, vegetable-microbial fuel cells, or microbial fuel cells with living hydrophytes (henceforth MFC-LH).

This technology offers a greater environmental contribution when a microbial fuel cell (henceforth MFC) is implemented in a constructed wetland for the wastewater treatment. Besides, it is possible to obtain simultaneously power and

Universidad Nacional Autónoma de México, Facultad de Química, Departamento de Ingeniería Química, Laboratorios de Ingeniería Química Ambiental y de Química Ambiental (UNAM-FQ-DIQ-LIQAyQA), Circuito de la Investigación Científica s/n. Ciudad Universitaria. Delegación Coyoacán, 04510 Ciudad de México, Mexico.
Email: maria.salinas@outlook.com

wastewater treatment, by means of a natural system that simulates the establishment of a terrestrial-aquatic ecosystem, contributing to the CO_2 emissions mitigation. The above-mentioned benefits make the combined bio-electrochemical system attractive to be implemented in remote rural areas. Remote regions in Mexico already utilize wood as the main energy source, due to the economic and technical difficulties for supplying power by conventional means (García-Ochoa 2010). Thus, the application of electrochemically assisted constructed wetlands in rural areas would supply the power needs and the wastewater treatment, and would also contribute with huge benefits in urban areas for the power supply from the wastewater depuration.

Electricity is an important energy source for the economic growth and the human development and its lack in remote communities constrains these development indicators. In México there are 139,156 small villages with less than 100 inhabitants which represent more than two percent of population (CONAGUA 2015); it is probable that these figures include the inhabitants who lack electricity at home. In general, these remote villages show a high dispersion degree and low economic capacity, hindering the power energy provision.

Incidentally, in 2014 the wastewater treatment in Mexico achieved just 48.68% of the domestic wastewater produced in all country (CONAGUA 2015); thus, in remote small villages, the wastewater treatment hardly exists. The constructed wetlands for wastewater treatment constitute an efficient system for the natural treatment of wastewater, require a minimal energy investment, have an operational and maintenance low cost, and are ease to operate. Due to their economic and operational advantages, constructed wetlands are a strongly recommended option for wastewater treatment in small and remote communities. This treatment system participates with barely nine percent of all domestic wastewater treatment plants in Mexico, cleaning less than one percent of the waste-flow produced in the entire country (CONAGUA 2015).

The combination of the two technologies, microbial fuel cells and constructed wetlands results in a novel system: the electrochemically assisted constructed wetlands. The novel system promises a potential opportunity so that distant communities can access electricity which could lead to social benefits such as life quality enhancement and the reduction of environmental impacts caused by the greenhouse gases emissions during electricity production.

Although, urban areas can count on electricity supply, using these bio-electrochemical systems in public gardens and green roofs could lead to great benefits. For example, a minimal environmental impact derived from the simultaneous wastewater treatment and electricity production, because the greenhouse gases emissions would be reduced in the process.

Electrochemically Assisted Constructed Wetlands

Microbial Fuel Cells with Living Hydrophytes

The microbial fuel cells with living hydrophytes or vegetal organisms include a bio-electrochemical system that generates power based on the principles of a galvanic electrochemical cell: two electrochemical half-reactions take place involving electric charge transference, an oxidation and a reduction reaction. The oxidation reaction occurs at anode and the reduction reaction at cathode. The released protons from oxidation migrate through an electrolyte that works as ionic conductor and the released electrons are transported by means of an external electric circuit, which works as electronic conductor, until electrons reach the cathode, where the reaction is completed with the oxygen reduction reaction, yielding water at the end.

Microbial fuel cells with living hydrophytes, MFC-LH, can be built with two compartments, one of them accommodates the anode, and the other one, accommodates the cathode. The two compartments can be joined by a cationic exchange membrane or a proton exchange membrane (Fig. 1a). The MFC-LH can be built with just one compartment receiving the two electrodes (Fig. 1b) (De Schamphelaire et al. 2008; Kaku et al. 2008). The system uses solar energy in order to produce electricity in a clean and efficient manner, including the plant roots in the anode compartment of a MFC (Strik et al. 2008, 2011). Photosynthesis plants secrete organic exudates, most of them are carbohydrates. The bacteria living at the rhizome take the carbohydrates as substrate to convert them in electrical energy which is conducted by the external circuit (Strik et al. 2008; Timmers et al. 2010a).

The current generation in a MFC-LH starts with the biofilm formation or a microbial consortium growth over the anode, where the electrochemically active bacteria capture and store energy as a result of their metabolic process. Bacteria obtain energy and electrons for its growth along a series of oxidation-reduction reactions. For that reason, bacteria requires an electron donor, which is the oxidized substance (with a lower potential), and an electron acceptor, that is the reduced substance (with a higher potential). In order to obtain a higher energy gain, microorganism take as electron donor the available substance with the lower potential, as well as, the acceptor electron with the higher potential, being this acceptor soluble or non-soluble. In the latter case, microorganisms appeal to the extracellular electron transference (Rabaey 2010).

Bacteria oxidize organic matter coming from the rhizodeposition and from the dead organic material of the plant. From oxidation the released electrons are transferred towards an electron acceptor and, at the same time, protons are released as well as CO_2. The anode of the electrochemical cell is the non-soluble electrons acceptor, and through the external circuit, electrons will be conducted to cathode, producing an electrical current. Simultaneously, the released protons are transported inside of a reactor, passing through the electrolyte in direction of cathode, crossing the cation exchange membrane, if this exists, or through the support media in the reactor, in which the plant is located. If an additional compartment exists for the cathode, this compartment contains the final electron acceptor, such as ferricyanide, or in most of the cases, oxygen is the final electron acceptor. If the bio-fuel cell does not contain a cation exchange membrane and is conformed by just one container, the cathode is located on the water surface or the electrolyte surface, next the atmospheric oxygen, which is the most used final electron acceptor. In the cathode, oxygen is reduced with the electrons and protons to yield water (Deng et al. 2011; Strik et al. 2008) (Figs. 1a and 1b).

The main phenomena involved in this technology are:

- Production of bio-fuel from the photosynthesis, in other words, rhizodeposition delivers organic compounds to the medium, where will the fuel that will feed the bio-electrochemical cell.

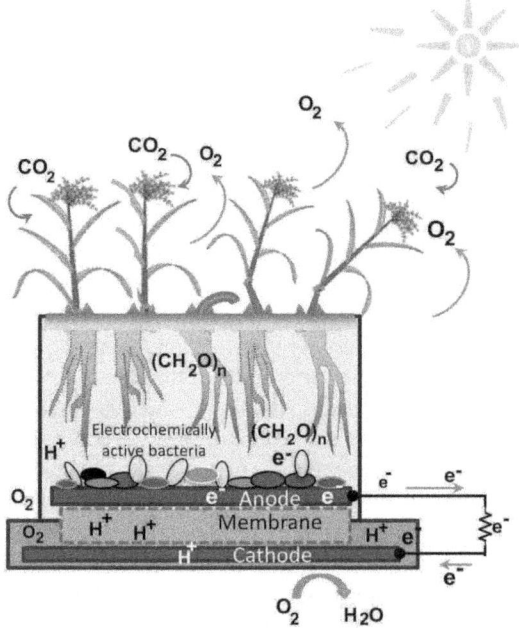

Fig. 1a. Microbial fuel cells with living plants, MFC-LP, built with two compartments, the anode and the cathode.

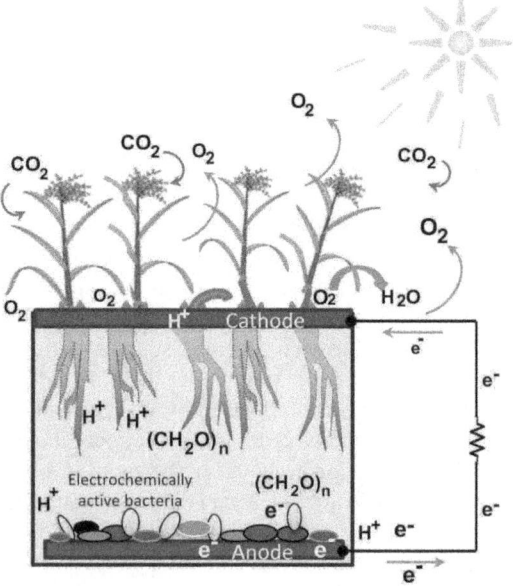

Fig. 1b. Microbial fuel cells with living plants, MFC-LH, built with one compartment.

- Anodic oxidation of organic matter by the electrochemically active bacteria (Strik et al. 2008).
- Oxygen reduction reaction at cathode (Strik et al. 2008; Rabaey 2010).
- Electricity generation from organic compounds by means of the MFC.

The most important benefits in this system are the continuous availability of fuel (organic matter from exudates released by the plant) and that electricity generation does not produce more pollutants than a conventional system. There is no fossil fuel combustion nor greenhouse gas emissions during power generation (Strik et al. 2008). The capacity of electricity production of MFC-LH has been estimated in 3.2 Wm^{-2} (28 $kWhm^{-2}yr^{-1}$), under the climatic conditions of west Europe (Helder et al. 2013a).

Reactions Involved in Electricity Generation in a MFC-LH

The voltage of the bio-electrochemical cell is yielded from the difference between the anode and cathode potential. The electrodes potential can be determined theoretically by means of the Nernst equation considering the oxidation and reduction reactions that take place in a bio-electrochemical cell.

The MFC-LH uses the subsurface biomass of a living plant in order to produce electricity. The biomass works as the support and feeding source for anodic consortiums and is made up of the roots, dead material of plant, as well as the exudates delivered. The diversity of organic compounds present and the non-uniform distribution of rhizodeposition provide a complex mixture of organic substances as substrate for bacteria, all of which are not used for electricity generation (Helder 2012). Considering that a great part of exudates contains acetate (Kaku et al. 2008), which is one of the most used substances as external carbon source in bio-fuel cells; thus, it is presented the oxidation of acetate as the reaction that occurs at anode, as in the following reaction:

$$CH_3COO^- + 4H_2O \rightarrow 2HCO_3^- + 9H^+ + 8e^-$$

From this reaction the anode potential it is determined applying the Nernst equation (equation 1):

$$E_{an} = E_{an}^0 - \frac{RT}{8F} ln \left(\frac{[CH_3COO^-]}{[H^+]^9 [HCO_3^-]^2} \right) \tag{1}$$

where:

E_{an} = anode potential (Volts, V)

E_{an}^0 = anode standard potential (298 K, 1 M), (Volts, V)

R = universal constant of gases (8.314 J/mol K)

T = temperature (K)

F = Faraday constant ($9.65*10^4$ C/mol)

8 = number of electrons involved in the reaction

$[CH_3COO^-]$ = acetate concentration (mol/L)

$[H^+]$ = protons concentration (mol/L)

$[HCO_3^-]$ = bicarbonate concentration (mol/L)

Under these conditions, the anode potential in open circuit normally is equal to –0.289 mV *versus* the hydrogen standard electrode, or –0.486 V *vs.* the Ag/AgCl electrode (Helder 2012).

In contrast, the cathode potential should be higher than the anode potential in order to produce electricity. If the bio-fuel cell includes the cathode in direct contact with oxygen, the cathode reaction consists of the oxygen reduction reaction as in the following reaction:

$$O_2 + 4H^+ + 4e^- \rightarrow 2H_2O$$

The Nernst potential for the oxygen reduction reaction is determined with equation 2:

$$E_{cat,O_2} = E^0_{cat} - \frac{RT}{4F} \ln\left(\frac{1}{pO_2[H^+]^4}\right) \tag{2}$$

where:

E_{cat,O_2} = cathode potential (Volts, V)

E^0_{cat} = cathode standard potential (298 K and atmospheric pressure), (V)

R = universal constant of gases (8.314 J/mol K)

T = temperature (K)

F = Faraday ($9.65*10^4$ C/mol)

4 = number of electrons involved in the reaction

pO_2 = oxygen partial presion (Pa)

$[H^+]$ = protons concentration (mol/L)

The cathode standard potential, with the oxygen reduction reaction, normally is equal to +0.0805 to 0.82 V *vs.* the hydrogen standard electrode or, 0.608 V *vs.* the Ag/AgCl electrode (Helder 2012; Rozendal et al. 2010).

Considering the theoretical potential of electrodes, the voltage of a MFC-LH with the oxidation reaction of acetate and the oxygen reduction reaction ideally would be equal to 1.1 V, according to the equation 3:

$$E_{cell} = E_{cat} - E_{an} \tag{3}$$

where:

E_{cell} = microbial fuel cell voltage (Volts, V)

E_{cat} = cathode potential (Volts, V)

E_{an} = anode potential (Volts, V)

In theory, the bio-fuel cell voltage it is determined with the reactions that take place on the electrodes, but in practice, the energy losses that occur in the system decrease the estimated voltage. These losses are originated by diverse factors involved with the internal resistance of the bio-fuel cell.

Electricity Generation from the Wastewater Treatment

The possibility of electric production in a bio-electrochemical device is the result of research carried out since the first decade of 2000 vis-à-vis the development and improvement of MFC. In this case, the components implementation of a conventional MFC with a vegetal species, it means, the integration of an anode and a cathode, an electrical circuit and a living hydrophyte, offers a novel device with promising results for the electricity generation. In fact, a huge interest in MFC was triggered by its environmental contribution, including the wastewater treatment occurring simultaneously to the electricity generation, transforming the chemical energy of organic matter (as contaminant) in electricity (Wu et al. 2015). In this way, the wastewater accomplishes two functions in a MFC: it is the fuel for the electrochemical cell and it is used as substrate for the electrochemically active bacteria at anode. It is estimated that the wastewater contains about nine times more chemical energy than that energy required for its treatment in wastewater treatment plants. However, this energy is not used nor recovered (Schröder 2008).

Among the wastewater treatment systems, constructed wetlands are economic systems; they consist of a submerged bed or a bed saturated with water that contains aquatic plants. The plants take nutrients from water and soil (support media) for its growth, taking advantage of the solar energy and the biological activity of the microorganisms that exist in the system, which responsible for organic matter degradation. Therefore, constructed wetlands simulate the physical, chemical, and biological processes that are performed in natural wetlands but in controlled environments (Salinas-Juárez 2011). The replicated natural processes by constructed wetlands lead to the physical-chemical and biochemical conditions that are needed for the current capture of electrons derived from the biological processes such in a microbial fuel cell.

The wetland study for the wastewater treatment began in the decade of 1950. One of the aims study was to improve the wastewater treatment performance in rural areas (Vymazal 2011). This depuration system was broadly used at the end of the 20th century on diverse geographical points. The applications above mentioned included different wastewater types, for example, sewage containing conventional pollutants such as organic matter, nutrients and microorganisms, and non-conventional pollutants such as heavy metals, hydrocarbons and emerging contaminants (EPA 2000; Vymazal 2011; Wu et al. 2015). In the present century, the artificial wetlands have been the focus of attention for diverse topics of research with the aim of achieving an effective and sustainable removal of the mentioned contaminants (Vymazal 2011; Wu et al. 2015). As a consequence, a variety of strategies and techniques have been proposed in order to improve the artificial wetlands performance; among them, the implementation of the components of a MFC in a wetland is suggested (Wu et al. 2015). From this implementation are derived the electrochemically assisted artificial wetlands (henceforth EAAW) for the concurrent wastewater treatment and the electricity generation.

The compatibility between constructed wetlands and MFC is based on the fact that both are biological systems that involve the organic matter degradation, and that a MFC requires a gradient of oxidation-reduction potential, which is naturally found in a constructed wetland as a consequence of the presence or absence of oxygen. The anode requires anaerobic conditions under which the electrons transference happens, from the electricigens bacteria to the anode; the oxygen presence interferes with this transference. On the contrary, the cathode requires a completely aerobic environment in which the oxygen reduction reaction should carry out with no restrictions (Doherty et al. 2015; Wu et al. 2015). The highest oxidation-reduction potential gradient it is generally found from the surface with aerobic regions, to the bottom, where anaerobic regions exist. Finally this gradient will depend on the flow direction, the design, and the operation conditions of the wetland (Doherty et al. 2015; Liu et al. 2016).

In constructed wetlands with surface flow, the water is over the soil and plants, emergent or submerged, plants are rooted in the sediment layer on the column base. Besides, a wetland can contain floating plants. In the sub-surface constructed wetlands, the water flows through the support media (porous media) such as gravel or any other stony material, in which the plants are rooted. A biofilm or microbial communities grow over the available surfaces in the roots of the plant, on sediments and/or on the support media; the microbial activity resulting is the main mechanism of contaminants removal, specially, the organic matter biodegradation (Kayombo et al. 2002; Liu et al. 2016). The oxygen concentration inside the wetland takes action in the microbial activity of the wetland and, therefore, on the contaminant degradation and on the oxidation-reduction potential gradient. There are three pathways in which the oxygen is introduced in a constructed wetland: oxygen that is included by means of the influent, the radial oxygen that the plant releases, and the atmospheric aeration (Fenoglio-Limón et al. 2001; Guido-Zárate and Durán-de-Bazúa 2008; Liu et al. 2016).

Depending on the characteristics and the type of selected wetland, it is possible to suggest different configurations of the electrochemical cell, in which the electrochemical components will be adapted with the aim of producing a higher electrical current and ensuring the contaminants removal. There is a diversity of designs and configurations among the studies conducted up until now. Some designs include materials that make the reactor more expensive, others establish a variety of operation parameters. In performed studies, small values of power have been obtained compared to those acquired by conventional MFC, where a generation of 2870 W/m³ (total reactor volume) have been achieved, with a coulombic efficiency of 83.5%; it means, that the total available electrons in the substrate or in the fuel, just 83.5% were used to produce power (Fan et al. 2012). These results were obtained under special operation arrangements and using specific materials.

The development of bio-fuel cells that include a vegetal species began about 2007, with no wastewater treatment, reporting the first results in 2008 (De Schamphelaire et al. 2008; Kaku et al. 2008; Strik et al. 2008). Since then, diverse studies have been done modifying the components and factors in the bio-electrochemical cell. For instance, with the aim of increasing the current and power density, a maximal current density of 469 mA m^{-2} was obtained and the highest power density of 211 mW m^{-2}.

The above values were produced from the modifications to the nutritive solution (electrolyte) that was formerly used (Helder et al. 2012a). Different vegetal species were proved, too, obtaining a maximum power density of 222 mW m^{-2} with the plant species *Spartina anglica* (Helder et al. 2010). Varying the reactor configuration, from tubular to a box (rectangular) reactor, with vertical electrodes, it was possible to obtain 1.6 A m^{-2} and 440 mW m^{-2} (Helder et al. 2012b). Reactors used in such experiments contained a cation exchange membrane and they were located in a greenhouse with temperature and light under control. In a more recent study the highest value of power in all the studies was obtained, producing 679 mW m^{-2}. This power generation was achieved with a bio-fuel cell that contained two cathodes, an anode, and two cation exchange membranes, and as current collectors, they used gold wires; in that bio-fuel cell the vegetal species *Spartina anglica* was used under controlled conditions of laboratory (Wetser et al. 2015). The same research team conducted an experiment under natural conditions of light, temperature and weather, obtaining variable values and a maximum power of 88 mW m^{-2}; in this latter case, a cation exchange membrane was not used between the electrodes (Helder et al. 2013b).

To obtain the mentioned results, except the last one, a cation exchange membrane was used, which significantly increases the reactors costs and makes less feasible its application in a real scale. In another investigation, the electricity generation was analyzed with the species *Lemna minuta* or *minor* that is a floating plant (known in Mexico since pre-Columbian times as *chichicaxtli*, from the Náhuatl or Aztec language *chichic*=sour and *achtli*=seed). In the installed reactor, neither nutrients nor organic matter was added, just 40 mL of tap water was introduced, obtaining an electricity generation of 1.62 ± 0.10 A m^{-2} and 380 ± 19 mW m^{-2} (Hubenova and Mitov 2012). However, it would be necessary to study the applicability of this species in a higher scale. Furthermore, in all the studies mentioned above, the current and power were calculated from the voltage measured and taking the value of the external resistance used. Thus, some losses are not considered; the losses interfere when the electrical parameters are directly measured with a multi-meter applying an external electric resistance to the bio-electrochemical system.

The research involving constructed wetlands use reactors with the own characteristics of a constructed wetland, such as the support media, vegetation, flow, depth, the hydraulic residence time, and the inclusion of wastewater, this being sewage domestic or industrial (from a farm, from a distillery or with textile dyes). In these investigations, the obtained results have been values of power lower than those achieved in MFC-LH. The maximum values accomplished of current were about 345 mA m^{-2} and of power, 179.78 mW m^{-2} in a 26 L container with four different emerging vegetal species and two submerged species in a reactor that would be a constructed wetland of surface flow, that was fed with two effluents, one coming from a hydrogen production process and other one originated in a treatment plant of water coming from a distillery and domestic wastewater (Chiranjeevi et al. 2013). In another artificial wetland, an electricity production of 87.79 mW m^{-2} was obtained with an upward flow reactor and using wastewater from a pig farm (Xu et al. 2016). A value of 85.2 mW m^{-2} was produced in artificial wetlands treating water with an azo dye, they obtained 93% of decoloration and an organic matter removal of 86%, demonstrating that the bio-electrochemical activity improved the removal processes (Fang et al.

2015). In the other studies, the values attained have been lower that those referred with coulombic efficiencies of less than 3.9% (Liu et al. 2014).

The differences between the MFC-LH and the EACW are the carbon source and that in the wetlands, the main objective is the wastewater treatment with the added value of the electricity generation in a simultaneous process. However, the EACW require a drastic improvement in terms of power generated and coulombic efficiency implicated for that power generation (Xu et al. 2015). A challenge in scaling the system up is to allocate the electrodes in an arrangement that optimize the potential redox gradient in the wetland, and the organic matter degradation feed the bacteria at the anode but does not interfere with the reduction reaction at cathode. An additional challenge regarding the design of the EAAW involves the optimal use of oxygen as final electron acceptor at the cathode and, besides that, ensuring that oxygen doesn't interfere with the anodic reactions of oxidation.

The EAAW technology is yet in an early stage of research, in which the main objective is increasing the power generation capacity. In spite of 28 kWh m^{-2} yr^{-1} of theoretical estimated generation in the MFC-LH, just 679 mWhm^{-2} were obtained in lab under control conditions (Wetser et al. 2015). The low coulombic efficiency, the energy losses, and the internal resistance of these bio-electrochemical systems demand to extend the research in pursuit of the factors involved. Even more, the implementation in a full scale faces challenges such as: finding the best materials technically and economically more suitable, and defining an optimal design for the electrochemical cell, in which every component be located in order to obtain the greatest advantage from the available fuel (organic matter) and the reactions occurring at the system, to convert that chemical energy into a usable current.

Electrochemically Assisted Constructed Wetlands Design and Materials

The electrical performance in a MFC depends on diverse factors that include the reactor design and planning as well as the operation conditions. It is necessary to determine the physical configuration (design) in order to boost the organic matter degradation and also the transport mechanisms for electrons and protons. It is also essential to procure the reduction of energy losses; this reduction is possible when the bio-fuel cell design reduce the ohmic resistance.

The materials selection is an important factor to achieve an optimal design. Since the importance of electrode materials in electricity generation, the materials selection for every component will be of special attention, specifically determining the electrical circuit components, the electrolyte composition, and in the case of the EAAW, the support media and vegetal species to use. In the base of the reported research, various materials are considered. The design activity should determine what the most suitable materials are in terms of economy and technique. It is important not forget the characteristics that a material must accomplish in order to promote the bio-electrochemical processes expected.

The former studies provide manifold observations that help to improve the MFC design. These considerations include: distance between electrodes and the surface area on both of them, the most suitable size for the reactor, the use of a cation exchange membrane, or a material for the separation between electrodes, the use of a catalysts, the operation conditions such as temperature, pH, operation time, hydraulic residence time, organic matter concentration, degradability of organic matter in the influent, and the likely presence of chemical toxic substances that could interfere in the bio-electrochemical processes (Kim et al. 2008; Logan et al. 2006).

Design—Physical Configuration

The design or physical configuration of a bio-electrochemical system determines the location and the position of every component of the biofuel cell. One of the factors that determine the configuration is the electrodes location and the type of separation that will be used between them. Furthermore, the task of design will determine if an ionic selective membrane or a material that act with a similar effect than a membrane will be used; it means that in this step should be determined if just a physical barrier will be used instead of a cation exchange membrane. According to the chosen option, the electrodes are to be located as well as the distance between them, determining the substance used as final electrons acceptor.

A bio-fuel cell with just one compartment takes the fact of the oxygen availability next to the water surface or the oxygen availability near the roots, which results from the oxygen transport by the aerial part of the plant. Therefore, the anode is introduced in a support matrix adjacent to the plant roots in order to obtain the exudates as fuel, or can be inserted in deepest part of reactor, where anaerobic conditions exist. The cathode is allocated in the superior layer, near the water surface, that is the region with the highest oxygen concentration available, or in an immediately contact with the rhizosphere.

The proximity between anode and cathode is related to the internal resistance of the bio-fuel cell. This closeness facilitates or hinders the protons transport through the membrane or through the electrolyte. The ohmic losses are reduced adjusting the space between the electrodes (Kim et al. 2008; Logan et al. 2006); however, if the distance between these components is reduced too much, it is feasible that oxygen invades the anode, affecting the anaerobic anodic bacteria (Kim et al. 2008).

Distinct configurations of MFC-LH have been developed and proved using a cationic or protonic exchange membrane. In the same way, bio-electrochemical systems with plants have been developed with no selective membrane. The first studies applied a design of bio-fuel cell that include a cation exchange membrane and the electrodes were allocated in two compartments (Helder et al. 2010; Strik et al. 2008; Timmers et al. 2010a). At the same time, other bio-electrochemical systems were developed in which, despite of the use of a membrane or a separation, the electrodes were located in an individual compartment (Bombelli et al. 2013; De Schamphelaire et al. 2008; Helder et al. 2012a; Hubenova and Mitov 2012; Timmers et al. 2013a,b). Bio-electrochemical systems were developed with no cation exchange membrane but the electrodes were put into a paddy field (Kaku et al. 2008; Takanezawa et al. 2010); thus, a sediment-MFC was simulated where the anode was buried in the sediment or in the plant support and the cathode was allocated on the water surface in such way that the cathode was in directly contact with the atmospheric oxygen (Chiranjeevi et al. 2012; De Schamphelaire et al. 2008; Helder et al. 2013b).

In the conducted studies in a paddy field, the cathode was put on the soil surface and the anode, 5 cm below the surface, buried into the soil, obtaining a maximal production of 6 mW m^{-2} (Kaku et al. 2008). In a similar experiment, two separation distances between anode and cathode were proved: the anode was put 2 and 5 cm below the soil surface and the cathode on the surface. The best result was obtained with 5 cm separation, obtaining 14 mW m^{-2} (Takanezawa et al. 2010). De Schamphelaire et al. 2008 used two different configurations. In the first one, a graphite felt was used as cathode, allocated on the water surface, and aerated with a pump, while the anode was composed by two pieces of felt graphite, allocated in different depths, buried into the soil. The second option consisted on a cathode made of graphite grains with ferricyanide, which required the use of a cation exchange membrane; the anode was composed by graphite granules. A significant difference was not observed between the obtained results with the air cathode and with the submerged cathode in ferricyanide. The difference was on the type of anode used, having a better result with graphite felt in comparison to the anode composed by graphite granules. In other experiments, the physical configuration was similar among them (Helder et al. 2010; Strik et al. 2008; Timmers et al. 2010a,b): the reactors were formed by two compartments, the anode compartment was contained by the cathode compartment; the two compartments were joined by a cation exchange membrane. In this configuration the distance between electrodes was estimated at 1 to 5 cm, the high values for electricity generation achieved 67 through 220 mW m^{-2}.

In the case of the EAAW, the experiments were performed using the vertical flow-constructed wetland design as physical configuration, utilizing cylindrical reactors of 50 cm in depth, with an upward flow and a huge separation between electrodes, compared to the separation in MFC-LH. In EACW the support media material is used as separation between electrodes and it is common to allocate the cathode near the roots so that the cathode receives the higher oxygen concentration. The distance between electrodes became more than 30 cm and the anode was located at bottom of the reactor with the aim to take advantage of the anaerobic conditions (Venkata et al. 2011; Yadav et al. 2012; Fang et al. 2013; Liu et al. 2013; Zhao et al. 2013). All of these systems have been built with an individual compartment allocating in the same recipient the cathode and the anode. Results obtained with EACW vary between 9.4 mW m^{-2} (Zhao et al. 2013) and 57 mW m^{-2} (Venkata et al. 2011).

Materials

Container (reactor). It is essential to ensure anaerobic conditions at the anode section, which are necessary to the electrochemically active bacteria perform the chemical energy transformation into electric energy (Kim et al. 2006; Strik et al. 2008). On the contrary, in the cathode section it is not necessary to preserve anaerobic conditions; if oxygen is the final electron acceptor, a higher oxygen availability is required, as higher as possible. Varied materials have been used in the performed experiments for the container or the reactor, being the recipient that contains most of the bio-fuel cell components. Materials as glass, plastic, acrylic, and plexiglas were used (De Schamphelaire et al. 2008; Helder et al. 2010; Strik et al. 2008; Timmers et al. 2010a), in the experiments at the paddy field the soil itself was used (Kaku et al. 2008; Takanezawa et al. 2010). Timmers et al. (2010b) designed a bio-fuel cell that used a glass tube as anode compartment, a glass cylinder of 30 cm in high and 3.5 cm in diameter; Helder et al. (2010) used a plexiglass cylinder of 9.9 cm in diameter; De Schamphelaire et al. (2008) employed two types of reactor, the first one with plastic containers 16.86 cm high and the second one, with cylindrical glass recipients 20 cm in high and 14 cm in diameter. In EAAW experiments, reactors generally were formed by an individual recipient made of plastic material as acrylic or PVC. The containers generally shaped cylindrical with 50 cm in high and at 10 to 30 cm in diameter (Fang et al. 2013; Liu et al. 2013; Yadav et al. 2012; Zhao et al. 2013).

Anode. With the aim of supporting the electron transference, the main characteristics that an anode material should have are: high electrical conductivity, high surface area, being non-corrosive, chemically stable in the electrolyte, and biocompatible with the microorganisms, especially with the electrochemically active bacteria and with the plants, that is, this material should not be toxic. In practice the anodic material mostly used is graphite due to its versatility (Kim et al. 2008; Logan et al. 2006). Graphite has diverse presentations such as compact plates, granules or powder, rods, paper, and fibers as felt or cloth. Graphite is relatively economic, easy to handle, and has a high defined surface area (Logan et al. 2006). Graphite felt and carbon granules are considered suitable material to conform the anode in a MFC-LH. From a comparison conducted among different anodic materials regarding their electric efficiency, resulted that carbon felt avoid clogging and over time carbon felt is not reduced in thickness; furthermore, the graphite granules (less than 2 m in diameter) offer a suitable support media for the roots penetration. The carbon felt advantage is that its compact state let a high contact with soil. On the contrary, carbon granules require the optimal quantity to ensure the electric contact among granules due to the movement caused by the roots growth or by the water flow, these phenomena can provoke the connection loss between anode and current collector. However, granules or powder are more suitable in a reactor, these materials shape according to the availability of space and the particles will be forced to contact each other (Arends et al. 2012). One of the major factors on the MFC performance is the biocatalysts quantity at anode, that is, the electrochemically active bacteria quantity on the anode surface, which is benefited by an anodic high surface area. The higher surface the area at anode is, the higher the contact between electricigens and the electrode is. This high contact reduces the internal resistance and increases the power density generated (Deng et al. 2011; Kim et al. 2008). A higher surface area it is achieved in graphite felt electrodes or by using layers of carbon granules (Logan et al. 2006).

Cathode. Cathodic material requires the same features than those needed by the anode. In the same way, the cathodic material significantly affects the MFC performance. Graphite felt (3 mm thickness) is the material most commonly used for cathode in MFC-LH. The difference among the cathode structure in different experiments is the cathode location. In some studies, the cathode is exposed to air or the graphite felt was aerated directly, while in two compartment reactors, the cathode (graphite felt) was submerged into ferricyanide and/or near the anode with a proton exchange membrane between them. Cathodic materials used in EACW vary among graphite discs or plates, as well as activated carbon and even a stainless steel mesh has been used (Venkata et al. 2011; Chiranjeevi et al. 2013; Fang et al. 2013; Liu et al. 2013; Zhao et al. 2013).

Electrodes Separation

The main objective of installing an ion selective membrane is enclosing the electrochemical reactions. Ideally, it is expected that organic matter remains at anode, but not at cathode, and it is expected too, that oxygen remains abundantly at cathode, but in zero concentration at anode. These aerobic conditions will ensure the oxidation and reduction reactions take place at the right location. However, even with a selective membrane, the reactants crossover is present in fuel cells (Yang and Pitchumani 2006). Furthermore, in a sediment-MFC or with a support matrix, it is common to find a natural redox barrier as separation between electrodes (Krieg et al. 2014). In this way, the oxygen concentration decreases with the reactor depth, and it is possible to avoid oxygen diffusion in the anodic region with no membrane (Deng et al. 2011). Conversely, the ohmic resistance exponentially increases using an ion selective membrane with a low electrolyte concentration in the wastewater (Harnisch and Schröder 2009). In experiments with and without membrane, the membrane-less reactors showed the best electrical performance, suggesting that the membrane influenced the results obtained (Helder et al. 2013b). Finally, using a cation exchange membrane significantly increases the reactor cost (Deng et al. 2011).

Electrolyte

In a MFC, the anolyte or anodic electrolyte consists of a solution that contains one or more organic compounds, which are the substrate for microorganisms. When wastewater is included, this water contains the organic matter that will be oxidized by electrochemically active bacteria. In a MFC-LH, the anolyte consists of a nutritive solution or the growth media from where plants take the required nutrients to its metabolic functions. In a constructed wetland, plants take the nutrients they need from wastewater and microorganisms take as carbon source the organic matter.

A reduced organic content in the electrolyte is associated with a low power generation, therefore a supplement or an extra substrate is recommended in order to obtain better electrical results (Kaku et al. 2008).

Besides that, it has been proved the power generated from different substrates. The organic compounds and substances more used to feed the microorganisms are several among glucose, glutamate, processed starch, wastewater, river water, marine sediments, ethanol, acetate, propionate, and butyrate. In many experiments acetate has been used with positive results, even better than those obtained with glucose or butyrate (Min and Logan 2004; Kaku et al. 2008; Kim et al. 2008). Acetate is one of the best organic compounds to stimulate the roots exudates and improves the power generation, even in darkness in MFC-LH (Kaku et al. 2008).

Pollutants present in domestic wastewater mostly comprise organic substances, and in a few cases could contain other inorganic contaminants; some of them are degradable and some non-biodegradable. Most organic pollutants are used by microorganisms as an energy source and to synthesize biomass (Kim et al. 2008). Synthetic wastewater to use should provide a carbon source, nitrogen, phosphorus, and trace of metals; all these substances are those required for the microbial reproduction and maintenance (Ng and Hermanowicz 2005).

Plants require primary nutrients as nitrogen, phosphorus, and potassium, secondary micronutrients as calcium, magnesium, and sulfur, and micronutrients needed in lower concentrations, as iron, manganese, chlorine, copper, zinc, boron, and molybdenum. In performed studies with MFC-LH, these nutrients are supplied by means of a "modified Hoagland nutritive solution" (Helder et al. 2012a; Strik et al. 2008; Timmers et al. 2010a,b), that contains every element needed for the plant growth under hydroponic conditions. In case of experiments in paddy fields, nutrients and part of organic matter were added by the soil in which the plants are located (Kaku et al. 2008).

In EAAW, organic matter and nutrients are supplied by the wastewater that is introduced with the aim of treatment, and simultaneously, to feed the microorganisms involved in power generation. Organic load vary according to each experiment reported; variation was used to observe the influence of different organic concentrations on power production and on contaminants removal, as well as to eventually adapt the system to an increasing organic load. The experiments with EACW have used different pollutants in wastewater as azo dyes (Yadav et al. 2012; Fang et al. 2013), domestic wastewater and distillery wastewater (Venkata et al. 2011; Chiranjeevi et al. 2012), likewise swine wastewater (Zhao et al. 2013). In every experiment, components and organic loads have been different and removals and power production showed diverse implications, due to the diversity of substances used.

In the case of two compartments MFC-LH, a catholyte or cathodic electrolyte was used. The catholyte increases the internal resistance and decreases the fuel cell efficiency. However, the objective of two compartments is to confine the half-reactions in order to produce a high power.

The aim of the use of two compartments in MFC-LH is to study the applicability of different final electron acceptor at cathode besides oxygen, such as potassium ferricyanide, hydrogen peroxide, or nitrates. Despite the results obtained with these electron acceptors, oxygen is considered as the more suitable final electron acceptor due to its abundance, non-cost availability, null toxicity, and its high oxidation potential; oxygen use is sustainable and does not produce chemical or pollutant wastes (Deng et al. 2011; Kim et al. 2008; Logan et al. 2006). Apart from that, the most important cathodic reaction to electricity production in MFC is the oxygen reduction reaction, yielding a higher potential at cathode than ferricyanide (Rozendal et al. 2010).

Electrons and protons react at the cathode with oxygen to obtain water; then, the oxygen reduction at cathode is one of the fundamental mechanisms in power generation in MFC-LH (Kim et al. 2008). In studies with MFC-LH and in EACW, oxygen is a final electron acceptor used frequently, especially where there are no two compartments. Ferricyanide has been used in two compartment MFC-LH, achieving similar results than with oxygen (De Schamphelaire et al. 2008; Timmers et al. 2010b).

Packed Bed or Support Matrix

In a bio-electrochemical system with living hydrophytes the support media must accomplish certain features such as: it should allow the roots growth as well as the aerial part development, and should have a high electric conductivity (Arends et al. 2012). Arends et al. 2011 compared the efficiency of electric charge transport using carbon felt and powder with different features finding that the material with better electric transfer are granules 1–5 mm in diameter, followed by carbon felt 18 mm thickness. Besides, the experiment was carried out with the aim of determine the best proportion mixing carbon powder with soil and carbon felt with soil. They observed that the best results were obtained with a mixture of 100 and 67% volume reactor of carbon powder and soil. In the case of felt, the highest results were yielding by the mixture of 24% of carbon felt (reactor volume). The results suggest that soil hinder the transport mechanisms as proton migration from anode to cathode and the exudates diffusion on the anode surface. These phenomena increase the ohmic resistance and the mass transfer resistance (Arends et al. 2011; Deng et al. 2011).

In MFC-LH research, the material mostly used as support media is graphite powder or graphite granules, in which the plant develops its roots, achieving the growth of five different vegetal species and proving the low graphite toxicity (De Schamphelaire et al. 2008; Strik et al. 2008; Helder et al. 2010; Timmers et al. 2010a,b). In other experiments, soil was used as support media and growth media (De Schamphelaire et al. 2008; Kaku et al. 2008; Takanezawa et al. 2010). Most of studies with EACW include gravel, although some of them include lake sediments and dewatered alum sludge from a water treatment plant (Venkata et al. 2011; Chiranjeevi et al. 2013).

The ion transport requires the minimum electric resistance, as minimum as possible (Harnisch and Schröder 2009); however, soil blocks the proton migration through the support matrix. For that reason, in diverse experiments a graphite bed is employed as plant support. In other experiments, *tezontle* was used as support media, determining that a graphite bed gives higher stability in power produced (Salinas-Juárez et al. 2016). *Tezontle* is a volcanic slag commonly used in Mexico for constructed wetlands (*tezontle* is also a Náhuatl or Aztec name, *tetl*=stone and *tzontli*=light as hair).

Vegetal Species

The vegetal species for bio-electrochemical systems with living hydrophytes should have particular features among them. The roots should be capable of living in anaerobic environments, which are the required conditions for the anode. Aquatic plants survive in anoxic or anaerobic conditions; therefore they are suitable for this type of MFC. Moreover, a huge variety of aquatic plants are used in constructed wetlands for wastewater treatment.

Perennial plants avoid the periodic change of electric installation and allow the power production in longer periods. Aquatic plants as *Glyceria maxima, Oriza sativa, Spartina anglica, Anisogramma anomala,* and *Arundo donax* have been used in experiments with MFC-LH. All of these plants produced electricity although the highest values were obtained with *Glyceria maxima* and *Spartina anglica* (Strik et al. 2008; Helder et al. 2010; Takanezawa et al. 2010; Timmers et al. 2010a,b). In experiments with EACW, the aquatic vegetal species used were *Eichhornia crassipe, Canna indica, Bryophyllum pinnatum, Solanum lycopersicum, Oriza sativa, Lycopodium* and *Adiantum, Hydrilla verticillata, Myriophyllum,* as submerged plants, and *Ipomoea aquatic, Phragmites australis,* and *Typha latifolia* as emerging plants (Yadav et al. 2012; Chiranjeevi et al. 2013; Fang et al. 2013; Liu et al. 2013; Zhao et al. 2013; Salinas-Juárez 2016).

The emerging species *Phragmites australis* is broadly used in constructed wetlands for wastewater treatment, due to the fact that its rhizome is capable of living in an aerobic-anaerobic environment. *Phragmites australis* proliferate without pests problems and is a resistant species upon weather or extreme conditions of temperature.

Inoculum

Electrochemically active bacteria oxidize organic matter in a MFC and transfer the released electrons to the anode. In an anode biofilm, it is possible to find a great diversity of microbial consortia. It is believed that electrochemically active microorganisms exist wherever in nature. Electrochemically active bacteria are well known as electricigens, anodophilic organisms, or exoelectrogens (Kim et al. 2008; Logan and Regan 2006). This type of organisms is characterized by the ability of transferring exocellular electrons towards the anode without any exogenous mediator, and the electricigens are capable of oxidizing organic compounds using the anode as the sole electron acceptor, although the inoculum could contain microorganisms capable of living at the anode, but not taking the anode as the final electron acceptor.

These organisms vary according to the inoculum used, the type of substrate (Kim et al. 2008; Logan and Regan 2006) and in the case of MFC-LH, the vegetal species included influences the microbial communities, because the plant rhizome can contain some electroactive microorganisms (Lu et al. 2015). Furthermore, the substrate applied in the bio-electrochemical system will define the species at anode.

An inoculum for a MFC could be taken from marine sediments, marine plankton, fresh water sediments, or anaerobic sludge. If a MFC is inoculated with a mixed culture it is expected that the electrochemically active bacteria survive as well as other microorganisms that live in association with them. Most of the microbial electroactive species are constituted by *Geobacter* or *Shewanella*, but a dominant species has not been found, it is suggested that electrochemically active bacteria need the functions that other microorganisms realized, for example the biodegraders that complete the oxidative activity (Kim et al. 2008; Logan and Regan 2006; Lu et al. 2015).

Among the microorganisms found in MFC are: *Shewanella putrefaciens*, *Geobacteraceae sulfurreducens*, *Geobacter multireducens*, *pseudomonas*, *acidobacteria*, *proteobacteria* α, β, and δ, as well as *firmicutes* (Kim et al. 2008; Logan and Regan 2006). The species considered as electrochemically active belong to the *Geobacter* and *Shewanella* species, being both iron-reducing bacteria; the family *Geobacteraceae*, especially, can transfer electrons directly to the anode. However, distinct microbial consortium participation, just like the respective metabolic pathways, are necessary to efficiently convert the stored energy from organic matter into electricity (Kim et al. 2008; Logan et al. 2006; Lu et al. 2015). Hence, the results are more promising when using a mixed culture for the inoculum than a pure culture due to the synergistic interactions among microorganisms which benefit power generation, despite the microorganisms involved be unknown.

Thus, the electrochemically active microorganisms depend and need the degrader organisms to produce the suitable electron donor from more complex organic compounds such as glucose. For that reason, the variety and composition of the bacterial community is broad in a MFC feed with glucose. In a MFC feed with simpler substances, the microbial variety is less diverse (Kim et al. 2008; Logan et al. 2006).

Populations related to *Rhizobiales* bacteria were found in the rice paddy anode, suggesting that this type of microorganisms would be associated with power generation, particularly, *Rhizobiales bacterium* A48 is likely to use the anode as electron acceptor after consuming the organic exudates from rice plant roots (Kaku et al. 2015).

External Electric Circuit

Materials used in the electric circuit should allow the electrons to flow; thus, the internal MFC resistance and the ohmic losses are reduced. For the external circuit, here exist a variety of materials such as copper wires or golden Teflon wires. The highest power values have been obtained using golden wires (Helder et al. 2010; Timmers et al. 2010a,b).

In the external circuit of MFC-LH, graphite rods are used and the external resistances vary from 10 to 1000 ohm (De Schamphelaire et al. 2008; Kaku et al. 2008; Strik et al. 2008; Takanezawa et al. 2010; Timmers et al. 2010a,b; Chiranjeevi et al. 2012). In the EACW, the material used as wire is copper and in some studies, graphite rods are employed and the external electric resistances to measure polarization vary from 5 to 100,000 Ω (Yadav et al. 2012; Fang et al. 2013; Liu et al. 2013; Zhao et al. 2013).

Conclusions

Electrochemically assisted constructed wetlands are a promising technology to obtain electricity in a sustainable and affordable way in addition to the wastewater treatment. However, the challenges that these devices face are diverse and require future research in order to develop and apply this technology in a real scale. The major difficulties are to find the materials for the components and the design that lead to the highest harvesting of power from this natural system.

Acknowledgements

The author acknowledges her graduate student scholarships from the Mexican Science and Technology Council (*CONACYT*, in Spanish *Consejo Nacional de Ciencia y Tecnología*) and the UNAM Coordination for Graduate Studies. Materials were provided by *UNAM DGAPA* projects *PAPIME EN103704, PE101709*, and *PE100514* and to *UNAM Programa de Apoyo a la Investigación y el Posgrado de la Facultad de Química, PAIP 5000-9067.* The author also appreciates the collegial support in the review of the first draft by Prof. Dr. Pedro Roquero-Tejeda, from Mexico's National Autonomous University of Mexico (*UNAM, Universidad Nacional Autónoma de México in Spanish*).

References Cited

Arends, J.B.A., E. Blondeel, S. Tennison, N. Boon and W. Verstraete. 2011. Anode materials for sediment microbial fuel cells. Comm. Appl. Biol. Sci. Ghent University BE 76(2): 47–50.

Arends, J.B., E. Blondeel, S.R. Tennison, N. Boon and W. Verstraete. 2012. Suitability of granular carbon as an anode material for sediment microbial fuel cells. J. Soils Sediments 12(7): 1197–1206.

Bombelli, P., D.M.R. Iyer, S. Covshoff, A.J. McCormick, K. Yunus, J.M. Hibberd et al. 2013. Comparison of power output by rice (*Oryza sativa*) and an associated weed (*Echinochloa glabrescens*) in vascular plant bio-photovoltaic (VP-BPV) systems. Appl. Microbiol. Biotechnol. 97(1): 429–438.

Chiranjeevi, P., G. Mohanakrishna and S.V. Mohan. 2012. Rhizosphere mediated electrogenesis with the function of anode placement for harnessing bioenergy through CO_2 sequestration. Bioresour. Technol. 124: 364–70.

Chiranjeevi, P., R. Chandra and S.V. Mohan. 2013. Ecologically engineered submerged and emergent macrophyte based system: An integrated eco-electrogenic design for harnessing power with simultaneous wastewater treatment. Ecol. Eng. 51: 181–190.

CONAGUA. 2015. Estadísticas del agua en México 2015. Comisión Nacional del Agua. Secretaría de Medio Ambiente y Recursos Naturales. México D.F. In Spanish.

Deng, H., Z. Chen and F. Zhao. 2011. Energy from plants and microorganisms: Progress in plant-microbial fuel cells. Chem. Sus. Chem. 5(6): 1006–1011.

De Schamphelaire, L., L.V.D. Bossche, H.S. Dang, M. Höfte, N. Boon, K. Rabaey et al. 2008. Microbial fuel cells generating electricity from rhizodeposits of rice plants. Environ. Sci. Technol. 42(8): 3053–3058.

Doherty, L., Y. Zhao, X. Zhao, Y. Hu, X. Hao, L. Xu et al. 2015. A review of a recently emerged technology: Constructed wetland—Microbial fuel cells. Water Res. 85: 38–45.

EPA. 2000. Folleto informativo de tecnología de aguas residuales, humedales de flujo libre superficial. EPA 832-F-00-024. 2000. Washington, US. In Spanish.

Fan, Y., S.K. Han and H. Liu. 2012. Improved performance of CEA microbial fuel cells with increased reactor size. Energy Environ. Sci. 5(8): 8273–8280.

Fang, Z., H.L. Song, N. Cang and X.N. Li. 2013. Performance of microbial fuel cell coupled constructed wetland system for decolorization of azo dye and bioelectricity generation. Bioresour. Technol. 144: 165–171.

Fang, Z., H.L. Song, N. Cang and X.N. Li. 2015. Electricity production from Azo dye wastewater using a microbial fuel cell coupled constructed wetland operating under different operating conditions. Biosens. Bioelectron 68: 135–141.

Fenoglio-Limón, F.E., J. Genescá-Llongueras and C. Durán-de-Bazúa. 2001. Construcción y evaluación de electrodos de medición de potenciales de óxido-reducción para la evaluación indirecta de las condiciones de aerobiosis en sistemas que simulan humedales artificiales. Tecnol. Ciencia Ed. (IMIQ) 16(2): 61–68 In Spanish.

García-Ochoa, R. 2010. Hacia una perspectiva de sustentabilidad energética. pp. 337–372. *In*: Lezama, J.L. and B. Graizbord [eds.]. Los grandes problemas de México; Colegio de México. Mexico City, Mexico.

Guido-Zárate, A. and C. Durán-de-Bazúa. 2008. Remoción de contaminantes en un sistema modelo de humedales artificiales a escala de laboratorio/Pollutants removal in a lab-scale constructed wetlands model system. Tecnol. Ciencia Ed. (IMIQ) 23(1): 15–22. In Spanish.

Harnisch, F. and U. Schröder. 2009. Selectivity versus mobility: Separation of anode and cathode in microbial bioelectrochemical systems. Chem. Sus. Chem. 2(10): 921–926.

Helder, M., D. Strik, H. Hamelers, J. Kuhn, C. Blok and C. Buisman. 2010. Concurrent bio-electricity and biomass production in three P-MFCs using *Spartina anglica, Arundinella anomala* and *Arundo donax*. Bioresour. Technol. 101(10): 3541–3547.

Helder, M. 2012. Design criteria for the Plant-Microbial Fuel Cell Electricity generation with living plants—from lab to application. PhD thesis, Wageningen University, NL.

Helder, M., D.P.B.T.B. Strik, H.V.M. Hamelers, R.C.P. Kuijken and C.J.N. Buisman. 2012a. New plant-growth medium for increased power output of the Plant-Microbial Fuel Cell. Bioresour. Technol. 104: 417–23.

Helder, M., D.P. Strik, H.V. Hamelers and C.J. Buisman. 2012b. The flat-plate plant-microbial fuel cell: the effect of a new design on internal resistances. Biotechnol. Biofuels 5(1): 70.

Helder, M., W. Chen, E.J.M.V.D. Harst and D.P.B.T.B. Strik. 2013a. Electricity production with living plants on a green roof: Environmental performance of the plant-microbial fuel cell. Biofuels, Bioprod. Biorefin. 7: 52–64.

Helder, M., D.P.B.T.B. Strik, R.A. Timmers, S.M.T. Raes, H.V.M. Hamelers and C.J.N. Buisman. 2013b. Resilience of roof-top Plant-Microbial Fuel Cells during Dutch winter. Biomass Bioenergy 51(0): 1–7.

Hubenova, Y. and M. Mitov. 2012. Conversion of solar energy into electricity by using duckweed in Direct Photosynthetic Plant Fuel Cell. Bioelectrochem. 87: 185–191.

Kaku, N., N. Yonezawa, Y. Kodama and K. Watanabe. 2008. Plant/microbe cooperation for electricity generation in a rice paddy field. Appl. Microbiol. Biotechnol. 79(1): 43–49.

Kayombo, S., T.S.A. Mbwette, J.H.Y. Katima, N. Ladegaard and S.E. Jørgensen. 2002. Waste Stabilization Ponds and Constructed Wetlands Design Manual. Danish University of Pharmaceutical Sciences. DK.

Kim, I.S., K. Chae, M. Choi and W. Verstraete. 2008. Microbial fuel cells: recent advances, bacterial communities and application beyond electricity generation. Environ. Eng. Res. 13(2): 51–65.

Krieg, T., A. Sydow, U. Schröder, J. Schrader and D. Holtmann. 2014. Reactor concepts for bioelectrochemical syntheses and energy conversion. Trends Biotechnol. 32(12): 645–655.

Liu, S., H. Song, X. Li and F. Yang. 2013. Power generation enhancement by utilizing plant photosynthate in microbial fuel cell coupled constructed wetland system. Int. J. Photoenergy (Online) Article ID 172010, 10 pages.

Liu, S., H. Song, S. Wei, F. Yang and X. Li. 2014. Bio-cathode materials evaluation and configuration optimization for power output of vertical subsurface flow constructed wetland—microbial fuel cell systems. Bioresour. Technol. 166: 575–583.

Liu, H., Z. Hu, J. Zhang, H.H. Ngo, W. Guo, S. Liang et al. 2016. Optimizations on supply and distribution of dissolved oxygen in constructed wetlands: A review. Bioresour. Technol. 214: 797–805.

Logan, B.E. and J.M. Regan. 2006. Microbial fuel cells—challenges and applications. Environ. Sci. Technol. 40(17): 5172–5180.

Logan, B.E., B. Hamelers, R. Rozendal, U. Schröder, J. Keller, S. Freguia et al. 2006. Critical review microbial fuel cells: methodology and technology. Environ. Sci. Technol. 40: 5181–5192.

Lu, L., D. Xing and Z.J. Ren. 2015. Microbial community structure accompanied with electricity production in a constructed wetland plant microbial fuel cell. Bioresour. Technol. 195: 115–121.

Min, B. and B.E. Logan. 2004. Continuous electricity generation from domestic wastewater and organic substrates in a flat plate microbial fuel cell. Environ. Sci. Technol. 38(21): 5809–5814.

Ng, H.Y. and S.W. Hermanowicz. 2005. Membrane bioreactor operation at short solids retention times: performance and biomass characteristic. Water Res. 39: 981–992.

Rabaey, K. 2010. Bioelectrochemical systems a new approach towards environmental and industrial biotechnology. pp. 1–14. *In*: Rabaey, K., L. Angenent, U. Schröder and J. Keller [eds.]. Systems: From Extracellular Electron Transfer to Biotechnological Application. Bioelectrochemical IWA Publishing, London, UK.

Rozendal, R.A., F. Harnisch, A.W. Jeremiasse and U. Schröder. 2010. Chemically catalyzed cathodes. pp. 263–304. *In*: Rabaey, K., L. Angenent, U. Schröder and J. Keller [eds.]. Systems: From Extracellular Electron Transfer to Biotechnological Application. Bioelectrochemical IWA Publishing, London, UK.

Salinas-Juárez, M.G. 2011. Evaluation of a system of biofiltration and a wetland for wastewater treatment from the textile industry. Master Thesis, Universidad Nacional Autónoma de México, Mexico City, Mexico.

Salinas-Juárez, M.G. 2016. Study of power generation in an electrochemically assisted constructed wetland for wastewater treatment. Ph.D. Thesis, Universidad Nacional Autónoma de México, Mexico City, Mexico.

Salinas-Juárez, M.G., P. Roquero and M.C. Durán-Domínguez-de-Bazúa. 2016. Plant and microorganisms support media for electricity generation in biological fuel cells with living hydrophytes. Bioelectrochem. 112: 145–152.

Schröder, U. 2008. From wastewater to hydrogen: Biorefineries based on microbial fuel-cell technology. Chem. Sus. Chem. 1(4): 281–282.

Strik, D.P.B.T.B., H.V.M.H. Bert, J.F.H. Snerl and C.J.N. Buisman. 2008. SHORT COMMUNICATION Green electricity production with living plants and bacteria in a fuel cell. Int. J. Energy Res. 86(3): 973–981.

Strik, D., R. Timmers, M. Helder, K. Steinbusch, H. Hamelers and C. Buisman. 2011. Microbial solar cells: applying photosynthetic and electrochemically active organisms. Trends Biotechnol. 29(1): 41–49.

Takanezawa, K., K. Nishio, K. Souichiro, K. Hashimoto and K. Watanabe. 2010. Factors affecting electric output from Rice-Paddy Microbial Fuel Cells. Biosci. Biotechnol. Biochem. 74(6): 1271–1273.

Timmers, R., D. Strik, H. Hamelers and C. Buisman. 2010a. Long-term performance of a plant microbial fuel cell with *Spartina anglica*. Environ. Biotechnol. 86: 973–981.

Timmers, R.A., D.P.B.T.B. Strik, H.V.M. Hamelers and C.J.N. Buisman. 2010b. Characterization of the internal resistance of a plant microbial fuel cell. Electrochim. Acta 72: 165–171.

Timmers, R.A., D.P.B.T.B. Strik, H.V.M. Hamelers and C.J.N. Buisman. 2013a. Electricity generation by a novel design tubular plant microbial fuel cell. Biomass Bioenergy 51: 60–67.

Timmers, R.A., D.P.B.T.B. Strik, H.V.M. Hamelers and C.J.N. Buisman. 2013b. Increase of power output by change of ion transport direction in a plant microbial fuel cell. Int. J. Energy Res. 37: 1103–1111.

Venkata, M., S.G. Mohanakrishna and P. Chiranjeevi. 2011. Sustainable power generation from floating macrophytes based ecological microenvironment through embedded fuel cells along with simultaneous wastewater treatment. Bioresour. Technol. 102(14): 7036–7042.

Vymazal, J. 2011. Constructed wetlands for wastewater treatment: five decades of experience. Environ. Sci. Technol. 45: 61–69.

Wetser, K., E. Sudirjo, C.J.N. Buisman and D.P.B.T.B. Strik. 2015. Electricity generation by a plant microbial fuel cell with an integrated oxygen reducing biocathode. Appl. Energy 137: 151–157.

Wu, H., J. Fan, J. Zhang, H.H. Ngo, W. Guo, S. Liang et al. 2015. Strategies and techniques to enhance constructed wetland performance for sustainable wastewater treatment. Environ. Sci. Pollut. Res. 22(19): 14637–14650.

Xu, L., Y. Zhao, L. Doherty, Y. Hu and X. Hao. 2015. The integrated processes for wastewater treatment based on the principle of microbial fuel cells: A review. Critical Reviews in Environ. Sci. Technol. 46(1): 1–32.

Xu, L., Y. Zhao, L. Doherty, Y. Hu and X. Hao. 2016. Promoting the bio-cathode formation of a constructed wetland-microbial fuel cell by using powder activated carbon modified alum sludge in anode chamber. Sci. Rep. 6: 26514.

Yadav, A.K., P. Dash, A. Mohanty, R. Abbassi and B.K. Mishra. 2012. Performance assessment of innovative constructed wetland-microbial fuel cell for electricity production and dye removal. Ecol. Eng. 47: 126–131.

Yang, F. and R. Pitchumani. 2006. Transport and electrochemical phenomena. pp. 69–163. *In*: Sammes, N. [ed.]. Fuel Cell Technology: Reaching Towards Commercialization. Springer-Verlag, London, UK.

Zhao, Y., S. Collum, M. Phelan, T. Goodbody, L. Doherty and Y. Hu. 2013. Preliminary investigation of constructed wetland incorporating microbial fuel cell: Batch and continuous flow trials. Chem. Eng. J. 229: 364–370.

15

Predictive Models to Assess the Uptake of Organic Microcontaminants and Antibiotic Resistant Bacteria and Genes by Crops

Josep M. Bayona,[1,*] *Stefan Trapp,*[2] *Benjamín Piña*[1]
and *Fabio Polesel*[2]

INTRODUCTION

The uptake of organic contaminants by plants is an area of growing interest during the 21st century. Pioneering studies were devoted to the application of vegetation to environmental monitoring of organic persistent pollutants in the atmosphere. More recently, the interest has been focusing on contaminants of emerging concern, which include many classes of contaminants such as pharmaceuticals and personal care products (PPCPs). In fact, a large number of experimental studies at different experimental scales have been carried out including cell culture, *in vitro*, hydroponics, controlled conditions, and field studies. These studies have showed that plants grown on contaminated soil or irrigated with contaminated water can take up a number of organic contaminants through the root and a fraction of the uptaken contaminants can be translocated to other plant organs.

---

[1] Department of Environmental Chemistry, IDAEA-CSIC, c/JordiGirona, 18-26, E-08034, Barcelona, Spain.
Email: benjamin.pina@idaea.csic.es
[2] DTU Environment, Technical University of Denmark, DK-2800 Kongens Lyngby, Bygningstorvet 115, Denmark.
Email: sttr@env.dtu.dk; fabp@env.dtu.dk
* Corresponding author: josep.bayona@idaea.csic.es

For most xenobiotic organic contaminants, root uptake is a passive process driven by plant transpiration (McFarlane 1994). Once the contaminant diffuses through the root membrane, it can be transported to other parts of the plant via xylem and phloem channels depending on its properties by the apoplastic or symplastic routes. While xylem channels conduct unidirectionally aqueous solutions of nutrients from roots to the plant photosynthetic section, phloem is a bidirectional flow that distributes sugars and other photosynthetized products throughout the plant (Marschner and Marschner 2012). However, within the root cortex and stem, lateral movement to adjacent cells occurs and may provide a pathway for contaminants to partition into the phloem. Xylem transport rates are in the range of 10 cm min^{-1}, while in the phloem, they are one order of magnitude lower (Lang 1990). Due to the high water volume transpired by leaves, the transpiration process can be a very effective pathway for chemicals from soil to leaves if they are not biodegraded.

Polar, non-volatile, and persistent organic compounds can have rather high accumulation in leaves, *ca.* several hundred times above chemical equilibrium (Doucette et al. 2005; Trapp 2007). Moreover, the exposure of organics in aboveground vegetation might also occur through the gas-vapor exchange and wet-dry deposition. The chemical contaminant properties and the environmental conditions will dictate the dominant pathway (McLachlan 2011). Gas exchange is thought to be the most important pathway for volatile organic chemicals, being faster than dry deposition. Wet deposition is important for water-soluble compounds and when aerosol compounds are trapped in precipitation (McLachlan 2011). After contact through gas exchange and deposition, the lipophilic cuticle is thought to be the main plant component governing air-to-shoot transfer, although the stomatal route of entry might be important for some gas phase chemicals (Barber et al. 2004). Chemicals that accumulate in aboveground tissues during periods of high atmospheric concentration can be released to the atmosphere when concentrations decrease (McLachlan 2011).

A variety of predictive models to assess the uptake of organic contaminants by plants from atmospheric and soil compartments have been developed over the last two decades. Usually, they are of great interest in risk assessment, ecotoxicology, environmental biotechnology, and plant physiology (Trapp 2004). In fact, these estimates are much faster and cost effective than conventional analytical methods. Pesticides and other organic priority organic pollutants (i.e., polychlorinated biphenyls, PCBs, polycyclicaromatic hydrocarbons, PAHs, polychlorinated dibenzodioxins, PCDDs, polychlorinated dibenzofurans, PCDFs), industrial contaminants, and more recently, contaminants of emerging concern (CEC), including several PPCPs, have been assessed with a variety of predictive models (Trapp et al. 1990; Li et al. 2005; Kulhanek et al. 2005; McKone and Maddalena 2007; Dettenmaier et al. 2009; Juraske et al. 2009; Legind and Trapp 2009; Collins and Finnegan 2010; Collins et al. 2010; Cropp et al. 2010; Rein et al. 2011; Fantke et al. 2011, 2012, 2013; Trapp and Eggen 2013; Prosser et al. 2014; Takaki et al. 2014; Polesel et al. 2015; Yang et al. 2016).

Predictive models can be classified according to their approach and complexity in two large categories, namely (a) **empirical** and (b) **mechanistic**. While the former methods are rather simplistic and based on correlations between physical-chemical properties of target contaminants and/or plant compartments and bioconcentration factors, mechanistic models rely on plant physiology and demand a large number of experimental parameters. Both approaches will be presented in this book chapter.

Another model classification can be based on the targeted **plant organ**, that is, aerial (phyllosphere: shoot and leaves) or underground (rhizosphere). The number of compartments in each exposure media can be variable. For instance, two-compartment root uptake might include the pore water in contact with roots and soil. In the phyllosphere, up to three compartments have been considered a part from the aerial plant compartments, namely vapour or gas phase, suspended particles, and wet deposition.

Finally, predictive models can be classified depending on whether the concentration of contaminant, which is exposed to the plant is in **steady-state** conditions (no change with time) or **dynamic** (time dependant) conditions.

Regulatory authorities have incorporated some of the steady-state or equilibrium models (i.e., RAIDAR, EUSES, CSOIL, CLEA and CALTOX) into the chemical exposure assessment tools, as presented in Table 1.

Steady-state models are commonly used to predict the uptake of organic contaminants in plants. However, the steady-state assumption may introduce errors when complex dynamic processes (e.g., growth), temperature fluctuations, and variable environmental concentrations significantly affect the major chemical uptake and elimination processes (Undeman et al. 2009).

In particular, steady-state models are not relevant when the chemical input is pulsed, as occurs for vertical flow wetlands, pesticide spraying, sewage sludge amendments, or crop irrigation, or when input data such as plant properties undergo changes over time. Under these conditions, the assumption of steady-state may cause an unacceptable deviation between the predicted and measured concentrations. For those cases, a dynamic model needs to be applied.

Table 1. Plant uptake models used in assessment tools for human contaminated soil (Takaki et al. 2014).

Model name	Approach	Compartments	Processes	Institution
RAIDAR[1]	multimedia fugacity	soil, air, aerosol	root, shoot, air, aerosol uptake; soil volatilization-deposition; air-aerosol partitioning	Environment Canada
EUSES[2]	multimedia	soil, air, aerosol	root, shoot, air uptake; soil volatilization; air-aerosol partitioning	European Union
CSOIL[3]		soil, air, aerosol	root, shoot, air uptake; soil volatilization-deposition; air-aerosol partitioning	RIVM
CLEA[4]		soil	root, shoot, air uptake	Environment Agency, UK
CALTOX[5]	multimedia fugacity	multimedia-multiexposure	root, shoot, air, aerosol uptake; soil volatilization-deposition; air-aerosol partitioning	California, US

[1]Risk Assessment IDentification And Ranking; [2]European Union System for the Evaluation of Substances; [3]An exposure model for human risk assessment of soil contamination; [4]Contaminated Land Exposure Assessment; [5]CALifornia TOXic Substances control

As expected, the model complexity increases with the number of compartments and exposure media considered and also when the contaminant's concentration in plants is variable, often leading to the applicability of only numerical model solutions (not analytical). However, Rein et al. (2011) presented mathematical solution methods based on systems theory for dynamic models with non-stationary chemical input and time-variant plant data (e.g., logistic growth of annual plants).

As opposed to the predictive models for persistent organic contaminants, PPCP models should take into account the extent of **dissipation-metabolization** in the different compartments (i.e., rhizosphere and phyllosphere) to fit predictions against experimental data (Jacobsen et al. 2015; Hurtado et al. 2016).

In addition, many CEC behave as weak electrolytes (i.e., acidic or basic) and several chemical species may coexist depending on the plant compartment (i.e., apoplast, cytoplasm and plant vacuole) and the prevailing pH. In this case, the "**ion trap**" effect can occur when the contaminant is neutral outside the cell and ionized inside (Trapp 2004). This can lead to strong phloem transport of weak acids (Briggs et al. 1987; Kleier 1988).

Moreover, protein-mediated transport of contaminants through membranes can lead to an enhanced transport for some CECs whose structure is closely related to biogenic molecules such as aminoacids from soil (Miller et al. 2016). Electrochemical attraction of ionic compounds seems to be relevant, in particular for cations (Fu et al. 2009).

Despite the big progress in the development of predictive mechanistic models over the last decades, especially for pesticide plant application (Legind et al. 2011; Fantke et al. 2014; Jacobsen et al. 2015) or pharmaceuticals along the food production chain (Chitescu et al. 2014), one of the main barriers encountered in the model development is their validation with experimental data. Obviously, it demands high quality experimental data and standardized experimental setup.

Plant physiology, which depends on the phenotype, can widely affect the uptake of organic compounds as well as the chemical properties (Trapp 2015). Unfortunately, this information is rarely available and as a consequence the application of default values in many assessments becomes unavoidable. In this regard, it has been highlighted that the application of complex mechanistic models does not improve the accuracy of simple empirical estimates usually based on a number of experimental data (Collins et al. 2007).

In order to circumvent this limitation, several plant modellers have highlighted the need of including at least a minimum data set in experimental plant studies that could allow for a proper model calibration or validation (Fantke et al. 2016; Trapp et al. 2016).

An example of the scattering of the experimental data can be illustrated with regard to the transpiration stream concentration factor (*TSCF*) (compound concentration ratio between xylem and the solution adjacent to roots) reported in 30 referred publications.

For 115 compounds, no trends are observed for *TSCF* plotted versus log K_{ow} = –2 to 6 (Dettenmaier et al. 2009). Similarly, significant variability of measured bioconcentration factors (BCFs) exists for the same substance, as shown, for example, for triclosan based on the results from seven publications (Polesel et al. 2015).

This chapter will be primarily focused on the soil-water-root(-leaf) pathway. Pesticide dissipation models following their foliar application will not be covered.

However, the aerial part of the plant can be exposed to contaminants by volatilization from soil, which is a relevant process in case of volatile contaminants (log K_{AW} > –3 and log K_{OA} < 9) (i.e., K_{AW}: air-water partition constant; K_{OA}: octanol-air partition constant) (Collins and Finnegan 2010). Similarly, particle-bound transport from soil to leaves is of great importance for mostly adsorbed compounds (Kulhanek et al. 2005). Many models, however, can include both foliar and root uptake pathways (Trapp and Matthies 1995; Trapp 2007).

In the following sections, a short description of the application of predictive uptake models for organic contaminants is presented. The criteria for model selection include their generic applicability to multiple crops and their validation with independent data sets.

One way to validate models is to calculate the ratio (C_r) between the predicted concentration in the model (C_m) in shoot or root (mg/kg fresh mass) and the experimental (C_e) concentration in shoot or root (mg/kg fresh mass). The C_r value is usually reported as logarithm.

Different models have been developed to predict the prevalence of **antibiotic resistance genes** (ARGs) and **antibiotic resistant bacteria (ARBs)** in the environment. Biological systems are subjected to selective pressures from the environment, and environmental stressors (like antibiotics) favour genetic setups able to cope with each given stressor and reduce the fitness of sensitive species and strains.

While the effect of selective pressure on biological communities has long been studied by **Population Genetics**, there are very few models devoted to the particular problem of antibiotic resistance. In addition, these approaches have been mainly focused on optimizing medical treatments in terms of efficiency and/or economy (Levin et al. 1997; Laxminarayan and Brown 2001; MacLean et al. 2010; Stewart et al. 1998).

These models could in principle be applied to the prediction of the antibiotic resistance in soil or food, but they usually lack any specific treatment for horizontal gene transfer or the presence of different ARGs in the same genetic element. A recent detailed model has been developed and used to predict the frequency of ARG-encompassing integrons as a response to the presence of antibiotics (Engelstadter et al. 2016). The model includes not only the selective advantage due to the presence of the integron but also the physiological cost of maintaining an active integrase and the added mutation rate. The authors generated a series of testable hypothesis on the stability of integrons in bacterial populations, still to be experimentally confirmed. A second level of complexity arises from the interactions between plant and soil microbiomes, and between different plant/soil compartments (soil, rhizosphere, roots, stem/leaves, and fruits).

The pathways for uptake of bacteria and ARGs into the different plant organs and their translocation follow mechanisms completely different from the uptake of chemical contaminants, and predictive models are far less developed.

Endophytic bacteria colonize plants internally without any apparent adverse effects on the host, either by entering the host from the surrounding soil through wounds in the roots or through root hairs (Upreti and Thomas 2015). They can reach the different plant organs through the vascular system or through the apoplast, the space outside the plasma membrane. Once thought to be present in small and occasional populations, the use of culture-independent technology (imaging, direct DNA sequencing) revealed large, well established endophytic microbiomes, likely playing important roles in plant physiology, more as symbions than as parasitic (Reinhold-Hurek and Hurek 2011). Unfortunately, our knowledge of endophytic bacterial populations and of their interactions with the host is still too imperfect to draw useful predictive methods.

The bacterial (and ARG) load in plants may vary by several orders of magnitude, and, while undoubtedly related to soil microbiome, its composition greatly depends on plant physiology (Reinhold-Hurek et al. 2015; Asakura et al. 2016). Even less known are the factors determining the fraction of ARBs in these microbiomes, but some reports indicate that their composition and the prevalence of the associated ARG are strongly influenced by the quality of water used in the field (Zhu et al. 2017).

However, these factors have not been characterized until recently, and are not adequately modelled yet.

Empirical Models

Empirical models rely on the prediction of concentration factors (=ratio of concentrations in plant and in the surrounding environmental compartment, e.g., soil) based on properties of the target contaminant (e.g., distribution coefficients) and, in some cases, of plant compartments. Regressions are typically identified from the results of dedicated experiments.

The existing models are only applicable within their applicability domain, which typically is neutral molecules. Consequently, they are limited to not ionized CECs subject to limited degradation in soil and plant compartments.

Nevertheless, empirical models will be presented in this chapter since, in case of lack of data required in high complexity models, their output provides a preliminary assessment of the feasibility of contaminant uptake. An overview of the empirical models described in this section is given in Table 2.

Soil-water-root

Topp et al. (1986)

A simple regression model derived in is valid for closed microcosms dealing with the uptake of 16 radioactively labelled organic compounds with molecular mass (*MM*) ranging from 75 to 600 g mol^{-1} (that is, benzene, atrazine, pentachlorophenol, chlorobenzenes, and DDT) by barley and cress seedlings from soil and separately from air after volatilization. The following equation was proposed based on the *MM* as contaminant descriptor (equation 1):

$$\log CF = 5.943 - 2.385 \cdot \log MM \tag{1}$$

where *CF* is the concentration factor (dimensionless).

Application: molecules of moderate lipophilicity (log $K_{ow} \approx 2\text{--}4$).

Input: molecular mass of contaminant (g mol^{-1}).

Output: bioconcentration factor above ground plant [mg kg^{-1} FM (fresh mass) plant over mg kg^{-1} DM (dry mass) soil].

Table 2. Summary of empirical models for the prediction of bioconcentration factors, related to plant uptake of organic contaminants in plants.

Reference	Uptake pathway	Input							Output	Applicability
		Contaminant properties				Soil properties				
		MW	K_{OW}	K_{AW}	K_{OC}	f_{OC}	Bulk density	Water content		
Briggs et al. (1982)	Soil-water-root		X						Root bioconcentration factor	Neutral molecules ($\log K_{OW} = -1$ to 5)
Topp et al. (1986)	Soil-water-root	X							Bioconcentration factor in aboveground plant	Neutral molecules, moderate lipophilicity
Travis and Arms (1988)	Soil-water-root		X						Bioconcentration factor in aboveground plant	Neutral molecules, high lipophilicity (chloro-organic pesticides)
Briggs et al. (1982)	Soil-water-root(-shoot)		X						Transpiration stream concentration factor	Neutral molecules ($\log K_{OW} = -1$ to 5)
Dettenmaier et al. (2009)	Soil-water-root(-shoot)		X						Transpiration stream concentration factor	Neutral molecules in a wide lipophilicity range ($\log K_{OW} = -2$ to 6)
Ryan et al. (1988)	Soil-water-root(-stem)		X		X	X	X	X	Root and stem bioconcentration factors	Neutral molecules ($\log K_{OW} = 0$ to 4)
Bacci et al. (1990)	Air-leaf-shoot		X	X					Leaf/air bioconcentration factor	Neutral lipophilic ($\log K_{OW} = 1$–7) semivolatile compounds (pesticides)

Validation: benzene and pentachlorophenol do not fit to this model.

Applicability domain: The data used were derived in a small closed system (exicator) with radiolabeled compounds.

Travis and Arms (1988)

The model is a simplified relationship taking into account only the octanol-water partition coefficient (K_{ow}) as descriptor of the target contaminant. This model was developed for 29 contaminants (mostly neutral pesticides) based on published data (equation 2):

$$\log B_V = 1.588 - 0.578 \cdot \log K_{ow} \tag{2}$$

where B_V is the bioconcentration factor in the plants (dimensionless).

Application: neutral molecules with a log K_{ow} ranging from 1 to 9.

Input: octanol water-partition coefficient (K_{ow}).

Output: bioconcentration factor above ground plant (mg kg^{-1} FM plant over mg kg^{-1} DM soil).

Applicability domain: The data were derived from published field studies, mostly for lipophilic chloro-organic pesticides applied to farmland and meadows. The results are rather close to prediction models considering growth dilution, for example, the potato model (Trapp et al. 2007) and the root model (Trapp 2002).

More recently, fast sorption rates of emerging and priority pollutants (i.e., toluene, *p*-xylene, naphthalene, bisphenol A and 4-bromodiphenyl ether) have been reported in leafy vegetables with a high lipid content including leafy rape, Chinese mustard and Chinese cabbage (Yang et al. 2016). The authors found a strong correlation between the plant lipid content and the compound's hydrophobicity ($R^2 = 0.92$) suggesting the predominance of a sorption equilibrium process.

Transpiration Stream Concentration Factors (TSCF)

The transpiration stream concentration factor (TSCF) has been measured for few hundreds of compounds and modelled from the contaminant's log K_{ow}. Two types of distribution have been reported for the TSCF as a function of log K_{ow}. One of them is the bell shape with maxima at the log $K_{ow} = 1$–3 (Briggs et al. 1982; Burken and Schnoor 1998; Hsu et al. 1990). The other one is a sigmoidal pattern with TSCF decreasing at hydrophobicity higher than log K_{ow} of 2. To describe this relationship, the following equation has been proposed by Dettenmaier et al. (2009) (equation 3):

$$TSCF = \frac{11}{11 + 2.6^{\log K_{ow}}} \tag{3}$$

Application: neutral molecules with a log K_{ow} ranging from –2 to 6.

Input: octanol water-partition coefficient (K_{ow}).

Output: transpiration stream concentration factor (TSCF), i.e., compound concentration ratio between xylem and the solution adjacent to roots.

Applicability domain and validation: only for neutral organic compounds, which are not degraded during translocation.

The sigmoidal shape of the curve and the decline of translocation for lipophilic compounds are confirmed by theoretical considerations (Trapp 2007).

Ryan et al. (1988)

This screening model is based on semi-mechanistic understanding of the plant uptake of neutral contaminants from soil. It is based on the experimental work of Briggs et al. (1982) by applying a correction factor due to the contaminant availability from soil compared to an aqueous nutrient solution (hydroponic setup). Briggs et al. (1982) undertook an experimental study on the uptake of two series of non-ionized chemicals (o-methylcarbamoyloximes and substituted phenylureas) by barley roots and shoots from hydroponic solution. The passive uptake of each chemical during transpiration was measured over 96 h and it was observed that root and stem bio-concentration factors were proportional to lipophilicity of the chemical estimated as K_{ow}.

Application: neutral molecules with a log K_{ow} ranging from 0 to 4.

Input:

- Octanol water-partition coefficient K_{ow},
- Soil bulk density (g DM/cm^3),
- Soil-water content by volume (cm^3/cm^3),
- Organic carbon-water partition coefficient for the contaminant (cm^3/DM)K_{oc},
- Fraction of organic carbon in the soil (dimensionless) f_{oc}.

Output:

- RCF is the calculated as soil-to-plant root concentration factor (μg/g FM plant over μg/g DM soil).
- SCF is the calculated as soil-to-plant stem concentration factor (μg/g FM plant over μg/g DM soil). Due to the model complexity, the reader is referred to the original publication for further information on the underlying equations.

Applicability domain and validation: The approach uses the empirical regressions of Briggs et al. (1982) which were derived in hydroponic studies with barley seedlings. The model can be applied to neutral chemicals only and was not validated on field scale.

Air-leaf-shoot

The air-shoot or air-leaf concentration ratios have been used to estimate plant concentrations, using the contaminant concentrations in air as input to the model. Based on the assumption that the lipophilic cuticle is the major plant component governing the air-plant interactions, simple regression models were developed that relate air-shoot *BCFs* to n-octanol-air partition coefficients (K_{OA}) (Tolls and McLachlan 1994) or to a combination of air-water partition coefficient (K_{AW}) and K_{OW} (Bacci et al. 1990). In the second case, a simple regression model was derived from 10 neutral pesticides with log K_{OW} ranging from 1 to 7 (equation 4):

$$\log BCF = -1.95 + 1.14 \log K_{ow} - \log K_{AW} \tag{4}$$

Input: octanol-water (K_{OW}) and air-water (K_{AW}) partition coefficients.

Output: plant-air concentration ratio (ng/L of wet leaf)/(ng/L air), equal to *BCF*.

Applicability domain and validation: semivolatile neutral lipophilic compounds.

Model Extension

Since cuticle consists of several lipid or lipid-like waxy fractions (i.e., cutin, cutan, and extractable waxes) that exhibit different affinities for organic contaminants, cuticle lipid speciation has been proposed to improve more accurately predict the air-partition *BCF* (Barber et al. 2004).

Mechanistic Models

Mechanistic mass balance models for estimating plant tissue concentrations from chemical exposure to air and soils include one or more compartments. Rates of input, output, and accumulation in each compartment are expressed using equations describing partitioning, degradation, and diffusion rates (Gobas et al. 2015).

Parameters for mechanistically modelling the uptake of organic contaminants into plants fall into three major categories: (1) properties of the organic contaminants, (2) properties of the plant and (3) properties of the environment.

Chemical properties include K_{OW}, K_{AW}, K_{OA}, K_{OC} (organic carbon normalized-water partition coefficient in soil or sediment), pK_a (acid dissociation constant), aqueous solubility, vapour pressure, and chemical (bio)transformation half-lives or rates within plant.

Plant properties include dimensions, masses, and volumes of all compartments, their growth rates, transpiration rates, xylem and phloem flow rates, lipid equivalent contents, and volume fractions of water and air.

Environmentally relevant parameters include soil and air composition and characteristics, temperature, relative humidity and (for field studies) rainfall. Unfortunately, most of the reported data do not provide the sort of information above mentioned and default values need to be used, possibly leading to biased results.

Partition-Limited Models

Chiou et al. (2001) proposed a mechanistic partition-limited model for passive root uptake of contaminants from soil, taking into consideration the chemical concentration in soil and plant composition. The approach involved establishing the upper (equilibrium) limit for the level of contaminant in a plant compared to that in soil, against which the actual equilibrium at the time of analysis can be estimated.

The authors assumed that passive root uptake is the dominant process for chemical accumulation by plant from soil. The organic chemical is dissolved in water, uptaken into the plant during transpiration, and partition from the water to plant tissue in contact with solution. The chemical concentration in external solution is assumed to drive chemical uptake through a series of partitioning processes that may or not come to equilibrium with the external solution (Chiou et al. 2001).

A quasi-equilibrium factor (α_{pt}) that reflects the extent to which equilibrium is established was introduced. The α_{pt} term is the concentration ratio in plant tissue to the soil pore water. If a passive plant uptake from the soil is assumed, the concentration in the plant tissues will be equal to the concentration in the pore water (i.e., $\alpha_{pt} = 1$). If α_{pt} is below 1, the equilibrium is not yet reached or never will be reached and if higher than 1, the plant uptake process is active. In practice, the α_{pt} value is assumed to be 1 (worst-case scenario) for passive uptake. The input parameters of the model are: (i) K_{ow} of contaminants, (ii) the weighted average of the lipid fraction, (iii) water and carbohydrate content, and (iv) fraction of soil organic matter (SOM). In case of plants with high lipid content (leafy rape, Chinese mustard, lettuce and Chinese cabbage), only the lipid fraction has been considered (Yang et al. 2016), showing a faster sorption kinetics for contaminants of medium hydrophobicity (log K_{ow} = 2.7–4.8). This approach is based on a strong correlation between log K_{lip} (lipid-water partition coefficient) and log K_{ow}.

Input:

- Octanol-water partition coefficient (K_{ow}).
- Soil organic matter normalised chemical concentration in soil (mg kg^{-1} DW soil) (C_{som}).
- Chemical partition coefficient between soil organic matter and water (dimensionless) (K_{som}).
- Total mass fraction of the organic matter in the plant (g g^{-1}) (f_{pom}).
- Chemical partition coefficient between plant organic matter and water (dimensionless) (K_{pom}).
- Mass fraction of water in the plant (g g^{-1}) (f_{pm}).

Output:

- C_{pt} is the calculated mass of chemical per unit mass of plant (mg kg^{-1} FM plant).

Applicability domain and validation: Neutral compounds. Notably, this type of models is of limited use for forward predictions because key input parameters are fitted to the experimental results but can help to interpret the experimental results.

Fugacity Models

Hung and Mackay (1997) proposed a model based on the concept of **chemical fugacity**, defined as the tendency of a chemical to migrate from one phase into another. In this model, the plant is subdivided into three homogeneous compartments (i.e., roots, stem, leaves) with surrounding compartments (air and soil). Chemical concentrations are modelled using in- and outflow for each compartment and processes including transport in the xylem and phloem, air-leaf exchange, passive root uptake, growth dilution, and metabolism. Steady-state conditions are established by using partition coefficients derived from all compartments with respect to water partition coefficients, which are based on experimental measurements or derived from correlation with other physical-chemical properties. The major assumptions are the following: (1) chemical transport is unidirectional; (2) air and soil concentrations are constant over the growth period; (3) all model parameters are constant with time; (4) aerosol deposition is not taken into account. RAIDAR, EUSES, and CALTOX are models used in several regulatory agencies based on the fugacity approach (Table 1). Nevertheless, their applicability is limited to non-ionizable compounds.

Application: high intensity in the input parameters, which limits its wide application in crops.

Input data:

- three compartment (roots, stem, leaf) volumes,
- seven partition coefficients,
- three half-lives for growth and three for metabolism,
- eight transport parameters (air-leaf exchange, xylem and phloem flow rate, diffusive and bulk flow rate from soil to root).

A fugacity model has been designed for assessing the fate of biosolids-derived chemicals in amended soil and concentrations in plant, vertebrate and mammal receptors can be predicted (Hughes and Mackay 2011). Up to four levels of complexity are available depending on the intensity of input data.

Applicability domain and validation: Neutral compounds only. Ionic substances have no fugacity (see Trapp et al. 2010).

One-Compartment Steady State Concentration

Trapp and Matthies (1995) developed a mechanistic, generic one-compartment model for the uptake of organic chemicals by leafy vegetation. It considers passive uptake from soil through the root system, translocation to shoots, air-leaf exchange, metabolism, and dilution by plant growth. The uptake process is based on the partition coefficient between the plant tissue and water taking into account the lipid composition in roots and leaves. Flow exchange in the different plant compartments from the surrounding media (air or soil) is solved by a mass balance. Dilution and metabolism are accounted for by using reported first-order half-lives. Passive root uptake is calculated from the equilibrium pore water concentration using a root/water partition coefficient. To estimate the concentrations in the plant shoots, the model takes into account the inflow from the roots and leaves and losses to the air and through metabolism along with growth dilution. Chemical concentrations in the transpiration stream are estimated by using the relationship of Briggs et al. (1982) and Hsu et al. (1990). Transpiration stream flow rates are used to calculate the flux into plant shoots via root uptake (Trapp and Matthies 1995). Leaf-air exchange is modelled as diffusive gaseous flux across a concentration gradient from the shoot to air. The net flux between air and leaf is determined using a leaf-air partition coefficient, foliar surface, and leaf conductance constant. The loss mechanisms are described by using first-order rate constants. Once uptake and loss fluxes have been calculated, an analytical solution is applied to find the steady-state concentrations in plant shoots.

This approach has been adopted by the European Union System for the Evaluation of New and Existing Substances (EUSES) for screening risks from plant uptake (EU 2003). In addition, it has been adopted in the Netherlands in the CSOIL model.

Application:

- Non-ionizable compounds.
- Plants with continuous exponential growth.

Input data:

- Plant parameters: water, lipid and air content of the plant, plant tissue density, transpiration stream flow rate, leaf conductance, leaf area and volume, harvest time, and growth constant.
- Chemical properties: octanol-water partition coefficient (K_{ow}), dimensionless air-water partition coefficient (K_{AW}).
- Soil parameters: soil bulk density, water filled soil porosity, and fraction of organic carbon.

Applicability domain and validation: a couple of validation studies have been performed because the model is suggested for chemical risk assessment tools (TGD, Technical Guidance Document, REACH, Registration, Evaluation, Authorisation and Restriction of Chemicals, CSOIL, CLEA).

Two-Compartment Models

Cropp et al. (2010) developed a predictive model to estimate maximum plant concentration and time to reach it. It is a two-compartment (soil/water and plant) model and has an analytical solution. First-order equations are further used to describe losses by degradation in soil/water and exchange rates between soil/water and plant. The model has been applied and validated for the antibiotic norfloxacin in soybean, with good agreement between the modelled concentrations and the experimental data. An extension of the model (Hawker et al. 2013), with chemical activity as the main state variable, was tested to predict the uptake of norfloxacin, oxytetracycline, and chlortetracycline in rice (*Oryza sativa* L.). The root concentration was predictable from the soil-water concentrations and was not related to hydrophobicity (D_{ow}).

Crop-Specific Models

Legind and Trapp (2009) extended the widely used analytical plant uptake model (Trapp and Matthies 1995; Kulhanek et al. 2005) to a variety of crops types, such as root crops, leafy vegetables, fruits (and also milk and meat). The model system was applied to predict the dietary exposure of Danish children and adults to a variety of chemical contaminants.

Applicability domain and validation: Neutral lipophilic compounds, field scale, steady-state.

Multi-Compartment Models

Multimedia Activity Models for Ionic and Non-ionic Molecules

Activity models are based on thermodynamics and, as a consequence, exact equations to describe the behavior of neutral and ionizable molecules in non-ideal systems can be derived. Activity drives the diffusion and thermodynamic equations describe the exchange between environmental compartments. Activity models can be applied to neutral, carboxylic acids, bases, amphoters, and zwitterions (Trapp 2004). This approach has been tested for regional fate prediction of the 2,4-dichlorophenoxic acids (2,4-D), aniline, and trimethoprim (TMP). Regional model predictions (i.e., concentrations in different environmental compartments) were validated in a realistic scenario and provided better results than conventional fugacity methods (Franco and Trapp 2010).

Applicability domain and validation: The multimedia model of Franco and Trapp (2010) is a regional model and does not specifically calculate plant uptake.

Dynamic Cascade Models

When the concentration of a contaminant, to which the plant is exposed, is not constant in time, models need to be developed accordingly. This scenario is typical of different agriculture practices such as biosolids or pesticide application. In addition, the contaminant concentration in the irrigation water and the irrigation volume are also variable. In this context, a **multi-cascade model** approximating the logistic growth and coupling transpiration to growing mass has been developed (Rein et al. 2011). The underlying differential equation systems are solved analytically using the well-known solutions for triangular matrices ('cascades'). Non-stationary input and conditions are calculated by using a series of such solutions ('multi-cascade model'). This model was set up, parameterized, and tested for the uptake into growing crops (i.e., carrots and wheat) and the outcome compared with a numerical solution. The same model or solution method was used in a number of other studies, namely Legind et al. (2011); Trapp and Eggen (2013); Prosser et al. (2014); Polesel et al. (2015); and Jacobsen et al. (2015).

Applicability domain and validation: Neutral and ionizable compounds, greenhouse and field studies.

Extension for Ionizable Compounds

About 50% of REACH chemicals are of ionic nature (Franco and Trapp 2010), and so are 80% of all pharmaceuticals (Manallack 2007). These ionic organics have unique properties and undergo processes that are different from those of neutral compounds:

(i)   Ionizable compounds occur in (at least) two species, namely the neutral and the ionized molecule(s).

(ii) The concentration ratios of these two (or more) species change with pH.

(iii) Ions are attracted or repelled by electrical fields, while neutral compounds are unaffected. Hence, different permeabilities of neutral and ionic molecules may lead to accumulation of electrolytes inside living cells, which is known as ion trap effect.

(iv) Ions, and in particular multivalent ions, are subject to larger changes in their "active concentration" (the activity) with ionic strength (in sea water or in body fluids) than neutral compounds.

(v)   Ions are much more polar than the corresponding neutral molecules.

(vi) Ions have no measurable vapor pressure and thus do not tend to volatilize. If formed in atmosphere they thus partition irreversibly to aerosol particles, fog, rain, or snow.

Due to the unique properties of ionic organics, a concept based on **chemical activity** instead of fugacity was considered as a feasible way to describe transport and partitioning of ionics (Trapp et al. 2010). In particular, partition processes between soil solution and plant cells, or air and plant cells, can no longer be described with the popular lipophilic partitioning ad the parameter log K_{ow}. Instead, dissociation and speciation (dependent on pH), different membrane permeabilities of the occurring molecule species (leading to ion trap effects), and electrical attraction by charged, living cells need to be considered. These effects can be calculated with the "cell model", which has its main application in medicine (Trapp and Horobin 2005; Trapp et al. 2008). The cell model was coupled to the described plant uptake models and replaces there the partition coefficients between plant cells and solution (e.g., RCF, LCF). This approach is described in detail elsewhere (Fu et al. 2009). Subsequently, it has been used to describe fate and plant uptake of pharmaceuticals and personal care products, but also of pesticides (Prosser et al. 2014; Polesel et al. 2015; Buchholz et al. 2015; Buchholz and Trapp 2016). Studies with this type of models are still ongoing. A major problem is the complexity of processes and the large number of input data—chemical, plant, environment—which are not always known with the required precision for good predictions.

Applicability domain and validation: Ionisable and neutral compounds, greenhouse and field studies. Available at https://homepage.env.dtu.dk/stt/.

Dynamic Multicrop Model

Fantke et al. (2011) presented a dynamic plant uptake model for various crops based on a flexible set of interconnected compartments. The purpose is characterisation of health impacts of pesticides applied to food crops in life cycle impact assessment. The model was mostly applied to pesticides and food crops and showed promising results. The model was parameterized for pesticide exposure via crop consumption (Fantke et al. 2012) and the system dynamics was analysed in detail (Fantke et al. 2013, 2014). The authors also provide a database of measured dissipation half-lives in plants (for pesticides) and provide a database in excel (Fantke and Juraske 2013).

Applicability domain and validation: Pesticides, field scale. Available at http://dynamicrop.org.

Integrated Models

ACC-HUMAN

Models including terrestrial and aquatic food chains have also been developed for lipophilic contaminants (Czub and McLachlan 2004). A **fugacity-based, dynamic, mechanistic mass balance model** was developed to describe the bioaccumulation of lipophilic contaminants through the agricultural food chain (i.e., leafy vegetables, root fruits, aerial fruits, and tubers) and aquatic food web. A human daily dose may be calculated from the daily intake rates of the different food items and water/inhalation rate of air. Contaminant uptake via dietary sources of persistent lipophilic contaminants was addressed. It was validated with PCBs in Sweden.

Applicability domain and validation: Neutral lipophilic compounds, field scale, dynamic.

Combined Soil Transport and Plant Uptake (Tipping Buckets Model)

A dynamic physiological plant uptake model based on multicascade approach coupled to tipping bucket soil transport has been applied to model the uptake of organophosphates, a insect repellant (N,N-diethyltoluamide) and a plasticizer (*n*-butyl benzenesulfonamide).

A field scenario was simulated with sewage sludge application and high uptake of the polar, low volatile compounds with the highest concentration in straw (leaves and stem) was found (Trapp and Eggen 2013). The combined soil transport and plant uptake model was further used for a field biosolids application scenario and validated for the uptake of plant uptake of pharmaceuticals and personal care products (Prosser et al. 2014).

Recently, the combined model has been extended to include phloem transport from stem back to roots (available at https://homepage.env.dtu.dk/stt/2017Release_Plant_Model/index.htm). A conceptual representation of the combined model, considering contaminant input to the soil/plant system via soil irrigation and extensions for phloem transport, is given in Fig. 1.

Applicability domain and validation: Neutral lipophilic compounds, ionizable compounds, greenhouse studies and field scale. A version for heavy metals was published by Legind et al. (2012).

Biosolids Amended-Soil

Plant uptake of PPCPs from biosolids has been reported by a dynamic plant uptake based on multicascade approach coupled with a tipping bucket model (Prosser et al. 2014). Partitioning and advection with water are the transport process considered in

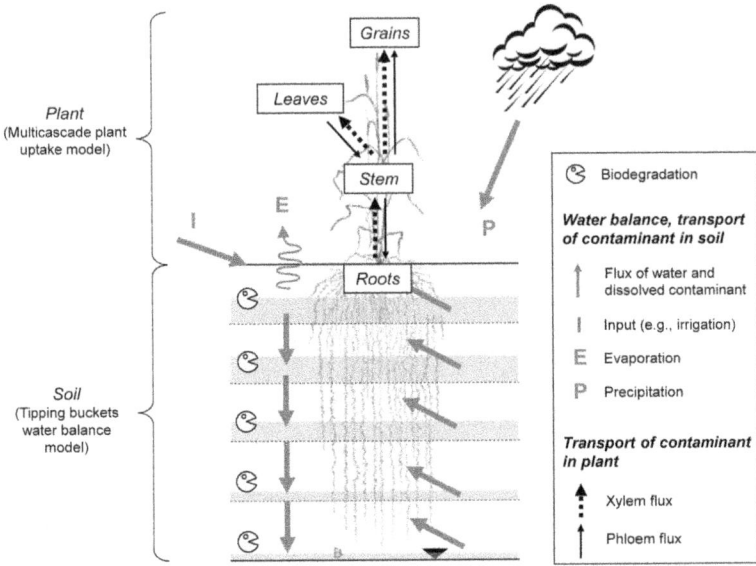

Fig. 1. Conceptual representation of the dynamic plant uptake model, coupling the tipping buckets model (for soil transport) and the multicascade model (for plant uptake and translocation). Inputs, outputs, and material fluxes considered in mass balances are given in the legend. The scenario considered in the figure is the release of contaminant to the soil/plant system via irrigation to soil.

the model to describe movement of the chemical throughout the plant. A cell model is coupled with the dynamic uptake model. The cell model is used to calculate partition coefficients (e.g., cytosol to soil, cytosol to xylem, and xylem to root) that are then used in the dynamic plant uptake model to estimate transport into different tissues of the plant.

The biosolids-amended soil fugacity approach (Hughes and Mackay 2011) was to predict the concentration of eight PPCPs. This model over-predicted the concentrations in root and shoot tissue by 2–3 orders of magnitude. In a model comparison study with field data, the fugacity approach was inferior to the cascade/buckets activity model approach described above (Prosser et al. 2014).

Pharmaceuticals Through Food Production System

The transfer of pharmaceuticals from contaminated soil, through plant uptake into dairy food production chain has been developed (Chitescu et al. 2014). The scenarios and model parameters refer to contaminant emission slurry production, storage time, emission to the soil, plant uptake, bioaccumulation in the animal's body, and transfer to the meat and milk.

Modelling results suggest the possibility of contamination of dairy's meat and milk due to the ingestion of contaminated feed by cattle. However, the estimated concentrations for oxytetracycline, sulfamethoxazole, and ketoconazole in milk and meat indicate a minor risk to human health.

Model Uncertainty, Sensitivity, Accuracy. Intercomparison and Validation of Model Data

The procedure to assess model uncertainty is commonly based on the comparison between different models for a defined chemical and scenario and on the use of the divergence in endpoint prediction as a measure of uncertainty. This approach has been used by McKone and Maddalena (2007) and Takaki et al. (2014), as described below.

McKone and Maddalena (2007) found high experimental variability (CV = 170%) when investigating soil-to-plant bioconcentration factors for organic contaminants but the variability between models used to predict the plant bioconcentration factors was even greater (CV = 1400%). The variability arises also by the very different scenarios and experimental set-ups that were considered, which would require an adaption of models and input parameters (Fantke et al. 2016; Trapp et al. 2016).

Takaki et al. (2014) evaluated the accuracy of plant uptake models for neutral hydrophobic organic contaminants ($1 < \log K_{OW} < 9$; $-8 < \log K_{AW} < 0$) used in regulatory exposure assessment tools using uncertainty and sensitivity analyses. The models considered were RAIDAR, EUSES, CLEA, CSOIL, and CALTOX (Table 1). CSOIL demonstrated the best performance among the five exposure assessment tools for root uptake from polluted soil in comparison with experimental data, but no model predicted accurately the shoot uptake. Recalibration of the transpiration and volatilization parameters improved the performance of some models (CSOIL and CLEA).

Alternatively, the model comparison can be performed by identifying the most important process for a wide range of chemicals and then evaluate the models' treatment of these models based on "current science" (Undeman and McLachlan 2011). By applying this approach, the EUSES and the ACC-HUMAN (steady-state version) were evaluated. In this regard, the equilibrium assumption for root crops in EUSES caused overestimations in daily intake of super hydrophobic chemicals ($\log K_{OW} > 11$, $\log K_{OA} > 10$). Uptake of hydrophilic chemicals from soil was identified as important research areas to enable further model uncertainty reduction (Undeman and McLachlan 2011).

The approaches above are used to evaluate the **structural uncertainty** of plant uptake models. However, there is another type of uncertainty that requires consideration—the one associated to model input. Model parameters do not exist as single values, but can be variable (thus uncertain). Accounting for **uncertainty in model parameters** and propagating it to model predictions can be useful to: (i) capture the variability of plant uptake measurements, and (ii) support decision makers in risk assessment by identifying worst-case scenarios, for example.

A common approach for **uncertainty analysis** is the definition of probability distributions for relevant model parameters and the propagation of uncertainty to model predictions using the Monte Carlo method. This approach has been used for uncertainty analysis of models used in risk assessment (Takaki et al. 2014) and a dynamic mechanistic model for ionisable PPCPs (Polesel et al. 2015). These studies revealed that up to two-order of magnitude variability can be expected in predicted BCFs for the same chemical, independently of the model considered. Such variability can be attributed to environmental (e.g., soil f_{OC}) and chemical properties (half-life in soil, partition coefficients). In the context of plant uptake of wastewater-borne

PPCPs, additional variability in BCFs will be likely associated to: (i) chemical input pulses (via, e.g., irrigation), which are region-, season- and plant-dependent; and (ii) degradation in plants, which is to date largely unknown (Hurtado et al. 2016).

Model Validation

The typical last step in modelling studies is its validation, whereby the model performance is assessed by comparing predictions with independent sets of measured data. This step is one of the main challenges for plant uptake models. Concentration factors (TCSF, BCF) rather than measured concentrations are commonly used as endpoint for model validation, given the wide range of chemical concentrations used in plant uptake tests. However, empirical data on concentration factors are themselves characterized by significant variability. As previously described, this has been clearly shown for TSCF values (Dettenmaier et al. 2009). For the same chemical (the biocide triclosan), variations of factor 100–1000 were shown for BCFs in root, stem, and leaf (Polesel et al. 2015). Such variability raises questions as to the repeatability and comparability of experiments and the impact of testing conditions (environmental properties, experimental set-ups) on empirical observations.

Another challenge in model validation is the existence of different approaches for BCF calculation. BCF is calculated as the ratio between the concentration in a plant compartment at harvest and the concentration in soil either: (i) at harvest or (ii) at the time of input to soil (for irrigation, the time of the latest irrigation pulse). While the first approach is more convenient and more commonly used, it may lead to overestimation of the actual translocation to plants by neglecting, for example, dissipation in soil. Therefore, consistent calculation methods are required for the reliable comparison of predicted and measured BCFs for model validation purposes.

To overcome these challenges, recommendations have been made as to the **minimum data reporting for experimental plant studies** (Fantke et al. 2016; Trapp et al. 2016). Data requirements include details on soil and plant characteristics, input loads and pathways of tests substances and their properties, and environmental conditions at the test site. In this way, improvements are expected for: (i) model predictions by using scenario- and site-specific (emission rates, transpiration rates) parameters, as exemplified in Trapp (2015); and (ii) interpretation of empirical data, with the possibility of selecting only the most relevant data sets for model validation. Benefits from the adoption of minimum data recommendations are illustrated in Fig. 2. Care must be taken to use empirical methods (e.g., regressions) solely within their applicability domain.

Recommendations and Conclusions

As described in this chapter, a wide variety of approaches and models exist to describe plant uptake of organic microcontaminants from soil, irrigation water, and air into plants and crops. The approach that fits best depends critically on the dynamics (equilibrium versus steady-state versus non stationary), the design of the study (laboratory, greenhouse, field experiments) and the crops under consideration (root crops, leafy vegetables, cereals, fruits). Besides, the properties of the chemicals play a major role,

Modelling plant uptake of PPCPs:

(i) using default input

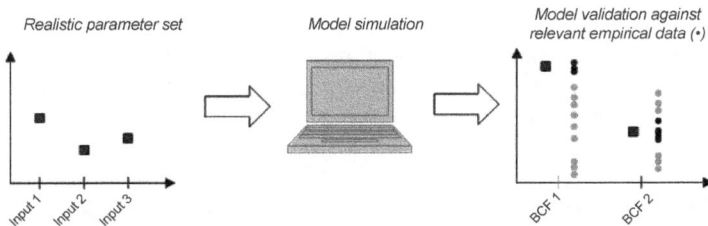

(ii) considering input uncertainty

(iii) using realistic input and relevant empirical data for validation (based on minimum data reporting)

Fig. 2. Summary of three approaches used for predicting and validating plant uptake of PPCPs against (highly variable) empirical data. The first approach stems from the lack of measured input data and, by relying on default parameter values as model input, appears more suitable for initial screening studies. In the second approach, the uncertainty in the model input is propagated to the model predictions (box plots), allowing researchers to partly capture the variability of empirical data. The third approach relies on "minimum data reporting" recommendations and, through the use of realistic input (scenario- and site-specific) and the selection of relevant empirical data, allows for refined model predictions and validation.

and models need to be differentiated for polar neutral compounds, lipophilic neutral compounds, ionizable and ionic compounds, and non-persistent compounds. There is no model approach that is optimal for all scenarios and all chemicals.

Pharmaceuticals and personal care products are often ionizable, and this deserves special attention in the formulation of processes and models. Furthermore, many new parameters and input data are required, and metabolism and degradation become far more important than for lipophilic compounds (which do not degrade in adsorbed form). When it comes to realistic, full-scale field scenarios, these parameters are often missing or only known with large uncertainty. The models then fail to deliver accurate predictions, but may still be valuable for an interpretation of the chemicals' behaviour. It is worth mentioning here that—as shown above—experimental results show large variability, and replicate studies may fail to reproduce findings obtained earlier or by others.

In order to progress towards better, more accurate prediction models for the plant uptake of pharmaceuticals and personal care products under real situations, studies might help where these relevant input parameters are identified, controlled, varied, or at least determined (Fantke et al. 2016; Trapp et al. 2016).

Finally, despite predictive plant uptake models have not been specifically designed for constructed wetland systems, some of the predictive models could be applied with minor modifications to the variety of wetland configurations to get further insight in the role of plants in treatment processes and to get further insight within water-plant interactions leading to contaminant removal.

Acknowledgements

The authors would like to acknowledge the financial support provided by COST-European Cooperation in Science and Technology, to the COST Action ES1403: New and emerging challenges and opportunities in wastewater reuse (NEREUS). Authors are indebted to Mrs. Professor Gladys Vidal from EULA (Chile) to review this chapter.

Disclaimer

References Cited

Asakura, H., M. Tachibana, M. Taguchi, T. Hiroi, H. Kurazono, S.I. Makino et al. 2016. Seasonal and growth-dependent dynamics of bacterial community in radish sprouts. J. Food Safety 36(3): 392–401.

Bacci, E., D. Calamari, C. Gaggi and M. Vighi. 1990. Bioconcentration of organic-chemical vapors in plant leaves—Experimental measurements and correlation. Environ. Sci. Technol. 24: 885–889.

Barber, J.L., G.O. Thomas, G. Kerstiens and K.C. Jones. 2004. Current issues and uncertainties in the measurement and modelling of air-vegetation exchange and within-plant processing of POPs. Environ. Pollut. 128: 99–138.

Briggs, G.G., R.H. Bromilow and A.A. Evans. 1982. Relationships between lipophilicity and root uptake and translocation of non-ionised chemicals by barley. Pest. Sci. 13: 495–504.

Briggs, G.G., R.L.O. Rigitano and R.H. Bromilow. 1987. Physico-chemical factors affecting uptake by roots and translocation to shoots of weak acids in barley. Pest. Sci. 19: 101–112.

Buchholz, A., A.C. O'Sullivan and S. Trapp. 2015. What makes a good compound against sucking pests? pp. 93–109. *In*: Maienfisch, P. and Th. M. Stevenson [eds.]. Discovery and Synthesis of Crop Protection Products. Chapter 8. ACS Symposium Series. Vol. 104. Chapter doi: 10.1021/bk-2015-1204.ch008.

Buchholz, A. and S. Trapp. 2016. How active ingredient localisation in plant tissues determines the targeted pest spectrum of different chemistries. Pest. Manage. Sci. 72(5): 929–939, doi: 10.1002/ps.4070.

Burken, J.G. and J.L. Schnoor. 1998. Predictive relationships for uptake of organic contaminants by hybrid poplar trees. Environ. Sci. Technol. 32: 3379–3385.

Chiou, C.T., G.Y. Sheng and M. Manes. 2001. A partition-limited model for the plant uptake of organic contaminants from soil and water. Environ. Sci. Technol. 35: 1437–1444.

Chitescu, C.L., A.I. Nicolau, P. Römkens and H.J. van der Fels-Klerx. 2014. Quantitative modelling to estimate the transfer of pharmaceuticals through the food production system. J. Environ. Sci. Health, Part B 49: 457–467.

Collins, Ch., J.C. White and S. Rock. 2007. Plant uptake of organic chemicals: Current developments and recommendations for future research. Environ. Toxicol. Chem. 26: 2465–2466.

Collins, Ch.D. and E. Finnegan. 2010. Modelling the plant uptake of organic chemicals, including the soil-air-plant pathway. Environ. Sci. Technol. 44: 998–1003.

Cropp, R.A., D.W. Hawker and M. Boonsaner. 2010. Predicting the accumulation of organic contaminants from soil by plants. Bull. Environ. Contam. Toxicol. 85: 525–529.

Czub, G. and S. McLachlan. 2004. A food chain model to predict the levels of lipophilic organic contaminants in humans. Environ. Toxicol. Chem. 23: 2356–2366.

Dettenmaier, E.M., W.J. Doucette and B. Bugbee. 2009. Chemical hydrophobicity and uptake by plant roots. Environ. Sci. Technol. 43: 324–329.

Doucette, W.J., T.J.K. Chard, B.J. Moore, W.J. Staudt and J.V. Headley. 2005. Uptake of sulfolane and diisopropanolamine (DIPA) by cattails (*Typha latifolia*). Microchem. J. 81: 41–49.

EC (European Commission). 2003. Technical Guidance Document on Risk Assessment in support of Commission Directive 93/67/EEC on Risk Assessment for new notified substances, Commission Regulation (EC) No 1488/94 on Risk Assessment for existing substances, and Directive 98/8/EC of the European Parliament and of the Council concerning the placing of biocidal products on the market. European Communities, Italy.

Engelstadter, J., K. Harms and P.J. Johnsen. 2016. The evolutionary dynamics of integrons in changing environments. ISME J. 10(6): 1296–1307.

Fantke, P., R. Juraske, A. Anton, R. Friedrich and O. Jolliet. 2011. Dynamic multicrop model to characterize impacts of pesticides in food. Environ. Sci. Technol. 45: 8842–8849.

Fantke, P., P. Wieland, R. Juraske, G. Shaddick, G.S. Itoiz, A. Anton et al. 2012. Parameterization models for pesticide exposure via crop consumption. Environ. Sci. Technol. 46: 12864–12872.

Fantke, P., P. Wieland, C. Wannaz, R. Friedrich and O. Jolliet. 2013. Dynamics of pesticide uptake into plants: From system functioning to parsimonious modeling. Environ. Model. Software 40: 316–324.

Fantke, P. and R. Juraske. 2013. Variability of pesticide dissipation half-lives in plants. Environ. Sci. Technol. 47: 3548–3562.

Fantke, P., B.W. Gillespie, R. Juraske and O. Jolliet. 2014. Estimating half-lives for pesticide dissipation from plants. Environ. Sci. Technol. 48: 8588–8602.

Fantke, P., J.A. Arnot and W.J. Doucette. 2016. Improving plant bioaccumulation science through consistent reporting of experimental data. J. Environ. Manag. 181: 374–384.

Franco, A. and S. Trapp. 2010. A multimedia activity model for ionizable compounds: Validation study with 2,4-dichlorophenoxyacetic acid, aniline, and trimethoprim. Environ. Toxicol. Chem. 29: 789–799.

Fu, W., A. Franco and S. Trapp. 2009. Methods for estimating the bioconcentration factors of polar and ionizable compounds in plants. Environ. Toxicol. Chem. 28(7): 1372–1379.

Gobas, F.A.P.C., L.P. Burkhard, W.J. Doucette, K.G. Sappington, E.M.J. Verbruggen, B.K. Hope et al. 2015. Review of existing terrestrial bioaccumulation models and terrestrial bioaccumulation modeling needs for organic chemicals. Integr. Environ. Assess. Manag. 12: 123–134.

Hawker, D.W., R. Cropp and M. Boonsaner. 2013. Uptake of zwitterionic antibiotics by rice (*Oryza sativa* L.) in contaminated soil. J. Hazard. Mater. 263: 458–466.

Hsu, F.C., R.L. Marxmiller and A.Y.S. Yang. 1990. Study of the root uptake and xylem translocation of cinmethylin and related compounds in detopped soybean roots using a pressure chamber technique. Plant Physiol. 93: 1573–1578.

Hughes, L. and D. Mackay. 2011. Model of the fate of chemicals in sludge amended soils with uptake in vegetation and soil dwelling organisms. Soil Sediment Contam. 20: 938–960.

Hung, H. and D. Mackay. 1997. A novel and simple model of the uptake of organic chemicals by vegetation from air and soil. Chemosphere 35: 959–977.

Hurtado, C., S. Trapp and J.M. Bayona. 2016. Inverse modelling of the biodegradation of emerging organic contaminants in the soil-plant system. Chemosphere 156: 236–244.

Jacobsen, R.E., P. Fantke and S. Trapp. 2015. Analyzing half-lives for pesticide dissipation in plants. SAR/QSAR Environ. Res. 26: 325–342.

Juraske, R., F. Castells, A. Vijay, P. Muñoz and A. Antón. 2009. Uptake and persistence of pesticides in plants: Measurements and model estimates for imidacloprid after foliar and soil application. J. Hazard. Mat. 165: 683–689.

Kleier, D.A. 1988. Phloem mobility of xenobiotics. Plant Physiol. 86: 803–810.

Kulhanek, A., S. Trapp, M. Sismilich, J. Janků and M. Zimová. 2005. Crop-specific human exposure assessment for polycyclic aromatic hydrocarbons in Czech soils. Sci. Total Environ. 339: 71–80.

Lang, A. 1990. Xylem, phloem and transpiration flows in developing apple fruits. J. Exp. Bot. 41: 645–651.

Laxminarayan, R. and G.M. Brown. 2001. Economics of antibiotic resistance: A theory of optimal use. J. Environ. Econom. Manag. 42: 183–206.

Legind, C.N. and S. Trapp. 2009. Modeling the exposure of children and adults via diet to chemicals in the environment with crop-specific models. Environ. Pollut. 157: 778–785.

Legind, C.N., C.M. Kennedy, A. Rein, N. Snyder and S. Trapp. 2011. Dynamic plant uptake model applied for drip irrigation of an insecticide to pepper fruit plants. Pest Manage Sci. 67: 521–527.

Legind, C.N., A. Rein, J. Serre, V. Brochier, C.-S. Haudin, P. Cambier et al. 2012. Simultaneous simulations of uptake into plants and leaching to groundwater of cadmium and lead for arable land amended with organic waste. PLoS One 7(10): e47002.

Levin, B.R., M. Lipsitch, V. Perrot, S. Schrag, R. Antia, L. Simonsen et al. 1997. The population genetics of antibiotic resistance. Clin. Infect. Diseases 24: S9–S16.

Li, H., G. Sheng, C.T. Chiou and O. Xu. 2005. Relation of organic contaminant equilibrium sorption and kinetic uptake in plants. Environ. Sci. Technol. 39: 4864–4870.

MacLean, R.C., A.R. Hall, G.G. Perron and A. Buckling. 2010. The population genetics of antibiotic resistance: Integrating molecular mechanisms and treatment contexts. Nat. Rev. Genet. 11(6): 405–414.

Manallack, D.T. 2007. The pK$_a$ distribution of drugs: Application to drug discovery. Perspect. Medicin. Chem. 1: 25–38.

Marschner, H. and P. Marschner. 2012. Marschner's Mineral Nutrition of Higher Plants. Waltham, MA. Elsevier.

McFarlane, J.C. 1994. Anatomy and physiology of plant conductive systems. pp. 13–36. *In*: Trapp, S. and J.C. McFarlane [eds.]. Plant Contamination: Modeling and Simulation of Organic Chemical Processes. CRC Press. Boca Raton, FL, US.

McKone, T.E. and R.L. Maddalena. 2007. Plant uptake of organic pollutants from soil: Bioconcentration estimates based on models and experiments. Environ. Toxicol. Chem. 26: 2494–2504.

McLachlan, M.S. 2011. Mass transfer between the atmosphere and plant canopy systems. pp. 137–158. *In*: Thibodeaux, L.J. and D. Mackay [eds.]. Handbook of Chemical Mass Transport in the Environment. CRC Press Boca Raton (FL) US.

Miller, E.L., S.L. Nason, K.K.G. Arthikeyan and J.A. Pedersen. 2016. Root uptake of pharmaceutical and personal care product ingredients. Environ. Sci. Technol. 50: 525–541.

Polesel, F., B.G. Plósz and S. Trapp. 2015. From consumption to harvest: Environmental fate prediction of excreted ionizable pharmaceuticals. Water Res. 84: 85–98.

Prosser, R.S., S. Trapp and P.K. Sibley. 2014. Modeling uptake of selected pharmaceuticals and personal care products into food crops from biosolids-amended soil. Environ. Sci. Technol. 48: 11397–11404.

Rein, A., C.N. Legind and S. Trapp. 2011. New concepts for dynamic plant uptake models. SAR and QSAR Environ. Res. 22: 191–2015.

Reinhold-Hurek, B. and T. Hurek. 2011. Living inside plants: Bacterial endophytes. Curr. Opin. Plant Biol. 14: 435–443.

Reinhold-Hurek, B., W. Bunger, C.S. Burbano, M. Sabale and T. Hurek. 2015. Roots shaping their microbiome: Global hotspots for microbial activity. Ann. Rev. Phytopathol. 53: 403–424.

Ryan, J.A., R.M. Bell, J.M. Davidson and G.A. O'Connor. 1988. Plant uptake of non-ionic chemicals from soils. Chemosphere 17: 2299–2323.

Stewart, F.M., R. Antia, B.R. Levin, M. Lipsitch and J.E. Mittler. 1998. The population genetics of antibiotic resistance II: Analytic theory for sustained populations of bacteria in a community of hosts. Theoret. Populat. Biol. 53(2): 152–165.

Takaki, K., A.J. Wade and Ch.D. Collins. 2014. Assessment of plant uptake models used in exposure assessment tools for soil contaminated with organic pollutants. Environ. Sci. Technol. 48: 12073–12082.

Tolls, J. and M.S. McLachlan. 1994. Partitioning of semivolatile organic-compounds between air and *Lolioum-multiflorum* (Welsch ray grass). Environ. Sci. Technol. 28: 159–166.

Topp, E., I. Scheunert, A. Attar and F. Korte. 1986. Factors affecting to the uptake of C-14 labeled organic-chemicals by plants from soil. Ecotoxicol. Environ. Safety 11: 219–228.

Trapp, S., M. Matthies, I. Scheunert and E.M. Topp. 1990. Modeling the bioconcentration of organic chemicals. Environ. Sci. Technol. 24: 1246–1252.

Trapp, S. and M. Matthies. 1995. Generic one-compartment model for uptake of organic chemicals by foliar vegetation. Environ. Sci. Technol. 29: 2333–2338.

Trapp, S. 2002. Dynamic root uptake model for neutral lipophilic organics. Environ. Toxicol. Chem. 21: 203–206.

Trapp, S. 2004. Plant uptake and transport models for neutral and ionic chemicals. Environ. Sci. Pollut. Res. 11: 33–39.

Trapp, S. and R.W. Horobin. 2005. A predictive model for the selective accumulation of chemicals in tumor cells. Eur. Biophys. J. 34: 959–966.

Trapp, S. 2007. Fruit tree model for uptake organic compounds from soil and air. SAR QSAR Environ. Res. 18: 367–387.

Trapp, S., A. Cammarano, E. Capri, F. Reichenberg and P. Mayer. 2007. Diffusion of PAH in potato and carrot slices and application for a potato model. Environ. Sci. Technol. 41: 3103–3108.

Trapp, S., G.R. Rosania, R.W. Horobin and J. Kornhuber. 2008. Quantitative modeling of selective lysosomal targeting for drug design. Europ. Biophys. J. 37: 1317–1328.

Trapp, S., A. Franco and D. Mackay. 2010. Activity-based concept for transport and partitioning of ionizing organics. Environ. Sci. Technol. 44: 6123–6129.

Trapp, S. and T. Eggen. 2013. Simulation of the plant uptake of organophosphates and other emerging pollutants for greenhouse experiments and field conditions. Environ. Sci. Pollut. Res. 11: 33–39.

Trapp, S. 2015. Calibration of a plant uptake model with plant- and site-specific data for uptake of chlorinated organic compounds into radish. Environ. Sci. Technol. 49(1): 395–402.

Trapp, S., W.J. Doucette, P. Fantke and T. Eggen. 2016. A suggested minimum data list to be reported for experimental plant uptake studies. SETAC Europe, Nantes, France. http://orbit.dtu.dk/files/127839929/SETAC_Europe_Abstractbook_Nantes.pdf.

Travis, C.C. and A.D. Arms. 1988. Bioconcentration of organics in beef, milk, and vegetation. Environ. Sci. Technol. 22: 271–274.

Undeman, E., G. Czub and M.S. McLachlan. 2009. Addressing temporal variability when modelling bioaccumulation in plants. Environ. Sci. Technol. 43: 3751–3756.

Undeman, E. and S. McLachlan. 2011. Assessing model uncertainty of bioaccumulation models by combining chemical space visualization with a process-based diagnostic ratios. Environ. Sci. Technol. 45: 8429–8436.

Upreti, R. and P. Thomas. 2015. Root-associated bacterial endophytes from *Ralstonia solanacearum* resistant and susceptible tomato cultivars and their pathogen antagonistic effects. Frontiers in Microbiol. 6: 255. doi: 10.3389/fmicb.2015.00255.

Yang, Ch.-Y., M. Chang, S.Ch. Wu and Y. Shih. 2016. Sorption equilibrium of emerging and traditional contaminants in leafy rape, Chinese mustard, lettuce and Chinese cabbage. Chemosphere 154: 552–558.

Zhu, B.K., Q.L. Chen, S.C. Chen and Y.G. Zhu. 2017. Does organically produced lettuce harbor higher abundance of antibiotic resistance genes than conventionally produced? Environ. Internatl. 98: 152–159.

16

The Role of Macrophytes in the Removal of Organic Micropollutants by Constructed Wetlands

A. Dordio,[1,2,]* A.J.P. Carvalho,[1,3] M. Hijosa-Valsero[4,5] and E. Becares[5]

INTRODUCTION

The huge development of chemical and agrochemical industries over the last century have resulted in the creation of a large number of new chemical compounds and their eventual dissemination into the environment. In fact, a lot of different organic substances are currently in use by modern society and many of these are frequently being detected in numerous environmental monitoring studies (Jones and de Voogt 1999; Fent et al. 2006; El Shahawi et al. 2010; Pal et al. 2010; Kümmerer 2011; Lapworth et al. 2012; Du and Liu 2012; Luo et al. 2014; Clarke and Cummins 2015; van Mourik et al. 2016; Ebele et al. 2017; Greaves and Letcher 2017; Bandowe and Meusel 2017). The number of new organic molecules generated by different industries expands daily, and the huge diversity of organic pollutants is one of the challenges that are presented to the mission of protecting the environment. Over the latest decades there has been an increased concern with this set of harmful organic

[1] Chemistry Department, School of Sciences and Technology, University of Évora, Évora, Portugal.
[2] MARE – Marine and Environmental Research Centre, University of Évora, Évora, Portugal.
[3] CQE – Évora Chemistry Centre, University of Évora, Évora, Portugal.
[4] Center of Biofuels and Bioproducts, Agrarian Technological Institute of Castilla and León (ITACyL), Villarejo de Orbigo, 24358 León, Spain.
[5] University of León, Department of Biodiversity and Environmental Management, Faculty of Biological Sciences, 24071 León, Spain.
* Corresponding author: avbd@uevora.pt

chemicals, mostly xenobiotics (i.e., synthetic compounds of anthropogenic origin that do not exist naturally in biological systems), that are commonly tagged as organic micropollutants (OMPs) (Jones and de Voogt 1999; Boethling et al. 2009; El Shahawi et al. 2010; Matthies et al. 2016). These compounds are characterized for generally being persistent pollutants (as result of being hardly biodegradable) or pseudo-persistent (as result of their continued inputs into the environment that help maintain approximately steady concentration levels of the pollutants irrespective of their potential degradability), bioaccumulating throughout the food chain, and potentially presenting some bioactivity. Together these form a set of characteristics which, despite the usually low concentration levels that feature OMPs' presence in the environment, ultimately pose non-negligible risks of causing adverse effects to human health and to the environment (Jones and de Voogt 1999; El Shahawi et al. 2010; Matthies et al. 2016). These OMPs include pesticides, polycyclic aromatic hydrocarbons (PAHs), phenolic compounds, petroleum hydrocarbons (e.g., alkanes, cycloalkanes, benzene, toluene, ethylbenzene, and xylenes), dyes, chlorinated compounds (e.g., trichloroethylene (TCE), perchloroethylene (PCE) and polychlorinated biphenyls (PCBs)), explosives (e.g., 2,4,6-trinitrotoluene (TNT), hexahydro-1,3,5-trinitro-1,3,5-triazine (RDX)), and a new class of emergent pollutants, the pharmaceuticals and personal care products (PPCPs) (Jones and de Voogt 1999; Boethling et al. 2009; El Shahawi et al. 2010). While some of these classes of pollutants are known to be present in the environment for a long time already and their toxic effects are well studied, awareness of the level of contamination of aquatic environments by some OMPs such as pharmaceuticals has only arisen more recently. This is due to the fact that most of these OMPs are generally found in the environment at very low concentration levels, which may range from μg L^{-1} to ng L^{-1}. In fact, advances in chemical analysis techniques and instrumentation made available over the last couple of decades (which allowed significantly lowering detection and quantification limits for analyses of organic compounds in complex environmental matrices) have aided in detecting low levels of other toxic organics in the environment. Ensuing such advances, the list of compounds that have been detected over the last decade in raw and treated wastewater, biosolids and sediments, receiving waters, ground water and drinking water has been steadily growing (Bolong et al. 2009; Pal et al. 2010; Kümmerer 2011; Lapworth et al. 2012; Du and Liu 2012; Luo et al. 2014; Li 2014; Clarke and Cummins 2015; Gavrilescu et al. 2015; Gothwal and Shashidhar 2015; Cizmas et al. 2015; Ebele et al. 2017). The ecotoxicity associated with OMPs is also largely unassessed but potential for chronic effects caused by long term exposure and for cumulative and even synergistic action exists (Fent et al. 2006; Bolong et al. 2009; El Shahawi et al. 2010; Zoller and Hushan 2010; Du and Liu 2012; Blaha and Holoubek 2013; Cizmas et al. 2015; Backhaus 2016; Bandowe and Meusel 2017).

Many of these substances receive inefficient treatment in WWTPs (which were not designed to deal with this type of pollutants) which is the main cause for their eventual release in the environment (Fent et al. 2006; Bolong et al. 2009; Luo et al. 2014; Li 2014; Haman et al. 2015; Dordio and Carvalho 2015).

Constructed wetlands systems (CWS) have been increasingly used over the last decades for removing OMPs from wastewater. CWS may be considered by now as a mature technology for the removal of bulk pollutants such as suspended solids, organic matter, pathogens and nutrients (Vymazal et al. 1998; Kadlec and Wallace

2009; Vymazal and Kröpfelová 2009). Meanwhile, focus has been moving in the last decades towards the removal of more specific and recalcitrant compounds for which the conventional wastewater treatment systems are not effective. In fact, CWS are currently being used more frequently and with some success for the cleanup of more specific pollutant types, namely some OMPs, and new challenges have been emerging such as the removal of pharmaceuticals and other micropollutants which present new environmental problems to be solved (Haberl et al. 2003; Davies et al. 2005; Imfeld et al. 2009; García et al. 2010; Hijosa-Valsero et al. 2011; Matamoros et al. 2012; Reyes-Contreras et al. 2012; Dordio and Carvalho 2013; Zhang et al. 2014; Verlicchi and Zambello 2014; Li et al. 2014a; Vymazal and Brezinová 2015; Ávila and García 2015; Toro-Vélez et al. 2016; Hussein and Scholz 2017; A et al. 2017).

Processes for Organic Pollutants Removal in Constructed Wetlands Systems

CWS are engineered to take advantage, and as much as possible enhance the effects, of the natural processes involving wetland vegetation, support matrix and their associated microbial assemblages to assist in treating wastewater and, in particular, to target them for certain specific types of pollutants (Vymazal et al. 1998; Kadlec and Wallace 2009). Wastewater clean-up is attained in a CWS through the concerted action of three main components, through a variety of chemical, physical and biological processes as illustrated in the scheme in Fig. 1, namely by exploiting the ability of plants to adsorb, uptake, and concentrate or metabolize organic xenobiotics, as well as to release root exudates that enhance compound biotransformation and microbial degradation. Vegetation also increases the diversity of niches in CW and therefore enhances microbial biodiversity, an aspect directly related to pollutant degradation.

Figure 1 highlights how the removal of OMPs by CWS involves several inter-dependent processes which, to some extent, may have an enhanced overall effect in

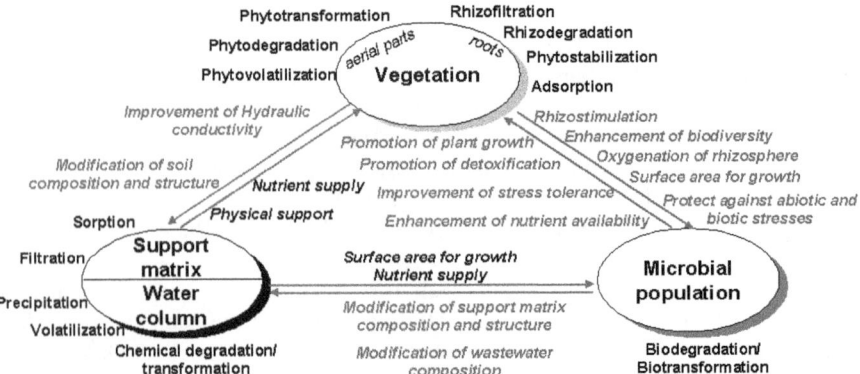

Fig. 1. Overview of organic micropollutant removal processes occurring within the domain of each CWS component and of the interactions and inter-dependences established between the components.

pollutants removal in these treatment systems as their concomitant action may lead to cooperative or synergistic effects.

Pollutant removal processes may be distinguished, under one of the possible criteria, as biotic (the class of processes carried out with an involvement of living organisms such as plants and microorganisms) or abiotic (processes which are strictly of physical or chemical nature, carried out in non-living components). Many of these removal processes are essentially the same that occur in natural wetlands environments (Reddy and DeLaune 2008). The way in which a CWS differs from a natural wetland is that an attempt is made in the CWS to enhance the effects of some of these processes (at least, for a specific set of target contaminants) through a thoughtful selection of the CWS components, careful system design, and optimizations of the most influential environmental and operation conditions within a controlled CWS environment.

Relevant abiotic and biotic processes that potentially may participate in removing OMPs from contaminated water in a CWS are described in Table 1.

CWS are biologically based systems in which biotic processes naturally play a key role in the removal of OMPs. The two biotic components which may be responsible in CWS for biological contributions to the removal of OMPs are the microbial populations and the wetland vegetation.

The Role of Aquatic Macrophytes in the Removal of Organic Micropollutants

The presence or absence of vegetation is one of the defining characteristics of wetlands and, as such, this is an indispensable component of natural and constructed wetlands systems that has several important roles, including one in wastewater treatment processes. Wetland plants are, in fact, the dominant structural component of most wetland treatment systems. Vegetation of CWS typically consists of aquatic macrophytes, also referred as hydrophytes or hydrophytic vegetation, which are the larger aquatic plants characteristic of wetland ecosystems. The term macrophytes includes aquatic vascular plants (angiosperms and fens), aquatic mosses, and some larger algae with tissues that are easily visible (Brix 1997; Vymazal et al. 1998; Kadlec and Wallace 2009). Aquatic macrophytes typically grow in or near regions that are seasonally or permanently water inundated. Hence, they are morphologically and physiologically adapted to living in such aquatic environments (saltwater or freshwater) or on a substrate that is at least periodically deficient in oxygen as a result of excessive water content (Vymazal et al. 1998; Kadlec and Wallace 2009). An important characteristic of these plants is that they developed elaborate structural mechanisms to avoid root anoxic conditions. Proper rhizosphere oxygenation is considered essential for active root function and also enables plants to counteract the effects of some soluble phytotoxins which may be present at high concentrations in anoxic substrates (Vymazal et al. 1998). A basic understanding of the growth requirements, characteristics and the roles performed by these wetland plants in CWS is essential for successful treatment wetland design and operation.

The wide variety of aquatic macrophytes naturally occurring in wetland environments may be distinguished in four classes (Brix 1997; Vymazal et al. 1998;

Table 1. Abiotic and biotic processes involved in organic micropollutants removal in CWS (Pilon-Smits 2005; Reddy and DeLaune 2008; Dordio and Carvalho 2011; Bulak et al. 2014).

Processes	Description
Abiotic	
Transfer/retention processes	*refers to processes that do not modify the compound but cause it to move among different compartments (water, biota, support matrix, atmosphere, etc.). Those processes lead to the pollutant storage in the support matrix, in the vegetation and microbial biomass or to losses of these compounds to the atmosphere.*
Sorption	The physical-chemical processes occurring at the surface of plants roots and of the supporting solid matrix which result in a short-term retention or long-term immobilization of the pollutants. The term refers broadly and without clear distinction to outer surface effects (adsorption) as well as diffusion into the inner surface of porous materials and incorporation in the sorbent (absorption).
Precipitation	For those compounds which can exist in several forms having distinct water solubilities, this corresponds to the conversion into the most insoluble forms. This removal process actually is just a transference between compartments (i.e., liquid to solid).
Filtration	Removal of particulate matter and suspended solids. It may lead to the removal of some dissolved organic pollutants as they are adsorbed to the filtered particles.
Volatilization	Release of some organic pollutants (or some smaller molecules that result from their prior decomposition through photodegradation, etc.) as vapors, which occurs when these compounds have significantly high vapor pressures.
Degradation/transformation processes	*refers to processes that lead to modifications of the structure of an organic pollutant, transforming it into one or several products with different behaviors and effects.*
Hydrolysis	The chemical breakdown of compounds by the action of water, a process which usually is pH-dependent.
Photodegradation	Decomposition of the pollutant molecules by the action of sunlight.
Redox reactions	Modification, which sometimes may be quite substantial, of the pollutant due to the action of oxidizing or reducing agents. Redox reactions are also frequently brought about by biotic agents (e.g., bacteria), or enzymatically catalyzed.
Biotic	
Transfer/retention processes	
Phytostabilization	Set of processes contributing to reduce the mobility of the organic pollutants and to attain their long-term immobilization in the rhizosphere.
Phytovolatilization	Uptake and transpiration of volatile organics through the aerial plant parts.
Rhizofiltration	Filtration through the mesh of roots and rhizomes to retain pollutants, by adsorption or deposition of particulates.

Table 1 contd. ...

...Table 1 contd.

Processes	Description
Degradation/transformation processes	
Biodegradation/ biotransformation	Decomposition of organic pollutants through metabolic or co-metabolic processes enacted by microorganisms, leading to complete decomposition (mineralization) or partial transformation.
Rhizodegradation	Enhancement of biodegradation of organic pollutants through the stimulation of microorganisms provided by substances (such as sugars, organic acids, enzymes, etc.) released in roots exudates.
Phytodegradation/ Phytotransformation	Direct uptake of organic compounds by plants and ensuing decomposition inside both root and shoot tissues. In addition to plant metabolism, endophytic microorganisms may be involved in the degradation of some compounds. Degradation products may be further subjected to the sequestration/lignification or undergo complete decomposition (mineralization). This process is usually much faster than biodegradation or rhizodegradation.

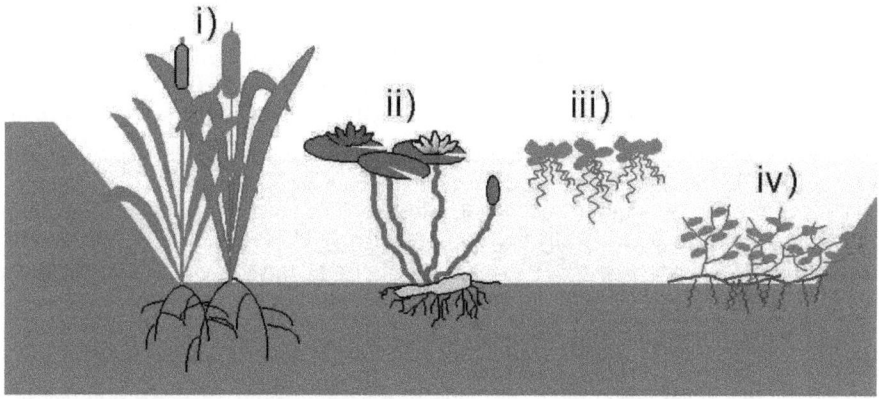

Fig. 2. Illustrative sketch outlining differences between classes of aquatic macrophytes.

Wetzel 2001; Stottmeister et al. 2003; Kadlec and Wallace 2009), according to their morphology and physiology (see also Fig. 2):

(i) **emergent aquatic macrophytes** are the dominating form of vegetation in natural wetlands. These plants grow on water-saturated or submersed soils from where the water table is about 0.5 m below the soil surface to where the sediment is covered with approximately 1.5 m of water. This type of macrophytes produce aerial stems and leaves which rise well above the water level and an extensive root and rhizome system in sediment, and comprise species like *Phragmites australis* (common reed), *Typha* spp. (cattail), *Scirpus* spp. (bulrush), *Juncus* spp. (rush), *Iris* spp. (blue and yellow flags), *Saggitaria latifolia* (duck potato), *Phalaris arundinocea* (reed canary grass), *Carex* spp. (sedges), *Zizania aquatica* (wild rice), *Eleocharis* spp. (spikerush), and *Glyceria* spp. (mannagrass);

(ii) **floating leaved macrophytes** are rooted in submersed sediments in water depths of approximately 0.5 to 3 m with stems submerged but possess either floating or slightly aerial leaves (e.g., *Nymphaea* spp. (water-lilies), *Nuphar* spp. (spatterdock), *Potamogeton* spp. (pondweeds), and *Hydrocotyle vulgaris* (pennyworth));

(iii) **free floating macrophytes** are not rooted to the substratum; they float freely on or in the water and are usually restricted to non-turbulent, protected areas (e.g., *Eichhornia crassipes* (water hyacinth), *Lemna* spp. (duckweed), *Spirodella* spp. (duckweed), *Pistia stratiotes* (water lettuce), *Azolla* spp. (water fern), and *Salvinia* spp. (watermoss));

(iv) **submerged macrophytes** occur at all depths within the photic zone and have their photosynthetic tissue entirely submerged with the flowers being exposed to the atmosphere. Two types of submerged aquatics are usually recognized: the elodeid type (e.g., *Elodea* spp. (waterweed), *Myriophyllum aquaticum* (parrot feather), *Ceratophyllum* spp. (coontail or hornwort)), and the isoetid (rosette) type (e.g., *Isoetes, Littorella, Lobelia*).

Some species may possibly develop in several of these forms; however, there is usually a dominant form and the plant species may be classified according to that form. In addition, emergent macrophytes are sometimes established/planted in floating mats, either with or without a supporting structure, and roots are kept submerged in the water instead of being anchored to the bottom substrate layer.

The plants that are most often used in CWS for OMPs treatment are persistent emergent plants, such as *Phragmites australis, Typha* spp., *Scirpus* spp., and *Juncus* spp.

Macrophytes growing in wetlands have several useful characteristics that allows them to contribute directly and indirectly to OMPs removal from wastewaters in a variety of ways (Brix 1997; Vymazal et al. 1998; Susarla et al. 2002; Kadlec and Wallace 2009; Shelef et al. 2013; Wang et al. 2014; Truu et al. 2015) (Fig. 1). Among the most relevant features, one can emphasize the effects of: (i) increasing the porosity of upper soil zones, improving the aeration/oxygenation of the rhizospere, supporting microbial respiration and aerobic microbial processes; (ii) stimulating microbial growth by the release of root exudates; (iii) being capable of accumulating, stabilizing, and degrading/transforming some OMPs; (iv) providing organic carbon substrates to adsorb some OMPs; (v) extracting available water and reversing the hydraulic gradient, thereby decreasing vertical and lateral migration of pollutants to groundwater.

In the participation of the plants in the operation of the CWS, different complementary roles are played by the different plant parts, according to the descriptions in Table 2.

Although the potential of plants to contribute significantly to remove organic pollutants from contaminated water has historically been underrated (in contrast with the widely recognized capability of some plants to uptake and accumulate heavy metals, for example), plants are, in fact, frequently capable of directly participating in the removal of OMPs, by adsorption to plant roots and/or uptake and subsequent

Table 2. Major roles of macrophytes in constructed wetlands (adapted from Brix (1997) and Shelef et al. (2013)).

Macrophyte plant part		Role in treatment process
Figure_table2	Aerial plant tissue	• Light attenuation → reduced growth of phytoplankton
		• Influence on microclimate → insulation from frost in winter and from radiation in summer
		• Reduced wind velocity → reduced risk of resuspension of solids
		• Aesthetic appearance
		• Transpiration → increased water loss
		• Nutrients and pollutants storage
		• Pollutants transformation/degradation or accumulation
		• Pollutants volatilization
	Plant tissue in water	• Filtering effect → filter out large debris
		• Reduced current velocity → increased rate of sedimentation, reduced risk of resuspension
		• Surface area for microbial attachment
		• Excretion of photosynthetic oxygen → increased aerobic degradation
		• Nutrient uptake
		• Provide surface area for pollutants adsorption
	Roots and rhizomes	• Stabilizing the sediment surface → less soil erosion
		• Prevents the medium from clogging in vertical flow systems
		• Release of oxygen, increases aerobic degradation and nitrification
		• Nutrients and pollutants uptake
		• Provide surface area for pollutants adsorption
		• Provide surface area for microbial attachment and development
		• Secretion of antibiotics for detoxification of root zone → pathogen removal
		• Release of exudates → stimulate microbial activity and biochemical transformations/degradation of pollutants in rhizosphere

translocation, metabolization, and accumulation/storage or volatilization (Hijosa-Valsero et al. 2016).

Several different species of macrophytes can be used in CWS but the selection of the more suitable ones to be used in each target case should be carefully made because, as has been stated before, plants play important roles, direct and indirect, in the global system efficiency. Precise treatment mechanisms vary based on the type of macrophyte, and the various available species and growth forms (e.g., annual or perennial, emergent, floating or submerged, woody or herbaceous) each fill different niches within the wetland and affect the final treatment results. Therefore, the selection of vegetation species has an important role in the optimization of CWS for particular types of target pollutants.

Specific requirements of plants characteristics will vary depending on the functional role of wetland plants in the treatment systems. This will be related to the type of CWS design and its mode of operation (continuous or batch), loading rate, and wastewater characteristics (Vymazal et al. 1998; Kadlec and Wallace 2009; Dhir et al. 2009; Reyes-Contreras et al. 2012; Shelef et al. 2013; Wang et al. 2014).

Other aspects of CWS design such as plant density and layout of the specimens (e.g., the way specimens of different species may be intermixed when planted in the beds) also have to be considered and their influence may range from subtle differences in the system's behavior to more pronounced impacts in the overall efficiency (Kadlec and Wallace 2009). In particular, the cycles of vegetative activity of some species in addition to variations of climate conditions may lead to significant seasonal changes in the system's efficiency, which in some cases may be mitigated by using polycultures of vegetation (Kadlec and Wallace 2009; Garfí et al. 2012). In this regard, recent studies have shown that, presumably desorption of emerging organic contaminants occurred from the rizhosphere during the senescence stage of plant biomass in a free water surface wetland (Ávila et al. 2017).

For the purpose of removing OMPs with well-known toxicity, plant species have to be selected based on the criteria of their tolerance to the pollutant (or cocktail of pollutants) and their capacity to lower pollutants concentrations in water. These characteristics may be assessed beforehand, either by testing with spiked influent directly in pilot-scale or fully-operating CWS, or by exposing plant specimens to aqueous solutions of the compounds in hydroponic assays.

Table 3 presents some recent examples of assessment studies conducted to determine the capacity of a variety of macrophyte species to remove certain types of OMPs from water, under common system setups and experimental conditions.

The examples of studies presented in Table 3 give an overview of the diversity of macrophyte species that have already been assessed and might be used for the removal of a wide variety of OMPs. It is also possible to have at least a rough idea of the direct influence of plants in diminishing the concentrations of the pollutants in the aqueous medium. Furthermore, in some studies the fate of the target pollutant in the plants is approached in more detail or an assessment is made of which processes contribute the most to its elimination. In addition, although the data is not presented in Table 3, it is common that in this type of studies also some assays are carried out with the aim of assessing phytotoxicity of the studied OMPs on the tested plant species.

While a few studies are conducted within at least mesocosm/pilot-scale CWS, assays conducted under hydroponic conditions are generally preferred. The rationale may be to keep the systems simpler, with minimal perturbations from other CWS components, when the goal is specifically to address plant–pollutant interactions.

A majority of the studies are conducted on single plant species systems and, in many cases, also with just a single pollutant substance. A much varied selection of plants species is studied in hydroponic experiments, but assays conducted in CWS typically employ a more restricted subset of those species, normally comprising the species that are more frequently studied in the hydroponic tests. Hydroponic

Table 3. Removal of selected organic micropollutants by several macrophytes.

Plant species	Compound (concentration, if available)	Assays conditions	% Removal (exposure time)	Observation of compound uptake by plant	Bibliographic reference
Acorus calamus (Sweet flag)					
Flame retardants					
	Polybrominated diphenyl ethers (3 mg/kg$_{soil}$)	Spiked soil	n.d.	Taken up, accumulation/metabolization (317.3 ± 10.7 ng/g in the roots) (60 d)	(Deng et al. 2016)
	PAHs	CW (lab scale)	27 (continuous-flow) 37 (intermittent-flow)	n.d.	(Jie-Ting et al. 2015)
Pesticides					
	Atrazine	Hydroponic conditions	62 (20 d)	Taken up, metabolization	(Wang et al. 2012)
	Triazophos (1 – 5 mg/L)	Hydroponic conditions	n.d.	Taken up in order of μg/(g$_{FW}$. d)	(Li et al. 2014b)
	Chlorpyrifos (1 – 16 mg/L)	Hydroponic conditions	65 – 92 (20 d)	Taken up	(Wang et al. 2016)
Armoracia rusticana (Horseradish)					
Pharmaceuticals					
	Diclofenac (31.8, 63.7 mg/L)	Hydroponic conditions	60 – 70 (8 d)	n.d.	(Kotyza et al. 2010)
	Ibuprofen (20.6, 41.3 mg/L)		40 – 94 (8 d)		
	Acetaminophen (15,30 mg/L)		100 (8 d)		
Arundo donax (Giant cane)					
	PAHs	CW (lab scale)	33 (continuous-flow) 41 (intermittent-flow)	n.d.	(Jie-Ting et al. 2015)

Table 3 contd. ....

...Table 3 cond.

Plant species	Compound (concentration, if available)	Assays conditions	% Removal (exposure time)	Observation of compound uptake by plant	Bibliographic reference
Azzola filicuolides (Water fern)					
Pharmaceuticals					
	Sulfadimethoxine (50 – 450 mg/L)	Hydroponic conditions	56 – 89 (35 d)	Taken up in the order of μg/g – mg/g	(Forni et al. 2002)
Baumea juncea					
PAHs					
	Phenanthrene (50, 250 mg/kg)	Spiked soil	71 – 97 (70 d) 99 – 100 (150 d)	Taken up and metabolization	(Zhang et al. 2010)
	Pyrene (50, 250 mg/kg)		37 – 50 (70 d) 65 – 97 (150 d)		
Boerhavia erecta (Erect spiderling)					
Pesticides					
	Imazalil (10, 100 μg/L, synthetic WW)	CW (mesocosms)	> 80 (5 d)	Taken up	(Lv et al. 2016b)
	Tebuconazole (10, 100 μg/L, synthetic WW)		> 50 (5 d)		
Cabomba aquatica					
Pesticides					
	Dimethomorph (0 – 1 mg/L)	Hydroponic conditions	3 (1 d)	Taken up in order of μg/ (g_{FW} d)	(Olette et al. 2008)
	Flazasulfuron (0 – 0.1 mg/L)		17 (4 d)		

Cabomba caroliniana			
Pharmaceuticals			
Codeine (24 ng/L, WW)	Hydroponic conditions	25 (4 d)	(Mackulak et al. 2015)
Methamphetamine (30 ng/L, WW)		71 (4 d)	
Benzoylecgonine (18 ng/L, WW)		28 (4 d)	
Tramadol (710 ng/L, WW)		48 (4 d)	
Venlafaxine (259 ng/L, WW)		41 (4 d)	
Oxazepam (83 ng/L, WW)		8 (4 d)	
Citalopram (84 ng/L, WW)		64 (4 d)	
Buprenorphine (11 ng/L, WW)		10 (4 d)	
Methadone (18 ng/L, WW)		71 (4 d)	
Canna indica (Purple arrowroot)			
Pesticides			
Triazophos (1 – 5 mg/L)	Hydroponic conditions	Taken up in order of $\mu g/(g_{FW} \cdot d)$ n.d.	(Li et al. 2014b)
Ceratophyllum demersum			
Pesticides			
MCPA (10 μg/L)	Hydroponic conditions	17 ± 5 (38 d) n.d.	(Matamoros et al. 2012)

Table 3 contd. ...

...Table 3 contd.

Plant species	Compound (concentration, if available)	Assays conditions	% Removal (exposure time)	Observation of compound uptake by plant	Bibliographic reference
	Pharmaceuticals and personal care products				
	Diclofenac (10 µg/L)	Hydroponic conditions	99 ± 1 (38 d)	Mainly photodegradation	(Matamoros et al. 2012)
	Naproxen (10 µg/L)		53 ± 5 (38 d)		
	Ibuprofen (10 µg/L)		52 ± 2 (38 d)	n.d.	
	Caffeine (10 µg/L)		81 ± 2 (38 d)		
	Clofibric acid (10 µg/L)		17 ± 3 (38 d)		
	Triclosan (10 µg/L)		99 ± 1 (38 d)		
Carex lurida					
	Organic solvents				
	Acetone (100 mg/L)	CW (microcosms)	84 (5 d) Summer 62 (5 d) Winter	n.d.	(Grove and Stein 2005)
	Tetrahydrofuran (100 mg/L)		88 (14 d) Summer 81 (14 d) Winter		
	1-butanol (100 mg/L)		87 (1 d) Summer 67 (1d) Winter		
Chrysopogon zizanioides (Vetiver grass)					
	Pharmaceuticals				
	Tetracycline (5 – 15 mg/L)	Hydroponic conditions	100 (40 d)	Taken up in order of µg/g	(Datta et al. 2013)

Cyperus alternifolius (Umbrella papyrus)				
Pesticides				
Triazophos (1 – 5 mg/L)	Hydroponic conditions	n.d.	Taken up in order of µg/(g_{FW} .d)	(Li et al. 2014b)
Pharmaceuticals				
Carbamazepine (0 – 500 µg/L)	CW (microcosms)	~50 – 64 (12 hr)	Taken up in order of ng/g	(Yan et al. 2016)
Sulfamethoxazole (0 – 500 µg/L)		~55 – 77 (12 hr)		
Ofloxacin (0 – 500 µg/L)		~75 – 91 (12 hr)		
Roxithromycin (0 – 500 µg/L)		~70 – 85 (12 hr)		
Diclofenac	CW (mesocosms)	69 (70 d)	Taken up and metabolization (21.4% of the total removed)	(Zhai et al. 2016)
Cyperus papyrus				
Flame retardants				
Polybrominated diphenyl ethers (3 mg/kg_{soil})	Spiked soil	n.d.	Taken up, accumulation/ metabolization (142.2 ± 11.4 ng/g in the roots) (60 d)	(Deng et al. 2016)
Cyperus rotundus (Coco-grass)				
Pesticides				
Triazophos (1 – 5 mg/L)	Hydroponic conditions	n.d.	Taken up in order of µg/(g_{FW} .d)	(Li et al. 2014b)

Table 3 contd. ...

...Table 3 contd.

Plant species	Compound (concentration, if available)	Assays conditions	% Removal (exposure time)	Observation of compound uptake by plant	Bibliographic reference
Eichhornia crassipes (Water hyacinth)					
PAHs					
	Naphthalene (15 mg/L)	Hydroponic conditions	53 (7 d)	Taken up and adsorption by plants	(Nesterenko-Malkovskaya et al. 2012)
Pesticides					
	Triazophos (1 – 5 mg/L)	Hydroponic conditions	n.d.	Taken up in order of µg/ ($g_{FW} \cdot d$)	(Li et al. 2014b)
Pharmaceuticals					
	Carbamazepine (0.8 mg/L)	Hydroponic conditions	36 ± 7 (12 d)	n.d.	(Lin and Li 2016)
	Ibuprofen (0.8 mg/L)		94 ± 5 (12 d)		
	Sulfadiazine (0.8 mg/L)		47 ± 4 (12 d)		
	Sulfamethoxazole (0.8 mg/L)		42 ± 7 (12 d)		
	Sulfamethazone (0.8 mg/L)		43 ± 6 (12 d)		
Elodea canadensis (Pondweed)					
Pesticides					
	Dimethomorph (0 – 1 mg/L)	Hydroponic conditions	5.5 (2 d)	Taken up in order of µg/ ($g_{FW} \cdot d$)	(Olette et al. 2008)
	Flazasulfuron (0 – 0.1 mg/L)		12 (4 d)		
	MCPA (10 µg/L)	Hydroponic conditions	11 ± 4 (38 d)	n.d.	(Matamoros et al. 2012)

Pharmaceuticals	Hydroponic conditions		Mainly photodegradation	(Matamoros et al. 2012)
Diclofenac (10 µg/L)		99 ± 1 (38 d)		
Naproxen (10 µg/L)		53 ± 3 (38 d)		
Ibuprofen (10 µg/L)		77 ± 2 (38 d)	n.d.	
Caffeine (10 µg/L)		94 ± 2 (38 d)		
Clofibric acid (10 µg/L)		23 ± 2 (38 d)		
Triclosan (10 µg/L)		97 ± 2 (38 d)		
Glyceria maxima (Great Manna Grass)				
Pesticides	CW (mesocosms)		Taken up and metabolization	(Chen et al. 2017)
Acetochlor (75, 260, 1100, 19 µg/L)		~100 (9 d)		
S-metolachlor (75, 260, 1100, 19 µg/L)		~99.8 (21 d)		
Metazachlor (75, 260, 1100, 19 µg/L)		>99 (21 d)		
Dimethachlor (75, 260, 1100, 19 µg/L)		>99 (21 d)		
Hydrilla verticillata (Esthwaite waterweed)				
PAHs	CW (lab scale)	34 (continuous-flow) 29 (intermittent-flow)	n.d.	(Jie-Ting et al. 2015)

Table 3 contd. ...

...Table 3 contd.

Plant species	Compound (concentration, if available)	Assays conditions	% Removal (exposure time)	Observation of compound uptake by plant	Bibliographic reference
Hordeum vulgare L. (Barley)					
Pharmaceuticals					
	Sulfadimethoxine (12 mg/L)	Hydroponic conditions	75 (2 d)	Taken up in order of µg/g (accumulated in roots)	(Ferro et al. 2010)
	Sulfamethazine (11 mg/L)		40 (2 d)		
	Diclofenac (31.8, 63.7 mg/L)		25 – 43 (8 d)	n.d.	(Kotyza et al. 2010)
	Ibuprofen (20.6, 41.3 mg/L)		17 – 58 (8 d)		
	Acetaminophen (15, 30 mg/L)		42 – 83 (8 d)		
Iris pseudacorus L. (Yellow flag)					
Pesticides					
	Acetone (100 mg/L)	CW (microcosms)	92 (5 d) Summer 61(5 d) Winter	n.d.	(Grove and Stein 2005)
	Tetrahydrofuran (100 mg/L)		82 (14 d) Summer 55 (14 d) Winter		
	1-butanol (100 mg/L)		82 (1 d) Summer 53 (1 d) Winter		
	Atrazine	Hydroponic conditions	76 (20 d)	Taken up and metabolization	(Wang et al. 2012)
	Chlorpyrifos (1 mg/L)	Hydroponic conditions	~ 88 (15 d)	Taken up (µg/g) and translocation	(Wang et al. 2013)
	Triazophos (1 – 5 mg/L)	Hydroponic conditions	n.d.	Taken up in order of µg/ (g_{FW}.d)	(Li et al. 2014b)

Compound	Conditions	Removal	Notes	Reference
Imazalil (10, 100 µg/L)	CW (mesocosms)	> 80 (5 d)	Taken up	(Lv et al. 2016b)
Tebuconazole (10, 100 µg/L)		> 60 (5 d)		
Imazalil (10 mg/L)	Hydroponic conditions	46 (24 d)	Taken up and metabolization	(Lv et al. 2016a)
Tebuconazole (10 mg/L)		25 (24 d)		
Pharmaceuticals				
Codeine (24 ng/L, WW)	Reactor (hydroponic conditions)	> 93 (4 d)	n.d.	(Mackulak et al. 2015)
Methamphetamine (30 ng/L, WW)		85 (4 d)		
Benzoylecgonine (18 ng/L, WW)		82 (4 d)		
Tramadol (710 ng/L, WW)		70 (4 d)		
Venlafaxine (259 ng/L, WW)		70 (4 d)		
Oxazepam (83 ng/L, WW)		58 (4 d)		
Citalopram (84 ng/L, WW)		94 (4 d)		
Buprenorphine (11 ng/L, WW)		> 70 (4 d)		
Methadone (18 ng/L, WW)		> 87 (4 d)		
Ibuprofen (10 mg/L)	Hydroponic conditions	~ 100 (24 d)	Taken up by roots and translocated to the aerial tissues	(Zhang et al. 2016)
Ihexol (10 mg/L)		13 (24 d)		
***Juncus acutus* L.** (Spiny rush)				
Pharmaceuticals				
Ciprofloxacin (0.1, 10 mg/L)	Hydroponic conditions	92 – 95 (28 d)	n.d.	(Christofilopoulos et al. 2016)
Sulfamethoxazole (50 mg/L)		61 (28 d)		

Table 3 contd. ....

...Table 3 contd.

Plant species	Compound (concentration, if available)	Assays conditions	% Removal (exposure time)	Observation of compound uptake by plant	Bibliographic reference
Juncus effusus L. (Common rush)					
	Organic solvents				
	Acetone (100 mg/L)	CW (microcosms)	89 (5 d) Summer 80 (5 d) Winter	n.d.	(Grove and Stein 2005)
	Tetrahydrofuran (100 mg/L)		95 (14 d) Summer 94 (14 d) Winter		
	1-butanol (100 mg/L)		82 (1 d) Summer 85 (1 d) Winter		
	Pesticides				
	Atrazine	Hydroponic conditions	~100 (8 d)	Adsorbed to roots, taken up and translocation	(Bouldin et al. 2006)
	Lambda-cyhalothrin		n.d. (8 d)		
	Imazalil (10, 100 µg/L)	CW (mesocosms)	> 80 (5 d)	Taken up	(Lv et al. 2016b)
	Tebuconazole (10, 100 µg/L)		> 60 (5 d)		
	Imazalil (10 mg/L)	Hydroponic conditions	67 (24 d)	Taken up and metabolization	(Lv et al. 2016a)
	Tebuconazole (10 mg/L)		41 (24 d)		
	Pharmaceuticals				
	Ibuprofen (10 mg/L)	Hydroponic conditions	~ 100 (24 d)	Taken up and translocation	(Zhang et al. 2016)
	Iohexol (10 mg/L)		31 (24 d)		

Juncus subsecundus (Fingered rush)

PAHs

Phenanthrene (50, 250 mg/kg)	Spiked soil	95 (70 d)	Taken up and metabolization	(Zhang et al. 2010)
Pyrene (50, 250 mg/kg)		13 – 39 (70 d)		

Landoltia punctata* and *Lemna minor (Duckweeds)

Pharmaceuticals

Ibuprofen (2 mg/L)	Hydroponic conditions	48 ± 4 (9 d)	n.d.	(Reinhold et al. 2010)
Fluoxetine (3 mg/L)		100 ± 4 (4 d)		
Clofibric acid (2 mg/L)		0 (9 d)		

Leersia oryzoides (Rice cutgrass)

Pesticides

Atrazine (2 ± 0.8 µg/L)	CW (mesocosms)	51 ± 6 (4 hr)	n.d.	(Moore et al. 2016)
Metolachlor (12 ± 0.4 µg/L)		54 ± 9 (4 hr)		
Diazinon (3.1 ± 0.2 µg/L)		55 ± 8 (4 hr)		

Lemna gibba (Gibbous duckweed)

Pesticides

Pesticide metabolites (0.1 mg/L)	Hydroponic conditions	n.d. (1, 2, 4 d)	Taken up and metabolization	(Fujisawa et al. 2006)

Pharmaceuticals

Acetaminophen (1 – 1000 µg/L)	Hydroponic conditions	66 ± 1 – 92 ± 2 (10 d)	n.d.	(Allam et al. 2015)
Diclofenac (1 – 1000 µg/L)		48 ± 2 – 84 ± 3 (10 d)		
Progesterone (1 – 1000 µg/L)		67 ± 2 – 95 ± 3 (10 d)		

Table 3 contd. ...

...Table 3 contd.

Plant species	Compound (concentration, if available)	Assays conditions	% Removal (exposure time)	Observation of compound uptake by plant	Bibliographic reference
	Tetracycline	CW (pilot scale)	73 (13 d)	n.d.	(Topal et al. 2015)
	4-epitetracycline		90 (13 d)		
	Anhydrotetracycline		63 (5 d)		
	4-epianhydrotetracycline		39 (13 d)		
Lemna minor (Common duckweed)					
Pesticides					
	Dimethomorph (0 – 1 mg/L)	Hydroponic conditions	12 (1 d)	Taken up in order of μg/ $(g_{FW}\cdot d)$	(Olette et al. 2008)
	Flazasulfuron (0 – 0.1 mg/L)	Hydroponic conditions	42 (4 d)		
	MCPA (10 μg/L)	Hydroponic conditions	6 ± 3 (38 d)	n.d.	(Matamoros et al. 2012)
Pharmaceuticals and personal care products					
	Diclofenac (10 μg/L)	Hydroponic conditions	99 ± 1 (38 d)	Mainly photodegradation	(Matamoros et al. 2012)
	Naproxen (10 μg/L)		40 ± 5 (38 d)		
	Ibuprofen (10 μg/L)		44 ± 3 (38 d)	n.d.	
	Caffeine (10 μg/L)		99 ± 1 (38 d)		
	Clofibric acid (10 μg/L)		16 ± 4 (38 d)		
	Triclosan (10 μg/L)		97 ± 3 (38 d)		

Limnophila sessiliflora (Dwarf Ambulia)

Pharmaceuticals

Codeine (24 ng/L, WW)	Hydroponic conditions	33 (4 d)	n.d. (Mackulak et al. 2015)
Methamphetamine (30 ng/L, wastewater)		76 (4 d)	
Benzoylecgonine (18 ng/L, WW)		33 (4 d)	
Tramadol (710 ng/L, WW)		59 (4 d)	
Venlafaxine (259 ng/L, WW)		62 (4 d)	
Oxazepam (83 ng/L, WW)		10 (4 d)	
Citalopram (84 ng/L, WW)		75 (4 d)	
Buprenorphine (11 ng/L, WW)		13 (4 d)	
Methadone (18 ng/L, WW)		63 (4 d)	

Linum usitatissimum (Common flax)

Pharmaceuticals

Diclofenac (63.6 mg/L)	Hydroponic conditions	100 (8 d)	n.d. (Kotyza et al. 2010)
Ibuprofen (41.3 mg/L)		100 (8 d)	
Acetaminophen (30.2 mg/L)		50 (8 d)	

Table 3 contd. ...

...Table 3 contd.

Plant species	Compound (concentration, if available)	Assays conditions	% Removal (exposure time)	Observation of compound uptake by plant	Bibliographic reference
Lupinus angustifolius (Narrowleaf lupin)					
Pesticides					
	Carbaryl (10 mg/L, 50 mg/L)	Hydroponic conditions	52 (16 d) at 10 mg/L 60 (16 d) at 50 mg/L	Taken up and metabolization	(Garcinuño et al. 2006)
	Linuron (10 mg/L, 50 mg/L)		59 (16 d) at 10 mg/L 45 (16 d) at 50 mg/L		
	Permethrin (10 mg/L, 50 mg/L)		73 (16 d) at 10 mg/L 47 (16 d) at 50 mg/L		
Lupinus luteolus (Pale yellow lupine)					
Pharmaceuticals					
	Diclofenac (31.8, 63.7 mg/L)	Hydroponic conditions	3 – 7 (8 d)	n.d.	(Kotyza et al. 2010)
	Ibuprofen (20.6, 41.3 mg/L)		4 – 11 (8 d)		
	Acetaminophen (15, 30 mg/L)		100 (8 d)		
***Lythrum salicaria* L.** (Spiked loosestrife)					
Pharmaceuticals					
	Flumequine (100 mg/L)	Hydroponic conditions	n.d. (10 – 30 d)	Taken up in order of µg/g	(Migliore et al. 2000)
	Flumequine (0.05 – 5 mg/L)		n.d. (35 d)		
Pesticides					
	Atrazine	Hydroponic conditions	66 (20 d)	Taken up and metabolization	(Wang et al. 2012)

Myriophyllum spicatum L. (Spiked water-milfoil)

Pesticides				
Linuron (~10 – 20 µg/L)	Spiked sediment	n.d. (7, 14, 21 d)	Taken up in order of µg/kg$_{DW}$	(Burešová et al. 2013)
Pharmaceuticals				
Tetracycine (5.0 mg/L)	Hydroponic conditions	70 (15 d)	n.d.	(Gujarathi et al. 2005)
Oxytetracycline (5.0 mg/L)		>90 (15 d)		
Phragmites australis (Common reed)				
Dyes				
Acid orange 7 (AO7) azo dye (129, 700 mg/L)	CW (mesocosms)	69 (n.d.) at 129 mg/L 74 (n.d.) at 700 mg/L	Active role in degradation	(Davies et al. 2005)
Flame retardants				
Polybrominated diphenyl ethers (3 mg/kg$_{soil}$)	Spiked soil	n.d.	Taken up, accumulation/ metabolization (103.1 ± 5.4 ng/g in the roots) (60 d)	(Deng et al. 2016)
PAHs	CW (lab scale)	29 (continuous-flow) 37 (intermittent-flow mode)	n.d.	(Jie-Ting et al. 2015)
Pesticides				
1,4-dichlorobenzene 1,2,4-trichlorobenzene γ-hexachlorocyclohexane	Hydroponic conditions	n.d. (7,14,21 and 28 d)	Taken up, translocation and phytovolatilization Taken up and translocation	(San Miguel et al. 2013)
Triazophos (1 – 5 mg/L)	Hydroponic conditions	n.d.	Taken up in order of µg/(g$_{FW}$.d)	(Li et al. 2014b)

Table 3 contd. ...

...Table 3 contd.

Plant species	Compound (concentration, if available)	Assays conditions	% Removal (exposure time)	Observation of compound uptake by plant	Bibliographic reference
	Imazalil (10, 100 µg/L)	CW (mesocosms)	> 80 (5 d)	Taken up	(Lv et al. 2016b)
	Tebuconazole (10, 100 µg/L)		> 60 (5 d)		
	Imazalil (10 mg/L)	Hydroponic conditions	96 (24 d)	Taken up and metabolization	(Lv et al. 2016a)
	Tebuconazole (10 mg/L)		31 (24 d)		
Pharmaceuticals					
	Diclofenac (31.8, 63.7 mg/L)	Hydroponic conditions	18 – 33 (8 d)	n.d.	(Kotyza et al. 2010)
	Ibuprofen (20.6, 41.3 mg/L)		49 – 60 (8 d)		
	Acetaminophen (15, 30 mg/L)		0 – 16 (8 d)		
	Enrofloxacin (10, 100 µg/L)	Hydroponic conditions	67 – 91 (7 d)	n.d.	(Carvalho et al. 2012)
	Tetracycline (10, 100 µg/L)		60 – 98 (7 d)		
	Enrofloxacin (100 µg/L, WW)		94 (7 d)		
	Tetracycline (100 µg/L, WW)		75 (7 d)		
	Ciprofloxacin (0.1 – 1000 µg/L)	Hydroponic conditions	n.d. (62 d)	Taken up in order of ng/g$_{DW}$	(Liu et al. 2013)
	Oxytetracycline (0.1 – 1000 µg/L)				
	Sulfamethazine (0.1 – 1000 µg/L)				
	Albendazole (2.65 mg/L)	Hydroponic conditions	n.d.	Taken up (in order of *pmol/g* – nmol/g) and transformation	(Podlipná et al. 2013)
	Flubendazole (3.13 mg/L)				
	Metformin (64.6, 129.2, 258.3 mg/L)	Hydroponic conditions	n.d.	Taken up by the roots and translocation	(Cui et al. 2015)

Carbamazepine (5 mg/L)	Hydroponic conditions	35 (1 d) 66 (4 d) 90 (9 d)	n.d.	(Sauvetre and Schrüder 2015)
Tetracycline 4-epitetracycline Anhydrotetracycline 4-epianhydrotetracycline	Field experiment	n.d.	Taken up in order of ppb	(Arslan Topal 2015)
Ibuprofen (10 mg/L)	Hydroponic conditions	~ 100 (24 d)	Taken up and translocation	(Zhang et al. 2016)
Iohexol (10 mg/L)		80 (24 d)		
Phenolic compounds				
Bisphenol A (BPA) (8.8 µg/L)	CWS (pilot scale)	71 ± 27 (1.8 d HRT)	n.d.	(Toro-Vélez et al. 2016)
Nonylphenols (1.67 ± 0.8 mg/L)		52 ± 37 (1.8 d HRT)		
Polychlorinated phenols				
Pentachlorophenol (10, 20 mg/kg)	Spiked soil	> 70 (75 d)	Taken up in order of mg/kg	(Hechmi et al. 2015)
Pistia stratiotes (Water cabbage)				
Pharmaceuticals				
Tetracycine (5.0 mg/L)	Hydroponic conditions	> 90 (6 d)	n.d.	(Gujarathi et al. 2005)
Oxytetracycline (5.0 mg/L)		> 90 (4 d)		
Carbamazepine (0.8 mg/L)	Hydroponic conditions	34 ± 5.9 (12 d)	n.d.	(Lin and Li 2016)
Ibuprofen (0.8 mg/L)		100 ± 1 (12 d)		
Sulfadiazine (0.8 mg/L)		77 ± 8 (12 d)		
Sulfamethoxazole (0.8 mg/L)		66 ± 5 (12 d)		
Sulfamethazone (0.8 mg/L)		89 ± 5 (12 d)		

Table 3 contd....

...Table 3 contd.

Plant species	Compound (concentration, if available)	Assays conditions	% Removal (exposure time)	Observation of compound uptake by plant	Bibliographic reference
Populus nigra L.					
	Pharmaceuticals				
	Ibuprofen (0.03, 3, 30 mg/L)	Hydroponic conditions	100 (21 d)	n.d.	(Iori et al. 2012)
Salvinia molesta (Giant salvinia)					
	Pesticides				
	MCPA (10 μg/L)	Hydroponic conditions	7 ± 3 (38 d)	n.d.	(Matamoros et al. 2012)
	Pharmaceuticals and personal care products				
	Diclofenac (10 μg/L)	Hydroponic conditions	99 ± 1 (38 d)	Mainly photodegradation	(Matamoros et al. 2012)
	Naproxen (10 μg/L)		43 ± 2 (38 d)		
	Ibuprofen (10 μg/L)		48 ± 5 (38 d)	n.d.	
	Caffeine (10 μg/L)		99 ± 1 (38 d)		
	Clofibric acid (10 μg/L)		18 ± 4 (38 d)		
	Triclosan (10 μg/L)		96 ± 1 (38 d)		
Salix alba L. (White willow)					
(clone SS5)	Ibuprofen (3, 30 mg/L)	Hydroponic conditions	37 − 81 (14 d)	n.d.	(Iori et al. 2013)
(clone SP3)	Ibuprofen (3, 30 mg/L)		30 − 77 (14 d)		

Salix fragilis (Crack willow)

Sulfadiazine (10 – 200 mg/kg$_{soil}$)	Soil spiked	n.d. (40 d)	Taken up in order of mg/kg, translocation and metabolization	(Michelini et al. 2012)

Salix nigra (Black willow)

Organochlorines

Chlorobenzene (10 mg/L)	Reactor	4 (1 d)	Taken up (0.048 mg) and translocation	(Gomez-Hermosillo et al. 2006)

PAHs

Phenanthrene (1 mg/L)	Reactor	5 (1 d)	Taken up (0.049 mg) and translocation	(Gomez-Hermosillo et al. 2006)

Scirpus olneyi (Olney bulrush)

Organochlorines

Chlorobenzene (10 mg/L)	Reactor	6 (1 d)	Taken up (0.067 mg) and translocation	(Gomez-Hermosillo et al. 2006)

PAHs

Phenanthrene (1 mg/L)	Reactor	4 (1 d)	Taken up (0.041 mg) and translocation	(Gomez-Hermosillo et al. 2006)

Scirpus triangulates Roxb.

Pesticides

Triazophos (1 – 5 mg/L)	Hydroponic conditions	n.d.	Taken up in order of µg/(g$_{FW}$·d)	(Li et al. 2014b)

Table 3 contd. ...

...Table 3 contd.

Plant species	Compound (concentration, if available)	Assays conditions	% Removal (exposure time)	Observation of compound uptake by plant	Bibliographic reference
Scirpus validus (Softstem bulrush)					
	Flame retardants				
	Polybrominated diphenyl ethers (3 mg/kg$_{soil}$)	Spiked soil	n.d.	Taken up, accumulation/ metabolization (223.5 ± 19.6 ng/g in the roots) (60 d)	(Deng et al. 2016)
	Pesticides				
	Chlorpyrifos (1 mg/L)	Hydroponic conditions	88 (15 d)	Taken up (μg/g) and translocation	(Wang et al. 2013)
	Pharmaceuticals				
	Diclofenac (0.5 – 2 mg/L)	Hydroponic conditions	84 – 87 (7 d) > 98 (21 d)	Taken up in the order of μg/g$_{FW}$; root uptake: 0.47 – 4.94%; shoot uptake: 0.28 – 2.07%	(Zhang et al. 2012)
	Carbamazepine (0.5 – 2 mg/L)	Hydroponic conditions	74 (21 d)	In the order of μg/g; 22% Carb. < 5% naproxen	(Zhang et al. 2013b)
	Naproxen (0.5 – 2 mg/L)		98 (21 d)		
	Clofibric acid (0.5 – 2 mg/L)	Hydroponic conditions	41 ± 3 – 56 ± 2 (7 d) 65 ± 3 – 78 ± 1(21 d)	Taken up and translocation In roots: 5.4 – 26.8 μg/g$_{FW}$ (21 d) In shoots: 7.2 – 34.6 μg/ g$_{FW}$ (21 d)	(Zhang et al. 2013a)

Sparganium americanum (American bur-reed)

Pesticides

Compound	System	Result		Reference
Atrazine (2 ± 0.8 µg/L)	CW (mesocosms)	25 ± 16 (4 hr)	n.d.	(Moore et al. 2016)
Metolachlor (12 ± 0.4 µg/L)		32 ± 16 (4 hr)		
Diazinon (3.1 ± 0.2 µg/L)		39 ± 18 (4 hr)		

***Typha* spp.** (Cattails)

Explosives

Compound	System	Result		Reference
Exahydro-1,3,5-trinitro-1,3,5-triazine (RDX) (0 – 10 mg/L)	CW (mesocosms)	89 – 96 (2 d)	Taken up	(Low et al. 2008)

Pesticides

Compound	System	Result		Reference
Metalaxyl (0.91 mg/L) Simazine (0.242 mg/L)	Hydroponic conditions	n.d. (1,3,5 and 7 d)	Taken up	(Wilson et al. 2000)
Chlorpyrifos (1 mg/L)	Hydroponic conditions	88 (15 d)	Taken up (mg/kg) and translocation	(Wang et al. 2013)
Triazophos (1 – 5 µg/L)	Hydroponic conditions	n.d.	Taken up (µg/g fw.day) and translocation	(Li et al. 2014b)
Imazalil (10, 100 µg/L) Tebuconazole (10, 100 µg/L)	CW (mesocosms)	> 80 (5 d) > 60 (5 d)	Taken up	(Lv et al. 2016b)
Imazalil (10 mg/L) Tebuconazole (10 mg/L)	Hydroponic conditions	53 (24 d) 27 (24 d)	Taken up and metabolization	(Lv et al. 2016a)
Atrazine (2 ± 0.8 µg/L) Metolachlor (12 ± 0.4 µg/L) Diazinon (3.1 ± 0.2 µg/L)	CW (mesocosms)	53 ± 2 (4 hr) 51 ± 2 (4 hr) 51 ± 1 (4 hr)	n.d.	(Moore et al. 2016)

Table 3 contd. ...

...Table 3 contd.

Plant species	Compound (concentration, if available)	Assays conditions	% Removal (exposure time)	Observation of compound uptake by plant	Bibliographic reference
Pharmaceuticals and personal care products					
	Clofibric acid (20 µg/L)	Hydroponic conditions	50 (within 1 – 2 d) 80 (21 d)	n.d.	(Dordio et al. 2009)
	Carbamazepine (1 mg/L, WW)	CW (microcosms)	97 (7 d) Summer 88 (7 d) Winter	n.d.	(Dordio et al. 2010)
	Clofibric acid (1 mg/L, WW)		75 (7 d) Summer 48 (7 d) Winter		
	Ibuprofen (1mg/L, WW)		96 (7 d) Summer 82 (7 d) Winter		
	Ibuprofen (20 µg/L)	Hydroponic conditions	60 (within 1 d) 99 (21 d)	n.d.	(Dordio et al. 2011b)
	Carbamazepine (0.5 – 2 mg/L)	Hydroponic conditions	$28 \pm 2 - 52 \pm 3$ (7 d) $39 \pm 3 - 66 \pm 2$ (14 d) $56 \pm 2 - 82 \pm 2$ (21 d)	Taken up and translocation	(Dordio et al. 2011a)
	Diclofenac (1 mg/L)	Hydroponic conditions	n.d. (7 d)	Taken up (in the order of $\mu g/g_{FW}$) and rapid metabolization	(Bartha et al. 2014)
	Metformin (64.6, 129.2, 258.3 mg/L)	Hydroponic conditions	n.d.	Taken up and potential translocation	(Cui et al. 2015)
	Metformin (6.5, 19.4, 32.3 mg/L)	Hydroponic conditions	$74 \pm 4 - 81 \pm 3$ (28 d)	Taken up (in the order of $\mu mol/g_{FW}$), translocation to shoots was restricted	(Cui and Schröder 2016)
	Ibuprofen (107.2 µg/L)	CW (mesocosm)	n.d. (342 d)	Taken up and metabolization	(Li et al. 2016)

Ibuprofen (10 mg/L)	Hydroponic conditions	~ 100 (24 d)	Taken up and translocation	(Zhang et al. 2016)
Iohexol (10 mg/L)		15 (24 d)		
Zannichellia palustris (Horned pondweed)				
Pharmaceuticals				
Sulfathiazole (200 µg/L, WW)	Hydroponic conditions	35 ± 12 (20 d)	n.d.	(Garcia-Rodriguez et al. 2013)
Sulfamethazine (200 µg/L, WW)		8 ± 6 (20 d)		
Sulfamethoxazole (200 µg/L, WW)		20 ± 2 (20 d)		
Sulfapyridine (200 µg/L, WW)		45 ± 3 (20 d)		
Tetracycline (200 µg/L, WW)		83 ± 5 (20 d)	Mainly photodegradation	
Oxytetracycline (200 µg/L, WW)		91 ± 1 (20 d)		
Zizania aquatic (Texas wild rice)				
PAHs	CW (lab scale)	28 (continuous-flow) 38 (intermittent-flow)	n.d.	(Jie-Ting et al. 2015)

d: days; hr: hours; n.d. not detailed; FW: fresh mass; DW: dry mass; WW: wastewater; HRT: hydraulic residence time studies thus seem to be approached essentially as exploratory work, with the goal of performing plant species screening among a wider range of possibilities, while CWS studies serve the purpose of testing more mature plant selections under more realistic conditions.

Most of these studies clearly show that in general plants have a significant capacity to contribute to the removal of several OMPs from contaminated waters. However, because of different assay conditions, the same plant may show different removals even for the same pollutants. In general, removal of OMPs with plants is not only related to the basic characteristics of the plants and of the pollutants, but also depends on other external factors, such as plant arrangement, environmental conditions (temperature, humidity), the support matrix characteristics, the exposure time, etc. Temperature is one of the important parameters which affects the removal of OMPs by plants, thereby, introducing seasonal effects in the system's performance. Some studies show that a high temperature has a significant positive impact on the removal of certain organic pollutants, most possibly due to the increased rates of the biodegradation, volatilization, photodegradation and, indirectly, of plant uptake due to the effects on evapotranspiration rate.

Although the contribution of plant uptake is not always verified, the studies where such data is available provide an indication that this process may in fact be relevant in a large number of cases.

Interesting results can also be appreciated in regard to the biochemical processes occurring as result of the plants exposure to OMPs since most of these compounds are xenobiotic to plants and, therefore, it is not expected that plants have metabolic mechanisms suited to deal with those substances. However, in many cases it is possible to conclude that the metabolic machinery of many plant species is able to cope with the toxicity of OMPs and can either eliminate (mineralize or volatilize) them or stabilize and sequester them in their tissues. However, currently, studies focusing on the metabolism of OMPs in plants are still scarce (Bartha et al. 2010; Dordio et al. 2011a; Huber et al. 2012; Podlipná et al. 2013; Bartha et al. 2014; Pietrini et al. 2015; Cui and Schröder 2016; Huber et al. 2016; Wu et al. 2016) in spite of the great interest that such data represents for phytoremediation applications such as CWS.

Conclusions

Over the latest years, constructed wetland systems have been gaining increasing importance as cost-effective wastewater treatment systems that are alternative or complementary to conventional wastewater treatment plants. They are more commonly employed nowadays as a polishing stage for removing pollutants that can not be efficiently removed by conventional processes, such as the organic micropollutants. This is a chemically vast class of contaminants which raise serious concerns due to significant ecotoxicity and potential human health risks that have been attributed to many of these substances. Constructed wetlands systems, through the concerted action of plant species adapted to water saturated environments, microorganisms characteristic of these environments, and minerals of the support matrix, have shown considerable success in removing a variety of these pollutants from wastewaters in an already sizeable number of research studies that have been conducted so far.

As the contents of this chapter hopefully illustrates, aquatic macrophytes that are commonly used in constructed wetlands can play important roles that can contribute directly and indirectly to the stabilization, sequestration, or degradation of these pollutants in these wastewater treatment systems. In fact, several different types of mechanisms may be involved in the pollutant removal processes, some of which may be of abiotic type but the potentially most important ones being of biotic nature, due either from direct action of plants or from degradation by microorganisms, stimulated by the plants in the rhizosphere or endophytic to plant tissues. In order to achieve higher efficiencies, a thorough understanding of the variety of processes involved in the action of plants as well as the conditions that affect them the most is, therefore, a valuable tool for optimization of these systems. Some studies have already been conducted with this focus on understanding how constructed wetlands work at the deepest levels (thereby surpassing a "black-box" approach) and in which ways they may be optimized. Already a significant base of knowledge exists about constructed wetlands components and the removal processes for which they are responsible, but there is substantial amount of work that can and still needs to be carried out in this field. Work on these hot topics show the possibilities that are still open to the development of phytotechnologies, and applications of constructed wetlands in particular, as low cost solutions to emergent environmental problems. Notwithstanding, over the latest years many studies have already showcased some important advantages in using these technologies. However, investigation of the mechanisms involved in the removal of these substances from the contaminated media, once again, may prove essential to attain the required maturity for wider application of these systems on a larger scale with reasonable results. Still, there is plenty of room for further research, as is the case of the most advanced genetic engineering technology applied both to the vegetation as to the microbial components, but also in the more traditional research subjects such as the advantages of introducing polyculture vegetation or employing composite materials for substrates. An additional area to explore may be a detailed characterization of how constructed wetland components may interact to provide synergistic enhancements for pollutants removal.

Although most researchers concur on the idea that plants are generally beneficial for wastewater treatment in constructed wetlands, the practical planning and maintenance of plantations is still incipient as appropriate knowledge to guide these endeavors remains scarce. This review hopefully highlights the need for widening the scope of knowledge regarding the roles of plants in constructed wetlands, which may lead to improving their contribution to the wastewater treatment processes.

Acknowledgements

The authors would like to express their gratitude to all those who reviewed the contents of this chapter, especially to Cristina Ávila, for the careful and helpful suggestions that greatly contributed to its final improvement.

References Cited

A, D., D. Fujii, S. Soda, T. Machimura and M. Ike. 2017. Removal of phenol, bisphenol A, and 4-tert-butylphenol from synthetic landfill leachate by vertical flow constructed wetlands. Sci. Total Environ. 578: 566–576.

Allam, A., A. Tawfik, A. Negm, C. Yoshimura and A. Fleifle. 2015. Treatment of drainage water containing pharmaceuticals using duckweed (Lemna gibba). Energy Procedia 74: 973–980.

Arslan Topal, E.I. 2015. Uptake of tetracycline and metabolites in *Phragmites australis* exposed to treated poultry slaughterhouse wastewaters. Ecol. Eng. 83: 233–238.

Ávila, C. and J. García. 2015. Pharmaceuticals and personal care products (PPCPs) in the environment and their removal from wastewater through constructed wetlands. Comprehensive Analytical Chemistry 67: 195–244.

Ávila, C., C. Pelissari, P.H. Sezerino, M. Sgroi, P. Roccaro and J. García. 2017. Enhancement of total nitrogen removal through effluent recirculation and fate of PPCPs in a hybrid constructed wetland system treating urban wastewater. Sci. Total Environ. 584: 414–425.

Backhaus, T. 2016. Environmental risk assessment of pharmaceutical mixtures: demands, gaps, and possible bridges. The AAPS Journal 18: 804–813.

Bandowe, B.A.M. and H. Meusel. 2017. Nitrated polycyclic aromatic hydrocarbons (nitro-PAHs) in the environment—A review. Sci. Total Environ. 581-582: 237–257.

Bartha, B., C. Huber, R. Harpaintner and P. Schröder. 2010. Effects of acetaminophen in Brassica juncea L. Czern.: investigation of uptake, translocation, detoxification, and the induced defense pathways. Environ. Sci. Pollut. Res. 17: 1553–1562.

Bartha, B., C. Huber and P. Schröder. 2014. Uptake and metabolism of diclofenac in *Typha latifolia*—How plants cope with human pharmaceutical pollution. Plant Sci. 227: 12–20.

Blaha, L. and I. Holoubek. 2013. Emerging Issues in Ecotoxicology: Persistent Organic Pollutants (POPs). Springer Netherlands, Dordrecht, pp. 429–436.

Boethling, R., K. Fenner, P. Howard, G. Klecka, T. Madsen, J.R. Snape et al. 2009. Environmental persistence of organic pollutants: guidance for development and review of POP risk profiles. Integr. Environ. Assess. Manag. 5: 539–556.

Bolong, N., A.F. Ismail, M.R. Salim and T. Matsuura. 2009. A review of the effects of emerging contaminants in wastewater and options for their removal. Desalination 239: 229–246.

Bouldin, J.L., J.L. Farris, M.T. Moore, J. Smith and C.M. Cooper. 2006. Hydroponic uptake of atrazine and lambda-cyhalothrin in Juncus effusus and Ludwigia peploides. Chemosphere 65: 1049–1057.

Brix, H. 1997. Do macrophytes play a role in constructed treatment wetlands? Water Sci. Technol. 35: 11–17.

Bulak, P., A. Walkiewicz and M. Brzezinska. 2014. Plant growth regulators-assisted phytoextraction. Biol. Plant. 58: 1–8.

Burešová, H., S.J.H. Crum, J.D. Belgers, P.I. Adriaanse and G.H.P. Arts. 2013. Effects of linuron on a rooted aquatic macrophyte in sediment-dosed test systems. Environ. Pollut. 175: 117–124.

Carvalho, P.N., M.C. Basto and C.M. Almeida. 2012. Potential of *Phragmites australis* for the removal of veterinary pharmaceuticals from aquatic media. Bioresour. Technol. 116: 497–501.

Chen, Z., Y. Chen, J. Vymazal, L. Kule and M. Koželuh. 2017. Dynamics of chloroacetanilide herbicides in various types of mesocosm wetlands. Sci. Total Environ. 577: 386–394.

Christofilopoulos, S., E. Syranidou, G. Gkavrou, E. Manousaki and N. Kalogerakis. 2016. The role of halophyte Juncus acutus L. in the remediation of mixed contamination in a hydroponic greenhouse experiment. J. Chem. Technol. Biotechnol. 91: 1665–1674.

Cizmas, L., V.K. Sharma, C.M. Gray and T.J. McDonald. 2015. Pharmaceuticals and personal care products in waters: occurrence, toxicity, and risk. Environmental Chemistry Letters 13: 381–394.

Clarke, R.M. and E. Cummins. 2015. Evaluation of "Classic" and emerging contaminants resulting from the application of biosolids to agricultural lands: A review. Hum. Ecol. Risk Assess. 21: 492–513.

Cui, H., B.A. Hense, J. Müller and P. Schröder. 2015. Short term uptake and transport process for metformin in roots of *Phragmites australis* and *Typha latifolia*. Chemosphere 134: 307–312.

Cui, H. and P. Schröder. 2016. Uptake, translocation and possible biodegradation of the antidiabetic agent metformin by hydroponically grown *Typha latifolia*. J. Hazard. Mater. 308: 355–361.

Datta, R., P. Das, S. Smith, P. Punamiya, D.M. Ramanathan, R. Reddy et al. 2013. Phytoremediation Potential of Vetiver Grass [Chrysopogon Zizanioides (L.)] for Tetracycline. Int. J. Phytorem. 15: 343–351.

Davies, L.C., C.C. Carias, J.M. Novais and S. Martins-Dias. 2005. Phytoremediation of textile effluents containing azo dye by using *Phragmites australis* in a vertical flow intermittent feeding constructed wetland. Ecol. Eng. 25: 594–605.

Deng, D., J. Liu, M. Xu, G. Zheng, J. Guo and G. Sun. 2016. Uptake, translocation and metabolism of decabromodiphenyl ether (BDE-209) in seven aquatic plants. Chemosphere 152: 360–368.

Dhir, B., P. Sharmila and P.P. Saradhi. 2009. Potential of aquatic macrophytes for removing contaminants from the environment. Crit. Rev. Environ. Sci. Technol. 39: 754–781.

Dordio, A. and A.J.P. Carvalho. 2015. Removal of pharmaceuticals in conventional wastewater treatment plants. pp. 1–44. *In*: Barret, L.M. [ed.]. Wastewater Treatment: Processes, Management Strategies and Environmental/Health Impacts. Nova Science Publishers, Hauppauge, NY, USA.

Dordio, A. and A.J.P. Carvalho. 2011. Phytoremediation: An option for removal of organic xenobiotics from water. pp. 51–92. *In*: Golubev, I.A. [ed.]. Handbook of Phytoremediation. Nova Science Publishers, Hauppauge, NY, USA.

Dordio, A.V., M. Belo, D.M. Teixeira, A.J.P. Carvalho, C.M.B. Dias, Y. Picó et al. 2011a. Evaluation of carbamazepine uptake and metabolization by *Typha* spp., a plant with potential use in phytotreatment. Bioresour. Technol. 102: 7827–7834.

Dordio, A.V. and A.J.P. Carvalho. 2013. Organic xenobiotics removal in constructed wetlands, with emphasis on the importance of the support matrix. J. Hazard. Mater. 252-253: 272–292.

Dordio, A., A.J.P. Carvalho, D.M. Teixeira, C.B. Dias and A.P. Pinto. 2010. Removal of pharmaceuticals in microcosm constructed wetlands using *Typha* spp. and LECA. Bioresour. Technol. 101: 886–892.

Dordio, A., R. Ferro, D. Teixeira, A.J.P. Carvalho, A.P. Pinto and C.M.B. Dias. 2011b. Study on the use of *Typha* spp. for the phytotreatment of water contaminated with ibuprofen. Intern. J. Environ. Anal. Chem. 91: 654–667.

Dordio, A.V., C. Duarte, M. Barreiros, A.J.P. Carvalho, A.P. Pinto and C.T. da Costa. 2009. Toxicity and removal efficiency of pharmaceutical metabolite clofibric acid by *Typha* spp.—Potential use for phytoremediation? Bioresour. Technol. 100: 1156–1161.

Du, L.F. and W.K. Liu. 2012. Occurrence, fate, and ecotoxicity of antibiotics in agro-ecosystems. A review. Agron. Sustain. Dev. 32: 309–327.

Ebele, A.J., M. Abou-Elwafa Abdallah and S. Harrad. 2017. Pharmaceuticals and personal care products (PPCPs) in the freshwater aquatic environment. Emerging Contaminants.

El Shahawi, M.S., A. Hamza, A.S. Bashammakh and W.T. Al Saggaf. 2010. An overview on the accumulation, distribution, transformations, toxicity and analytical methods for the monitoring of persistent organic pollutants. Talanta 80: 1587–1597.

Fent, K., A.A. Weston and D. Caminada. 2006. Ecotoxicology of human pharmaceuticals. Aquat. Toxicol. 76: 122–159.

Ferro, S., A.R. Trentin, S. Caffieri and R. Ghisi. 2010. Antibacterial sulfonamides: Accumulation and effects in Barley plants. Fresenius Environ. Bull. 19: 2094–2099.

Forni, C., A. Cascone, M. Fiori and L. Migliore. 2002. Sulphadimethoxine and Azolla filiculoides Lam.: a model for drug remediation. Water Res. 36: 3398–3403.

Fujisawa, T., M. Kurosawa and T. Katagi. 2006. Uptake and transformation of pesticide metabolites by Duckweed (Lemna gibba). J. Agric. Food Chem. 54: 6286–6293.

García, J., D.P.L. Rousseau, J. Morató, E. Lesage, V. Matamoros and J.M. Bayona. 2010. Contaminant removal processes in subsurface-flow constructed wetlands: A review. Crit. Rev. Environ. Sci. Technol. 40: 561–661.

Garcia-Rodríguez, A., V. Matamoros, C. Fontàs and V. Salvadò. 2013. The influence of light exposure, water quality and vegetation on the removal of sulfonamides and tetracyclines: A laboratory-scale study. Chemosphere 90: 2297–2302.

Garcinuño, R.M., P. Fernández Hernando and C. Cámara. 2006. Removal of Carbaryl, Linuron, and Permethrin by Lupinus angustifolius under hydroponic conditions. J. Agric. Food Chem. 54: 5034–5039.

Garfí, M., A. Pedescoll, E. Bécares, M. Hijosa-Valsero, R. Sidrach-Cardona and J. García. 2012. Effect of climatic conditions, season and wastewater quality on contaminant removal efficiency of two experimental constructed wetlands in different regions of Spain. Sci. Total Environ. 437: 61–67.

Gavrilescu, M., K. Demnerová, J. Aamand, S. Agathos and F. Fava. 2015. Emerging pollutants in the environment: present and future challenges in biomonitoring, ecological risks and bioremediation. New Biotech. 32: 147–156.

Gomez-Hermosillo, C., J.H. Pardue and D.D. Reible. 2006. Wetland plant uptake of desorption-resistant organic compounds from sediments. Environ. Sci. Technol. 40: 3229–3236.

Gothwal, R. and T. Shashidhar. 2015. Antibiotic pollution in the environment: A review. Clean Soil Air Water 43: 479–489.

Greaves, A.K. and R.J. Letcher. 2017. A review of organophosphate esters in the environment from biological effects to distribution and fate. Bull. Environ. Contam. Toxicol. 98: 2–7.

Grove, J.K. and O.R. Stein. 2005. Polar organic solvent removal in microcosm constructed wetlands. Water Res. 39: 4040–4050.

Gujarathi, N.P., B.J. Haney and J.C. Linden. 2005. Phytoremediation potential of myriophyllum aquaticum and pistia stratiotes to modify antibiotic growth promoters, tetracycline, and oxytetracycline, in aqueous wastewater systems. Int. J. Phytoremediat. 7: 99–112.

Haberl, R., S. Grego, G. Langergraber, R.H. Kadlec, A.R. Cicalini, S. Martins-Dias et al. 2003. Constructed wetlands for the treatment of organic pollutants. J. Soil Sediment. 3: 109–124.

Haman, C., X. Dauchy, C. Rosin and J.F. Munoz. 2015. Occurrence, fate and behavior of parabens in aquatic environments: A review. Water Res. 68: 1–11.

Hechmi, N., N. Ben Aissa, H. Abdenaceur and N. Jedidi. 2015. Uptake and bioaccumulation of pentachlorophenol by emergent wetland plant *Phragmites australis* (Common Reed) in cadmium co-contaminated soil. Int. J. Phytorem. 17: 109–116.

Hijosa-Valsero, M., V. Matamoros, R. Sidrach-Cardona, A. Pedescoll, J. Martín-Villacorta, J. García et al. 2011. Influence of design, physico-chemical and environmental parameters on pharmaceuticals and fragrances removal by constructed wetlands. Water Sci. Technol. 63: 2527.

Hijosa-Valsero, M., C. Reyes-Contreras, C. Domínguez, E. Bécares and J.M. Bayona. 2016. Behaviour of pharmaceuticals and personal care products in constructed wetland compartments: Influent, effluent, pore water, substrate and plant roots. Chemosphere 145: 508–517.

Huber, C., B. Bartha and P. Schröder. 2012. Metabolism of diclofenac in plants—Hydroxylation is followed by glucose conjugation. J. Hazard. Mater. 243: 250–256.

Huber, C., M. Preis, P.J. Harvey, S. Grosse, T. Letzel and P. Schröder. 2016. Emerging pollutants and plants—Metabolic activation of diclofenac by peroxidases. Chemosphere 146: 435–441.

Hussein, A. and M. Scholz. 2017. Dye wastewater treatment by vertical-flow constructed wetlands. Ecol. Eng. 101: 28–38.

Imfeld, G., M. Braeckevelt, P. Kuschk and H.H. Richnow. 2009. Monitoring and assessing processes of organic chemicals removal in constructed wetlands. Chemosphere 74: 349–362.

Iori, V., F. Pietrini and M. Zacchini. 2012. Assessment of ibuprofen tolerance and removal capability in Populus nigra L. by *in vitro* culture. J. Hazard. Mater. 229-230: 217–223.

Iori, V., M. Zacchini and F. Pietrini. 2013. Growth, physiological response and phytoremoval capability of two willow clones exposed to ibuprofen under hydroponic culture. J. Hazard. Mater. 262: 796–804.

Jie-Ting, Q.U., L.U. Shao-Yong, W.U. Xue-Yan, L. Ke, X. Wei and C. Fang-Xin. 2015. Impact of hydraulic loading on removal of polycyclic aromatic hydrocarbons (PAHs) from vertical-flow wetland. Toxicol. Environ. Chem. 97: 388–401.

Jones, K.C. and P. de Voogt. 1999. Persistent organic pollutants (POPs): state of the science. Environ. Pollut. 100: 209–221.

Kadlec, R.H. and S.D. Wallace. 2009. Treatment Wetlands. CRC Press, Boca Raton.

Kotyza, J., P. Soudek, Z. Kafka and T. Vanek. 2010. Phytoremediation of pharmaceuticals—preliminary study. Int. J. Phytoremediat. 12: 306–316.

Kümmerer, K. 2011. Emerging contaminants. pp. 69–87. *In*: Wilderer, P. [ed.]. Treatise on Water Science. Elsevier, Oxford.

Lapworth, D.J., N. Baran, M.E. Stuart and R.S. Ward. 2012. Emerging organic contaminants in groundwater: A review of sources, fate and occurrence. Environ. Pollut. 163: 287–303.

Li, W.C. 2014. Occurrence, sources, and fate of pharmaceuticals in aquatic environment and soil. Environ. Pollut. 187: 193–201.

Li, Y.F., G.B. Zhu, W.J. Ng and S.K. Tan. 2014a. A review on removing pharmaceutical contaminants from wastewater by constructed wetlands: Design, performance and mechanism. Sci. Total Environ. 468: 908–932.

Li, Y., J. Zhang, G. Zhu, Y. Liu, B. Wu, W.J. Ng et al. 2016. Phytoextraction, phytotransformation and rhizodegradation of ibuprofen associated with Typha angustifolia in a horizontal subsurface flow constructed wetland. Water Res. 102: 294–304.

Li, Z., H. Xiao, S. Cheng, L. Zhang, X. Xie and Z. Wu. 2014b. A comparison on the phytoremediation ability of triazophos by different macrophytes. J. Environ. Sci. 26: 315–322.

Lin, Y.L. and B.K. Li. 2016. Removal of pharmaceuticals and personal care products by Eichhornia crassipe and Pistia stratiotes. J. Taiwan Inst. Chem. Eng. 58: 318–323.

Liu, L., Y.h. Liu, C.x. Liu, Z. Wang, J. Dong, G.f. Zhu et al. 2013. Potential effect and accumulation of veterinary antibiotics in *Phragmites australis* under hydroponic conditions. Ecol. Eng. 53: 138–143.

Low, D., K. Tan, T. Anderson, G.P. Cobb, J. Liu and W.A. Jackson. 2008. Treatment of RDX using down-flow constructed wetland mesocosms. Ecol. Eng. 32: 72–80.

Luo, Y., W. Guo, H.H. Ngo, L.D. Nghiem, F.I. Hai, J. Zhang et al. 2014. A review on the occurrence of micropollutants in the aquatic environment and their fate and removal during wastewater treatment. Sci. Total Environ. 473-474: 619–641.

Lv, T., Y. Zhang, M.n.E. Casas, P.N. Carvalho, C.A. Arias, K. Bester et al. 2016a. Phytoremediation of imazalil and tebuconazole by four emergent wetland plant species in hydroponic medium. Chemosphere 148: 459–466.

Lv, T., Y. Zhang, L. Zhang, P.N. Carvalho, C.A. Arias and H. Brix. 2016b. Removal of the pesticides imazalil and tebuconazole in saturated constructed wetland mesocosms. Water Res. 91: 126–136.

Mackulak, T., M. Mosný, J. Škubák, R. Grabic and L. Birošová. 2015. Fate of psychoactive compounds in wastewater treatment plant and the possibility of their degradation using aquatic plants. Environ. Toxicol. Pharmacol. 39: 969–973.

Matamoros, V., L.X. Nguyen, C.A. Arias, V. Salvadó and H. Brix. 2012. Evaluation of aquatic plants for removing polar microcontaminants: A microcosm experiment. Chemosphere 88: 1257–1264.

Matthies, M., K. Solomon, M. Vighi, A. Gilman and J.V. Tarazona. 2016. The origin and evolution of assessment criteria for persistent, bioaccumulative and toxic (PBT) chemicals and persistent organic pollutants (POPs). Environ. Sci. Processes Impacts 18: 1114–1128.

Michelini, L., R. Reichel, W. Werner, R. Ghisi and S. Thiele-Bruhn. 2012. Sulfadiazine uptake and Effects on Salix fragilis L. and Zea mays L. Plants. Water Air Soil Pollut. 223: 5243–5257.

Migliore, L., S. Cozzolino and M. Fiori. 2000. Phytotoxicity to and uptake of flumequine used in intensive aquaculture on the aquatic weed, Lythrum salicaria L. Chemosphere 40: 741–750.

Moore, M.T., M.A. Locke and R. Kröger. 2016. Mitigation of atrazine, S-metolachlor, and diazinon using common emergent aquatic vegetation. J. Environ. Sci. (in press).

Nesterenko-Malkovskaya, A., F. Kirzhner, Y. Zimmels and R. Armon. 2012. Eichhornia crassipes capability to remove naphthalene from wastewater in the absence of bacteria. Chemosphere 87: 1186–1191.

Olette, R., M. Couderchet, S. Biagianti and P. Eullaffroy. 2008. Toxicity and removal of pesticides by selected aquatic plants. Chemosphere 70: 1414–1421.

Pal, A., K.Y.-H. Gin, A.Y.-C. Lin and M. Reinhard. 2010. Impacts of emerging organic contaminants on freshwater resources: Review of recent occurrences, sources, fate and effects. Sci. Total Environ. 408: 6062–6069.

Pietrini, F., D. Di Baccio, J. Acena, S. Pérez, D. Barceló and M. Zacchini. 2015. Ibuprofen exposure in Lemna gibba L.: Evaluation of growth and phytotoxic indicators, detection of ibuprofen and identification of its metabolites in plant and in the medium. J. Hazard. Mater. 300: 189–193.

Pilon-Smits, E. 2005. Phytoremediation. Annu. Rev. Plant Biol. 56: 15–39.

Podlipná, R., L. Skálová, H. Seidlová, B. Szotáková, V. Kubícek, L. Stuchláková et al. 2013. Biotransformation of benzimidazole anthelmintics in reed (*Phragmites australis*) as a potential tool for their detoxification in environment. Bioresour. Technol. 144: 216–224.

Reddy, K.R. and R.D. DeLaune. 2008. Toxic organic compounds. pp. 507–536. Biogeochemistry of Wetlands: Science and Applications. CRC Press, Boca Raton.

Reinhold, D., S. Vishwanathan, J.J. Park, D. Oh and F. Michael Saunders. 2010. Assessment of plant-driven removal of emerging organic pollutants by duckweed. Chemosphere 80: 687–692.

Reyes-Contreras, C., M. Hijosa-Valsero, R. Sidrach-Cardona, J.M. Bayona and E. Bécares. 2012. Temporal evolution in PPCP removal from urban wastewater by constructed wetlands of different configuration: A medium-term study. Chemosphere 88: 161–167.

San Miguel, A., P. Ravanel and M. Raveton. 2013. A comparative study on the uptake and translocation of organochlorines by *Phragmites australis*. J. Hazard. Mater. 244-245: 60–69.

Sauvetre, A. and P. Schrüder. 2015. Uptake of Carbamazepine by rhizomes and endophytic bacteria of *Phragmites australis*. Front. Plant Sci. 6.

Shelef, O., A. Gross and S. Rachmilevitch. 2013. Role of plants in a constructed wetland: current and new perspectives. Water 5: 405.

Stottmeister, U., A. Wiessner, P. Kuschk, U. Kappelmeyer, M. Kastner, O. Bederski et al. 2003. Effects of plants and microorganisms in constructed wetlands for wastewater treatment. Biotechnol. Adv. 22: 93–117.

Susarla, S., V.F. Medina and S.C. McCutcheon. 2002. Phytoremediation: An ecological solution to organic chemical contamination. Ecol. Eng. 18: 647–658.

Topal, M., G. Uslu Senel, E. Öbek and E.I. Arslan Topal. 2015. Removal of tetracycline and the degradation products by Lemna gibba L. exposed to secondary effluents. Environ. Prog. Sustainable Energy 34: 1311–1321.

Toro-Vélez, A.F., C.A. Madera-Parra, M.R. Peña-Varón, W.Y. Lee, J.C. Cruz, W.S. Walker et al. 2016. BPA and NP removal from municipal wastewater by tropical horizontal subsurface constructed wetlands. Sci. Total Environ. 542, Part A: 93–101.

Truu, J., M. Truu, M. Espenberg, H. Nõlvak and J. Juhanson. 2015. Phytoremediation and plant-assisted bioremediation in soil and treatment wetlands: A review. Open Biotechnol. J. 9: 85–92.

van Mourik, L.M., C. Gaus, P.E.G. Leonards and J. de Boer. 2016. Chlorinated paraffins in the environment: A review on their production, fate, levels and trends between 2010 and 2015. Chemosphere 155: 415–428.

Verlicchi, P. and E. Zambello. 2014. How efficient are constructed wetlands in removing pharmaceuticals from untreated and treated urban wastewaters? A review. Sci. Total Environ. 470-471: 1281–1306.

Vymazal, J. and T. Brezinová. 2015. The use of constructed wetlands for removal of pesticides from agricultural runoff and drainage: A review. Environ. Int. 75: 11–20.

Vymazal, J., H. Brix, P.F. Cooper, M.B. Green and R. Haberl. 1998. Constructed wetlands for wastewater treatment in Europe. Backhuys Publishers, Leiden, The Netherlands.

Vymazal, J. and L. Kröpfelová. 2009. Removal of organics in constructed wetlands with horizontal sub-surface flow: A review of the field experience. Sci. Total Environ. 407: 3911–3922.

Wang, C., S. Zheng, P. Wang and J. Qian. 2014. Effects of vegetations on the removal of contaminants in aquatic environments: A review. J. Hydrodyn. B 26: 497–511.

Wang, C., Q. Zhou, L. Zhang, Y. Zhang, E. Xiao and Z. Wu. 2013. Variation characteristics of chlorpyrifos in nonsterile wetland plant hydroponic system. Int. J. Phytorem. 15: 550–560.

Wang, Q., C. Li, R. Zheng and X. Que. 2016. Phytoremediation of chlorpyrifos in aqueous system by riverine macrophyte, Acorus calamus: toxicity and removal rate. Environ. Sci. Pollut. Res. 23: 16241–16248.

Wang, Q., W. Zhang, C. Li and B. Xiao. 2012. Phytoremediation of atrazine by three emergent hydrophytes in a hydroponic system. Water Sci. Technol. 66: 1282.

Wetzel, R.G. 2001. Limnology: lake and river ecosystems. Academic Press, San Diego, CA.

Wilson, P.C., T. Whitwell and S.J. Klaine. 2000. Metalaxyl and Simazine toxicity to and uptake by Typha latifolia. Arch. Environ. Contam. Toxicol. 39: 282–288.

Wu, X., Q. Fu and J. Gan. 2016. Metabolism of pharmaceutical and personal care products by carrot cell cultures. Environ. Pollut. 211: 141–147.

Yan, Q., G. Feng, X. Gao, C. Sun, J.s. Guo and Z. Zhu. 2016. Removal of pharmaceutically active compounds (PhACs) and toxicological response of Cyperus alternifolius exposed to PhACs in microcosm constructed wetlands. J. Hazard. Mater. 301: 566–575.

Zhai, J., M. Rahaman, J. Ji, Z. Luo, Q. Wang, H. Xiao et al. 2016. Plant uptake of diclofenac in a mesocosm-scale free water surface constructed wetland by Cyperus alternlifolius. Water Sci. Technol.

Zhang, D.Q., R.M. Gersberg, W.J. Ng and S.K. Tan. 2014. Removal of pharmaceuticals and personal care products in aquatic plant-based systems: A review. Environ. Pollut. 184: 620–639.

Zhang, D.Q., R.M. Gersberg, T. Hua, J. Zhu, W.J. Ng and S.K. Tan. 2013a. Assessment of plant-driven uptake and translocation of clofibric acid by Scirpus validus. Environ. Sci. Pollut. Res. 20: 4612–4620.

Zhang, D.Q., T. Hua, R.M. Gersberg, J. Zhu, W.J. Ng and S.K. Tan. 2012. Fate of diclofenac in wetland mesocosms planted with Scirpus validus. Ecol. Eng. 49: 59–64.

Zhang, D.Q., T. Hua, R.M. Gersberg, J. Zhu, W.J. Ng and S.K. Tan. 2013b. Carbamazepine and naproxen: Fate in wetland mesocosms planted with Scirpus validus. Chemosphere 91: 14–21.

Zhang, Y., T. Lv, P.N. Carvalho, C.A. Arias, Z. Chen and H. Brix. 2016. Removal of the pharmaceuticals ibuprofen and iohexol by four wetland plant species in hydroponic culture: plant uptake and microbial degradation. Environ. Sci. Pollut. Res. 23: 2890–2898.

Zhang, Z., Z. Rengel and K. Meney. 2010. Polynuclear aromatic hydrocarbons (PAHs) differentially influence growth of various emergent wetland species. J. Hazard. Mater. 182: 689–695.

Zoller, U. and M. Hushan. 2010. Synergistic ecotoxicity of APEOs-PAHs in rivers and sediments: is there a potential health risk? Rev. Environ. Health 25: 351–357.

17

Influence of Design and Operation Parameters in the Organic Load and Nutrient Removal in Constructed Wetlands

Jason A. Hale[1,2,]* and *Joan García*[2]

INTRODUCTION

An artificial or constructed wetland is an engineered system that removes contaminants from wastewater by taking advantage of the physical, chemical, and biological conditions which characterize natural wetlands. More than fifty years of research, design, operation, and monitoring have proven that this can be a low maintenance and low operating cost waste water treatment technology (Vymazal 2011).

The net removal of contaminants is the sum of many processes, including sedimentation, filtration, precipitation, volatilization, adsorption, plant uptake, and various microbial processes influencing carbon, nitrogen, and phosphorus cycles, among others (Vymazal 2007; Kadlec and Wallace 2009; Faulwetter et al. 2009; Wu et al. 2014). Most constructed wetlands are built to treat organic matter and nutrients. Within this context, the purpose of this paper is to present and discuss the principal processes, constraints, and design and operation parameters which control organic load and nutrient removal in the basic types of constructed wetlands. The goal is to contribute to the understanding of constructed wetlands as complex systems which

[1] Pandion Saudia Co. Ltd. P.O. Box 1751 Al Khobar 31952 Kingdom of Saudi Arabia.
[2] GEMMA-Environmental Engineering and Microbiology Group, Department of Civil and Environmental Engineering, Universitat Politècnica de Catalunya-Barcelona Tech, c/Jordi Girona 1-3, Building D1, E-08034 Barcelona, Spain.
  Email: joan.garcia@upc.edu
* Corresponding author: jhale@pandiontech.com

nonetheless behave in generally predictable ways when managed and operated within the limits of their design.

The first two sections of this chapter will introduce the basic features of constructed wetlands, and summarize forms of organic matter and nutrients, and the processes affecting them.

The rest of the chapter will expand the discussion of constructed wetlands as a wastewater treatment technology by describing how different design parameters and alternative operating activities reduce concentrations of organic matter and nutrients from wastewater.

Basic Features of Constructed Wetlands

While early development of constructed wetland technology for treatment of wastewater dates to the early 1900s, the period of modern systematic research and development of the technology began in the 1950s with research into different plant species for remediation of polluted surface water (Vymazal 2008). In the decades that followed research and implementation of the technology incorporated a wide range of plants, media and planting substrate, and directions of flow to optimize removal of contaminants from different wastewater streams around the world (Vymazal 2011).

Constructed wetlands combine hydrologic, substrate, and vegetation variables across many thousands of implemented full-scale projects (e.g., Fonder and Headley 2013; Vymazal 2014). For a given wastewater treatment need, a design is selected and specified based on a number of criteria, including wastewater composition, quantity, regulatory requirements, and available physical space (Kadlec and Wallace 2009).

A convenient starting point for conceptualizing constructed wetland design is to consider hydrology and the flow path of wastewater. Free-water surface (FWS) wetlands designs are characterized by wastewater which is exposed to the atmosphere.

Pre-treated influent enters at one end of a gently sloped basin (lined) and flows horizontally through a planted basin. The wastewater stays on top of the substrate, and is exposed to the atmosphere along the entire length of the system. In contrast, treatment processes in sub-surface flow (SSF) wetlands designs occur as wastewater passes through substrate, for example, sand or gravel, rather than on top of it.

This single distinction yields two basic differences between FWS and SSF designs. First, while many of the waste treatment processes in constructed wetlands are dominated by microbial activity, SSF designs provide a much higher surface area for microbial colonization on substrate grains than FWS, where only the basin sides, substrate surface, and macrophyte (plant) stems are available for colonization by microbes. Second, exposure of FWS wastewater stream to the atmosphere promotes gas exchange, and allows for a higher and more consistent supply of oxygen. These differences lead to significant implications in their capacities for specific wastewater treatment applications. For example, aerobic conditions are more common in FWS, and most organic matter is oxidized by microbes that use oxygen as an electron acceptor (aerobic treatment). In SSF, where anoxic conditions are more common, organic matter may be oxidized by anaerobic organisms that can use nitrate (denitrification) or sulfate as an electron acceptor in addition to oxygen.

There are two broad types of sub-surface constructed wetland designs, distinguished by the direction of flow. Wastewater in "horizontal" sub-surface flow

(HSSF) wetlands flows horizontally down a gentle gradient; again, unlike FWS systems, the wastewater stream in HSSF flows below the substrate surface, and is not exposed to the atmosphere after leaving the inlet zone. Under normal operation, a continuous supply of wastewater to HSSF wetlands floods substrate pore spaces, maintaining generally anoxic conditions. These conditions favor certain processes relevant to control of water pollution (e.g., denitrification, anaerobic metabolism), while others are suppressed or restricted. In contrast to these, water flow through "vertical" sub-surface flow (VSSF) wetlands is usually intermittent, or "pulsed". This practice allows water to drain from pore spaces, which may fully or partially re-aerate between flooding events, depending on the specific operation activity.

From the earliest research into constructed wetlands, the practice of combining several single constructed wetlands into a "hybrid" system has been recognized as an important tool to improve overall pollutant removal efficiency. In such systems, treated discharge from one system supplies influent to the next. Parameters such as the type, order, number, and size of individual units in a hybrid system are determined by the project needs.

The roles macrophytes play in the treatment process of constructed wetlands have been studied since the 1950s (Vymazal 2011), and quantifying these roles remains an important topic today. Macrophytes serve different functions in different constructed wetlands. Brix (1994) gave priority to the physical effects of macrophytes on constructed wetland treatment efficiency, while a survey by Vymazal (2013a) notes that some relationships between macrophyte taxa and treatment efficiency have been found. A number of studies have investigated the potential for removal of pollutants by removal (harvest) of macrophyte biomass. Kadlec and Wallace (2009) note that reviews of the effectiveness and efficiency of harvesting are mixed, though most studies of the effects of macrophytes in SSF constructed wetlands recognize that the influence of macrophytes increases in cases of lower nutrient loads.

Given the basic hydrology, flow-path direction, and roles of macrophytes as a starting point, other critical constructed wetland design criteria include:

- Treatment goals, including nutrient or organic matter reduction; heavy metal removal; primary, secondary or tertiary treatment.
- Hydraulic residence time (HRT), conceptualized as the ratio of constructed wetland volume and flow rate (Zahraeifard and Deng 2011).
- Choice of substrate media (for SSF), including physical size and shape, chemical composition, and capacity for chemical sorption (e.g., Brix et al. 2001; García et al. 2004a; Albuquerque et al. 2009).
- Physical dimensions of the system, particularly the depth of flooding and length of flowpath surface area (for FWS), and cross-sectional area (for SSF) (García et al. 2005; Kadlec and Wallace 2009).
- While not considered design criteria *per se*, local conditions (e.g., physical setting, meteorological environment) also play a role in constructed wetland design and operation (Kadlec and Wallace 2009).

Failure to appreciate the importance of these criteria, as well as the interactions between them, can lead to serious challenges in the operation of constructed wetlands for treatment of wastewater.

Organic Matter, Nutrients, and Transformation Processes

Biodegradable organic matter (often represented as biochemical oxygen demand, BOD), nitrogen, and phosphorus are common wastewater contaminants which must be addressed by treatment systems before re-use or discharge back to the environment. The purpose of this section is to provide an overview of these contaminants, and the processes which control their abundance in constructed wetlands.

While many physical processes act and interact to reduce and remove pollutants from wastewater, many of the nutrient and organic matter removal effects of constructed wetlands are dominated or influenced by microbial activity (Truu et al. 2009; Ligi et al. 2015), and research continues to quantify the effects of design and operations on the composition and activity of microbial communities in constructed wetlands (e.g., Adrados et al. 2014; Button et al. 2015; Murphy et al. 2016).

Organic Matter

"Organic matter" in the context of wastewater treatment wetlands refers to a complex set of organic compounds. Particulate organic matter, or POM, is defined as material which is large enough to be retained on 0.45 mm filter (Kadlec and Wallace 2009), and may include:

- Biomass of living or dead microbes;
- Extracellular polymeric substances (EPS), also known as biofilm, and sludge;
- Above and belowground biomass of macrophytes, and litter.

Influent POM is mainly retained by physical processes such as filtration and sedimentation. These particles accumulate and disintegrate (hydrolysis), producing dissolved organic compounds that can be degraded by different metabolic pathways, thus removing a portion of influent organic matter from constructed wetland as CO_2 (García et al. 2010).

The main mechanisms for dissolved organic matter removal from wastewater streams are off-gassing of metabolic products such as CO_2 and CH_4. The relative importance of these mechanisms for removing organic matter depends primarily on the redox conditions (García et al. 2004b; García et al. 2005). The main biological processes linked to dissolved organic matter transformation and fate are described below.

Aerobic transformation consumes organic matter in the presence of molecular oxygen producing carbon dioxide, water, and energy. In systems where oxygen is abundant, aerobic metabolism to carbon dioxide and subsequent off-gassing is the primary pathway for removal of dissolved organic matter from a constructed wetland system. Aerobic processes play a major role in contaminant removal from free water surface FWS wetlands and vertical flow VF wetlands systems.

As oxygen becomes less abundant, other metabolic pathways including denitrification and other anaerobic transformations become more common. Denitrification requires nitrate for metabolism of organic matter. The process is inhibited in the presence of oxygen and produces less energy than aerobic treatment;

but in cases where nitrate is available and oxygen is scarce (e.g., anoxic conditions), it is a common pathway for removal of organic matter from wastewater.

As the absence of oxygen becomes more extreme, sulfate reduction and methanogenesis (reduction of carbon to methane) are other metabolic pathways which remove organic matter. Both pathways occur in strictly anaerobic conditions of approximately equal Eh (reduction potential). The abundance of organic carbon to sulfur may determine which pathway is most prevalent (García et al. 2010). Facultative and strictly anaerobic processes are more prevalent in SSF than FWS.

For the purpose of wastewater treatment, biochemical oxygen demand (BOD) is used to describe the amount of oxygen required to support aerobic metabolism of readily degradable organic matter (mainly carbon) in wastewater. Historically, the test was known as BOD_5, the subscript representing a 5-day test though Jouanneau et al. (2014) describe more robust and rapid methods to determine BOD. Because nitrification also consumes oxygen, a method is needed to inhibit this process so that only the oxygen used during metabolism is accounted for. Thus, in conditions of high nitrification, the value of carbonaceous biochemical oxygen demand (CBOD) will be lower than BOD, and provide a complementary measure of labile organic matter. Chemical oxygen demand (COD) represents the sum of all carbonaceous material, and includes labile and recalcitrant organic matter. The relationship between BOD and COD is an important consideration for constructed wetland design and operation as it helps distinguish between organic compounds that may be easily degraded versus those that require longer time or more aggressive conditions (Kadlec and Wallace 2009).

Nitrogen

This section discusses the processes which control the relative concentration and abundance of nitrogen as a nutrient. This section also addresses the controls and constraints on these processes which can determine constructed wetland treatment efficiency.

Organic nitrogen, ammonia, and ammonium ion (NH_4^+ and NH_3), nitrate (NO^{3-}), and nitrite (NO^{2-}) are usually considered be the principal forms of nitrogen involved in the nitrogen cycle in constructed wetlands. Most of the biologically relevant nitrogen contained in domestic and municipal wastewater is present as ammonia and organic nitrogen (Kadlec and Wallace 2009), and the dominant form often has a substantial impact on constructed wetland design and operation.

Ammonification is the production of ammonia-nitrogen (the sum of ammonia NH_3 and ammonium ion NH_4^+) by microbial degradation of particulate organic nitrogen, urea, and uric acid. Ammonification occurs both aerobically and anaerobically, and the proportion of ammonia to ammonium ion in wastewater is a function of pH and temperature; ammonium ion is usually the dominant form at pH values typical of constructed wetlands (Kadlec and Wallace 2009). Ammonia volatilization may be a pathway of total nitrogen reduction from wastewater in conditions of higher pH; however, the principal pathway for removal of nitrogen from constructed wetlands is by nitrification-denitrification to dinitrogen gas (N_2), and its subsequent release to the atmosphere.

Nitrification is the conversion of reduced nitrogen in ammonium ion to oxidized states such as nitrite and nitrate. The process is completed during microbial metabolism in aerobic or mildly anoxic ($-50 <$ Eh < 200 mV) conditions. In many (if not most) cases, the paired processes of nitrification and denitrification represent the largest pathway for nitrogen removal from wastewater by constructed wetlands. However, total system performance must take ambient environmental conditions into account. For example, Kuschk et al. (2003), as well as a number of other researchers have noted that nitrification can be limited by temperature during all seasons.

Nitrification is facilitated by the presence of oxygen; thus, the efficiency of this process is greatly influenced by the basic design of a constructed wetland (e.g., FWS, SSF). Field trials of SSF by Murphy et al. (2016) were conducted to determine if the process of nitrification loss was reduced when mechanical aeration of a constructed wetland system was switched off; the study also measured the system recovery rate after the aeration is switched back on. The loss of dissolved oxygen was observed to be more rapid than loss of nitrification. Nitrate was observed in the effluent long after the aeration was suspended, continuing for more than 48 hours. A complementary modeling study predicted nitrate diffusion out of biofilm over a 48 h period. After two weeks of no aeration in the established system, nitrification recovered within two days, whereas nitrification establishment in a new system was previously observed to require 20 to 45 days. These results suggest that once established in constructed wetland systems, resident nitrifying microbial communities are robust.

Denitrification is a microbial-mediated metabolic process which converts nitrate to N_2; this conversion to gaseous phase removes nitrogen from the treatment system. As denitrification is also a metabolic process, heterotrophic bacteria metabolize organic matter, and are thus responsible for removal of additional carbon from the system in the form of CO_2. García et al. (2010) review how strong an effect denitrification can have on organic matter removal under certain conditions. Denitrification is most prevalent under slightly anaerobic conditions; thus, as with nitrification, this property becomes critically important when setting basic constructed wetland design parameters.

In their review of contaminant removal processes in subsurface constructed wetlands, García et al. (2010) noted that optimum pH level for denitrification is between pH 6 and 8. Denitrification occurs more slowly but still to a significant degree below pH 5 and is negligible or absent below pH 4. Critical biological processes such as photosynthesis and metabolism of organic matter influence pH and alkalinity, and may play a role in limiting or facilitating removal of nitrogen by exerting controls over different steps in the removal of contaminants under some conditions. Denitrification is also strongly dependent on temperature. However, Prost-Boucle et al. (2015) reported removal of total nitrogen from constructed wetland systems in mountain areas at slow but measurable rates, demonstrating that the nitrification-denitrification cycle proceeded even where ambient temperatures were near or below freezing for part of the year.

Dissimilatory nitrate reduction to ammonia (DNRA) is a heterotrophic process which conserves nitrogen within the system by converting nitrate to ammonia, in contrast to denitrification of nitrate to gaseous N_2. In their summary of microbial processes in constructed wetlands, Faulwetter et al. (2009) recognized the potential for DNRA to occur in wastewater treatment systems, particularly in systems where denitrification may be limited by either an abundance of oxygen, low nitrate, or low organic carbon. As the process retains nitrogen within the treatment system, other processes may be required to reduce total nitrogen concentration to required treatment standards.

Anaerobic ammonia oxidation (anammox) consumes both ammonia and nitrate to produce N_2 which is released from the system to the atmosphere. In pilot-project reactors, the anammox process produces less sludge per unit of nitrogen removed from the system (Fux et al. 2002). In the context of constructed wetlands, Saeed and Sun (2012) identify three potential benefits anammox may present over nitrification-denitrification for removal of nitrogen from wastewater: (a) anammox is an autotrophic process requiring no external organic carbon sources; (b) lower oxygen demand; and (c) low energy consumption. However, the authors point out that the anammox starts and proceeds at a slow rate, and is sensitive to several other environmental parameters, including concentrations of other compounds common in the wastewater treatment system, including nitrite, sulfate, and ammonia. Thus, while uncommon in most wastewater treatment systems, Zhai et al. (2016) describe how anammox may be an alternative to denitrification for removal of total nitrogen from the system in some cases. Hu et al. (2016) have quantified the contribution of the process to total nitrogen reduction in lab-scale planted VF mesocosms.

Phosphorus

The phosphorus cycle in constructed wetlands includes a large number of transfer and alteration processes: adsorption, desorption, precipitation, dissolution, plant and microbial uptake, fragmentation, leaching, mineralization, sedimentation or accretion, an increase by natural growth or by gradual external addition; growth in size or extent, and finally, long-term burial (Vymazal 2007).

Basic processes which remove phosphorus from wastewater are assimilated into biomass; sorption onto particles, filtration or burial into the substrate. Each of these processes is reversible, to an extent, meaning at least a portion of the phosphorus from each may be returned to the wastewater stream (e.g., decay of biomass). Of these, burial represents the longest-term potential removal (Kadlec and Wallace 2009). Harvest of above-ground macrophyte biomass may represent an additional removal mechanism in some cases (Vymazal 2008).

In comparison with characteristic removal rates of organic matter and nitrogen, constructed wetland removal of phosphorus seems low. The major differences in mechanisms driving removal processes of carbon, nitrogen, and phosphorus are: (1) the significance of microbial interactions which conserve phosphorus, and (2) the lack of a significant gaseous phase in the phosphorus cycle, both in contrast with carbon and nitrogen cycles. While metabolism of organic matter to CO_2 or CH_4, and the ultimate denitrification of nitrogen compounds to N_2 gas provide removal of carbon and nitrogen from the treatment system, the long-term "removal" of phosphorus is through the internal incorporation into soil (FWS) or stabilization-complexation onto reactive media (SSF). In the context of interactions with sediment, García et al. (2010) adds a third characteristic of phosphorus cycling in constructed wetlands to consider: most substrates used as granular media have low P-sorption and P-complexation capacities.

Reddy and D'Angelo (1994) described how phosphorus solubility may be influenced by the following ambient conditions:

- in low pH conditions, phosphorus may be fixed by iron or aluminum, if present;
- in higher pH conditions, phosphorus may be fixed by calcium or magnesium, if present;
- in reducing conditions (e.g., anaerobic), iron and iron-bound phosphorus may be solubilized and returned to the water phase.

Considering these three tendencies, some generalities within constructed wetland systems may be made. While conditions in FWS are usually aerobic and may tend to higher pH as a result of photosynthesis, SSF conditions are usually anaerobic and reducing, though still generally around or above pH 7. Soluble reactive phosphorus (as HPO_4^{2-}) becomes more abundant at higher pH (pK = 7.2), and thus, available either for biological assimilation or sorption onto soil and sediments. Therefore, incorporation into sediment may be the most common pathway for long-term sequestration of phosphorus in constructed wetlands (Kadlec and Wallace 2009), though the process can be very sensitive to (1) internal conditions and (2) the tendency for phosphorus to be conserved in biological systems.

The degree of phosphate adsorption by the granular medium depends mainly on its texture (effective surface area) and grain size distribution as well as chemical affinity for phosphorus, chiefly measured by iron content in some conditions, and aluminum and calcium content in others. García et al. (2010) noted that gravel is the most widespread medium in horizontal SSF CWs because it does not clog easily. However, it only adsorbs small amounts of phosphorus because it has a coarse texture (small surface-to-volume ratio) and generally contains low levels of Fe (iron) and Al (aluminum). Drizo et al. (1999) conducted screening experiments with a range of materials to identify materials with optimal properties for treating phosphorus in wastewater. Their study identified shale (a carbonate) as exhibiting the best combination of a number of characteristics including pH, cation exchange capacity, and hydraulic conductivity.

In a survey of substrates used in vertical flow, VF, constructed wetlands, Molle et al. (2003) tentatively identified crystal formation as a mechanism of phosphorus removal by calcareous media from wastewater, and many studies conducted in the following years have continued to identify and refine substrate capacity for phosphorus removal by chemical reaction (e.g., Molle et al. 2003; Prigent et al. 2013; Troesch et al. 2016).

Properties Controlling Transformations

As described above, biological cycling of organic matter, nitrogen, and phosphorus within the constructed wetland itself must be taken into account when determining design parameters. The following section describes some of the major physical and chemical controls on organic matter and nutrient cycling and removal from wastewater. These properties both influence and are influenced by organic matter and nutrient abundance and cycling, and constructed wetland design parameters (e.g., depth, HRT) can be selected so that some level of control is gained on each.

Oxidation-reduction (or "redox") potential describes the tendency for a given system to oxidize or reduce available substrates (Kadlec and Wallace 2009). Closely related to the availability of oxygen, measurement of the redox potential quantifies the strength of the tendency to exchange electrons and reduce oxidation state; this property influences physical, chemical, and biological processes.

Because of the large surface area FWS CW present for atmospheric exchange of oxygen, FWS often maintain relatively well oxygenated, and thus a high, but narrow range of redox potential. SSF CWs are distinguished from other wastewater treatment processes by the juxtaposition of areas with different redox status at the small and large spatial scale (García et al. 2010).

Another distinction between FWS and SSF designs is the emphasis of different parts of the sulfur cycle. The abundance and dominant forms of sulfur can have substantial effects on several important bio-physical processes in constructed wetlands. Given the generally aerobic conditions found in FWS constructed wetlands, the oxidized forms of sulfur are usually dominant. However, sulfate can support microbial metabolism in the strongly reducing conditions which characterize some HSSF systems, increasing the importance of reduced sulfur compounds.

Temperature affects microbial metabolic activity, dissolution of gaseous compounds in water (e.g., oxygen, ammonia, carbon dioxide), and, in turn, temperature can influence many other processes related to nutrient cycling and contaminant removal. Individual processes respond to high and low temperatures differently, and this section considers a number of different processes involved in removal of organic matter and nutrients from wastewater.

In a review of nutrient removal processes in constructed wetlands, Vymazal (2007) noted that ammonification, nitrification, and denitrification are all sensitive to temperature, and each exhibits a different optimal range. These differences may be attributed to environmental preferences of the microbial taxa. In this summary, the minimum temperature required for growth of nitrifying bacteria taxa is between 4 and 5°C; denitrification was recorded below 5°C, but at a slower rate.

Yet the sum of the biological and physical interactions occurring in constructed wetlands may create microclimates or other conditions which extend the basic ambient environmental limits of microbes. Given the sensitivity of nitrification to low temperatures, Prost-Boucle et al. (2015) conducted a detailed study of several hybrid-design constructed wetlands operating year-round in mountain areas. Air temperatures during the study ranged from approximately –5 to 25°C, yet total nitrogen removal by the system was relatively unaffected during the winter season. The study identified locations within the system where nitrification exhibited different sensitivities to the low temperatures during winter. Their results from full-sized systems indicate that the combination of conditions present in some constructed wetland designs may play a large role in extending the capacity of microbe-mediated processes to remove contaminants.

Alkalinity and pH are important properties in aquatic chemistry, and are associated with many critical wastewater treatment processes in constructed wetlands. The fundamental distinctions between conditions in FWS and SSF constructed wetlands described above also impact the typical values and variability of these properties.

The presence of photosynthetic algae or submerged aquatic vegetation (SAV) in FWS wastewater promotes the potential for daily increases in pH as observed in natural water bodies by consuming carbon dioxide, and shifting the bicarbonate-carbonate-carbon dioxide equilibrium from lower to higher pH (Kadlec and Wallace 2009). Respiration of photosynthesizing algae at night produces carbon dioxide, adding to the volume produced by the host of other microbes present in wastewater treatment systems. Nonetheless, constructed wetlands treating municipal wastewater produce effluent of mean pH of 7.18 (Kadlec and Wallace 2009). The seasonal mean pH during summer may be slightly elevated, at 7.24, which may indicate internal buffering capacity against large seasonal swings in effluent pH due to macrophyte photosynthesis.

Photosynthesis does influence ambient processes in SSF constructed wetlands, though the pathways and mechanisms and products are predictably different. In most cases, the major photosynthetic elements in SSF constructed wetlands are rooted macrophytes; thus, CO_2 exchange for photosynthesis occurs in the atmosphere, not the aquatic medium. However, Kadlec and Wallace (2009) noted that changes in pH from HSSF influent to effluent may reflect microbial processes more than the activities or presence of macrophytes citing two reports based on un-planted gravel beds (Bavor et al. 1988; Theis and Young 2000).

While constructed wetland effluent is usually near neutral (Kadlec and Wallace 2009), changes in pH within the constructed wetland system influence, and are influenced by, many other critical wastewater treatment processes. Nitrification is most efficient at pH 7.2 and higher, while the companion process denitrification is most efficient between 6.5 and 7.5, possibly a reflection of oxidation of organic matter required for this process. Value of pH affects the phosphorus cycle at several points in wastewater treatment, as noted above.

Hydraulic loading rate (HLR) is defined as the total flow rate (e.g., volume per day) per unit of wetted land area (e.g., square meters). HLR is an important design parameter for constructed wetlands because it exerts a strong influence on the amount of a contaminant that can be removed by a constructed wetland system. García et al. (2004b) found significant differences in COD:BOD, dissolved reactive phosphorus removal related to HLR.

Hydraulic residence time (HRT) is a critical design criterion as it represents the period of time wastewater is exposed to the range of processes described earlier. COD, BOD, and total suspended solids (TSS) removal efficiencies are related to HRT (Kadlec and Wallace 2009).

In the simplest terms, HRT can be described as the ratio of constructed wetland volume to flow rate (Zahraeifard and Deng 2011).

In practice, HRT is a function of many other design criteria, including aspect ratio (width/length), macrophytes, substrate porosity (hydraulic conductivity), wastewater depth and flow, and bed slope (Revitt et al. 1999).

HRT is only one of many interrelated design variables which must be optimized to suit project requirements. In general, HSSF and FWS systems exhibit longer HRT, while VF systems exhibit shorter HRT, though some wetland treatment types are more suited to some waste treatment needs. However, the rate at which some biological and chemical processes can proceed (e.g., denitrification) may dictate the range of HRT which can be considered. Across all constructed wetland types and wastewater treatment needs, HRT can range from hours to weeks.

Efforts to model HRT and related processes must consider areas of variable flow within certain constructed wetland designs.

Zahraeifard and Deng (2011) noted a number of examples where other researchers identified three different HRTs within a single system, making accurate modeling of system-wide HRT under multiple treatment scenarios a challenge.

Further, constructed wetland designs may need to anticipate changes in HRT due to biofilm accumulation, plant colonization, and evapotranspiration (ET) rates (Langergraber et al. 2009).

CW Designs and Alternative Operational Strategies

To this point, only broad constructed wetland designs and the basics of nutrient and organic matter roles and cycling in wastewater treatment have been presented. As it was discussed, the net removal of contaminants within a constructed wetland is the sum of many processes, including sedimentation, filtration, precipitation, volatilization, adsorption, plant uptake, and various microbial processes influencing carbon, nitrogen, and phosphorus cycles, among others.

The following part of the chapter describes basic constructed wetland designs and alternative operation activities, and how these influence treatment of organic matter and nutrients in wastewater.

Free Water Surface Constructed Wetland Design

FWS constructed wetlands are characterized by wastewater flow that is exposed to the atmosphere. Along with the aspect ratio (length: width), the proportion of planted area to wetted area, and the size and shape of each may have significant implications on constructed wetland cell capacity and efficiency (Kadlec and Wallace 2009).

FWS constructed wetland treatment systems have been composed of single or multiple cells ranging in sizes from a few square meters to many hectares. The largest FWS systems are composed of multiple cells totaling many square kilometers in size. Because of their exposure to the atmosphere and the variability associated with daily meteorological conditions, design of FWS of all sizes must include environmental parameters such as evapotranspiration rate and relative humidity to account for potential changes in total volume of water in the system.

FWS basins are usually lined, either with artificial impermeable membranes (e.g., PVC, polyvynil chloride, or HDPE, high density polyethylene), or with clay, bentonite, or similar materials, especially in cases of larger sizes (e.g., greater than 0.5 ha). A 30 to 40 cm layer of soil or other media may be used to provide rooting substrate for macrophytes (Kadlec and Wallace 2009). The design water depth is often less than 0.5 m, dictated to some extent by the species of emergent macrophytes present, though a range of depths from centimeters to permanent pools have been implemented under different conditions (Kadlec and Wallace 2009).

A literature survey conducted by Vymazal (2013a) identified more than 150 species of rooted macrophytes occurring in FWS designs worldwide; the five most common are *Typha*, *Scirpus* (*Schoenoplectus*), *Phragmites*, *Juncus*, and *Eleocharis*. The choice of plant(s) for a constructed wetland project may be a function of climate and latitude, cost, water quality, required maintenance, and other factors (Kadlec and Wallace 2009).

In addition to rooted emergent macrophytes, FWS constructed wetlands may include floating or submerged taxa. Free-floating taxa may include small duckweed (Lemnaceae) or larger water lettuce (*Pistia* sp.) (Vymazal 2008), and the effects of each type of plant must be considered during the design phase. For example, floating species may reduce light penetration into the wastewater column, and so greatly impact photosynthesis by phytoplankton and epiphyton. Gas exchange (e.g., ammonia, oxygen, carbon dioxide) may also be reduced. SAV taxa used in full-scale FWS constructed wetlands may include *Hydrilla* and *Ceratophyllum* (Kadlec and Wallace 2009).

When FWS receive wastewater with little or no pre-treatment, processes such as settling of suspended solids occur near the inlet. Some of these suspended solids represent organic matter, which is composed of carbon, nitrogen, and phosphorus components.

As described above, the principal pathways for organic carbon and nitrogen removal from FWS system are associated with microbial metabolism. The concentration of organic nitrogen in the form of ammonia may be reduced by aerobic metabolism to nitrite and nitrate, off-gassing during periods of higher pH (e.g., due to photosynthesis by phytoplankton or periphyton).

It should be noted that because the wastewater is exposed in FWS, new sources of ammonia (in the form of uric acid or urea) may be introduced by birds, fish, or mammals. This exposure also limits the application of FWS in some areas where disease vectors are a concern (e.g., secondary treatment of domestic wastewater).

As particulate organic nitrogen and ammonia are usually the most abundant forms of inorganic nitrogen in domestic and municipal wastewater, rates of ammonification and nitrification are often highest early in the FWS treatment stream. However, removal of total nitrogen from the waste stream is usually considered to be limited by denitrification. Denitrification is inhibited when enough oxygen is available for aerobic treatment of organic matter; thus, the process requires either substantial differences in water chemistry across the site to support nitrification-denitrification, or microsites within a FWS system to support anoxic conditions and subsequent denitrification. These microsites may include deeper areas of the constructed wetland, or collections of leaf litter.

The most common application of FWS constructed wetlands is for secondary and tertiary treatment of municipal sewage. However, treatment of wastewater from agricultural and industrial facilities is becoming more common. For example, storm water runoff from a range of land-uses (Vymazal 2008) and oil-field produced water have been treated with FWS (Breuer et al. 2012).

Sub-Surface Flow Constructed Wetlands

As noted earlier, the substantial difference between basic constructed wetland designs stems from whether the wastewater flows is exposed to the atmosphere, or flows "underground".

The material through which wastewater flows in SSF design constructed wetlands is referred to as support media, and variables to consider include the size and shape, effects of media characteristics on HRT, surface area for biofilm development, chemical reactivity, and the effects of media on macrophyte (e.g., capacity to support roots). Like many other characteristics of constructed wetland design and operation, optimizing all variables for particular wastewater treatment needs can be a challenge. For example, while smaller diameter media present greater surface area for microbial colonization or physical or chemical reaction, hydraulic conductivity (the ease with which fluid moves through the medium) is reduced.

García et al. (2005) compared treatment performance in experimental constructed wetlands given four design variables: aspect ratio, granular size of the media, water depth, and HLR. Of those variables, only water depth and HLR had a significant impact on removal of nutrients and organic matter. In contrast, Akratos and Tsihrintzis (2007) studied pilot-scale constructed wetland performance using three different media, alongside a number of other variables (e.g., plant, HRT). Their study did identify a difference between the influence of media size and possible chemical composition on constructed wetland efficiency.

Some SSF designs include layers of different media arranged in strategic locations to enhance particular processes of physical removal of contaminant through sedimentation and filtration, while other layers maybe designed to retain water to increase residence time and maintaining anaerobic conditions (e.g., Morvannou et al. 2015).

Macrophytes play somewhat different roles in SSF constructed wetlands compared with FWS. Because the roots of emergent macrophytes such as *Phragmites* and *Typha* come into direct contact with wastewater, there are many pathways for interactions with microbial communities, and additional potential to influence treatment performance.

Photosynthesis impacts ambient processes in SSF wetlands. However, the pathways and mechanisms and products are predictably different. In most cases, the major photosynthetic elements in SSF wetlands are rooted macrophytes; thus CO_2 exchange for photosynthesis occurs in the atmosphere, not the aquatic medium. Here, the products of photosynthesis, oxygen and organic carbon compounds, are available to the SSF system. As observed by Nivala et al. (2013), subsurface oxygen availability tends to be one of the main rate-limiting factors for removal of carbon and nitrogen compounds in SSF constructed wetlands.

Horizontal Sub-Surface Flow Constructed Wetlands

As for FWS designs, HSSF constructed wetland basins are often lined with an impermeable material. Unlike some FWS constructed wetlands, the maximum size of HSSF is constrained by the hydraulic properties of the media.

HSSF designs typically include a latitudinal channel or layer of coarse stone at the inlet to distribute flow across the cross-sectional area of the constructed wetland; a similar channel at the outlet may promote even flow through the substrate (Kadlec and Wallace 2009). While uncommon in practice, layers or sections of different sized substrate may be used in HSSF to control the flow of water through the wetland cell and reduce the effect of accumulated solids. The practice of layering media of different sizes and properties is more common in VF designs (e.g., Molle et al. 2005).

Carballeira et al. (2016) reviewed the differences between HSSF beds in a pilot-project across three years of operations based on a number of important variables. The study reviewed responses in effluent total suspended solids (TSS) and BOD$_5$ in beds planted with *Phragmites australis*, *Typha latifolia*, *Iris pseudocorus*, and *Juncus effusus*, and discovered differences between planted beds based on organic matter loading when loading rates were higher (design load of average 4.7 g BOD L^{-1} d^{-1}) than when they were diluted to lower concentration (avg. 2.5 g BOD L^{-1} d^{-1}). The authors concluded that differences in constructed wetland performance may be influenced by plant productivity and the relationship between productivity and evapotranspiration rate. Thus, while an early summary by Brix (1997) found that treatment-related functions of macrophytes such as nutrient uptake were of significant value only in lower-loaded systems, the results of Carballeira et al. (2016) suggest the differences between macrophyte taxa may be evident only at higher loads.

Among the most useful aspects of HSSF design is the juxtaposition of a range of redox-conditions (García et al. 2010). Given adequate nitrate and organic matter, the tendency toward anaerobic conditions that characterizes HSSF wetlands may maintain conditions for extremely effective denitrification.

In their review, Vymazal and Kröpfelová (2009) found HSSF to be most commonly used type of constructed wetland around the world, being often used to treat domestic wastewater and municipal sewage from small settlements and villages, producing effluent water to secondary or tertiary treatment standards. Their review identified other applications for a range of industrial and agricultural wastewaters, as well as some systems treating landfill leachate, though this is uncommon for HSSF.

Vertical Flow Constructed Wetlands

In contrast to both FWS and HSSF, VF designs are most commonly operated by intermittent flooding of wastewater influent across the entire surface of the bed. This water is allowed to drain through the media (often layers of sand and gravel), and is then collected from the bottom of the bed. The bed is usually allowed to drain completely which allows air to refill the bed; the period of time allowed for reaeration of pore spaces is a topic of research (Murphy et al. 2016). VF beds operated this way provide greater oxygen transfer into the bed, thus producing a nitrified effluent which is (ideally) low in ammonium.

This most common method of VF operation provides a large surface area for remineralization of organic matter. Over time, from months to 1 year (Molle et al. 2005) the accumulation of this sludge layer promotes distribution of wastewater across the entire surface (e.g., not concentrated at the inlets), and demineralizes recalcitrant organic matter.

As noted earlier, the roles of macrophytes between FWS and SSF constructed wetlands designs differ. There are also differences in the effects macrophytes have between HSSF and VF designs. In their review of VF constructed wetlands for sludge treatment, Uggetti et al. (2010) include the effect of macrophytes evapotranspiration through roots and rhizomes, which contributes to drying and dewatering of the sludge; transport of oxygen to support mineralization of sludge organic matter around roots and rhizomes; and maintenance of cracks in the drying upper layers of sludge through stem movement, which contributes to dewatering and remineralization, as well.

Molle et al. (2006) studied the effect of reeds before and after harvest on wastewater infiltration into VF media. They determined that part of the interaction between reed stems and rhizomes (when present) and accumulated sludge layer lay in increased initial flooding of lower filter layers with wastewater via reed root and rhizomes. In the absence of reed stems, flooding through lower VF filter layers was slowed, and controlled by the development of the sludge layer, which increases over time.

Media in VF is often layered, where finer media are at the top of the bed, and coarser material at the bottom (Molle et al. 2005; Vymazal 2008). The composition, distribution, and thickness of the layers have a great impact on performance and other operational activities. For example, Prigent et al. (2013) studied a "compact" VF system including several layers of expanded schist and different sized gravel. The schist provided three times more effective surface area available for microbial community development as biofilm, and served as a subsurface control on water percolation to subsequent layers. Their study determined that use of their configuration could meet national standards for measures of organic matter (BOD, COD, TSS) while occupying less area (only 1.2 m^2 pe^{-1} where pe = population equivalent, a standard measure of wastewater treatment capacity).

In his broad-based review of constructed wetland designs, Vymazal (2008) identified at least nine different wastewater applications where VF wetlands were used (in contrast to their use in hybrid designs, see below). These included several agricultural (livestock and pesticides) and industrial (petroleum hydrocarbons) wastewater types.

Vymazal (2008) also noted the significant difference in sizes between VF and other constructed wetland designs. VF systems generally required less land per unit of wastewater treatment capacity (1–3 m^2 pe^{-1} compared to 5–10 m^2 pe^{-1} of HF systems). However, VF systems may generally require more maintenance and operation efforts because of the use of pumps, timers, and other electric and mechanical devices.

Hybrid Constructed Wetland Designs

"Hybrid" constructed wetland designs include combinations of free water surface, horizontal sub-surface, and/or vertical sub-surface flow constructed wetlands. In a recent survey of hybrid CW examining applications in 24 countries, Vymazal (2013b) reported that all types of hybrid constructed wetlands are more efficient in total nitrogen removal than single HF or VF constructed wetlands. Out of 60 surveyed hybrid systems, 38 were designed for municipal sewage while 22 hybrid systems were designed to treat various industrial and agricultural wastewaters. His survey revealed that HF-VF design was used to treat municipal sewerage only, while the most commonly used hybrid system, VF-HF arrangement, had been applied to a wider range of wastewater types, including sewage and industrial wastewaters.

The selection, sequencing, and dimensions of each component in a hybrid constructed wetland system is to be determined by quality and quantity of wastewater, as well as the requirements of effluent. For example, nitrification of ammonia requires an adequate supply of oxygen; thus, FWS and VF components are more efficient for nitrification than HSSF. However, anaerobic conditions are needed for denitrification (in addition to nitrate and organic matter to complete the nitrification-denitrification process and reduce total nitrogen from the wastewater). HSSF systems provide far greater capacity for anaerobic processes. Thus, hybrid designs which include VF followed by HSSF are very efficient at removal of total nitrogen, as well as recalcitrant organic matter. Alternatively, when reduction of ammonia from effluent is the discharge requirement, upstream treatment may be followed by VF stages, which provide conditions for nitrification.

Molle et al. (2005) completed a review of 20 years of systematic research, application, and follow-up monitoring of a group of two-stage VF, which has formed the basis for hybrid designs and practice known as the "French System". After 30 years of development of systems for treatment of domestic waste produced by small, rural communities, current practice for this system usually includes two treatment stages, with three units in parallel on the first stage and two for the second, with successive periods of feeding (3.5 days), and rest periods (7 days at the first stage and 3.5 at the second stage).

This feeding-resting practice contributes to the maintenance of permeability, oxygen content, and control the biomass growth (Morvannou et al. 2015). When combined with other design specification (e.g., media), this schedule of feeding reduces (or eliminates) the requirement for pre-treatment. Often, these systems are fed with raw wastewater (after screening for debris), in contrast to most other designs which are fed with wastewater that has undergone some settling to reduce the load of biosolids delivered to the system.

Alternative Operational Strategies

There are a number of optional strategies that can be applied to the operation of the basic constructed wetland designs. These optional operations activities may be implemented as part of a constructed wetland design to enhance system performance, retard or reverse development of clog matter and resultant conditions (e.g., Kadlec and Wallace 2009; Knowles et al. 2011). Alternative operations may also be considered in cases where unintended influent or environmental conditions must be managed within an existing constructed wetland design. For example, very high initial organic matter concentration (BOD higher than 250 mg L^{-1}), presence of fats and oils, and management of macrophyte abundance can present significant challenges to maintenance of effective treatment wetland performance.

The basic range of operational activities available to wastewater treatment operators included here is surprisingly similar across different constructed wetland designs; however, the specific application and effect of different activities varies greatly across designs.

Feeding

Variables involved in supply of wastewater to a constructed wetland system differ in when or how frequently wastewater is introduced; the composition of the influent wastewater; and where wastewater is introduced.

There are many possible strategies to consider when determining the timing of wastewater introduction to a constructed wetland system, from continuous flooding to allowance of long "resting" periods when a system is allowed to dry out.

FWS and HSSF are normally operated in continuous feeding (i.e., flooded) mode, though water levels can be changed to maintain a number of methods (e.g., standpipe). In both cases, ambient conditions between aerobic and anaerobic conditions can be manipulated by managing water depth (e.g., García et al. 2004b). Allowing FWS and SSF constructed wetlands to drain completely for months at a time is an operational strategy to reduce clogging with biosolids, and could be considered maintenance (Knowles et al. 2011).

"Feeding mode" describes how often and how much wastewater is delivered to the treatment system; the period of time it is held; and the rate at which it is drained. Examples of alternative feeding modes include continuous, batch (or "batch-fill-drain"), intermittent, pulse, sporadic, and event-driven (e.g., sporadic, Kadlec and Wallace 2009; Nivala et al. 2013). While the distinction between these terms is sometimes semantic, the critical variables to consider are the timing, volume, water quality (e.g., combine sewer overflow, CSO), and degree to which water drains from the support media and provides time for re-aeration of media pore spaces.

Timing between feeding events is a critical parameter in the design of staged VF treatment systems which characterize the "French System" described above. As noted, long resting periods between feeding events early in the process are important for separation and dewatering of biosolids. Three beds compose the first stage of the French System, giving longer resting periods between each feeding. The second stage of the French system includes only two beds, thus reducing the period of time between feeding. This improves conditions for denitrification as the pore spaces within the media have less time to drain, thus some fluid remains.

Based on their review of microbial characteristics of constructed wetlands, Faulwetter et al. (2009) noted that the redox condition can be managed by selecting a suitable mode of operation, especially in subsurface flow systems. According to their review, the most important operating factors are *feeding mode*, hydraulic loading rate, and retention time. Feeding mode can be classified into a number of distinct categories depending on the wastewater management strategy.

Recirculation returns a portion of constructed wetland-treated water back into the treatment system. This method has been used to provide nitrified wastewater from vertical flow treatment stages to support pre-treatment denitrification, and to provide some degree of stabilization to the system as a whole (Brix and Arias 2005). Wu et al. (2015) added that recirculation can be used to dilute influent concentration of some industrial wastewaters.

Step feeding refers to introduction of additional wastewater to the interior of a constructed wetland, as opposed to only the upstream end of a system (Kadlec and Wallace 2009). It contrasts with recirculation in that step-feeding introduces a waste stream of previously untreated water into the system. Descriptions of this strategy (e.g., Kadlec and Wallace 2009; Wu et al. 2014) acknowledge there are few full-size examples, thus there is a paucity of design and monitoring data.

Step-feeding describes addition of wastewater to different points within a constructed wetland stage. Kadlec and Wallace (2009) give several examples of the step-feeding alternative used to distribute wastewater (and degradation processes) across a larger area, and to avoid overload of the initial inlet. An alternative application, according to the authors, may be to introduce new wastewater of different characteristics which may promote or enhance desirable contaminant removal characteristics (unpublished).

Aeration

As described above, oxygen availability is one property that distinguishes basic constructed wetland designs from each other. Given the influence that oxygen has over many treatment processes, particularly in sub-surface flow designs, some treatment systems pre-aerate influent wastewater, while others introduce active aeration within the wetland by means of blowers. Clearly, the location of increased aeration and oxygen supply in the treatment system has a critical impact on the contaminant removal processes, and many studies have been conducted to determine the optimal volume and timing of aeration (e.g., Nivala et al. 2013; Murphy et al. 2016).

Wallace (1998) and other authors have developed methods for introducing and distributing air into HSSF. The purpose of increased aeration is to increase and manipulate oxygen supplied to the HSSF treatment systems in a way that enhances aerobic-dependent microbial processes, and promotes thicker and deeper macrophyte root growth. For example, aeration may increase the capacity for nitrification in HSSF; following nitrification, the natural conditions within HSSF are available in close proximity to continue nitrogen treatment with denitrification and then off-gassing of N_2 which leaves the system. Wallace and Kadlec (2005) implemented a full-scale hybrid-design constructed wetland for benzene-toluene-ethylbenzene-xylene, BTEX-contaminated groundwater. The enhanced aeration in this project occurred before the wastewater was introduced to the wetland system for the purpose of iron oxidation; this step contributed to removal of iron as precipitates of iron oxide in the FWS cell which followed.

Aeration may also be applied to VF constructed wetlands. To characterize the capacity for increased coupling of nitrification-denitrification in VF constructed wetland designs, Boog et al. (2014) applied intermittent and continuous active aeration to pilot-scale VF constructed wetlands. In their trials, intermittent aeration produced conditions which enhanced denitrification over continuously aerated systems, and reduced total nitrogen removal from the wastewater.

Conclusions

It is clear that many aspects must be considered in the design and operation of constructed wetlands for effective removal of organic load and nutrients.

This chapter can be concluded with this advice for the reader: a thorough and holistic review of this and the accompanying chapters, as well as the recent literature, must be performed to decide which are the most important considerations to be addressed for each case.

Acknowledgements

M. Hartl provided comments to this manuscript. Mr. Hartl is a PhD Fellow in the SuPER-W "European Joint Doctorate" programme at UPC Barcelona and Ghent University. His project, titled "Coupling bio-electrochemical systems and phytotechnologies for wastewater treatment used in rural areas towards economic recovery of (micro)nutrients, biomass & water reuse" integrates several themes within the discipline of ecological engineering, especially constructed wetlands and bio-electrochemical systems; sustainable sanitation; and other application of the study of biogeochemical cycles.

References Cited

Adrados, B., O. Sanchez, C.A. Arias, E. Becares, L. Garrido, J. Mas et al. 2014. Microbial communities from different types of natural wastewater treatment systems: Vertical and horizontal flow constructed wetlands and biofilters. Water Res. 55: 304–312.

Akratos, C.S. and V.A. Tsihrintzis. 2007. Effect of temperature, HRT, vegetation and porous media on removal efficiency of pilot-scale horizontal subsurface flow constructed wetlands. Ecol. Eng. 29(2): 173–191.

Albuquerque, A., J. Oliveira, S. Semitela and L. Amaral. 2009. Influence of bed media characteristics on ammonia and nitrate removal in shallow horizontal subsurface flow constructed wetlands. Bioresource Technol. 100: 6269–6277.

Bavor, H.J., D.J. Roser, S.A. McKersie and P. Breen. 1988. Treatment of secondary effluent. Report to Sydney Water Board, Sydney. New South Wales, Australia.

Boog, J., J. Nivala, T. Aubron, S. Wallace, M. van Afferden and R.A. Muller. 2014. Hydraulic characterization and optimization of total nitrogen removal in an aerated vertical subsurface flow treatment wetland. Bioresource Technol. 162: 166–174.

Breuer, R., B. Breuer, B. Al Sharji, T.R. Headley and Y. I. Thaker. 2012. The first year operation of the Nimr Water Treatment Plant in Oman: Sustainable produced water management using wetlands. International Conference on Health, Safety and Environment in Oil and Gas Exploration and Production, p.7.

Brix, H. 1994. Functions of macrophytes in constructed wetlands. Water Sci. Technol. 29: 71–78.

Brix, H. 1997. Do macrophytes play a role in constructed treatment wetlands? Water Sci. Technol. 35(5): 11–17.

Brix, H., C.A. Arias and M. del Bubba. 2001. Media selection for sustainable phosphorus removal in subsurface flow constructed wetlands. Water Sci. Technol. 44(11-12): 47–54.

Brix, H. and C.A. Arias. 2005. The use of vertical flow constructed wetlands for on-site treatment of domestic wastewater: New Danish guidelines. Ecol. Eng. 25(5): 491–500.

Button, M., J. Nivala, K.P. Weber, T. Aubron and R.A. Müller. 2015. Microbial community metabolic function in subsurface flow constructed wetlands of different designs. Ecol. Eng. 80: 162–171.

Carballeira, T., I. Ruiz and M. Soto. 2016. Effect of plants and surface loading rate on the treatment efficiency of shallow subsurface constructed wetlands. Ecol. Eng. 90: 203–214.

Drizo, A., C.A. Frost, J. Grace and K.A. Smith. 1999. Physico-chemical screening of phosphate-removing substrates for use in constructed wetland systems. Water Res. 33(17): 3595–3602.

Faulwetter, J.L., V. Gagnon, C. Sundberg, F. Chazarenc, M.D. Burr, J. Brisson et al. 2009. Microbial processes influencing performance of treatment wetlands: A review. Ecol. Eng. 35(6): 987–1004.

Fonder, N. and T. Headley. 2013. The taxonomy of treatment wetlands: A proposed classification and nomenclature system. Ecol. Eng. 51: 203–211.

Fux, C., M. Boehler, P. Huber, I. Brunner and H. Siegrist. 2002. Biological treatment of ammonium-rich wastewater by partial nitrification and subsequent anaerobic ammonium oxidation (anammox) in a pilot plant. J. Biotechnol. 99(3): 295–306.

García, J., J. Chiva, P. Aguirre, E. Alvarez, J.P. Sierra and R. Mujeriego. 2004a. Hydraulic behaviour of horizontal subsurface flow constructed wetlands with different aspect ratio and granular medium size. Ecol. Eng. 23(3): 177–187.

García, J., P. Aguirre, R. Mujeriego, Y. Huang, L. Ortiz and J.M. Bayona. 2004b. Initial contaminant removal performance factors in horizontal flow reed beds used for treating urban wastewater. Water Res. 38(7): 1669–1678.

García, J., P. Aguirre, J. Barragana, R. Mujeriegoa, V. Matamoros and J.M. Bayona. 2005. Effect of key design parameters on the efficiency of horizontal subsurface flow constructed wetlands. Ecol. Eng. 25(4): 405–418.

García, J., D.P.L. Rousseau, J. Morato, E. Lesage, V. Matamoros and J. Bayona. 2010. Contaminant removal processes in subsurface-flow constructed wetlands: A review. Crit. Rev. Env. Sci. Tec. 40(7): 561–661.

Hu, Y., F. He, L. Ma, Y. Zhang and Z. Wu. 2016. Microbial nitrogen removal pathways in integrated vertical-flow constructed wetland systems. Bioresource Technol. 207: 339–345.

Jouanneau, S., L. Recoules, M.G. Durand, A. Boukabache, V. Picot, Y. Primault et al. 2014. Methods for assessing biochemical oxygen demand (BOD): A review. Water Res. 49(1): 62–82.

Kadlec, R.H. and S.D. Wallace. 2009. Treatment wetlands, Second Edition. CRC Press, Boca Raton, London, New York.

Knowles, P., G. Dotro, J. Nivala and J. Garcia. 2011. Clogging in subsurface-flow treatment wetlands: Occurrence and contributing factors. Ecol. Eng. 37(2): 99–112.

Kuschk, P., A. Wiessner, U. Kappelmeyer, E. Weissbrodt, M. Kästner and U. Stottmeister. 2003. Annual cycle of nitrogen removal by a pilot-scale subsurface horizontal flow in a constructed wetland under moderate climate. Water Res. 37(17): 4236–4242.

Langergraber, G., D. Giraldib, J. Menac, D. Meyerd, M. Peñae, A. Toscano et al. 2009. Recent developments in numerical modelling of subsurface flow constructed wetlands. Sci. Total Environ. 407(13): 3931–3943.

Ligi, T., M. Truu, K. Oopkaup, H. Nõlvak, Ü. Mandera, W.J. Mitsch et al. 2015. The genetic potential of N2 emission via denitrification and ANAMMOX from the soils and sediments of a created riverine treatment wetland complex. Ecol. Eng. 80: 181–190.

Molle, P., A. Lienard, A. Grasmick and A. Iwema. 2003. Phosphorus retention in subsurface constructed wetlands: Investigations focused on calcareous materials and their chemical reactions. Water Sci. Technol. 48(5): 75–83.

Molle, P., A. Lienard, C. Boutin, G. Merlin and A. Iwema. 2005. How to treat raw sewage with constructed wetlands: An overview of the French systems. Water Sci. Technol. 51(9): 11–21.

Molle, P., A. Lienard, A. Grasmick and A. Iwema. 2006. Effect of reeds and feeding operations on hydraulic behaviour of vertical flow constructed wetlands under hydraulic overloads. Water Res. 40(3): 606–612.

Morvannou, A., N. Forquet, S. Michel, S. Troesch and P. Molle. 2015. Treatment performances of French constructed wetlands: Results from a database collected over the last 30 years. Water Sci. Technol. 71(9): 1333–1339.

Murphy, C., A.R. Rajabzadeh, K.P. Weber, J. Nivala, S.D. Wallace and D.J. Cooper. 2016. Nitrification cessation and recovery in an aerated saturated vertical subsurface flow treatment wetland: Field studies and microscale biofilm modeling. Bioresource Technol. 209: 125–132.

Nivala, J., S. Wallace, T. Headley, K. Kassa, H. Brix, M. van Afferden et al. 2013. Oxygen transfer and consumption in subsurface flow treatment wetlands. Ecol. Eng. 61: 544–554.

Prigent, S., G. Belbeze, J. Paing, Y. Andres, J. Voisin and F. Chazarenc. 2013. Biological characterization and treatment performances of a compact vertical flow constructed wetland with the use of expanded schist. Ecol. Eng. 52: 12–18.

Prost-Boucle, S., O. Garcia and P. Molle. 2015. French vertical-flow constructed wetlands in mountain areas: How do cold temperatures impact performances? Water Sci. Technol. 71(8): 1219–1228.

Reddy, K.R. and E.M. D'Angelo. 1994. Soil process regulating water quality in wetlands. pp. 309–324. *In*: Mitsch, W.J. (ed.). Global Wetlands: Old World and New. Elsevier, Amsterdam, The Netherlands.

Revitt, M., B. Shutes and L. Scholes. 1999. The use of constructed wetlands for reducing the impacts of urban surface runoff on receiving water quality. IAHS-AISH P. 259(259): 349–356.

Saeed, T. and G. Sun. 2012. A review on nitrogen and organics removal mechanisms in subsurface flow constructed wetlands: Dependency on environmental parameters, operating conditions and supporting media. J. Environ. Manage. 112: 429–448.

Theis, T. and T. Young. 2000. Subsurface flow wetland for wastewater treatment at Minoa. Final Report to the New York State Energy Research and Development Authority: Albany. New York.

Troesch, S., D. Esser and P. Molle. 2016. Natural rock phosphate: A sustainable solution for phosphorous removal from wastewater. Procedia Eng. 138: 119–126.

Truu, M., J. Juhanson and J. Truu. 2009. Microbial biomass, activity and community composition in constructed wetlands. Sci. Total Environ. 407(13): 3958–3971.

Uggetti, E., I. Ferrer, E. Llorens and J. Garcia. 2010. Sludge treatment wetlands: A review on the state of the art. Bioresource Technol. 101(9): 2905–2912.

Vymazal, J. 2005. Horizontal sub-surface flow and hybrid constructed wetlands systems for wastewater treatment. Ecol. Eng. 25: 478–490.

Vymazal, J. 2007. Removal of nutrients in various types of constructed wetlands. Sci. Total Environ. 380(1–3): 48–65.

Vymazal, J. 2008. Constructed Wetlands for Wastewater Treatment: A Review. Proc. Taal 2007: The 12th World Lake Conference. pp. 965–980.

Vymazal, J. 2011. Constructed Wetlands for Wastewater Treatment: Five decades of Experience. Environ. Sci. Technol. 45(45): 61–69.

Vymazal, J. 2013a. Emergent plants used in free water surface constructed wetlands: A review. Ecol. Eng. 61: 582–592.

Vymazal, J. 2013b. The use of hybrid constructed wetlands for wastewater treatment with special attention to nitrogen removal: A review of a recent development. Water Res. 47(14): 4795–4811.

Vymazal, J. 2014. Constructed wetlands for treatment of industrial wastewaters: A review. Ecol. Eng. 73: 724–751.

Vymazal, J. and L. Kröpfelová. 2009. Removal of organics in constructed wetlands with horizontal sub-surface flow: A review of the field experience. Sci. Total Environ. 407(13): 3911–3922.

Wallace, S.D. 1998. System for removing pollutants from water. US Patent # 6,200,469.

Wallace, S. and R. Kadlec. 2005. BTEX degradation in a cold-climate wetland system. Water Sci. Technol. 51(9): 165–171.

Wu, S., P. Kuschk, H. Brix, J. Vymazal and R. Dong. 2014. Development of constructed wetlands in performance intensifications for wastewater treatment: A nitrogen and organic matter targeted review. Water Res. 57: 40–45.

Wu, S., S. Wallace, H. Brix, P. Kuschk, W.K. Kirui, F. Masi et al. 2015. Treatment of industrial effluents in constructed wetlands: Challenges, operational strategies and overall performance. Environ. Pollut. 201: 107–120.

Zahraeifard, V. and Z. Deng. 2011. Hydraulic residence time computation for constructed wetland design. Ecol. Eng. 37(12): 2087–2091.

Zhai, J., M.H. Rahaman, X. Chen, H. Xiao, K. Liao, X. Li et al. 2016. New nitrogen removal pathways in a full-scale hybrid constructed wetland proposed from high-throughput sequencing and isotopic tracing results. Ecol. Eng. 97: 434–443.

18

Design and Construction of an Artificial Subsuperficial Wetland of Double Cell

An Experience in Palmillas, Querétaro Mexico

Salvador Alejandro Sánchez-Tovar

INTRODUCTION

This paper presents the joint experience of a private wastewater treatment company and the National Autonomous University of Mexico, modifying the traditional design conditions of a horizontal subsurface flow artificial wetland system.

A small wastewater treatment plant was built for the cafeteria of a tourist inn at km 193 of the Mexico-Querétaro highway in the town of El Marqués, state of Querétaro, Mexico. It consists essentially of a **septic tank, a degasser, a grease trap, an upflow anaerobic bioreactor, followed by a hydraulic regulator tank and a horizontal subsurface flow artificial wetland** in which the purpose of varying the hydraulics was for avoiding the formation of preferential roads that gradually tends to occur with the flow of wastewater and in the non-degradation of pollutants.

This paper aims to convey the experience gained during this innovative construction for those who want to apply the knowledge that is the result of it freely,

Universidad Nacional Autónoma de México, Facultad de Química, Departamento de Ingeniería Química, Laboratorios de Ingeniería Química Ambiental y de Química Ambiental (UNAM-FQ-DIQ-LIQAyQA), Ciudad Universitaria, 04510 Ciudad de México, México; INVENTEC, S.A. de C.V. Calle Lauro Villar 39, Col. Providencia, 02440 Ciudad de México, México.
Email: salvadorinvestigador@live.com.mx

especially to small communities where sanitation is non existing and with scarce funds to install a full conventional wastewater treatment system (Dallas and Ho 2005).

Artificial Wetlands

An artificial wetland (AW) can be defined as an area that is saturated by surface or groundwater with such frequency and duration that it is sufficient to maintain these saturation conditions. They have three basic advantages that make them an attractive potential for wastewater treatment:

- Physically fix pollutants on soil surface and organic matter.
- Use and transform the elements in water by microorganisms.
- Achieve consistent water treatment levels with low energy consumption and minimal maintenance.

Artificial wetlands can be used as a supplementary system in an existing water treatment plant to improve water quality (polish) and can also be used as the main treatment system in small communities (Durán-de-Bazua 2004; IWA 2000).

Fundamentals of Horizontal Subsurface Flow Artificial Wetland Design

In order to provide an overview of the "El Marqués" sewage treatment plant operation, the flow diagram of the wastewater treatment processes can be seen in Fig. 1. Figure 2 shows its location within the property and Fig. 3 depicts the general arrangement of the plant.

The treatment process begins with a fat and oil trap (TG-101). As the name implies, the fats and oils, being of lower density, float on the wastewater and are trapped. The trap also has a screen at the entrance where the largest solids are trapped, preventing occlusions or blockages in pipes and equipment.

At the outlet, in the TG-101 trap exit, a submersible pump BC-101 that feeds the digester or anaerobic deaerator DA-101 is located. This equipment removes 50% of the organic load measured as a biochemical oxygen demand. This means that the flow of partially treated water reaches the wetland with an average concentration of 170 mgL^{-1} ± 10%. The design flow is 0.2 L s^{-1}. The effluent of the anaerobic reactor is the influent of the artificial wetland.

This effluent flows by gravity from the anaerobic reactor DA-101 to the Regulatory Tank TR-101. This tank has an intermediate bulkhead so that sludge and suspended solids that can be entrained with the effluent from the digester are retained and reinjected to the reactor. This regulatory tank feeds the horizontal subsurface flow artificial wetland subsystem, which functions as a tertiary treatment for wastewater.

The artificial wetland is composed of two units, HA-101 and HA-102, so that each receives half of the effluent from the anaerobic reactor. In case one of the units requires maintenance, the full flow can be sent to the second unit.

Finally, the effluent from the artificial wetland system HA-101 and HA-102 flows by gravity into the TCL-101 chlorination tank. After this disinfection process, the water is either discharged or reused.

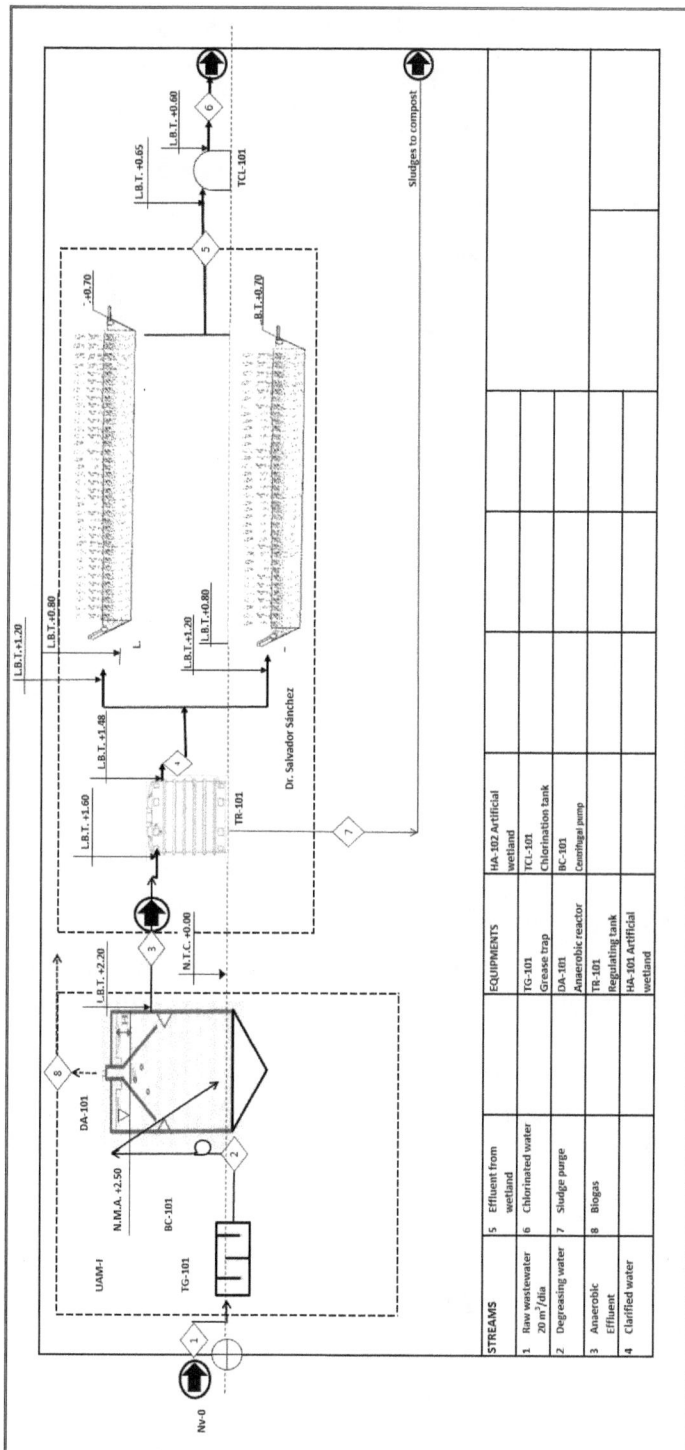

Fig. 1. Flow diagram of the wastewater treatment plant "El Marqués".

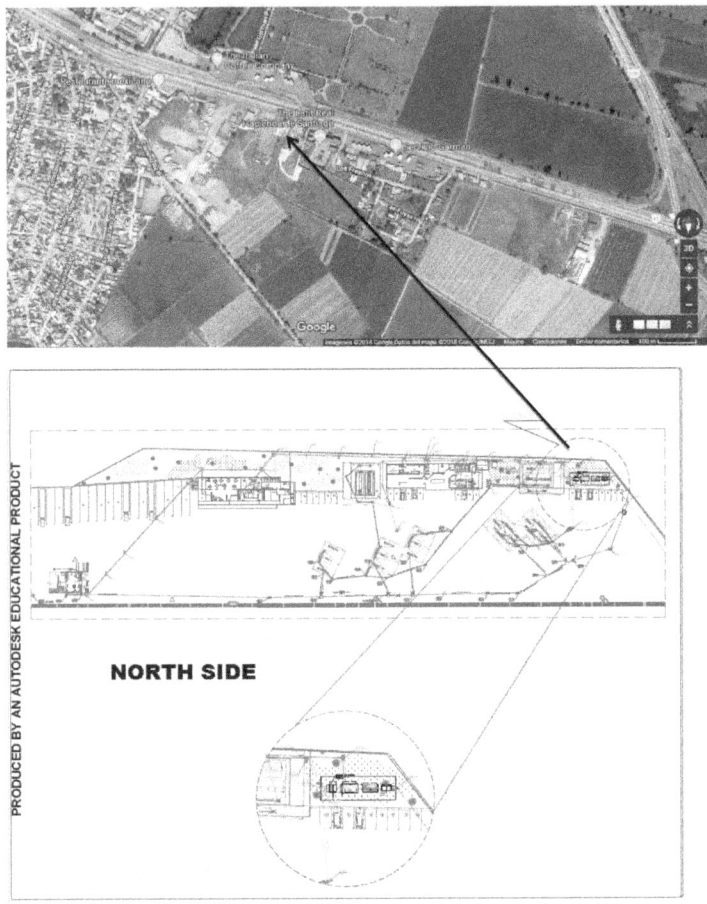

Fig. 2. Location of the wastewater treatment plant inside the facilities.

The design details and development of the project are described below by means of the next points, and is described how they were developed in the construction of the referred wastewater treatment plant.

- The type of terrain, as this will mean the type of civil and earthwork work.
- The waterproofing of the subsurface layer of the ground as a function of the permeability of the same.
- Primary treatment.
- The anaerobic degrader.
- The hydraulic distributor.
- Design of the wetland according to the classic model of a horizontal flow subsurface wetland.
- The vegetation used.
- The support media used.

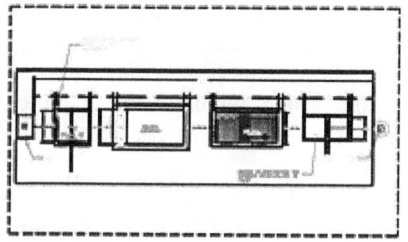

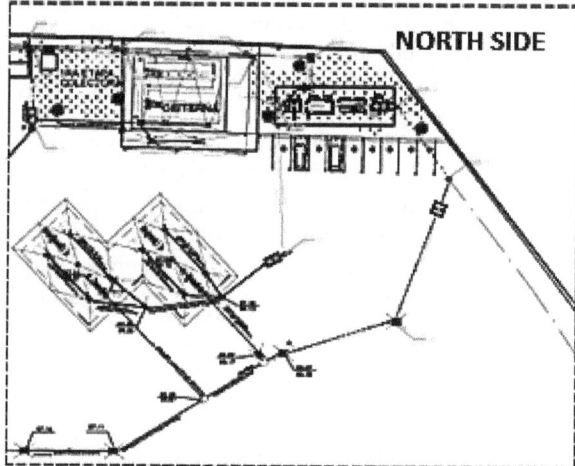

Fig. 3. General arrangement of the wastewater treatment plant inside the building.

- Selection and placement of the granular medium for subsurface systems.
- Disinfection.

Type of Terrain

Basically it can be hard or soft, spongy or compact, crumbly, sandy or stony, or even a mixture of the above. The advice of a professional or skilled craftsman of the construction is of total importance to define the type of machinery and manual force that will be needed for the practical work.

The hard rock terrains such as basalt, slabs and volcanic rock, are difficult to excavate and require more specialized machinery and personnel, so they can increase construction costs to more than three hundred percent.

It is highly recommended never to use trash or compost as fillers, as these materials have very low density, and are susceptible to irreversible sinking.

Neither should the wetland be built on debris fill, since the latter's low density causes it to sink and, in the long run, does not support the hydraulic load.

In some silt terrains, "piloting" work may be required. This practice, however, increases the budget allotted to the project but it is sometimes necessary.

It can be concluded that for the construction of artificial wetlands, highly compactable and free clay soils, or pebbles are the best option.

Artificial wetlands built above the soil line should have an adequate foundation. Masonry stone treated with asphalt is recommended. Concrete and steel rod foundation is not recommended without proper waterproofing, as it is susceptible to chemical corrosion (saltpetre) and may require cathodic protection, which is not always effective. Besides, concrete as alkaline material usually needs a pretreatment to protect it from any acid attack.

Artificial wetlands built below the ground line generally require excavation machinery (backhoes, tamping machines, etc.). The water table, the sand, the humidity, the feasibility of sinking, and the capacity to withstand the hydraulic load must be taken into account to choose either construction.

In the case of the artificial wetland of the "Parador Turistico El Marquéz", it was built in the backyard of a tourist inn at kilometer 193.12 of the Mexico-Querétaro highway, north of Mexico City. A study of soil mechanics was carried out at 10 meters depth and a rocky sandy soil was found, so it was determined to make concrete piloting structures of 14710 kPa (150 $kg_f cm^{-2}$), with a 50 cm masonry foundation for the sandblast. For the anaerobic reactor and for the artificial wetland system, a 60 cm masonry foundation was built. The concrete was reinforced with 1.2502 cm (0.5") thick steel rod.

In the following series of figures (Figs. 4–7), different photographs of the artificial wetland system of the tourist lodging of "El Marqués", Palmillas, Querétaro, Mexico, are shown.

Waterproofing

It is sometimes present naturally made by a layer of clay or silicates clays *in situ*, which can be compacted to a state close to a waterproof layer. Plasticizing chemical treatments can also used. Even a layer of bentonite or asphalt may be added. So far, the best and most durable option is the use of geotextiles and geomembranes. The latter requires equipment (welding machines, staplers, thermofusion devices) as well as specialized personnel for its installation.

For this project, the use of a commercial chemical additive as a pre-treatment for the concrete, which closes the micropores and produces water impermeability, was chosen and the artificial wetlands basins were covered with a geomembrane of high density polyethylene of 2.0 mm in thickness.

Primary Treatment, Septic Tank of Multiple Screens (TG-101)

The characterization of wastewater to be treated should be considered as the first design parameter. Firstly, solids content of wastewater should be measured. These can be divided into:

- Macrosolids. These are organic and inorganic rocks, sand and rubbish that must be removed by sieving, meshing, and desanding channels as they can clog the system, and
- Microsolids. These can be dissolved or suspended and can be divided into fixed and volatile solids. Some of them are eliminated in the preliminary treatment and the rest are degraded in the anaerobic reactor and in the artificial wetland.

Fig. 4. "Parshall" channel for flow measurement.

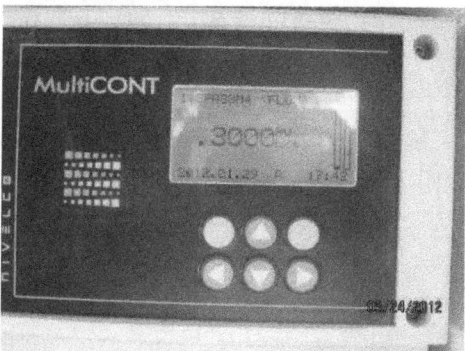

Fig. 5. Electronic flow meter.

Fig. 6. Primary treatment of the wetland.

This first container must be designed properly, since in addition to the function of receiving the raw wastewater, it acts as a first separator of macrosolids (using flotation and settling), converting the turbulent flow into a laminar one.

Fig. 7. Double-cell wetland with mature plant species.

It is advisable to place a removable sieve or mesh of plastic or stainless steel at the entrance to this device. In the case of urban wastewater, an inclined sieve formed by "medium" mesh is recommended, the separations varying between 1.5 and 5 cm, refining the sieving with a mesh of 0.5 to 1.5 cm, so that only the sands pass. The use of a manual, removable, and easy maintenance brush screen is recommended for low flow rates and low macrosolids contents. In this project, a sieve of 1 cm in diameter was installed. This sieve retains plastic, glass, metal, and paper objects.

To measure the volume of a hydraulic device, the hydraulic residence time or TRH must be considered, which is calculated by the equation

$$TRH = V/Q \tag{1}$$

where

V = Reactor volume, m^3

Q = Flow rate, m^3 s^{-1}

The number of septic tank screens will depend on the number of separation steps intended to be made therein. In the first bulkhead, the coarse solids (gravel, coarse sand, fibers and heavy grease—the light greases float-) are settled. The experience recommends a hydraulic residence time, TRH, of 2 hr (7,200 s). For this case, the restaurant has between 50 to 100 diners per day. The facilities had four toilets, four urinals, and four washbasins. The kitchen had a two-tier facility. Considering a flow rate Q = 0.5 L s^{-1}, the volume V may be calculated

$$V = TRH * Q \tag{1a}$$

$$V = 7,200 \ s * 0.5 \ Ls^{-1} = 3600 \ L = 3.8 \ m^3$$

Considering a safety margin of 100% and having an original volume of 1.2 m^3, if there are four separation screens, the separation volume will be 9.2 m^3, which is close to the volume rates proposed by Official Mexican Standard NOM-006-CONAGUA-1997 which are expressed in Table 1.

A relationship between length (L) and width (b), $2 \leq L/b \leq 4$ is recommended. If a minimum internal width (b) = 0.80 m and a minimum useful height (h) = 1.20 m. Width (b) and useful height (h), will be $b \leq 2$ m. The total length of the pit and the bulkhead for a volume of 9.2 m^3 will be 9.58 m.

Table 1. Mexican recommendation for the number of septic tanks depending on the number of users in dwellings or apartment buildings (NOM-006-CONAGUA-1997).

Number of people served	Working volume, m³	
	Rural sites	**Urban sites**
Up to 5	0.60	1.05
6 to 10	1.15	2.10
11 to 15	1.75	3.10
16 to 20	2.30	4.15
21 to 30	3.50	6.25
31 to 40	4.65	8.30
41 to 50	5.80	10.40
51 to 60	6.95	12.45
61 to 80	9.25	16.60
81 to 100	11.55	20.75

Anaerobic Bioreactor or Digester (DA-101)

The anaerobic bioreactor was designed by Dr. Oscar Armando Monroy-Hermosillo (*Universidad Autónoma Metropolitana - Iztapalapa* Unit in Mexico City) and it consists of an upflow anaerobic sludge blanket type, UASB, which reduces the initial load of pollutants measured as chemical oxygen demand, COD, biochemical oxygen demand, BOD, and total Kjeldahl nitrogen by half.

Hydraulic Regulator Tank (TRH-101)

The hydraulic regulator tank was constructed in order to avoid variations of flow from the digester to the wetland, controlling the possible entrainment of microorganisms from the reactor and maintaining a constant laminar flow to the wetland. Figure 8 shows a commercial plastic tank of 1,100 liters to which a high density polyethylene bulkhead is welded to its center with a height over the cylinder 0.25 m producing a U-type flow, first descending and then ascending, further obtaining a second sedimentation of solids.

Hydraulic Distribution

The hydraulic distributor established is a "logarithmic" system of flow distribution, with small circular holes having a diameter of 1.0 cm, which were first spaced 30 cm in the center of the distributor and from there the distance was decreasing to the next hole in a "logarithmic" scale that can be base 3 to base 10. This type of distribution forces the flow to move to the edges of the distributor and not to concentrate in the system, obtaining a better distribution along the bed of the artificial wetland (Fig. 9).

It could be affirmed that the mathematical models most used in the design of artificial wetlands are: The model of Reed et al. (1995), the Crites and Tchobanoglous model (Crites et al. 1988), and the Kadlec and Knight model (1996). Any of the three

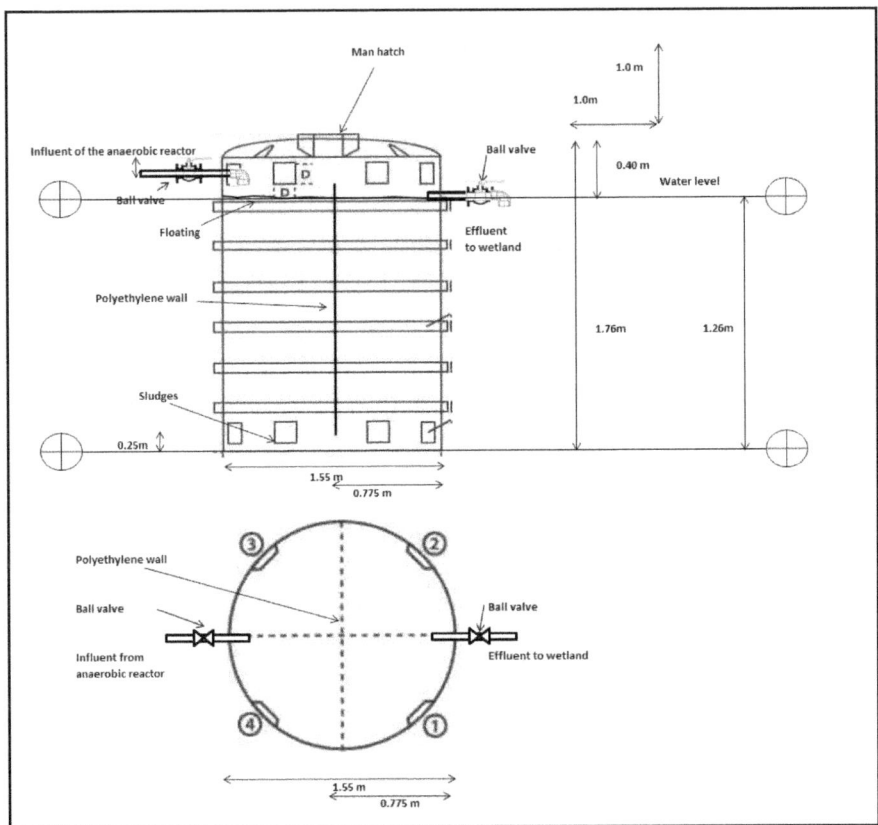

Fig. 8. Hydraulic regulator tank after the digester (Influent, effluent).

mentioned models permit to calculate the design dimensions in a range of 3.0% to approximately 5.0% according to the practical experience of the author. Although chemical engineering bases such as thermodynamics and heat transfer are considered, none of the three models take into account hydraulic factors.

From that practical experience in the construction of wastewater treatment plants, WWTP, the lesson learned is that hydraulic models are very important to avoid the so called dead zones that may reduce the overall performance of any artificial wetland system.

The first model was chosen considering the hydraulic properties. As these biochemical reactors in general follow a first order kinetics for removing organic biodegradable matter, particularly carbon and nitrogen, the basic equation of a piston flow reactor is (Durán-de-Bazúa 1997; EPA 2000; Langergraber et al. 2008; Libmaber and Orozco-Jaramillo 2013):

$$\frac{Ce}{Co} = e^{-KTt} \qquad (2)$$

where:

Ce = Concentration of the pollutant in the effluent, mg L^{-1}

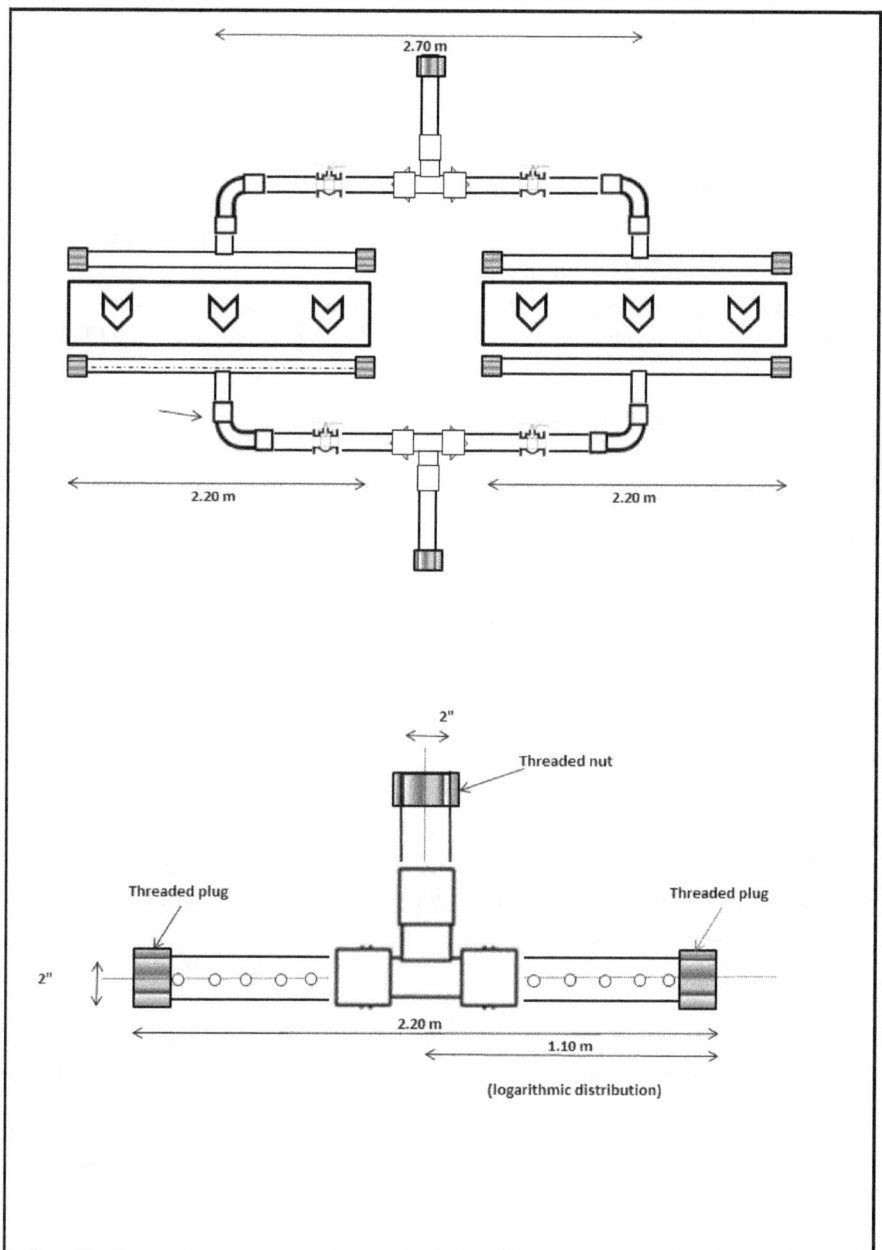

Fig. 9. Hydraulic inlet and outlet distributors to the artificial wetland system.

Co = Concentration of the pollutant in the influent, mg L^{-1}

K_T = First order reaction constant, dependent on temperature, d^{-1}

t = Hydraulic residence time, d.

Hydraulic residence time may be calculated using the following equation:

$$t = \frac{LWyn}{Q} \tag{3}$$

where:

L: Length of the artificial wetland cell, m

W: Width of the artificial wetland cell, m

y: Depth of the artificial wetland cell, m

n: Porosity—available space for water flow through the support media of the artificial wetland cell left by vegetation roots and other solids-, generally expressed between 0 and 1.0.

Average flow rate is calculated as follows:

$$Q = \frac{Q_e + Q_o}{2} \tag{4}$$

where:

Q: Average flow rate through the reactor, $m^3\ s^{-1}$

Q_e: Flow rate at the exit of the reactor, $m^3\ s^{-1}$

Q_o: Flow rate at the entrance of the reactor, $m^3\ s^{-1}$

With equations (2) and (3) surface area of the artificial wetland is calculated:

$$A_s = LW = \frac{Q \cdot \ln(C_o/C_e)}{K_T yn} \tag{5}$$

where A_s = Surface area of the artificial wetland, m^2.

The value of K_T depends on the pollutant to be eliminated. In case of not having values from the literature, the following equation can be applied, considering its value at 20°C, K_{20}:

$$K_T = K_{20}(1.1)^{(T-20)} \tag{6}$$

Table 2 shows the values for K_{20} and other physical characteristics, such as porosity and hydraulic conductivity, most commonly found in the literature for calculating artificial wetlands as a function of the support media utilized.

According to the calculation reports and applying equation 5, an area of 65 m^2 was found, considering a width:length ratio of 2:1, a cell of 6 × 10.5 m would be needed. Due to space availability, and given the design of the hydraulic distributor, two cells of 3.0 x 12.0 m were considered, as shown in Fig. 10. Figure 11 shows the application of the hydraulic distributor and Fig. 12 shows the cross-section of a cell.

The Vegetation Used

As mentioned before, hydrology is the most important design factor in an artificial wetland as it brings together all the wetland's functions and it is often the primary

Table 2. Commonly used values for K_{20} and other physical characteristics for media materials to be used in the calculations of artificial wetlands (Reed et al. 1995).

Material	Average diameter of media, mm	Porosity, %	Hydraulic conductivity, k_s, $m^3m^{-2}d^{-1}$	K_{20}
Medium size sand	3.2	36–40	10,000–50,000	1.84
Gross size sand	5–7	28–32	1000–10000	1.35
Gravel	8	30–55	500–5000	0.6
*"Tezontle"** (volcanic slag)	20–30	40–50	20000–25000	2.1

*Aztec derived word, meaning stone as light as hair, *tetl* = stone, *tzontli* = hair

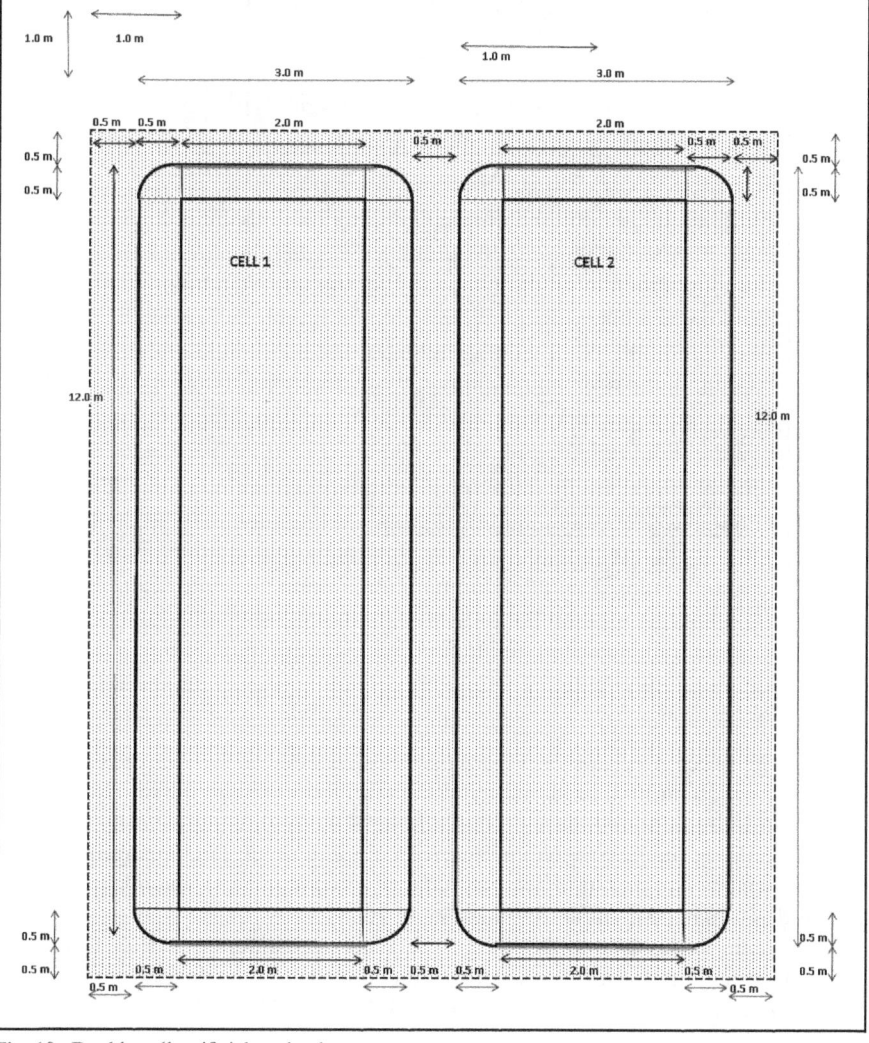

Fig. 10. Double-cell artificial wetland system.

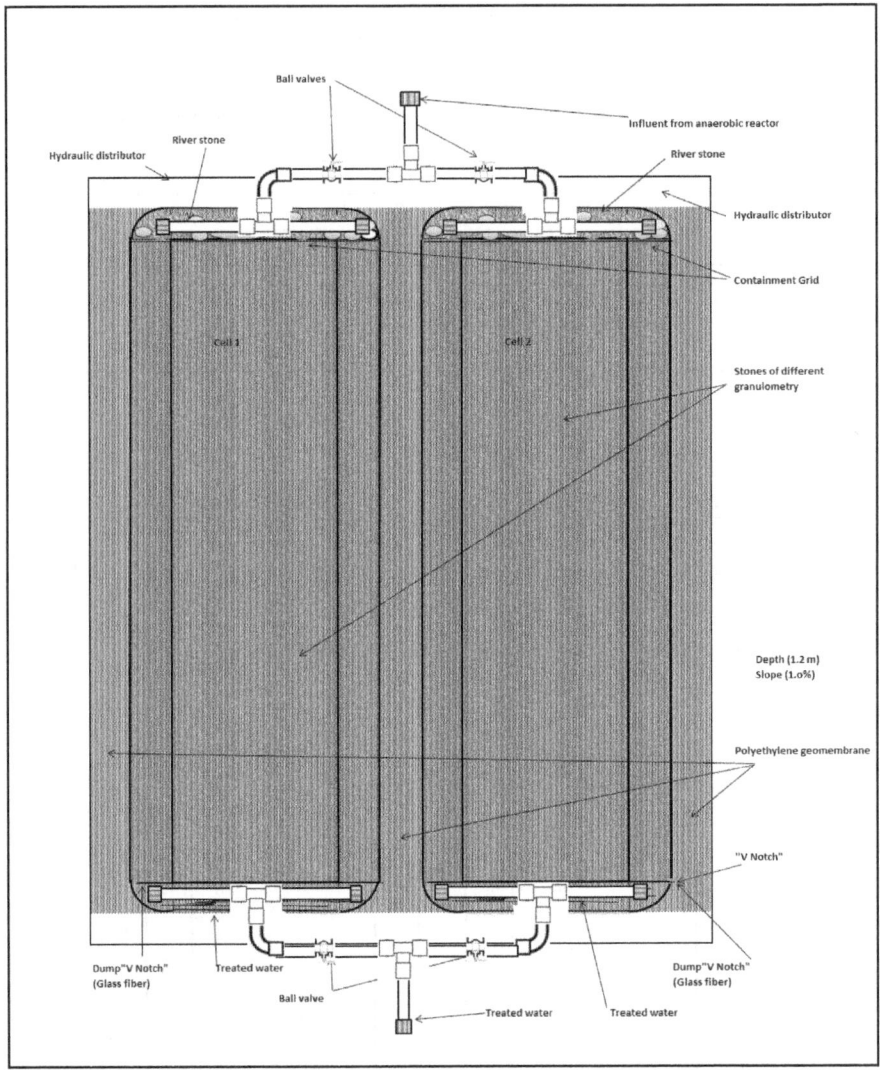

Fig. 11. Dual-cell artificial wetland system with the hydraulic distributor (glass fiber).

factor in its success or failure. Hydrology can be affected by the density of the vegetation which causes sinuous movements of the water through the roots, rhizomes, stems, and loose decaying leaves. The fact that a wetland is permanently saturated with water allows the formation of an anaerobic zone in which certain biogeochemical processes can be carried out to help the removal of contaminants and other functions (Durán-de-Bazúa 2004; Durán-de-Bazúa et al. 2008).

The main importance of the use of vegetation in artificial wetlands lies in the oxygen transfer to the root zone. As already mentioned at the beginning, the functions of vegetation for wastewater treatment are as follows (ITRC 2003):

• Stabilizes the support media and limits flow channeling.

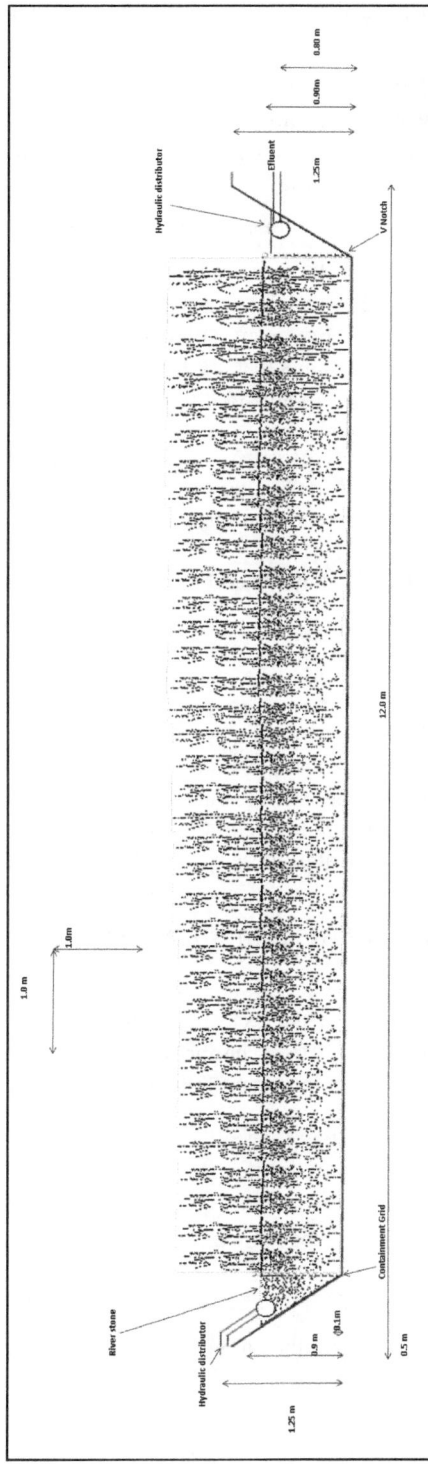

Fig. 12. Cross-section of the horizontal subsurface flow artificial wetland system considered.

- Slows down water velocities to allow suspended materials to settle and remain in the reactor.
- Takes nutrients and trace elements by incorporating them into its tissues.
- Transfers gases between the atmosphere and sediments.
- Releases oxygen through photosynthesis from surface structures of plants oxygenating the spaces within pores of support media.
- Stem and root systems give rise to sites for the attachment of microorganisms.

The selection of vegetation will depend on local climatic conditions, water depth, wetland design (whether it is a surface, subsurface, or floating-type wetland) and the characteristics of the wastewater to be treated. Emergent, submerged, floating, and floating root type plants can be used. Emergent plants that are frequently used in the construction of artificial wetlands for wastewater treatment are bulrushes, reeds, among many others (IWA 2000).

Plants or macrophytes selected for the assembly of this system were the reeds (*Phragmites australis*) and the Egyptian papyrus (*Cyperus papyrus*) for their adaptability and resistance to climate changes. In the end, the system was dominated by the second plant.

Every six months, the plants are cut to a height of 60 cm and the biological mass is chopped and sold for cattle feeding.

The Support Media Used

The types of support media commonly used in the installation of artificial wetlands, as shown in Table 2 before, include soil, sand, gravel, rock, plastic materials, and even organic materials such as compost or corks or residues such as shredded plastic from soft drinks bottles. Some sediment and debris can accumulate in the wetland because of the low water velocity and high productivity typical of these systems.

Selection and Placement of the Granular Medium for Subsurface Systems

Selection of the granular medium for the case of subsurface systems is also important (EPA 2000; IWA 2000; ITRC 2003; Amábilis et al. 2016). The factors by which the support media, sediments, and vegetation remains are important follow:

- They act as support for many living organisms in the system.
- Its permeability can affect the movement of water through the artificial wetland system.
- Many chemical and biological transformations (especially of the microbial type) can occur.
- Provides storage for many contaminants.
- The accumulation of vegetation remains increases the amount of organic matter in the wetland.

Organic matter gives rise to the exchange of matter, fixation of microorganisms, and is a source of carbon, which is the source of energy for some of the most important biological reactions being carried out.

The physical and chemical characteristics of the support media can be altered by flooding, since water replaces the atmospheric gases in the pores and the microbial metabolism consumes the available oxygen producing the formation of an anoxic milieu, which will be important for the removal of contaminants such as metals and the production of electricity by anaerobic bacteria (Amábilis-Sosa et al. 2016; Salinas-Juárez et al. 2016).

Disinfection

The disinfection of the treated effluent is carried out by means of a 5% sodium hypochlorite commercial solution system injected before disposal or reuse of the effluent to achieve a dilution of 1.0 to 2.0 mg L^{-1}.

System Starting-up and Stabilization

Just as the operation of the anaerobic reactor requires a gradual start, the development of the vegetation and the biomass that degrades the organic matter dissolved in the residual water in the wetland, entails a gradual process.

Once the reeds are planted, they may partially or totally lose their foliage. The development of new shoots depends on the environmental conditions of the geographical region where the artificial wetland is to be built and the capacity for natural adaptation of the plants in the sowing, as well as the correct contribution of nutrients to them, and this may take from some days to several months.

The flow of the influent to be treated should be gradually increased. The rate will depend on the results of periodic analyses using as common parameters the chemical oxygen demand, pH, and electric conductivity (Soto-Esquivel et al. 2013).

Once the system is stable, considering variations of the chemical oxygen demand with less than ± 10 percent in the effluent, it is given to the responsible person that has to take care of the growth of the vegetation, and periodically measure influent and effluent parameters.

Final Remarks

The systems byproducts, such as fats and oils and suspended solids coming out of the primary settler tanks are sent to a container where lime is added to disinfect them, and once the pH value is constant, *Eisenia foetida* is added to promote its stabilization. The final product is used as a soil improver in the green area around the inn.

Thus, no wastes are accumulated from the whole system.

The Appendix presents two tables with the calculations data, useful for any community wishing to install a similar system.

Conclusions

Considering the intrinsic goodness of the small system presented in this contribution, the artificial wetlands system as it was planned together with the other parts of it, such as the fats and oil trap, the primary settler, and the anaerobic reactor have been operating without any problem since its stabilization period.

The costs involved will depend upon the factors mentioned in this contribution but definitely are much lower than a conventional wastewater treatment system.

Appendix. Memory of Calculation

Table A.1. Water composition and flow design data.

COMPOSITION	UNITS	RAW WASTEWATE INFLUENTE	ANAEROBIC REACTOR EFFLUENT	WETLAND EFFLUENT
DQO	(mg/L)	336	235.2	23.52
DBO$_5$	(mg/L)	155	108.5	10.85
SST	(mg/L)	157	109.9	10.99
SSV	(mg/L)	93	65.1	6.51
FAST & OILS	(mg/L)	36	25.2	2.52
NTK	(mg/L)	29	20.3	2.03
P	(mg/L)	8	5.6	0.56

FLOW DESIGN DATA								
Population			200 Habitants					
Q= Flow average(population):			17 m³/d-population					
Flow (per habitant)=			85 L / d-habitant					
Total flow (L/d-población)=			17,000 L /d-population					
Total Flow (L/min-population)=			11.81 L / min-population					
Flow numerical convertions								
Qmed =	0.20	L/s	=	17.00	m³/d =	0.71	m³/h =	0.0118	m³/min
Q max inst =	0.26	L/s	=	22.10	m³/d =	0.92	m³/h =	0.0153	m³/min
							Q max inst =	0.0003	m³/s

Media: Red Tezontle (Red stone)		25	mm	n=	0.38
Vegetation:		Papyrus (Papyrus egiptian)			
Depth of the wetland SFSF:		0.7 m			
Porosity of the wetland FWS		0.6 %			
Critical temperature in winter:		5 °C		25	
Average annual water temperature at the entrance:		25 °C			

Table A.2. Dimensioning of a horizontal subsurface flow artificial wetland.

CALCULATIONS AND PREDIMENSIONING			
Calculation of the wetland area			

$$A_s = LW = \frac{Q \cdot \ln(C_o / C_e)}{K_T y n} \qquad .(1)$$

Y = Depth of wetland cell, m
n = Porosity, or space available for water flow through the wetland

$$K_T = K_{20}*(1.06)^{(T-20)}$$

$$K_{20} = 1.104 * d^{-1}$$

$$K_T = 1.104 * (1.06)^{(T-20)} \qquad\qquad K_T = \qquad 1.5256 \quad d^{-1}$$

$$As = \quad 65 \quad m^2$$

TRH = (Area (m2) * Depth (m) * Porosity) / Flow (m3 / d)
TRH =   1.24 d.
 Calculation of the average temperature of the water,
Calculation of the average temperature of the water,

Winter temperatures	10	°C
Water ionlet temperature	25	°C
Clobal coefficient of heta transfer		
It is assumed		
A layer of vegetation residues	15	cm
A layer of dry gravel	0.8	cm
A layer of medium saturated gravel	60	cm

$$U = \quad \cfrac{1}{\dfrac{0.15}{0.05} \quad + \quad \dfrac{0.0008}{1.5} \quad + \quad \dfrac{0.6}{2}} \qquad 0.30$$

Using Tc

$$Tc = \quad \frac{(To-Ta)*U*\alpha*As*t}{Cp*\delta*As*y*n} \qquad = \quad 0.3 \qquad °C$$

$$Te = \quad To - Tc \qquad = \quad 24.7 \qquad °C$$

$$Tw = \quad \frac{To - Te}{2} \qquad = \quad 24.9 \qquad °C$$

T, assumed   =   T of design

Calculation of length by width

$$W = \frac{1}{y}\left[\frac{(Q)(A_s)}{(m)(k_s)}\right]^{0.5} \qquad .(2)$$

W: Width of a wetland cell, m
As: Wetland surface area, m2
L: Length of wetland cell, m

$$L = \left[\frac{A_s y^{8/3} m^{1/2} \cdot 86400}{a \cdot Q}\right]^{2/3} \qquad .(3)$$

m: Bed bottom slope,% expressed as decimal.
Y: depth of water in the wetland, m
ks: Hydraulic conductivity of a unit of wetland area
Perpendicular to the flow direction, m3 / m2 / d.

L:W	3:1		
It is designed for two units		32.7 m2 each one It proposes 2 of 200 m2 in parallel	
As =	65 m²		
As$_1$ = As$_2$ =	33 m²		
	L= $\quad$ =	12.1	m
W= $\quad \dfrac{33}{12.1}$	=	2.7	m
L : W $\quad$ 3:1		36.2	m²

PROPOSAL 1		PROPOSAL 1	
Reactors	1	Reactors	2
Width (m)	6	Width (m)	3
Length (m) =	12	Length (m) =	12
Maximum depth of grave (n	1	Maximum depth of grave (ı	1
Maximum depth of water (n	0.9	Maximum depth of water (ı	0.9
Slope 1-2% according to surveyor		Slope 1-2% according to surveyor	

Acknowledgements

The author acknowledges the collegial peer revision from Dr. Maria Guadalupe Salinas-Juárez that greatly improved this contribution.

References Cited

Amábilis-Sosa, L.E., C. Siebe, G. Moeller-Chávez and M.C. Durán-Domínguez-de-Bazúa. 2016. Remoción de mercurio por *Phragmites australis* empleada como barrera biológica en humedales artificiales inoculados con cepas tolerantes a metales pesados. Rev. Int. Contam. Ambie. 32(1): 47–53. In Spanish.

Crites, R., D. Gunther, A. Kruzic, J. Pelz and G. Tchobanoglous. 1988. Design manual: constructed wetlands and aquatic plant systems for municipal wastewater treatment. EPA/625/1-88/022.U.S. Environmental Protection Agency, Office of Research and Development. Washington, D.C. US.

Dallas, S. and G. Ho. 2005. Subsurface flow reed beds using alternative media for the treatment of domestic greywater in Monteverde, Costa Rica, Central America. Wat. Sci. Technol. 51(10): 119–128.

Durán-de-Bazúa, C. 1997. Tratamiento biológico de aguas residuales. Pub. UNAM, Facultad de Química, Programa de Ingeniería Química Ambiental y de Química Ambiental. 7th edition. Mexico City, Mexico. In Spanish.

Durán-de-Bazúa, C. 2004. Tratamiento sostenible de aguas de suministro y residuales para países con economías emergentes/Sustainable treatment of water and wastewaters for emerging countries. *In*: Durán-de-Bazúa, C. and L.I. Ramírez Burgos [eds.]. Proceedings of the Third International Minisymposium on Removal of Contaminants from Wastewaters, Atmosphere, and Soils. Compact disk, July 7–10, UNAM, Facultad de Química, PIQAyQA, México D.F. Mexico. In Spanish.

Durán-de-Bazúa, C., A. Guido-Zárate, T. Huanosta, R.M. Padrón-López and J. Rodríguez-Monroy. 2008. Artificial wetlands performance: nitrogen removal. Water Sci. Technol. 58(7): 1357–1360.

EPA. 2000. Folleto informativo de tecnología de aguas residuales. Humedales de flujo subsuperficial. Agencia de Protección al Ambiente. Washington, D.C. US. In Spanish.

ITRC. 2003. Technical and regulatory guidance document for constructed treatment wetlands. The Interstate Technology and Regulatory Council Wetlands Team. 128 pp. http://www.itrcweb.org/GuidanceDocuments/WTLND-1.pdf.

IWA. 2000. IWA Specialist Group on Use of Macrophytes in Water Pollution Control. Constructed wetlands for pollution control: processes, performance, design, and operation. Scientific and Technical Report No.8, London, United Kingdom.

Kadlec, R.H. and R.L. Knight. 1996. Treatment Wetlands, CRC Lewis Publishers. New York, US.

Langergraber, G., K. Leroch, A. Pressl, R. Rohrhofer and R. Haberl. 2008. A two-stage subsurface vertical flow constructed wetland for high-rate nitrogen removal. Water Sci. Technol. 57(12): 1881–1887.

Libmaber, M. and A. Orozco-Jaramillo. 2013. Sustainable treatment of municipal wastewater. Water 21. IWA. London, UK. http://www.iwapublishing.com/books/9781780400167/sustainable-treatment-and-reuse-municipal-wastewater.

NOM-006-CONAGUA-1997. NORMA Oficial Mexicana Fosas sépticas prefabricadas - Especificaciones y métodos de prueba. Diario Oficial de la Federación, 29 de enero de 1999. Mexico City, Mexico. In Spanish.

Reed, S.C., R.W. Crites and E.J. Middlebrooks. 1995. Natural systems for waste management and treatment. McGraw-Hill. Second Edition. New York, US.

Salinas-Juárez, M.G., P. Roquero-Tejeda and M.C. Durán-Domínguez-de-Bazúa. 2016. Plant and microorganisms support media for electricity generation in biological fuel cells with living hydrophytes. J. Bioelectrochemistry. DOI 10.1016/j.bioelechem.2016.02.007.

Soto-Esquivel, M.G., A. Guido-Zárate, S. Guzmán-Aguirre, A.G. Mejía-Chávez, R.S. García-Gómez, T. Huanosta et al. 2013. Algunos aspectos interesantes de sistemas de humedales a escala de laboratorio y de banco en México. Química Central (Ecuador) 3(2): 53–65. In Spanish.

19

Strategies of the Constructed Wetlands Operation under the Perspective of the Global Change Scenario

Gladys Vidal, * *Daniela López, Ana María Leiva, Gloria Gómez, William Arismendi* and *Sujey Hormazábal*

INTRODUCTION

Constructed wetlands are engineered systems that have been designed and constructed to treat water or biosolids under different operational conditions.

These systems take advantage of natural processes involving wetland vegetation, soils and microbial assemblages to remove organic matter, nutrients, and specific compounds contained in domestic or industrial wastewater, sludge and diffuse emissions of runoff, and drainage from agricultural and forest soil (Vymazal 2007).

Moreover, they are a cost-effective natural alternative to conventional wastewater systems and a feasible technology for the treatment of point or non-point source of water pollution (Plaza de Los Reyes and Vidal 2007).

Constructed wetlands are commonly used for the treatment of domestic and industrial wastewater. According to hydraulic characteristics, they are classified in three types of wetlands (Kadlec and Knight 1996):

(a) surface flow (SF),

(b) horizontal subsurface (HSSF), and

(c) vertical subsurface constructed wetlands (VSSF).

Engineering and Biotechnology Environmental Group, Environmental Science Faculty & Center EULA–Chile, University of Concepción. P.O. Box 160-C, Concepción-Chile.
* Corresponding author: glvidal@udec.cl

Also, various types of these systems may be combined and known as hybrid constructed wetlands.

Liquid waste generated by anthropogenic activities present high concentrations of organic matter and nutrients. For these reasons, they cannot be discharged to water bodies due to the possible environmental impact that may cause. The conventional treatment for removing these compounds consists on partial degradation or total mineralization by biological and chemical process (Kadlec and Wallace 2009).

However, under the global change scenario, these effluents can be used for resource recovery. In this perspective, wastewater treatment technology has to apply different strategies for reducing global climate change such as:

(a) nutrient and energy recovery,

(b) water reclamation and reuse,

(c) biodiversity conservation and

(d) greenhouse gas (GHG) reduction.

Therefore constructed wetlands could be an alternative for dealing this challenge.

Recently, these systems have been applied for wastewater reuse in agriculture. Many researches had determined that constructed wetlands effluents have very good physical-chemical properties for water reuse. However, there are divided opinions about the disinfection efficiencies of these systems (Melián et al. 2010).

Despite this application, constructed wetlands can be sources of three important GHG: carbon dioxide (CO_2), methane (CH_4), and nitrous oxide (N_2O) (Mander et al. 2008). One strategy for reducing GHG emissions is to combine these systems with anaerobic digestion technology (Plaza de Los Reyes and Vidal 2014; López et al. 2015). The advantage of this treatment sequence is the energy recovery.

Regarding nutrients recovery, phosphorus is contained (60–80%) as phosphate (PO_4^{-3}-P) in sewage, beside nitrogen, solids and organic matter. The potential phosphorus recovery from sewage could be a good alternative for agriculture in rural areas using constructed wetland and natural compounds with high level of adsorption as support medium (Vera et al. 2014). On the other hand, constructed wetlands can contribute to decreasing the negative impacts on the environment and to improve biodiversity by the production of fibers and flowers (Vidal and Hormazabal 2016).

Macrophyte plants as *Phragmites* spp., *Schoenoplectus* spp., and *Typha* spp. assimilate nutrients and then use these components for fiber production. In the same way, species of ornamental plants as *Zantedeschia* spp. and *Iris* spp. can produce flowers from nutrients presented in wastewater (López et al. 2016; Burgos et al. 2017).

The goal of this chapter is to describe the strategy of the constructed wetlands for treating domestic and/or industrial wastewater until now and to describe the use of constructed wetlands under a perspective of the global change scenario.

Design and Operational Characteristics of Different Constructed Wetlands

There exist several types of constructed wetlands classification. Considering Kadlec and Knight (1996), they could be classified according to the wetland hydrology into surface (SF) and subsurface constructed wetland (SSF). At the same time, SSF can

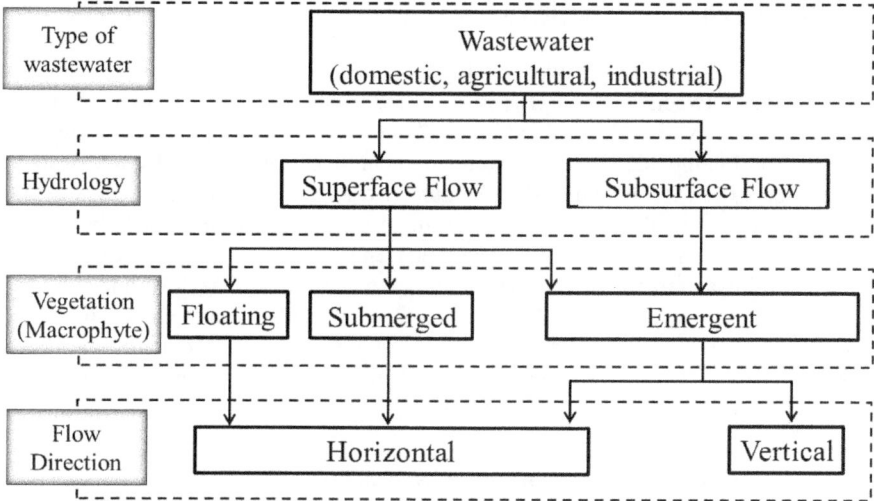

Fig. 1. Constructed wetland classification (adapted from Morales et al. (2014); Plaza de Los Reyes et al. (2011)).

be divided according to the flow direction (horizontal and vertical). Moreover, these types of constructed wetlands may be combined to achieve a higher treatment effect by forming hybrid constructed wetlands.

Other classification based on the life form of the dominating macrophyte divides constructed wetlands into:

(a) free floating macrophyte system, with vegetation which do not present rooting in the deep zone (e.g., *Lemna* spp. and *Eichornia crassipes*),

(b) submerged macrophyte system, with plants which have their foliar system submerged in the water column (e.g., *Isoetes lacustris, Lobelia dortmanna, Egeria densa,* and *Elodea canadensis*), and

(c) emergent macrophyte system compounded by rooted vegetation on the deep zone but their leaves and stems over the surface (e.g., *Phragmites* spp., *Typha* spp., *Schoenoplectus* spp., and *Carex* spp.) (Sundaravadivel and Vigneswaran 2001; Plaza de Los Reyes et al. 2011). These classifications can be combined as shown Fig. 1.

Surface Flow Constructed Wetlands (SF)

SF wetlands are a type of constructed wetlands where the wastewater is exposed to the atmosphere. This system mimics natural wetland in appearance and function because water flows horizontally across the surface and it is filtered through a dense support of aquatic plants (USEPA 2000). As the wastewater moves slowly through the wetland, it is treated by the process of sedimentation, filtration, oxidation, reduction, adsorption, and precipitation (Halverson 2004; Kadlec and Wallace 2009). Moreover, it is identified the aerobic zone near surface layer of water while the anaerobic zone is located in deeper water and in the substrate (Halverson 2004). Typical design and operational characteristics of SF are: (a) 0.2–0.3 m of rooting soil; (b) water depths between 0.2–0.6 m; (c) hydraulic loading rates between 12–50 mm d^{-1}; (d) organic loading rates between 1–11 g BOD$_5$ m^{-2}·d^{-1}, and (e) hydraulic residence time (HRT) between 5–15 d (Vymazal and Kröpfelová 2008). As shown in Fig. 1, SF's vegetation may be free floating, submerged, or/and emergent macrophytes (Kadlec and Wallace 2009). Because of the potential risk for human exposure to pathogens, FS wetlands are commonly used for advanced treatment of effluent from secondary or tertiary treatment. At the same time, they have been employed for urban agricultural and industrial rainwater treatment due to their ability to work with pulse flows and changing water levels (Vymazal 2013). FS wetlands are suitable in all climates, including cold zones. However, they are considered more sustainable in warm climates (> 10°C). Furthermore, FS systems provide significant benefits in the form of human uses and wildlife habitat (Kadlec and Knight 1996).

Horizontal Subsurface Constructed Wetlands (HSSF)

HSSF wetland is a system where wastewater circulates through the support medium, rhizomes, and roots plants. The typical design consists of a rectangular (bed slope between 0.5–1%) filled with gravel or soils and planted with emergent vegetation (Kadlec and Wallace 2009). Wastewater (pre-treated) are fed in at the inlet and flow horizontally and slowly under the surface through the support medium until the outlet. Finally, they are collected via level control arrangement. During the passage of wastewaters through the wetland, these makes contact with a network of aerobic, anaerobic, and anoxic zones (Vymazal 2005). The presence of roots plants and support medium favors the development of biofilm which promotes the removal of organic matter through the anaerobic degradation.

Nonetheless, nutrients removal (nitrogen and phosphorous) usually reaches lower levels (Kadlec and Knight 1996; Akratos and Tsihrintzis 2007; Kadlec and Wallace 2009). Regarding operational characteristics of HSSF systems, the hydraulic loading rate recommended varies between 23–50 mm d^{-1} with HRT between 2–10 d.

The organic loading rate may be between 3–15 gBOD$_5$ m^{-2}·d^{-1} and the area of bed request is between 5–10 m^2 P.E^{-1} (P.E.: population equivalent) (Vymazal and Kröpfelová 2008; Plaza de Los Reyes et al. 2011).

HSSF wetlands are probably the most commonly operated of constructed wetlands over the world. They are frequently used for secondary treatment for single-family homes or for small communities (Wallace and Knight 2006).

In addition, there are many other applications like treatment of industrial and landfill wastewater (Vymazal and Kröpfelová 2008).

Vertical Subsurface Flow Constructed Wetlands (VSSF)

VSSF systems have achieved a growing interest because they are an alternative for producing nitrified effluents as a consequence of their greater oxygen transfer capacity

(Green 1996; Cooper 1999); thus, allowing the partial removal of reduced nitrogen. In a VSSF wetland, wastewater is fed intermittently onto the bed surface flooding the entire surface and then by gravity, percolates and drains through the medium support (Stefanakis et al. 2014). This characteristic allows that air fills the void spaces from the atmosphere to the system (Prochaska et al. 2007) achieving oxygen transfer rate between 10 to 24 mg O_2 $m^{-2}·d^{-1}$ (Cooper 2005). This way of feeding is important for the treatment because of the enhanced aerobic conditions for the oxidation of ammonia nitrogen (NH_4^+-N) (nitrification) and decomposition of organic matter. However, these conditions do not favor denitrification due to lack of convertion from nitrate to gaseous nitrogen (N_2) (Vymazal 2008; Plaza de Los Reyes et al. 2011). In VSSF systems, the treatment of effluents is made by physical, chemical, and biological interactions between the different components of wetlands (wastewater, support medium, vegetation, and microorganisms). The biofilm attached to the support medium and to the rhizosphere consist of a dense and rich microbiota that carries out removal mechanism by microorganism in parallel of filtration, sedimentation, absorption, and adsorption processes (García-Serrano and Corzo-Hernández 2008; Kadlec and Wallace 2009; Vymazal 2011a). Regarding operational characteristics of VSSF wetlands, some authors determine that the hydraulic loading rate have to be close to 50 mm/d and the organic loading rates may be between 20–30 $gBOD_5$ $m^{-2}·d^{-1}$ (Stefanakis et al. 2014). Furthermore, the nitrogen loading rate and HRT can create variations between 2.8–30 g TN $m^{-2}·d^{-1}$ and 1–2 d, respectively (Plaza de Los Reyes et al. 2011). Compared to HSSF systems, VSSF systems require less area for treatment which ranges between 1–3 m^2 P.E^{-1} (Cooper 1999). The most common use of VSSF wetlands is for municipal and domestic wastewater treatment. Due to their ability to oxidize NH_4^+-N, they have also been employed for the treatment of wastewater with high NH_4^+-N concentrations like landfill leachates, food processing wastewaters, and dairy wastewater (Kadlec and Wallace 2009).

Hybrid Constructed Wetlands

Hybrid constructed wetlands are systems which operation and performance consists in a combination of various constructed wetlands operated sequentially (Zhi and Ji 2014). The concept is to use advantages of VSSF and HSSF wetlands for enhancing the transformation and removal of nitrogen (Ayaz et al. 2012). In addition to provide more diverse redox condition for microorganism, they achieve higher potential for removal of multiple pollutants (Meng et al. 2014). VSSF systems which are intermittently fed provide suitable conditions for nitrification but denitrification does not take place (Vymazal 2008). On the other hand, HSSF systems are predominantly anaerobic/anoxic and therefore, can provide good conditions for denitrification if nitrate is present. Thus, hybrid systems generate good conditions for nitrification and denitrification at the same time. The most commonly combinations of hybrid systems are VSSF-HSSF, HSSF-VSSF, and VSSF-VSSF (Ye and Li 2009; Vymazal 2013). Furthermore, multiple stage systems have been studied where more than two constructed wetland in sequence were used (Vymazal 2013).

Effluent Treatment by Constructed Wetlands

Constructed Wetlands for Sewage Treatment

During the last decades, constructed wetlands have been used for treatment of domestic sewage, especially for low population density areas (Steinmann et al. 2003; Ciria et al. 2005). These systems offer an environmental and sustainable technology which generates economic savings in terms of electricity and human labor and provide a wildlife habitat. Likewise, they have lower construction and maintenance costs (Cooper 2006).

Table 1 compares the performances of wastewater treatment using SF, HSSF, and VSSF systems. For organic matter removal, HSSF and VSSF systems show efficiencies of up to 95% while SF wetlands only achieve 85%. As SF systems are exposed to the atmosphere, the rate of organic matter removal is reduced directly when temperatures decreased (USEPA 2000). However, removal efficiencies of total nitrogen (TN) in HSSF and VSSF wetlands are lower than 67%. In fact, nitrogen is transformed through nitrification (VSSF system) and denitrification (HSSF system) but it is not eliminated by nitrification-denitrification pathway (Vymazal 2007; Saeed and Sun 2012). For this reason, there exist an increasing interest in the development and evaluation of hybrid constructed wetlands which generate aerobic and anaerobic conditions for promoting TN removal (Vymazal and Kröpfelová 2015). Actually, hybrid systems are in operation in many countries worldwide (Vymazal 2013). Nevertheless, the most commonly used hybrid system for treating domestic or municipal sewage is VSSF-HSSF (Zapater-Pereyra et al. 2015). For this system, the removal efficiencies of BOD_5, COD were between 86–92.5% and 78–84%, respectively.

Some studies have been shown that hybrid systems remove between 76–80% of TN compared with a single HSSF or VSSF system. Contrary to nitrogen performance, efficiencies of BOD_5, COD, TSS, and TP show no significant difference between hybrid constructed wetlands and single stage HSSF or VSSF (Vymazal 2013).

Constructed Wetlands for Industrial Wastewater Treatment

Industrial wastewaters have used constructed wetlands technology for their treatment. Agricultural and dairy effluents contain high organic matter rates, therefore these systems are frequently combined with a physical, chemical, and/or biological pretreatment (Kadlec and Wallace 2009). Table 2 shows examples of industrial wastewaters treated by constructed wetlands. Regarding the swine and dairy industry, HSSF and VSSF systems have been implemented using anaerobic lagoons or sedimentation tanks as a pretreatment.

The most commonly used macrophytes in these systems were *Phragmites australis*, *Typha latifolia*, and *Schoenoplectus* ssp. (Stone et al. 2004; Sun et al. 2005). For BOD_5 and nutrients removal, these values ranged between 57–93% and 26–60% respectively. Likewise, HSSF and VSSF wetlands have been studied for

Table 1. Operating, design, and depuration of sewage depending on constructed wetlands type.

| Type of wetlands | Design parameters | | | Operational Characteristics | | Removal parameters | | | | | | | |
| | Depth (m) | Unit area (m² E.I.⁻¹ᵃ) | Size ratio | HRT (d) | Hydraulic Load (mm d⁻¹) | Organic Matter | | Nitrogen | | Phosphorus | | Pathogens | |
						Loading rate (gBOD₅ m⁻²·d⁻¹)	Removal efficiency (%)	Loading rate (gTN m⁻²·d⁻¹)	Removal efficiency (%)	Loading rate (gTP m⁻²·d⁻¹)	Removal efficiency (%)	Bacteria[a] (n°/100 mL)⁻¹	Removal efficiency (%)
SF	0.06–1.2	5.0–20	2:1–4:1	5–15	12–50	1–11	54–88	0.02–4.2	20–52	0.1	3–66	1,800–556,000	79–99
HSSF	0.3–1.0	1.2–12	0.25:1–4:1	2–10	23–50	3–15	65–95	1–10	39–59	0.04–0.6	15–58	5×10^7–6.5×10^7	98–99
VSSF	0.5–1.4	0.85–5.5	3:1–14:1	1–2	27–110	10–25	30–95	2.8–30	23–67	0.06	23–63	1.0×10^6–1.7×10^6	99–99

Notes: ᵃEquivalent inhabitant (E.I.), Hydraulic Residence Time (HRT), Total nitrogen (TN), Total phosphorus, 5-day Biochemical Oxygen Demand (BOD₅).
References: USEPA 2000; García et al. 2004; Vymazal 2005; Vohla et al. 2005; Cooper et al. 2005; Wallace and Knight 2006; Crites et al. 2006; Akratos and Tsihrintzis 2007; García-Serrano and Corzo-Hernández 2008; Vymazal and Kröpfelová 2008; Kadlec and Wallace 2009; Zhao et al. 2010; Vera et al. 2010; Plaza de Los Reyes et al. 2011; Liu et al. 2012; Vera et al. 2013; Vera et al. 2014; Zhu et al. 2014; Pelissari et al. 2014; Rai et al. 2015

Table 2. Industrial wastewater treated by constructed wetlands.

Country	Constructed Wetlands	Industry	Pre-treatment	Macrophyte	F (m³ d⁻¹) HRT(d)	Influent (mg L⁻¹)						Effluent (mg L⁻¹)						Ref.
						BOD₅	COD	TSS	TN	NH₄⁺-N	TP	BOD₅	COD	TSS	TN	NH₄⁺-N	TP	
Tanzania	HSSF	Winery	TPFBB	*Phm*	0.02/9	-	922	-	-	-	-	-	86	-	-	-	-	a
Portugal	SF	Swine	-	*Ips, Csp*	-	1,207	4,608	2,350	1,420	1615	99	265	1,814	95	232	216	24	b
Slovenia	VSSF	Textiles	-	*Pa*	1.4/0.3	99–350	276–1,379	-	7–82	0.2–4.5	-	43–95	122–487	-	10–19	7–16	68	c
US	M-P-M	Swine	AL, FT	*Tl, Sch*	8.3/13–15	-	445	160	66	-	71	-	287	105	30	-	-	d
Portugal	HSSF	Tannery	FT	*Tl, Ips, Csp, Sts, Pa*	0.03/6.8	875	1,966	75	143	-	0.3	454	869	19	105	-	0.5	e
US	M-P-M	Swine	AL, FT	*Tl, Sch*	12.6/10–18	-	808	363	116–174	86	56–73	-	313	160	30–70	53	38–48	f
US	SF	Swine	AL	*Je, Sci, Spar, Tl, Ta*	2.6/12–14	-	-	-	134	118	30	-	-	-	26	20	22	g
Ireland	VSSF	Swine	FT	*Pa*	0.4/0.17	2,157	-	444	110	104	-	918	-	192	80.8	76	-	h
Italy	HSSF	Dairy	ST	*Pa*	4.4/10	451	1,219	690	65	22.4	12.8	28	98	60	33.3	24.5	5	i
Italy	HSSF, VSSF, SF	Winery	ST	*Tl, Pa, El, Cer, Ny, Ips, Je*	10–35/3.5–13	425	4,0459	222	15	-	4.9	29	91	24	2.6	-	1.3	j,k
Spain	VSSF-HSSF	Winery	HUSB	*Pa, Je*	-	2,950	4,238	580	-	171	2.3	279	448	17	-	12.5	1.9	l
Bangladesh	VSSF-HSSF-VSSF	Tannery	-	*Pa*	4.8/12.5/2.4	4,200	11,500	27,600	-	111	30	80	200	12400	-	15	3	m

Notes: TS: SepticTank; B: Biofilter, ST: Sedimentation Tank, AL: Anaerobic lagoon, FT: Feed Tank, TPFBB: Three Phase Fluidized Bed Bioreactor, F: Flow, HRT: Hydraulic Residence Time, *Tl:Typha latifolia, Ta:Typha angustifolia, Phm:Phragmites mauritianus, Pa: Phragmites australis, Sch: Schoenoplectus sp., Ny: Nymphaea sp., Je: Juncus effusus; Ips: Iris pseudacorus; Sci: Scirpus sp.; Sts: Stenotaphrum secundatum; Csp: Canna sp.; El: Elodea canadensis; Cer: Ceratophyllum demersum; Spar: Sparganium americanum*, M-P-M = March-Pond-March, COD: Chemical Oxygen Demand, SST: Total Suspended Solids.
References: a: Renalda et al. 2006; b: Rodrigues et al. 2006; c: Bulc et al. 2006; d: Poach et al. 2007; e: Calheiros et al. 2007; f: Poach et al. 2004; g: Stone et al. 2004; h: Stone et al. 2000; i: Sun et al 2005; j: Mantovi et al. 2003; k: Masi et al. 2002; l: Serrano et al. 2011; m: Saeed et al. 2012

treating effluents of winery industry. Using a septic tank as a pretreatment, these systems achieved performances of 94% and 78% for BOD_5 and nutrients removal, respectively. Recently, hybrid constructed wetland have been applied for industrial wastewater treatment (Vymazal 2013). As shown Table 2, VSSF-HSSF and VSSF-HSSF-VSSF combinations were used for winery and tannery industry. In both cases, removal efficiencies of BOD_5 and NH_4^+-N were between 90–98%.

Constructed Wetlands Under the Global Change Scenario

Generation of Methane and Nitrous Oxide in Constructed Wetlands

Constructed wetlands have been suggested to be an attractive solution for wastewater treatment in rural areas for small units and small villages (Mander et al. 2014).

Despite their advantages as a technology with low operational and maintenance costs, constructed wetlands have been found to be sources of GHG such as CO_2, CH_4, and N_2O (Zhang et al. 2014; Zhao et al. 2016).

Table 3 shows CH_4 emissions of different HSSF wetlands used for wastewater treatment. CH_4 is among the most important GHG. In constructed wetlands, CH_4 is produced in soil and/or anoxic-anaerobic sediments (–100 to –350 mV) and it comes from the anaerobic degradation of organic matter (Zhao et al. 2016). About 70% of CH_4 is produced from acetate while the other 30% is generated from CO_2 and hydrogen (Wang et al. 2013). The methanogenesis is a complex phenomenon accomplished by coordinated action of three trophic groups of bacteria (hydrolytic-fermentative, acetogenic, and methanogens) (Shah et al. 2015). Specifically, the main organic matter removal processes in HSSF system correspond to anaerobic microbiological mechanisms (greater than 94.7%) because anoxic-anaerobic conditions prevail in this system (–100 to –500 mV) (Vasudevan et al. 2011). HSSF wetlands have been shown to produce CH_4 emissions of 1,068 $mgCH_4 \ m^{-2}{\cdot}d^{-1}$ with ranges between 2–6,480 $mgCH_4/m^{-2}{\cdot}d^{-1} \ m^2{\cdot}d$ (Mander et al. 2014; Corbella and Puigagut 2015; López et al. 2015; Casas et al. 2017). Nevertheless, in VSSF, CH_4 is produced in anaerobic regions which are located in the bottom zones of this system. This information is corroborated by Wang et al. (2013) who showed an increased development of methanogens (archaea) in deep zones (5–40 × 10^7 cells g $m^{-2}{\cdot}d^{-1}$ soil to –40 cm), with respect to shallow zones (5–15 × 10^7 cells g^{-1} soil to –10 cm). Thus, CH_4 emissions in VSSF wetland are less significant with an average of 72 mg $CH_4 \ m^2{\cdot}d^{-1}$ (ranges: 7.2–140 mg $CH_4 \ m^2 d^{-1}$) (Wang et al. 2013; Mander et al. 2014).

Particularly, CH_4 production in constructed wetlands will depend on substrate characteristics (Vymazal 2008). Elmitwalli et al. (2001) determined that the maximum conversion to methane was the highest for the colloidal fraction (86%), followed by suspended fraction (78%), and finally, the dissolved fraction (62%). Moreover, constructed wetland depth influences redox conditions and therefore, the microbiology of the system. Due to redox conditions above –244 mV, methanogens bacterial activity increased (Mitsch and Gosselink 2007).

Likewise, the type of vegetation plays a remarkable role in the emissions, oxidation, and transport of CH_4 (Wang et al. 2013). The soil-plant systems can promote consumption of approximately 20–50% of the CH_4 by methanotrophs inhabiting the root zone (Wang et al. 2013).

Table 3. Methane emissions in urban wastewater treatment by HSSF wetlands.

Country	Climate	Variation factor	Area (ha)	Support medium	Macrophyte	COD removal efficiency (%)	Methane emission (mg CH$_4$ m^{-2}·d^{-1})
Estonia	T/B	Summer	0.03	Coarse Sand	Phr, TL y SS	89	454.1 ± 320,5
		Winter					2.0 ± 0,40
		Summer	0.04	Gravel/Silt	Phr	55	213.7 ± 50,8
		Winter					14.7 ± 6,00
Norway	B	Summer	0.0001	Coral Sand	UP	40	173.6 ± 57,4
		Winter					−2.00 ± 9,2
Poland	T/O	Summer	0.56	Sand/ Gravel	Phr	-	894.8 ± 293,8
		Winter					58.76 ± 45,40
Chile	T/M	Different species	0.00045	Gravel	Phr	59	1,455 ± 482
			0.00045	Gravel	Sch	58	1,305 ± 27
Spain	T/M	210 mgDQO L^{-1}	0.0055	Grava	Phr	71	28.1
		420 mgDQO L^{-1}				76	21.4
Czech Republic	T	0 m^{a}	0.07	Coarse stones/ Gravel	Phr	74	416 − 2,980
		2 − 14 m^{b}					352

Table 3 contd. ...

...Table 3 contd.

Country	Climate	Variation factor	Area (ha)	Support medium	Macrophyte	COD removal efficiency (%)	Methane emission (mg CH$_4$ m^{-2} d^{-1})
		UP			UP	73 – 95	267.1 – 654.2
Japan	T/W	Different species	0.0002	Gravel/Sand	Phr	75 – 95	1,063 – 1,697
					ZL	77 – 98	1,621 – 6,487
					TL	55 – 98	432.6 – 2,540
Spain	T/M	ST	0.00004	Gravel	Phr	64	20.2 – 161.2
		UASB				55	14.2 – 571.6
Japan	T/W	-	0.0012	Sand	Phr	99	62.4 – 240

Notes: T: Temperate, M: Mediterranean, B: Boreal, O: Oceanic, W: Warm, Phr: *Phragmites australis*, TL: *Typha latifolia*, SS: *Scirpus silvaricus*, UP: unplanted; Sch: *Schoenoplectus californicus*; ZL: *Zizania latifolia*, UASB: Upflow Anaerobic Sludge Blanket reactor, ST: Sedimentation tank, a: from 0 m of wetland input, b: between 2–14 m of wetland input.

References: Sovik et al. 2006; Garcia et al. 2007; Picek et al. 2007; Wang et al. 2008; Liu et al. 2009; López et al. 2015; Corbella and Puigagut 2015

On the other hand, recent research has shown that constructed wetlands can produce N_2O through a different pathways, both chemical and biochemical or during aerobic and/or anaerobic conditions (Chuersuwan et al. 2014). It has been identified that N_2O can be generated as a nitrification minor product or as an intermediary of denitrification process. Under aerated constructed wetlands, N_2O can be emitted during nitrification by ammonia-oxidizing bacteria during the oxidation of hydroxylamine (NH_2OH) to nitrite (NO_2^-) and also via reducing NO_2^- to N_2O and N_2 under aerobic conditions by nitrification-denitrification (Inamori et al. 2007; Mander et al. 2014). Denitrification rates are influenced by nitrate and carbon availability, temperature, and pH (Mitsch and Gosselink 2007). The conversion of N_2O to N_2 is very sensitive to oxygen and redox conditions. The disruption of this last step can generate an incomplete denitrification and N_2O emissions (Mander et al. 2014).

Regarding N_2O emissions in constructed wetlands, Mander et al. (2008) observed that N_2 emissions are the most important component in nitrogen retention from HSSF (50% of TN removal) whereas only 1.2% of TN was removed as N_2O. Furthermore, typical HSSF systems (< 0.12 P.E m^{-2}) have N_2 and N_2O emissions rates of 400 and 10 kgN $ha^{-1} \cdot yr^{-1}$, respectively. For VSSF system, Jia et al. (2011) showed that intermittent feeding mode would promote the emission of N_2O. In this case, the N_2O flux ranged between 0.09–7.33 mg m^{-2} hr^{-1}. However, Mander et al. (2014) evaluated N_2O emissions of different constructed wetland and determined no significant differences between constructed wetlands. The lowest values were found in FS (0.13 mg m^{-2} hr^{-1}) followed by VSSF (0.14 mg m^{-2} hr^{-1}) and finally, HSSF systems (0.24 mg m^{-2} hr^{-1}). Furthermore, this research observed that all of constructed wetlands types showed a significant positive correlation between the inflow TN loading and N_2O emission values. Likewise, N_2O emissions is likely to be highly variable across a range of water and environmental conditions (Wu et al. 2009). This compound may be affected by dissolved oxygen, HRT, organic carbon, and nitrate available, C/N ratio (Teiter and Mander 2005; Inamori et al. 2007), and macrophyte species (Mitsch and Gosselink 2007; Wang et al. 2008; Mander et al. 2014).

Some investigations have noted that the design parameters of HSSF and VSSF systems are one of the most important factors that control CH_4 and N_2O emission. Specifically, the deeper water table could directly influence on GHG production (CH_4, N_2O) (Dong and Reddy 2010; Mander et al. 2014). This effect is observed by Aguirre et al. (2005) who determined that in the shallow HSSF system (30 cm), the organic matter removal was achieved by aerobic degradation (9.9%), denitrification (56.9%), sulfate-reduction (33.2%), and methanogenesis (0%). In contrast, in the deeper HSSF system (50 cm), methanogenesis was 4.9%. This information is corroborated by Dong and Sun (2007) who demonstrated that the deeper VSSF system (25 cm) enhanced TN removal and at the same time, promoted another nitrogen removal pathway such as ANNAMOX.

The application hybrid constructed wetlands should not just consider organic matter and nitrogen removal. Also, it is important to evaluate their CH_4 and N_2O emissions into the atmosphere. Even more, there exists limited information about strategies for GHG reduction in hybrid system. For this reason, it is necessary to study the contribution of these systems in GHG generation.

Advantages and Strategies for Decreasing Climate Change Impact through Constructed Wetlands Application

Global climate change is a global threat to the world. Increases in ambient temperature are directly linked to rising anthropogenic GHG concentrations in the atmosphere. The potential impacts of climate change are global warming, loss of biodiversity, and the widespread retreat of glaciers and icecaps which is related to water stress. By 2025, as much as two thirds of the world population may be subjected to moderate to high water stress (Nema et al. 2012). Under this global change scenario, it is necessary that the operation of technologies for wastewater treatment must consider these following parameters:

(a) low energy use and/or energy recovery from liquid waste,

(b) water reclamation and reuse,

(c) ecosystem biodiversity conservation,

(d) economic services,

(e) nutrients recovery, and

(f) mitigation of GHG emissions.

Constructed wetlands are a promising technology used to improve the quality of wastewater from point and not point sources of water pollution including domestic, industrial, and municipal wastewater (Vymazal et al. 2006; Vymazal 2007; García-Serrano and Corso-Hernández 2008; Vera et al. 2011). In addition to their effectiveness as wastewater treatment, these systems provide several ecosystem services: provisional (food, energy and fibers), supporting (biodiversity, nutrient cycling), and cultural (recreational, educational) services (Mitsch and Gosselink 2007).

In spite of the fact that the operation of constructed wetlands has low energy consumption and maintenance requirements, these systems operated with high organic loading rates can generate GHG (López et al. 2015; Sepúlveda et al. 2017). As opposed to conventional wastewater treatments as activated sludge treatment, constructed wetlands emit significantly lower levels of N_2O (Mander et al. 2014). Moreover, the life cycle assessment (LCA) observed that VSSF system emitted 3.18 $kg \cdot CO_2$-eq whereas conventional wastewater treatment emitted 7.3 $kg \cdot CO_2$-eq (Pan et al. 2011). One strategy for mitigating CH_4 emissions under global change scenarios is to change the hydrological regime of constructed wetlands (e.g., intermittent loading). Likewise, the use of combined systems such anaerobic treatment and constructed wetland is another strategy which can contribute to recovery of energy and at the same time, minimize GHG emissions.

Water reclamation from wastewater has also gained considerable importance due to water stress (Asano 1998; Vera et al. 2016). In this way, constructed wetlands offer great potential for polishing wastewater effluent to a standard suitable for reuse (Ghermandi et al. 2007). Particularly, water reuse from sewage offers two potential resources: water and nutrients. Both are essential for plant growth, agriculture, horticulture, forestry, golf courses, parks, and gardens (Greenway 2005). Nevertheless, the main risk with sewage effluent reuse is pathogens that pose potential risk to human health. In fact, removal of pathogens is the prime objective in treating wastewater for reuse. For this reason, there exist quality requirements for water use which establish maximum concentration of fecal coliforms, nematodes, *salmonella* spp., and enteric virus. The World Health Organization (WHO) establishes three water use categories. Category A is for irrigation of crops likely to be eaten uncooked, sports fields, and publics parks. Category B corresponds for irrigation of cereal crops, industrial crops, fodder crops, pasture, and trees. Finally, category C is used for localized irrigation of crops in category B if exposure to workers and the public does not occur. Regarding fecal coliforms standards, concentrations of 10^3 and 10^3–10^5 n°/100 mL are required for category A and B, respectively. In contrast to the other categories, category C has not restrictions (WHO 1989).

Marecos do Monte and Albuquerque (2010) studied the possibility of used final effluents of HSF system for agriculture. In term of physical-chemical quality, results show that effluent is suitable for reuse in agricultural and landscape irrigation. However, HSSF system had a low performance in terms of pathogenic removal (80% for FC removal). Other research has evaluated the use of hybrid constructed wetlands for the treatment and reuse for irrigation (Melian et al. 2010). Due to 90% of FC removal being achieved, the hybrid system could be a robust technology for wastewater treatment and reuse.

Based on different limits of international guidelines for wastewater reuse and good quality of effluent generated by constructed wetlands, it is necessary to study in detail the use of these systems as an efficient and low cost technology for sewage treatment and reuse for irrigation.

In order to improve the biodiversity of rural areas and the landscape, more than 150 plants species have been documented in constructed wetland. *Phragmites* spp., *Typha* spp., *Scirpus* spp., *Schoenoplectus* spp., and *Juncus* spp. are commonly used (Vymazal 2013; López et al. 2016). However, the use of ornamental species has recently increased and over 20 species have been studied (Vymazal 2011b). *Zantedeschia* spp., *Canna* spp., *Cyperus* spp., *Iris* spp., *Agapanthus* spp., *Strelitzia* spp., *Heliconia* spp. have been used mainly in SSF wetland to treat sewage (Zurita et al. 2009, 2011; Morales et al. 2014; García et al. 2013; Calheiros et al. 2015; Burgos et al. 2017). The introduction of plants and flowers in the rural areas promotes a greater capture of CO_2, improves the landscape, and flowers can boost the sector producing honey, by pollinating flowers.

Final Comments

Table 4 summarizes constructed wetlands' effects on factors affecting global climate change. These parameters are energy recovery, water reclamation, biodiversity, fibers production, nutrients recovery, and GHG emissions.

The combination of anaerobic degradation and constructed wetlands enhances resources recovery and simultaneously decreases GHG emissions.

Other positive effects on global climate change include the contribution of constructed wetlands to biodiversity conservation, fibers production, and water reclamation and reuse.

However, nutrients recovery is possible if constructed wetlands are combined with another treatment.

Table 4. Strategies for decreasing wastewater treatment impact on climate change using constructed wetlands.

Technology	Factors affecting global climate change						
	Wastewater application	Energy recovery	Water reuse	Biodiversity	Fibers production	Nutrient recovery	GHG emissions
SSF	Domestic	w.e.	+	+	+	w.e.	-
SSF + Lagoon	Domestic	w.e.	+	+	+	w.e.	-
SSF + HCFSV	Domestic	w.e.	+	+	+	w.e.	-
AT + VSSF	Domestic	+	+	+	+	-	+
AT + VSSF with adsorption	Domestic	+	+	+	+	+	+
AT + SF	Swine	+	+	+	+	+/-	-
HSSF, VSSF, SF	Winery	w.e.	+	+	+		-

Notes: AT: anaerobic treatment, (-): negative effect, (+): positive effect, w.e.: without effect

Therefore, constructed wetlands offer the possibility to be a suitable and sustainable technology to deal with global climate change. Thus, constructed wetlands investigations have to focus on mitigating the negative effect of climate change.

Acknowledgments

This work was supported by Grant No. 13.3327-IN.IIP INNOVA BIOBIO and CONICYT/FONDAP/15130015.

References Cited

Aguirre, P., E. Ojeda, J. García, J. Barragán and R. Mujeriego. 2005. Effect of water depth on the removal of organic matter in horizontal subsurface flow constructed wetlands. J. Environ. Sci. Health A Tox. Hazard. Subst. Environ. Eng. 40: 1457–1466.

Akratos, C. and V. Tsihrintzis. 2007. Effect of temperature, HRT, vegetation and porous media on removal efficiency of pilot-scale horizontal subsurface flow constructed wetlands. Ecol. Eng. 29: 173–191.

Araya, F., I. Vera, K. Sáez and G. Vidal. 2016. Enhanced ammonium removal from sewage in mesocosm-scale constructed wetland with artificial aeration and natural zeolite. Environ. Technol. 37: 1811–1820.

Asano, T. 1998. Wastewater Reclamation and Reuse. Technomic Publishing, Lancaster.

Ávila, C., M. Garfi and J. García. 2013. Three-stage hybrid constructed wetland system for wastewater treatment and reuse in warm climate regions. Ecol. Eng. 61: 43–49.

Ávila, C., J. García and M. Garfi. 2016. Influence of hydraulic loading rate, simulated storm events and seasonality on the treatment performance of an experimental three-stage hybrid CW system. Ecol. Eng. 87: 324–332.

Ayaz, S., Ö. Aktaş, N. Fındık, L. Akça and C. Kınacı. 2012. Effect of recirculation on nitrogen removal in a hybrid constructed wetland system. Ecol. Eng. 40: 1–5.

Bulc, T.G., A. Ojstrsek and D. Vrhovsek. 2006. The use of constructed wetland for textile wastewater treatment. IWA. Portugal. 1667–1675.

Burgos, V., F. Araya, C. Reyes-Contreras, I. Vera and G. Vidal. 2017. Performance of ornamental plants in mesocosm subsurface constructed wetlands under different organic sewage loading. Ecol. Eng. 99: 246–255.

Calheiros, C.S.C., O.S.S. Rangeland and M.L. Castro. 2007. Constructed wetland systems vegetated with different plants applied to the treatment of tannery wastewater. Water Res. 41: 1790–1798.

Calheiros, C.S.C., V.S. Bessa, R.B.R. Mesquita, H. Brix, A.O.S.S. Rangel and P.M.L. Castro. 2015. Constructed wetland with a polyculture of ornamental plants for wastewater treatment at a rural tourism facility. Ecol. Eng. 79: 1–7.

Casas, Y., A. Rivas, D. López and G. Vidal. 2017. Life-cycle greenhouse gas emission assessment and extended exergy accounting of horizontal flow constructed wetland for wastewater treatment: A case study in Chile. Energy Indicators 74: 130–139.

Chuersuwan, S., P. Suwanwaree and N. Chuersuwan. 2014. Estimating greenhouse gas fluxes from constructed wetlands used for water quality improvement. Songklanakarin J. Sci. Technol. 36: 367–373.

Ciria, M.P., M.L. Solano and P. Soriano. 2005. Role of macrophyte *Typha latifolia* in a constructed wetland for wastewater treatment and assessment of its potential as a biomass fuel. Biosyst. Eng. 92: 535–544.

Cordell, D., J. Drangert and S. White. 2009. The story of phosphorus: Global food security and food for thought. Glob. Environ. Chang. 19: 292–305.

Cooper, P. 1999. A review of the design and performance of vertical flow and hybrid reed bed treatment systems. Wat. Sci. Tech. 40: 1–9.

Cooper, P. 2005. The performance of vertical flow constructed wetland systems with special reference to the significance of oxygen transfer and hydraulic loading rates. Water Sci. Technol. 5: 81–90.

Cooper, P. 2007. The constructed wetland association UK data base of constructed wetland systems. Water Sci. Technol. 56: 1–6.

Corbella, C. and J. Puigagut. 2015. Effect of primary treatment and organic loading on methane emissions from horizontal subsurface flow constructed wetlands treating urban wastewater. Ecol. Eng. 80: 79–84.

Crites, R.W., E.J. Middlebrooks and S.C. Reed. 2006. Natural Wastewater Treatment Systems. Taylor and Francis Group, Boca Raton.

Dong, Z. and T. Sun. 2007. A potential new process for improving nitrogen removal in constructed wetlands-promoting coexistence of partial-nitrification and ANAMMOX. Ecol. Eng. 31: 69–78.

Dong, X. and G.B. Reddy. 2010. Soil bacterial communities in constructed wetlands treated with swine wastewater using PCR-DGGE technique. Bioresour. Technol. 101: 1175–1182.

Elmitwalli, T., J. Soellner, A. De Keizer, H. Bruning, G. Zeeman and G. Lettinga. 2001. Biodegradability and change of physical characteristics of particles during anaerobic digestion of domestic sewage. Water Res. 35: 1311–1317.

Fan, J., W. Wang, B. Zhang, Y. Guo, H.H. Ngo, W. Guo et al. 2013. Nitrogen removal in intermittently aerated vertical flow constructed wetlands: impact of influent COD/N ratios. Bioresour. Technol. 143: 461–466.

García, J., P. Aguirre, R. Mujeriego, Y. Huang, L. Ortiz and J.M. Bayona. 2004. Initial contaminant removal performance factors in horizontal flow reed beds used for treating urban wastewater. Water Res. 38: 1669–1678.

García, J., V. Capel, A. Castro, I. Ruiz and M. Soto. 2007. Anaerobic biodegradation tests and gas emissions from subsurface flow constructed wetlands. Bioresour. Technol. 98: 3044–3052.

García-Serrano, J. and A. Corzo-Hernández. 2008. Depuración con humedales construidos. Guía práctica de diseño, construcción y explotación de sistemas de humedales de flujo subsuperficial. Universidad Politécnica de Catalunya, Catalunya. In Spanish.

García, J., D. Paredes and J. Cubillos. 2013. Effect of plants and the combination of wetland treatment type systems on pathogen removal in tropical climate conditions. Ecol. Eng. 58: 57–62.

Ghermandi, A., D. Bixio and C. Thoeye. 2007. The role of free water surface constructed wetlands as polishing step in municipal wastewater reclamation and reuse. Sci. Total Environ. 380: 247–258.

Green, M.B. 1997. Experience with establishment and operation of reed bed treatment for small communities in the UK. Wetlands Ecol. Manag. 4: 147–158.

Greenway, M. 2005. The role of constructed wetlands in secondary effluent treatment and water reuse in subtropical and arid Australia. Ecol. Eng. 25: 501–509.

Halverson, N. 2004. Review of constructed subsurface flow *vs.* surface flow wetlands. Westinghouse Savanna River Company, Aiken, US.

Inamori, R., P. Gui, P. Dass, M. Matsumura, K.Q. Xu, T. Kondo et al. 2007. Investigating CH_4 and N_2O emissions from ecoengineering wastewater treatment processes using constructed wetland microcosms. Process Biochem. 42: 363–373.

Jia, W., J. Zhang, P. Li, H. Xie, J. Wu and J. Wang. 2011. Nitrous oxide emissions from surface flow and subsurface flow constructed wetland microcosms: Effect of feeding strategies. Ecol. Eng. 37: 1815–1821.

Kadlec, R. and R. Knight. 1996. Treatment Wetlands, Lewis Publishers, CRC Press.

Kadlec, R. and S. Wallace. 2009. Treatment Wetlands.Taylor and Francis Group, Boca Raton, US.

Liu, C., K. Xu, R. Inamori, Y. Ebie, J. Liao and Y. Inamori. 2009. Pilot-scale studies of domestic wastewater treatment by typical constructed wetlands and their greenhouse gas emissions. Front. Envir. Sci. Eng. China. 3: 477–482.

Liu, X., S. Huang, T. Tang, X. Liu and M. Scholz. 2012. Growth characteristics and nutrient removal capability of plants in subsurface vertical flow constructed wetlands. Ecol. Eng. 44: 189–198.

López, D., D. Fuenzalida, I. Vera, K. Rojas and G. Vidal. 2015. Relationship between the removal of organic matter and the production of methane in subsurface flow constructed wetlands designed for wastewater treatment. Ecol. Eng. 83: 296–304.

López, D., M. Sepúlveda and G. Vidal. 2016. *Phragmites australis* and *Schoenoplectus californicus* in constructed wetlands: Development and nutrient uptake. J. Soil Sci. Plant Nutr. 16: 763–777.

Mander, Ü., K. Lõhmus, S. Teiter, T. Mauring, K. Nurk and J. Augustin. 2008. Gaseous fluxes in the nitrogen and carbon budgets of subsurface flow constructed wetlands. Sci. Total Environ. 404: 343–353.

Mander, Ü., G. Dotro, Y. Ebie, S. Towprayoon, C. Chiemchaisri, S.F. Nogueira et al. 2014. Greenhouse gas emission in constructed wetlands for wastewater treatment: A review. Ecol. Eng. 66: 19–35.

Mantovi, P., M. Marmiroli, E. Maestri, S. Tagliavini, S. Piccinini and N. Marmiroli. 2003. Application of a horizontal subsurface flow constructed wetland on treatment of diary parlor wastewater. Bioresour. Technol. 88: 85–94.

Marecos do Monte, H. and A. Albuquerque. 2010. Analysis of constructed wetland performance for irrigation reuse. Water Sci. Technol. 61: 1699–1705.

Masi, F., G. Conte, N. Martinuzzi and B. Pucci. 2002. Winery high organic content wastewater treated by constructed wetlands in mediterranean climate. IWA. TZ. 1: 274–282.

Melián, J., A. Rodríguez, J. Arana and J. Henríquez. 2010. Hybrid constructed wetlands for wastewater treatment and reuse in the Canary Islands. Ecol. Eng. 36: 891–899.

Meng, P., H. Pei, W. Hu and Y. Shao. 2014. How to increase microbial degradation in constructed wetlands: influencing factors and improvement measures. Bioresour. Technol. 157: 316–326.

Mitsch, W. and J. Gosselink. 2007. Wetlands, John Wiley and Sons, Inc., New York, US.

Morales, G., D. López, I. Vera and G. Vidal. 2014. Constructed wetland with ornamental plants for the treatment of the organic matter and nutrient contain in domestic wastewater. Theoria. 22: 33–46.

Nema, P., S. Nema and P. Roy. 2012. An overview of global climate changing in current scenario and mitigation action. Renew. Sust. Ener. Rev. 16: 2329–2336.

Pan, T., X.D. Zhu and Y.P. Ye. 2011. Estimate of life-cycle greenhouse gas emissions from a vertical subsurface flow constructed wetland and conventional wastewater treatment plants: A case study in China. Ecol. Eng. 37: 248–254.

Pelissari, C., P. Sezerino, S. Decezaro, D. Wolff, A. Bento, O. de Carvalho Junior et al. 2014. Nitrogen transformation in horizontal and vertical flow constructed wetlands applied for dairy cattle wastewater treatment in southern Brazil. Ecol. Eng. 73: 307–310.

Picek, T., H. Čížková and J. Dušek. 2007. Greenhouse gas emissions from constructed wetland-plants as important sources of carbon. Ecol. Eng. 31: 98–106.

Plaza de Los Reyes, C. and G. Vidal. 2007. Humedales construidos: Una alternativa a considerar para el tratamiento de aguas servidas. Tecnología del Agua. 28: 34–48. In Spanish.

Plaza de Los Reyes, C., L. Vera, M. Salvato, M. Borin and G. Vidal. 2011. Perspectives of the nitrogen removal by constructed wetlands. Tecnología del Agua. 31: 40–49. In Spanish.

Poach, M.E., P.G. Hunt, G.B. Reddy, K.C. Stone, M.H. Johnson and A. Grubbs. 2004. Swine wastewater treatment by marsh-pond-marsh constructed wetlands under varying nitrogen loads. Ecol. Eng. 23: 165–175.

Poach, M.E., P.G. Hunt, G.B. Reddy, K.C. Stone, M.H. Johnson and A. Grubbs. 2007. Effect of intermittent drainage on swine wastewater treatment by marsh–pond–marsh constructed wetlands. Ecol. Eng. 30: 43–50.

Prochaska, C.A., A.I. Zouboulis and K.M. Eskridge. 2007. Performance of pilot-scale vertical-flow constructed wetlands, as affected by season, substrate, hydraulic load, and frequency of application of simulated urban sewage. Ecol. Eng. 31: 57–66.

Rai, U., A. Upadhyay, N. Singh, S. Dwivedi and R. Tripathi. 2015. Seasonal applicability of horizontal sub-surface flow constructed wetland for trace elements and nutrient removal from urban wastes to conserve Ganga River water quality at Haridwar, India. Ecol. Eng. 81: 115–122.

Renalda, M., K.N.N. Jau and J.H.Y. Katima. 2006. Performance of horizontal subsurface flow constructed wetland (HSSFCW) in the treatment of tannins wastewater. IWA. Portugal. 1603–1610.

Rodrigues, F., J. Catarino, A. Maia, E. Mendonca, A. Picado, Z. Figueiredo et al. 2006. Quality improvement of digested swine wastewater by aquatic macrophytes. IWA. Portugal. 1645–1651.

Rojas, K., I. Vera and G. Vidal. 2013. Influence of season and species *Phragmites australis* and *Schoenoplectus californicus* on the removal of organic matter and nutrients contained in sewage wastewater during the start up operation of the horizontal subsurface flow constructed wetland. Revista Facultad de Ingeniería, Universidad de Antioquia. 69: 230–240. In Spanish.

Saeed, T. and G. Sun. 2012. A review on nitrogen and organics removal mechanisms in subsurface flow constructed wetlands: Dependency on environmental parameters, operating conditions and supporting media. J. Environ. Manage. 112: 429–448.

Saeed, T., R. Afrin, A. Al Muyeed and G. Sun. 2012. Treatment of tannery wastewater in a pilot-scale hybrid constructed wetland system in Bangladesh. Chemosphere 88: 1065–1073.

Sepúlveda, M., L. López and G. Vidal. 2017. Methane production in horizontal subsurface flow constructed wetlands treating domestic wastewater. Ecol. Eng. (in press).

Serrano, L., D. De la Varga, I. Ruiz and M. Soto. 2011. Winery wastewater treatment in a hybrid constructed wetland. Ecol. Eng. 37: 744–753.

Shah, F., Q. Mahmood, N. Rashid, A. Pervez, I. Raja and M. Shah. 2015. Co-digestion, pretreatment and digester design for enhanced methanogenesis. Renew. Sustainable Energy Rev. 42: 627–642.

Sharma, P., I. Takashi, K. Kato, K. Tomita and T. Nagasawa. 2013. Seasonal efficiency of a hybrid sub-surface flow constructed wetland system in treating milking parlor wastewater at northern Hokkaido. Ecol. Eng. 53: 257–266.

Søvik, A., J. Augustin, K. Heikkinen, J. Huttunen, J. Necki, S. Karjalainen et al. 2006. Emission of the greenhouse gases nitrous oxide and methane from constructed wetlands in Europe. J. Environ. Qual. 35: 2360–2373.

Stefanakis, A., C.S. Akratos and V.A. Tsihrintzis. 2014. Vertical flow constructed wetlands: eco-engineering systems for wastewater and sludge treatment, Elsevier, Amsterdam.

Sundaravadivel, M. and S. Vigneswaran. 2001. Constructed wetlands for wastewater treatment. Crit. Rev. Env. Sci. Technol. 31: 351–409.

Steinmann, C.R., S. Weinhart and A. Melzer. 2003. A combined system of lagoon and constructed wetland for an effective wastewater treatment. Water Res. 37: 2035–2042.

Stone, K.C., M.E. Poach, P.G. Hunt and G.B. Reddy. 2004. Marsh-pond-marsh constructed wetland design analysis for swine lagoon wastewater treatment. Ecol. Eng. 23: 127–133.

Stone, K.C., P.G. Hunt, A.A. Szogi, F.J. Humenik and J.M. Rice. 2000. Constructed wetland design and performance for swine lagoon waste water treatment. Trans. ASAE 45: 723–730.

Sun, G., Y. Zhao and S. Allen. 2005. Enhanced removal of organic matter and ammonia–nitrogen in a column experiment of tidal flow constructed wetland system. J. Biotechnol. 115: 189–197.

Teiter, S. and Ü. Mander. 2005. Emission of N_2O, N_2, CH_4, and CO_2 from constructed wetlands for wastewater treatment and from riparian buffer zones. Ecol. Eng. 25: 528–541.

USEPA. 1992. Manual Wastewater Treatment/Disposal for Small Communities. United States Environmental Protection Agency. USEPA edition, Ohio.

USEPA. 2000. Manual: Constructed Wetlands Treatment of Municipal Wastewaters. United States Environmental Protection Agency. USEPA edition, Ohio.

Vasudevan, P., P. Griffin, A. Warren, A. Thapliyal and M. Tandon. 2011. Localized domestic wastewater treatment: Part I. Constructed wetlands. An overview. J. Sci. Ind. Res. 70: 583–594.

Vera, I., J. García, K. Sáez, L. Moragas and G. Vidal. 2011. Performance evaluation of eight years' experience of constructed wetland systems in Catalonia as alternative treatment for small communities. Ecol. Eng. 37: 364–371.

Vera, I., F. Araya, E. Andrés, K. Sáez and G. Vidal. 2014. Enhanced phosphorus removal from sewage in mesocosm-scale constructed wetland using zeolite as medium and artificial aeration. Environ. Technol. 35: 1639–1649.

Vera, I., C. Jorquera, D. López and G. Vidal. 2016. Constructed wetlands for wastewater treatment and reuse in Chile: reflections. Tecnología y Ciencias del Agua. 7: 19–35.

Vidal, G. and S. Hormazabal. 2016. Las fibras vegetales y sus aplicaciones: Innovación en su generación a partir de la depuración de agua. Ediciones Universidad de Concepción, Concepción, Chile.

Vohla, C., E. Poldvere, A. Noorvee, V. Kuusemets and Ü Mander. 2005. Alternative filter media for phosphorous removal in a horizontal subsurface flow constructed wetland. J. Environ. Sci. Health. 40: 1251–1264.

Vymazal, J. 2005. Horizontal sub-surface flow and hybrid constructed wetlands systems for wastewater treatment. Ecol. Eng. 25: 478–490.

Vymazal, J. 2007. Removal of nutrients in various types of constructed wetlands. Sci. Total Environ. 380: 48–65.

Vymazal, J. 2008. Constructed wetland for wastewater treatment: A review. pp. 965–980. *In*: Sengupta, N. and R. Dalwani [eds.]. Wetlands and Natural Resource Management. Proceedings of Taal 2007. The 12th World Lake Conference.

Vymazal, J. and L. Kröpfelová. 2008. Wastewater Treatment in Constructed Wetlands with Horizontal Sub-Surface Flow. Environmental Pollution. Springer, Heidelberg.

Vymazal, J. 2011a. Constructed wetlands for wastewater treatment: five decades of experience. Environ. Sci. Technol. 45: 61–69.

Vymazal, J. 2011b. Plants used in constructed wetlands with horizontal subsurface flow: A review. Hydrobiologia. 674: 133–156.

Vymazal, J. 2013. Emergent plants used in free water surface constructed wetlands: A review. Ecol. Eng. 61: 582–592.

Vymazal, J. and L. Kröpfelová. 2015. Multistage hybrid wetland for removal of nitrogen. Ecol. Eng. 84: 202–208.

Wang, Y., R. Inamori, H. Kong, K. Xu, Y. Inamori, T. Kondo et al. 2008. Influence of plant species and wastewater strength on constructed wetland methane emissions and associated microbial populations. Ecol. Eng. 32: 22–29.

Wallace, S. and R. Knight. 2006. Small-scale constructed wetland treatment systems: Feasibility, design criteria and OM requirements. IWA Publishing. London, UK.

Wang, Y., H. Yang, C. Ye, X. Chen, B. Xie, C. Huang et al. 2013. Effects of plant species on soil microbial processes and CH_4 emission from constructed wetlands. Environ. Poll. 174: 273–278.

WHO. 1989. Health guidelines for the use of wastewater in agriculture and aquaculture. World Health Organization. WHO Technical Report Series, Switzerland.

Wu, J., J. Zhang, W. Jia, H. Xie, R.R. Gu, C. Li et al. 2009. Impact of COD/N ratio on nitrous oxide emission from microcosm wetlands and their performance in removing nitrogen from wastewater. Bioresour. Technol. 100: 2910–2917.

Ye, F. and Y. Li. 2009. Enhancement of nitrogen removal in hybrid constructed wetland to treat domestic wastewater for small rural communities. Ecol. Eng. 35: 1043–1050.

Zapater-Pereyra, M., H. Ilyas, S. Lavrnić, J.J.A. van Bruggen and P.N.L. Lens. 2015. Evaluation of the performance and space requirement by three different hybrid constructed wetlands in a stack arrangement. Ecol. Eng. 82: 290–300.

Zhang, D.Q., K.B.S.N. Jinadasa, R.M. Gersberg, Y. Liu, W.J. Ng and S.K. Tan. 2014. Application of constructed wetlands for wastewater treatment in developing countries—a review of recent developments (2000–2013). J. Environ. Manage. 141: 116–131.

Zhao, Y.J., B. Liu, W.G. Zhang, Y. Ouyang and S.Q. An. 2010. Performance of pilot-scale vertical-flow constructed wetlands in responding to variation in influent C/N ratios of simulated urban sewage. Bioresource Technol. 101: 1693–1700.

Zhao, Z., X. Song, Y. Wang, D. Wang, S. Wang, Y. He et al. 2016. Effects of algal ponds on vertical flow constructed wetlands under different sewage application techniques. Ecol. Eng. 93: 120–128.

Zhi, W. and G. Ji. 2014. Quantitative response relationships between nitrogen transformation rates and nitrogen functional genes in a tidal flow constructed wetland under C/N ratio constraints. Water Res. 64: 32–41.

Zhu, H., B. Yan, Y. Xu, J. Guan and S. Liu. 2014. Removal of nitrogen and COD in horizontal subsurface flow constructed wetlands under different influent C/N ratios. Ecol. Eng. 63: 58–63.

Zurita, F., J. De Anda and M. Belmont. 2009. Treatment of domestic wastewater and production of commercial flowers in vertical and horizontal subsurface-flow constructed wetlands. Ecol. Eng. 35: 861–869.

Zurita, F., M. Belmont, J. De Anda and J. White. 2011. Seeking a way to promote the use of constructed wetlands for domestic wastewater treatment in developing countries. Water Sci. Technol. 63: 654–659.

20

Uptake and Detoxification of Organic Micropollutants by Macrophytes in Constructed Wetlands

A. Dordio,[1,2,*] *A.J.P. Carvalho,*[1,3] *M. Hijosa-Valsero*[4,5] and *E. Becares*[5]

INTRODUCTION

The ubiquitious presence of organic micropollutants (OMPs) in the aquatic environment has been causing worldwide concern over recent years. OMPs consist of a vast and ever expanding group of anthropogenic as well as natural substances which are characterized by their persistence (related with their non-biodegradability) or pseudo-persistence (as result of steady input loads) in the environment, tendency to bioaccumulate in living organisms, and potentially harmful bioactivity. The wide range of substances that comprise this class of pollutants includes pesticides, polycyclic aromatic hydrocarbons (PAHs), phenolic compounds, petroleum hydrocarbons, dyes, chlorinated compounds, explosives, and a new class of emergent pollutants, the pharmaceuticals and personal care products (PPCPs) (Jones and de Voogt 1999; Boethling et al. 2009; El Shahawi et al. 2010).

[1] Chemistry Department, School of Sciences and Technology, University of Évora, Évora, Portugal.
[2] MARE—Marine and Environmental Research Centre, University of Évora, Évora, Portugal.
[3] CQE—Évora Chemistry Centre, University of Évora, Évora, Portugal.
[4] Center of Biofuels and Bioproducts, Agrarian Technological Institute of Castilla and León (ITACyL), Villarejo de Orbigo, 24358 León, Spain.
[5] University of León, Department of Biodiversity and Environmental Management, Faculty of Biological Sciences, 24071 León, Spain.
* Corresponding author: avbd@uevora.pt

OMPs have been frequently detected at trace concentrations (ranging from a few ng/L to several μg/L) in raw and treated wastewater, biosolids and sediments, surface water, groundwater, and even in drinking water (Bolong et al. 2009; El Shahawi et al. 2010; Pal et al. 2010; Lapworth et al. 2012; Du and Liu 2012; Luo et al. 2014; Clarke and Cummins 2015; Gavrilescu et al. 2015; Gothwal and Shashidhar 2015; Ebele et al. 2017). These low concentrations and the chemical diversity of OMPs not only make the associated detection and quantification procedures necessarily more complex from an analytical perspective, but they also create new challenges for water and wastewater treatment processes. In fact, effluents of wastewater treatment plants (WWTPs) are generally regarded as the primary source of many OMPs in aquatic systems. OMPs' removal in conventional WWTPs is commonly incomplete and variable (Fent et al. 2006; Bolong et al. 2009; Luo et al. 2014; Li 2014; Haman et al. 2015; Dordio and Carvalho 2015), a difficulty that may be attributed to the diversity of OMPs physical and chemical properties (e.g., hydrophobicity and biodegradability) and to the fact that the type of removal processes that are more prominent in WWTP are not suitable for such low concentration levels.

Constructed wetlands (CWS) have been increasingly studied and applied to the removal of OMPs from wastewater. In these systems, the concerted action of macrophytes, microorganisms, and minerals of the support matrix, has in many cases already been effective in reducing levels of OMPs in wastewaters (Davies et al. 2005; Imfeld et al. 2009; Matamoros et al. 2012; Reyes-Contreras et al. 2012; Dordio and Carvalho 2013; Zhang et al. 2014; Shehzadi et al. 2014; Verlicchi and Zambello 2014; Li et al. 2014a; Vymazal and Brezinová 2015; Sauvêtre and Schröder 2015; Ávila and García 2015; Toro-Vélez et al. 2016; Syranidou et al. 2016; Hijosa-Valsero et al. 2016; Hussein and Scholz 2017; A et al. 2017; Hijosa-Valsero et al. 2017). In order to achieve higher efficiencies on these engineered systems, the operating conditions and selection of components can be optimized through a comprehensive understanding of OMPs' fate, the mechanisms involved in the removal processes, and the roles played by each CWS component.

An increasingly important role is being attributed to CWS macrophytes in the removal of poorly biodegradable organic pollutants. Indirectly, they may contribute to regulate and enhance biotic processes via the release of root exudates which can either participate and improve pollutants transformation or have an effect on microbial populations to aid their development and stimulate their degradation processes in the rhizosphere. In addition, plants may also participate directly, through adsorption, uptake, translocation and, finally, volatilization or detoxification (which involves functionalization and/or conjugation, followed by compartmentalization (Brix 1997; Vymazal et al. 1998; Macek et al. 2000; Susarla et al. 2002; Pilon-Smits 2005; Kvesitadze et al. 2006; Kadlec and Wallace 2009; Dordio and Carvalho 2011; Shelef et al. 2013; Wang et al. 2014; Carvalho et al. 2014; Truu et al. 2015; Hijosa-Valsero et al. 2016).

A potentially significant role (that may have been underestimated for a long time) in plants adaptation to polluted environments and in enhancing phytoremediation may be played by microorganisms that are endophytic to plants, that is, microorganisms that live symbiotically inside plants (Truu et al. 2009; Weyens et al. 2009; Li et al. 2012; Afzal et al. 2014; Shehzadi et al. 2014; Zhu et al. 2014; Sauvêtre and Schröder 2015; Truu et al. 2015; Syranidou et al. 2016; Ijaz et al. 2016; Calheiros et al. 2017;

Feng et al. 2017). Several benefits provided by endophytic microorganisms to the host plants have already been identified (Weyens et al. 2009; Li et al. 2012; Afzal et al. 2014; Ijaz et al. 2016; Arslan et al. 2017; Feng et al. 2017). In addition, there may also be relevant contributions of endophytes in the degradation of organic pollutants that are taken up by plants (Weyens et al. 2009; Li et al. 2012; Afzal et al. 2014; Ijaz et al. 2016; Arslan et al. 2017; Feng et al. 2017).

In this chapter, an overview is presented of the relevant aspects concerning OMPs interactions with aquatic macrophytes as the former are taken up by the latter in constructed wetlands systems employed for wastewater treatment. Firstly, an account is given of the main factors affecting the extent and rate of uptake of OMPs by plants and their subsequent translocation to different plant parts. The chapter then proceeds to describe the various metabolic stages enacted by plants to transform the pollutants into a form that is less toxic and that can possibly remain in plant tissues without compromising their survival. Finally, the conditions of abiotic stress that are provoked in plants by their exposure to OMPs and the response of the protective antioxidant systems (mainly enzymatic) are described briefly, thereby giving an overview of phytotoxic issues that may be involved in the use of plants to remove OMPs from wastewater after a long-term exposure or in cases where OMPs loads may be high (e.g., in some industrial wastewaters).

Uptake and Translocation of Organic Micropollutants by Plants

Plants play an important role in the biotic processes responsible for the removal of OMPs by CWS, involving several processes of which some details are still unclear and remain to be fully understood. Subjected to the direct or indirect action of plants, OMPs can be stabilized or degraded in the rhizosphere, adsorbed or uptaken and accumulated in the roots and rhizomes, transported to the aerial parts, and volatilized or transformed/degraded and accumulated inside plant tissues (Fig. 1).

Sorption onto their roots and rhizomes would be considered the first step of the plants' direct intervention in the removal of OMPs from the aqueous medium, occurring when dissolved pollutants come into contact with the extensive surface of the root system (Hijosa-Valsero et al. 2016). They may readily adsorb (in irreversible or relatively reversible form, depending on different variables), bind to the root structure and cell walls, or be taken up inside root tissues. Sorption of OMPs to plant roots has been reported for several macrophyte species and it can be estimated in control experiments using dead or inactivated biomass (Carvalho et al. 2012; Hijosa-Valsero et al. 2016).

The extent of sorption processes is usually estimated by the so-called root concentration factor (RCF), defined as the ratio between the amount of compound sorbed by the root/rhizome (per unit weight of fresh root/rhizome tissue) and the compound concentration within the aqueous medium (Briggs et al. 1982). The RCF describes the potential of a given organic substance to transfer to plant roots, without differentiating between accumulation on roots surface and uptake into root tissues (Schröder and Collins 2002). For a given compound, the value of RCF seems to be relatively dependent on its octanol-water partitioning coefficient (K_{ow}). In particular, some authors have found linear correlations between the logarithms of the two quantities, log RCF and log K_{ow} (Briggs et al. 1982; Burken and Schnoor 1998).

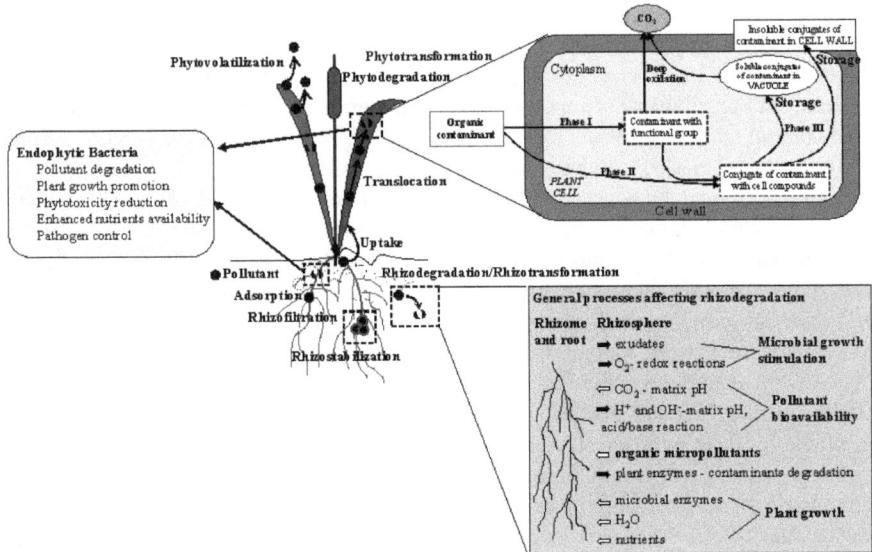

Fig. 1. Major removal processes and transformation pathways of organic micropollutants in plants (adapted from Sandermann (1994); Kvesitadze et al. (2006); Abhilash et al. (2009); Dordio and Carvalho (2011); Truu et al. (2015)).

Along with adsorption, uptake by plants roots is, for many OMPs, an important process with a significant contribution to removal of the pollutants from the aqueous medium. In fact, many organic substances can be readily taken up, but because most are xenobiotic to plants, there are no specific transporters for them in plant membranes. Therefore, xenobiotic organics move through cell walls and into the roots by simple diffusion (passive uptake). Root absorption is performed in two phases: in the first, faster phase, pollutants diffuse from the surrounding medium into the roots, while in the second phase, they more slowly accumulate in the tissue (Korte et al. 2000).

OMPs that are taken up by roots may subsequently be translocated into different parts of the plants as a result of the physiological processes involved in the transport of nutrients. The main force driving this transport is the water potential gradient created throughout the plant by the transpiration stream (the transport of water and dissolved substances from roots to shoots, passing through vessels and tracheids located in the xylem) which depends on plant characteristics and environmental conditions (Dietz and Schnoor 2001; Pilon-Smits 2005; Kvesitadze et al. 2006; Collins et al. 2006). The importance of the transpiration stream for the uptake and translocation of organics by plants is expressed in the following equation (Briggs et al. 1983; Dietz and Schnoor 2001):

$$U = (TSCF)(T)(C) \tag{1}$$

where U is the rate of organic compound assimilation, T is the rate of plant transpiration, C is the organic compound concentration in the water phase, and TSCF is the transpiration stream concentration factor (dimensionless ratio between the concentration of the organic compound in the liquid of the transpiration stream (xylem sap) and the bulk concentration in the root zone solution) (Dietz and Schnoor 2001; Doucette et al. 2005; Kvesitadze et al. 2006).

With possible exceptions of some hormone-like substances such as phenoxy acid herbicides, there is no evidence of active uptake (corresponding to a TSCF > 1) of xenobiotic organics (Doucette et al. 2005). Therefore, if xenobiotic uptake is not directly proportional to water uptake (TSCF = 1) the compound is said to be excluded (TSCF < 1), although there might still be a passive process. However, factors such as membrane permeability and xylem sap solubility of the pollutant may limit the extent or kinetics of passive uptake (Doucette et al. 2005). Sorption and rapid metabolism of pollutants within the plant may also reduce xylem concentrations and keep TSCF < 1 (Doucette et al. 2005).

Several empirical relationships have been deduced between TSCF and log K_{ow} that suggest a gaussian-like dependence (a bell-shaped form) (Briggs et al. 1983; Hsu et al. 1990; Sicbaldi et al. 1997; Burken and Schnoor 1998; Collins et al. 2006; de Carvalho et al. 2007; Paraiba 2007; Limmer and Burken 2014) pointing to optimal hydrophobicity properties (corresponding to the maxima of the gaussian curves) for uptake and translocation of the organic pollutants, in the range of moderately hydrophobic behaviors. In fact, a range of log K_{ow} values between 0.5 and 3.5 is frequently advanced as optimal (Dietz and Schnoor 2001; Pilon-Smits 2005; Pascal-Lorber et al. 2008), considering that it should be hydrophobic enough to allow movement through the lipid bilayer of membranes, and still water soluble enough to allow transference into the cell fluids. If organics are too hydrophilic (log K_{ow} < 0.5), they cannot pass membranes and never get into the plant; if they are too hydrophobic (log K_{ow} > 3) they get stuck in membranes and cell walls in the periphery of the plant and cannot enter the cell fluids. However, there are indications that the log K_{ow} alone is not an absolute predictor of the tendency for a compound to be taken up by plants. It has been observed that some compounds with low log K_{ow} have, nevertheless, still shown ability to penetrate plant cell membranes (Renner 2002). The tendency of a OMP to be removed from water by a given plant may also depend on other factors such as initial pollutant concentration, the plant's anatomy, and its root system (Chaudhry et al. 2002). Laboratory and field experiments with some OMPs such as 1,4-dioxane, MTBE (Methyl tertbuthyl ether), sulfolane, and diisopropanolamine also suggest that these predictive schemes may not be applicable for some non-ionizable, highly water soluble organics (Rubin and Ramaswami 2001; Groom et al. 2002; Doucette et al. 2005; Chard et al. 2006; Dettenmaier et al. 2009; Limmer and Burken 2014).

Very hydrophobic compounds (log K_{ow} > 3.5) that may not be significantly taken up by plants, can also be candidates for phytostabilization and/or rhizodegradation owing to their long residence times within the root zone (Dietz and Schnoor 2001; Pilon-Smits 2005).

As soon as OMPs penetrate plant cells, they can be exposed to a plant's metabolic transformations eventually leading to their partial or complete degradation, or through which they may be transformed into less toxic compounds that are subsequently sequestered in plant tissues (Korte et al. 2000; Kvesitadze et al. 2006). Metabolism of xenobiotic compounds in plant systems is generally a "detoxification" process that is analogous to the hepatic metabolism of xenobiotic compounds in humans; hence the name "green liver" that sometimes is applied to these systems (Sandermann 1994; Dietz and Schnoor 2001; Kvesitadze et al. 2006).

Detoxification of Organic Micropollutants by Plants

Once an organic xenobiotic is taken up and translocated, it undergoes one or more phases of metabolic transformation, according to the scheme illustrated in Fig. 1. There are three possible phases of metabolic transformation in higher plants which can be identified (Sandermann 1994):

- *Phase I: functionalization*—this phase consists of the conversion or activation (through oxidation, reduction, or hydrolysis) of lipophilic compounds (Komives and Gullner 2005; Eapen et al. 2007) by enzymatic transformations that result in the addition of a hydrophilic group (e.g., hydroxyl, amine, carboxyl, sulphydryl). As a result of these transformations, an increase of the polarity and water solubility of the compound is achieved. These transformations may also enhance the compound's affinity to enzymes that will catalyze further transformation (conjugation or deep oxidation (Korte et al. 2000; Kvesitadze et al. 2006)) by the addition or exposure of the appropriate functional groups. Some xenobiotics, if available in low concentrations, may be degraded by oxidative processes to common metabolites of the cell and to CO_2. Following this pathway, a plant cell can not only detoxify the compound but also assimilate the resulting carbon atoms for cell needs. Otherwise, if concentrations are high, full detoxification is not achieved and the transformation further proceeds to conjugation (Korte et al. 2000).

    During Phase I, several groups of enzymes are known to play important roles (Sandermann 1992; Sandermann 1994; Macek et al. 2000; Eapen et al. 2007). The oxidative metabolism of xenobiotics is mediated mainly by Cytochrome P450 monooxygenase. This enzyme is crucial in the oxidative processes that lead to the bioactivation of xenobiotics by forming reactive electrophilic compounds that will proceed to be conjugated in Phase II. Peroxidases, which help convert some xenobiotics, and peroxygenases, which may also participate in oxidative processes, are some other important enzyme groups. In addition, nitroreductase is involved in the degradation of nitroaromatics and laccase where it has the function of breaking up aromatic ring structures.

    The reactions involved in Phase I of xenobiotics' metabolism are the first step; its purpose is to render the xenobiotic into a less toxic form. After modifications during Phase I, the molecule should be ready for the reactions comprising Phases II and III, which further detoxify the pollutant. However, if it already has a functional group that is suitable for Phase II metabolism, the compound can directly enter Phase II without going through Phase I.

- *Phase II: conjugation*—this phase involves the conjugation of metabolites from Phase I (or of the xenobiotics themselves, when they already contain suitable functional groups) to endogenous molecules (proteins, peptides, amino acids, organic acids, mono-, oligo- and polysaccharides, pectins, lignin, etc.) (Coleman et al. 1997; Korte et al. 2000; Dietz and Schnoor 2001; Eapen et al. 2007). Compounds with higher molecular weights result from conjugation, which usually have lower mobilities and a greatly diminished biological activity. Hence, final products of Phase II metabolism are usually less toxic than the original xenobiotics or Phase I derivatives.

Conjugation in Phase II is catalyzed by transferases such as glutathione-S-transferases, glucosyl transferase, and N-malonyl transferases (Eapen et al. 2007). Conjugation of Phase I metabolites takes place in the cytosol, but it is harmful to accumulate these compounds there (Eapen et al. 2007).

- **Phase III: compartmentalization**—this phase involves getting the modified (conjugated) xenobiotics compartmentalized in vacuoles or getting them bound to cell wall components such as lignin or hemicellulose (Coleman et al. 1997; Dietz and Schnoor 2001; Eapen et al. 2007). The conjugates are, thus, removed from vulnerable sites in cytosol and transported to sites where they can not interfere with cellular metabolism: conjugates that are soluble (e.g., formed with peptides, sugars, amino acids, etc.) are accumulated in vacuoles, whereas insoluble conjugates (e.g., coupled with pectin, lignin, xylan, and other polysaccharides) are removed from inside the cell and accumulated at plant cell walls (Sandermann 1992; Sandermann 1994; Eapen et al. 2007). This is a potential final step in the non-oxidative utilization of xenobiotics.

Metabolic reactions of Phase III are unique to plants because, unlike animals, they do not have means to eliminate xenobiotics from their systems by excretion. Plants, therefore, need to somehow restrain the xenobiotic within their own system. ATP-driven vacuolar transporters are the main enzymes involved in this phase. Further processing of conjugates may occur in the vacuolar matrix or cell walls, where they are stored (Eapen et al. 2007). It is assumed that Phase III products are no longer toxic.

Although in several studies it has already been shown that some macrophyte species have the ability to uptake and translocate several OMPs into their aerial parts (Forni et al. 2002; Dordio et al. 2011a; Zhang et al. 2012; Wang et al. 2013; San Miguel et al. 2013a; Li et al. 2014b; Lv et al. 2016; Wang et al. 2016; Zhang et al. 2016; Chen et al. 2017; Pi et al. 2017), knowledge about the range of metabolic transformations endured by this type of compounds and their ultimate fate inside plants is still very scarce. Because plants are photo-autotrophic organisms, they did not evolve to metabolize organic contaminants as sources of energy or carbon. Nevertheless, they often metabolize or sequester organic compounds, transforming them in more water-soluble forms and/or immobilizing them, in order to detoxify them (Pilon-Smits 2005; Weyens et al. 2009; Truu et al. 2015; Feng et al. 2017).

Recently, some attention has been focused on the role of endophytes in the detoxification and degradation of organic contaminants inside plants (Weyens et al. 2009; Li et al. 2012; Afzal et al. 2014; Sauvêtre and Schröder 2015; Ijaz et al. 2016; Arslan et al. 2017; Feng et al. 2017).

Endophytes are a group of varied microorganisms, mainly bacteria or fungi, that live asymptomatically inside plants for at least part of their life cycle, establishing an array of interactions with host plants that affect plant growth, health, and survival (Hardoim et al. 2015; Wani et al. 2015; Compant et al. 2016; Feng et al. 2017). They can be found in nearly every plant species, with the highest densities observed in plant's roots, and decreasing from stems to leaves.

In a plant-endophyte symbiotic relationship, endophytes obtain from their host plants a growth medium that is rich in nutrients and a safe habitat that protects them from biotic and abiotic stressors (Bacon and Hinton 2006; Reinhold-Hurek and Hurek 2011; Li et al. 2012; Afzal et al. 2014; Feng et al. 2017). In return, endophytes can provide several benefits to their hosts, namely they produce a wide range of natural bioactive substances that promote plant growth (Bacon and Hinton 2006; Li et al. 2012; Santoyo et al. 2016; Feng et al. 2017), they increase plant nutrient uptake, inhibit plant pathogen growth, reduce disease severity, and enhance tolerance to environmental stresses (drought, salt, heat, contaminant) (Fig. 1) (Li et al. 2012; Hamilton and Bauerle 2012; Hamilton et al. 2012; Afzal et al. 2014; Feng et al. 2017).

Moreover, in recent studies, several plant endophytes have been isolated and shown to be capable of degrading organic pollutants such as petroleum derivatives, PAHs, organochlorines, pesticides, phenolic compounds or pharmaceuticals (Weyens et al. 2009; Kang et al. 2012; Andreolli et al. 2013; Bisht et al. 2014; Barman et al. 2014; Sauvêtre and Schröder 2015; Anudechakul et al. 2015; Syranidou et al. 2016; Zhu et al. 2016). In re-inoculation experiments, some of these strains had a clear effect of enhancing remediation, favoring the metabolism of pollutants as well as decreasing phytotoxicity and improving plant health (Afzal et al. 2014; Feng et al. 2017). Therefore, the latest findings strongly point to a view in which some endophytes participate with an important role in the detoxification and efficient degradation of organic contaminants that are taken up by the plant. However, until now there has not been a clear assessment of the relative importances of plants' own phytodegradation processes in comparison with endophytes contributions to the metabolization of organic pollutants inside plants. These contributions may be concurrent or complementary to plant's metabolic processes, and any cooperative or synergistic interplay between the two types of organisms and respective pollutant removal processes has yet to be studied in detail.

Notwithstanding all the recent research work, fate of most OMPs in plants is yet poorly understood, especially in regard to the characterization of transformed and sequestered products. The short- and long-term fate and potential toxicity of some of the intermediate products produced by plant metabolism remains an important issue, both in regard to the plant's viability but also in regard to harvesting strategies, implemented to avoid the reentry of contaminants into the environment through the decay of dead plant materials or the consumption of these plants by herbivores.

Xenobiotic Phytotoxicity

Because plants can not excrete potentially toxic compounds that are xenobiotic to them, all plants have special defense mechanisms to protect them from the negative effects caused by the exposure to small quantities of such foreign substances. Plants exposure to xenobiotics, therefore, will create stress conditions that are responsible for the activation of such defense mechanisms. The relative rates of organic xenobiotics uptake, translocation, and metabolism usually determines whether or not these compounds will induce the plant's response, which can be inferred from the plant's physiological, biochemical, and molecular responses.

Physiological responses to xenobiotic stress such as growth reduction, chlorosis, and necrosis of tissues usually can be inspected with relatively ease. An assessment of physiological toxic responses can also be based on quantitative parameters, namely the plants' relative growth rates (RGR) or the photosynthetic pigments concentrations in plant tissues. Alteration of these two parameters have been observed in plants, sometimes compromising plant viability, as consequence of their exposure to xenobiotics (Mishra et al. 2006).

The photosynthetic apparatus is one of the most important targets of stress in plants. Indeed, most of the metabolic responses induced by stress conditions have consequences on the plant's aptitude to maintain an efficient light energy conversion (Rmiki et al. 1999). In fact, it has been reported that plants subjected to stress conditions do show alterations in chlorophyll (total, a and b) and carotenoid contents (Ferrat et al. 2003; Gill and Tuteja 2010; Das and Roychoudhury 2014). In general, in stressed plants, carotenoid contents tend to increase to provide protection against the formation of free oxygen radicals. Conversely, a decrease in total chlorophyll and in the ratio chlorophyll/carotenoids are often observed (Ferrat et al. 2003). These variations in photosynthetic pigments under exposure to organic xenobiotics have been observed for various macrophyte species such as *Azolla filiculoides*, *Elodea canadensis*, *Lemna gibba*, *Lemna minor*, *Cabomba aquatica*, *Cyperus alternifolius,* or *Typha* spp. (Ferrat et al. 2003; Olette et al. 2008; Dordio et al. 2009; Dordio et al. 2011b; Vafaei et al. 2012; Pietrini et al. 2015; Yan et al. 2016).

The exposure to organic xenobiotics also induces biochemical alterations, one important outcome of which is the production of reactive oxygen species (ROS). These highly reactive substances are partially reduced forms of atmospheric oxygen (O_2). They are typically the result of the excitation of ordinary triplet O_2 to form the high-energy singlet oxygen ($^1O_2$) or of the transfer of one, two, or three electrons to O_2 to form, respectively, superoxide radicals ($\bullet O_2^-$), hydrogen peroxide (H_2O_2) or O_2^{3-} (which dismutates into water and hydroxyl radicals, $\bullet OH$) (Wojtaszek 1997; Mittler 2002; Blokhina et al. 2003; Apel and Hirt 2004; Smirnoff 2005; Gill and Tuteja 2010). These are usually produced by plants as by-products of various metabolic pathways (such as photosynthesis and respiration) localized in different cellular compartments (predominantly in chloroplasts, mitochondria, and peroxisomes) (Apel and Hirt 2004; Gill and Tuteja 2010; Das and Roychoudhury 2014). Under physiological steady state conditions, ROS are scavenged and kept at adequate levels by different antioxidative defense components that are often confined to particular cell compartments (Apel and Hirt 2004; Das and Roychoudhury 2014). However, their overproduction can be triggered by external stress factors such as xenobiotics exposure. In such conditions, plants activate pathways to metabolize these foreign compounds, but as a side-effect they produce large amounts of ROS that can perturb the normal equilibrium between ROS production and scavenging. These disturbances can cause the intracellular levels of ROS to rapidly rise, which may risk cell viability (Mittler 2002; Mittler et al. 2004; Masella et al. 2005). The rapid and transient production of high quantities of ROS results in what is called an "oxidative burst" (Wojtaszek 1997; Apel and Hirt 2004; Gill and Tuteja 2010). ROS, unlike atmospheric oxygen, are capable of unrestricted oxidation of various cellular components. If their levels are not properly controlled, damage to some biomolecules may result (e.g., membrane lipid peroxidation, protein oxidation, enzyme inhibition, and DNA and RNA damage) that can ultimately lead to the oxidative destruction of the cell (Mittler 2002; Mittler et al. 2004; Masella et al. 2005). It is, therefore, vital that ROS levels within cells are strictly controlled and kept within a narrow concentration range. Plants have developed mechanisms to monitor and scavenge excessive amounts of ROS and their tolerance to organic pollutants is related to their capacity to cope with ROS over-production.

Despite the above mentioned problems caused by excessive levels of ROS, a steady-state is still required within the cells because modifications of ROS levels also act as a signal for the activation of stress response and defense pathways. Thus, ROS can be viewed as cellular indicators of stress, and as secondary messengers involved in the stress-response signal transduction pathway. As such, the measurement of antioxidant enzymes activities (that counteract rising ROS levels) has been frequently used to assess environmental stress induced in plants by various pollutants (Mittler 2002; Apel and Hirt 2004; Gill and Tuteja 2010; Das and Roychoudhury 2014).

There are two types of ROS level control, involving two different mechanisms, one that enables fine modulation of low levels of ROS for signaling purposes, and one that will enable the detoxification of excess ROS, especially during stress. Mechanisms of ROS detoxification exist in all plants and can be categorized either as non-enzymatic, provided by anti-oxidant agents such as flavanones, anthocyanins, carotenoids, and ascorbic acid (AsA), or as enzymatic (Table 1). Major enzymatic ROS scavengers in plants include superoxide dismutase (SOD), ascorbate peroxidase (APX), glutathione peroxidase (GPX), and catalase (CAT), which are present in different cell compartments (Table 1). Their expression is genetically controlled and regulated both by developmental and environmental stimuli, according to the needs for removing excess ROS produced in cells (De Gara et al. 2003).

The first line of defense against ROS is provided by the action of SOD that dismutates superoxide ($\cdot O_2^-$) to H_2O_2, which is subsequently detoxified by the action of other enzymes, namely CAT and peroxidases such as APX, GPX, and GPOX, each acting on different cell compartments. In contrast to CAT, APX requires an ascorbate and glutathione (GSH) regeneration system, the ascorbate-glutathione cycle, involving the MDHAR, DHAR, and GR enzymes. Like APX, GPX also detoxifies H_2O_2 to H_2O, but uses GSH directly as a reducing agent. GPOX prefers phenolic substrates such as guaiacol or pyragallol as electron donors (Wojtaszek 1997; Apel and Hirt 2004; Gill and Tuteja 2010; Das and Roychoudhury 2014).

A summary of the several enzymes and antioxidants and respective reactions involved in ROS control and detoxification processes (both enzymatic and non-enzymatic) is presented in Table 1.

In the processes of detoxification and response against oxidative stress, phenolic compounds, which are ubiquitous secondary metabolites in plants, may have an important role both in enzymatic as well as non-enzymatic mechanisms. Many phenolic compounds are used as substrates for antioxidant enzymes (in particular in the case of GPOX) (Smirnoff 2005; Gill and Tuteja 2010; Das and Roychoudhury 2014). In addition, phenolic compounds have anti-oxidant properties on their own because their chemical structure enables them to quench radicals, which also makes them important players in the non-enzymatic control of the oxidative burst (Smirnoff 2005; Gill and Tuteja 2010; Das and Roychoudhury 2014). Therefore, alteration of phenolic compounds levels can provide a quantitative measure of oxidative stress that is complimentary to the evaluation of anti-oxidant enzymes activities.

Toxic effects induced in plants by some classes of OMPs have already been studied, especially in the case of pesticides (for motives that go beyond the application to phytoremediation), but also, to a lesser extent, by other classes as well (Wilson et al. 2000; Amaya-Chavez et al. 2006; Olette et al. 2008; Faure et al. 2012; Raja et al. 2012; San Miguel et al. 2013b; Garanzini and Menone 2015; Prasad et al. 2015).

Table 1. Detoxifying enzymes and non-enzymatic antioxidants and respective ROS scavenging reactions (Wojtaszek 1997; Mittler 2002; Blokhina et al. 2003; Ashraf 2009; Gill and Tuteja 2010; Das and Roychoudhury 2014).

Antioxidant enzymes	EC number	Location	Reaction catalyzed
Superoxide dismutase (SOD)	1.15.1.1	Chl, Cyt, Mit, Per, Apo	$O_2^{\cdot-} + O_2^{\cdot-} + 2H^+ \rightarrow 2\ H_2O_2 + O_2$
Catalase (CAT)	1.11.1.6	Per	$2\ H_2O_2 \rightarrow O_2 + 2H_2O$
Ascorbate peroxidase (APX)	1.11.1.11	Chl, Cyt, Mit, Per, Apo	$2Asa + H_2O_2 \rightarrow 2MDA + 2H_2O$ $(2MDA \rightarrow AsA + DHA)$
Guaiacol peroxidase (GPOX)	1.11.1.7	Cyt	Donor + $H_2O_2 \leftrightarrow$ oxidized donor + $2H_2O$
Glutathione peroxidase (GPX)	1.11.1.12	Cyt	$2GSH + H_2O_2 \leftrightarrow GSSG + 2H_2O$
Glutathione S-transferase (GST)	2.5.1.18	Cyt, Mit	$RX + GSH \leftrightarrow HX + R\text{-}S\text{-}GSH*$
MDA reductase (MDHAR)	1.6.5.4	Chl, Mit, Per, Cyt	$NADPH + MDA \leftrightarrow NAD(P)^+ + AsA$
DHA reductase (DHAR)	1.8.5.1	Chl, Mit, Per	$2GSH + DHA \leftrightarrow GSSG + AsA$
Glutathione reductase (GR)	1.6.4.2	Chl, Cyt, Mit	$2NADPH + GSSG \leftrightarrow 2NADP^+ + 2GSH$

Non-enzymatic antioxidants	Location	Function
Ascorbatic acid (AA)	Chl, Cyt, Mit, Per, Apo, Vac	Detoxifies H_2O_2 via action of APX
Reduced Glutathione (GSH)	Chl, Cyt, Mit, Per, Apo, Vac	Acts as a detoxifying co-substrate for enzymes like peroxidases, GR and GST
α-Tocophenol	Mostly in membranes	Guards against and detoxifies products of membrane LPO
Carotenoids	Chl and other non-green plastids	Quenches excess energy from the photosystems, LHCs
Flavonoids	Vac	Direct scavengers of H_2O_2 and $^1O_2$ and OH
Proline	Chl, Cyt, Mit	Efficient scavenger of OH and $^1O_2$ and prevent damages due to LPO

Abbreviations—Apo: apoplast; AsA: ascorbate; Chl: chloroplast; Cyt: cytosol; DHA: dehydroascorbate; GSH: glutathione; GSSG: oxidized glutathione; LHC: light harvesting complex; LPO: lipid peroxidation; MDA: monodehydroascorbate; Mit: mitochondria; Per: peroxisome; Vac: Vacuole; *R may be an aliphatic, aromatic or herocyclic group; X may be a sulfate, nitrite or halite group

However, there are still entire classes of OMPs (e.g., pharmaceuticals) for which the oxidative stress induced by them on plants is still poorly studied and characterized, thus impairing a more adequate design of phytoremediation solutions for such types of pollutants (Pomati et al. 2004; Dordio et al. 2009; Dordio et al. 2011b; Liu et al. 2013; Iori et al. 2013; Zezulka et al. 2013; Bartha et al. 2014; Nunes et al. 2014; Yan et al. 2016; Christofilopoulos et al. 2016).

Conclusions

In constructed wetlands systems, plants can contribute directly to remove organic micropollutants via uptake, translocation, accumulation, and/or transformation/degradation. A deeper understanding of these processes and a more complete information regarding bioaccumulation or transformation/degradation fate of organic micropollutants in aquatic macrophytes can be useful for assessing the potential value of phytotreatment technologies, namely of constructed wetlands systems, for removing pollutants from wastewaters. In addition, the information regarding uptake and elimination kinetics of organic micropollutants in aquatic macrophytes may aid in bioaccumulation modeling and environmental risk assessment initiatives addressing this type of contaminants.

Metabolic fate undergone by organic micropollutants inside plants is a relevant topic which, historically, has been studied mainly on crops and a few ornamental plants, whereas plants with value for phytoremediation have been the subject of only a scarce number of reports. Conjugative metabolism in species that are most frequently used in phytoremediation has not been investigated in significant depth so far. There is some relevance in expanding this knowledge in the context of constructed wetlands operation and maintenance. For instance, the awareness of pollutants sequestration in plant tissues may advise the adoption of educated practices of harvesting that prevents pollutant reintroduction resulting from decaying plant leaves, or the construction of adequate of barriers that deter herbivores from feeding on these plants. Furthermore, a better knowledge of the mechanisms involved in organic xenobiotics metabolism in plants may additionally present opportunities for enhancing metabolic rates for this type of compounds. Yet another subject that is emerging in recent studies as an important piece of the plant system and that needs better characterization and understanding is the community of endophyte microorganisms, which may be responsible for a major contribution to metabolic transformations of organic micropollutants. This component also provides further possibilities of additional system enhancement, for example, through the inoculation of plants with selected or genetically modified strains with more efficient characteristics.

In summary, the fate of organic micropollutants inside plants is currently an understudied and poorly understood topic, but which is fundamental for optimization and management of constructed wetlands designed for an efficient treatment of wastewaters contaminated with organic micropollutants.

Acknowledgements

The authors would like to express their gratitude to all those who reviewed the contents of this chapter, especially to Cristina Ávila, for the careful and helpful suggestions that greatly contributed to its final improvement.

References Cited

A, D., D. Fujii, S. Soda, T. Machimura and M. Ike. 2017. Removal of phenol, bisphenol A, and 4-tert-butylphenol from synthetic landfill leachate by vertical flow constructed wetlands. Sci. Total Environ. 578: 566–576.

Abhilash, P.C., S. Jamil and N. Singh. 2009. Transgenic plants for enhanced biodegradation and phytoremediation of organic xenobiotics. Biotechnol. Adv. 27: 474–488.

Afzal, M., Q.M. Khan and A. Sessitsch. 2014. Endophytic bacteria: Prospects and applications for the phytoremediation of organic pollutants. Chemosphere 117: 232–242.

Amaya-Chavez, A., L. Martinez-Tabche, E. Lopez-Lopez and M. Galar-Martinez. 2006. Methyl parathion toxicity to and removal efficiency by Typha latifolia in water and artificial sediments. Chemosphere 63: 1124–1129.

Andreolli, M., S. Lampis, M. Poli, G. Gullner, B. Biró and G. Vallini. 2013. Endophytic Burkholderia fungorum DBT1 can improve phytoremediation efficiency of polycyclic aromatic hydrocarbons. Chemosphere 92: 688–694.

Anudechakul, C., A.S. Vangnai and N. Ariyakanon. 2015. Removal of chlorpyrifos by water hyacinth (Eichhornia crassipes) and the role of a plant-associated bacterium. Int. J. Phytorem. 17: 678–685.

Apel, K. and H. Hirt. 2004. Reactive oxygen species: Metabolism, oxidative stress, and signal transduction. Annu. Rev. Plant Biol. 55: 373–399.

Arslan, M., A. Imran, Q.M. Khan and M. Afzal. 2017. Plant-bacteria partnerships for the remediation of persistent organic pollutants. Environ. Sci. Pollut. Res. 24: 4322–4336.

Ashraf, M. 2009. Biotechnological approach of improving plant salt tolerance using antioxidants as markers. Biotechnol. Adv. 27: 84–93.

Ávila, C. and J. García. 2015. Pharmaceuticals and personal care products (PPCPs) in the environment and their removal from wastewater through constructed wetlands. Compr. Anal. Chem. 67: 195–244.

Bacon, C.W. and D.M. Hinton. 2006. Bacterial endophytes: The endophytic niche, its occupants, and its utility. pp. 155–194. *In*: Gnanamanickam, S.S. [ed.]. Plant-Associated Bacteria. Springer Netherlands, Dordrecht.

Barman, D.N., M. Haque, S.M. Islam, H.D. Yun and M.K. Kim. 2014. Cloning and expression of ophB gene encoding organophosphorus hydrolase from endophytic *Pseudomonas* sp. BF1-3 degrades organophosphorus pesticide chlorpyrifos. Ecotox. Environ. Safe. 108: 135–141.

Bartha, B., C. Huber and P. Schröder. 2014. Uptake and metabolism of diclofenac in Typha latifolia— How plants cope with human pharmaceutical pollution. Plant Sci. 227: 12–20.

Bisht, S., P. Pandey, G. Kaur, H. Aggarwal, A. Sood, S. Sharma et al. 2014. Utilization of endophytic strain *Bacillus* sp. SBER3 for biodegradation of polyaromatic hydrocarbons (PAH) in soil model system. Eur. J. Soil Biol. 60: 67–76.

Blokhina, O., E. Virolainen and K.V. Fagerstedt. 2003. Antioxidants, oxidative damage and oxygen deprivation stress: a review. Ann. Bot. 91: 179–194.

Boethling, R., K. Fenner, P. Howard, G. Klecka, T. Madsen, J.R. Snape et al. 2009. Environmental persistence of organic pollutants: Guidance for development and review of POP risk profiles. Integr. Environ. Assess. Manag. 5: 539–556.

Bolong, N., A.F. Ismail, M.R. Salim and T. Matsuura. 2009. A review of the effects of emerging contaminants in wastewater and options for their removal. Desalination 239: 229–246.

Briggs, G.G., R.H. Bromilow, A.A. Evans and M. Williams. 1983. Relationships between lipophilicity and the distribution of non-ionized chemicals in barley shoots following uptake by the roots. Pestic. Sci. 14: 492–500.

Briggs, G.G., R.H. Bromilow and A.A. Evans. 1982. Relationships between lipophilicity and root uptake and translocation of non-ionised chemicals by barley. Pestic. Sci. 13: 495–504.

Brix, H. 1997. Do macrophytes play a role in constructed treatment wetlands? Water Sci. Technol. 35: 11–17.

Burken, J.G. and J.L. Schnoor. 1998. Predictive relationships for uptake of organic contaminants by hybrid poplar trees. Environ. Sci. Technol. 32: 3379–3385.

Calheiros, C.S.C., S.I.A. Pereira, H. Brix, A.O.S.S. Rangel and P.M.L. Castro. 2017. Assessment of culturable bacterial endophytic communities colonizing Canna flaccida inhabiting a wastewater treatment constructed wetland. Ecol. Eng. 98: 418–426.

Carvalho, P.N., M.C.P. Basto, C.M.R. Almeida and H. Brix. 2014. A review of plant-pharmaceutical interactions: from uptake and effects in crop plants to phytoremediation in constructed wetlands. Environ. Sci. Pollut. Res. 21: 11729–11763.

Carvalho, P.N., M.C. Basto and C.M. Almeida. 2012. Potential of *Phragmites australis* for the removal of veterinary pharmaceuticals from aquatic media. Bioresour. Technol. 116: 497–501.

Chard, B.K., W.J. Doucette, J.K. Chard, B. Bugbee and K. Gorder. 2006. Trichloroethylene uptake by apple and peach trees and transfer to fruit. Environ. Sci. Technol. 40: 4788–4793.

Chaudhry, Q., P. Schröder, D. Werck-Reichhart, W. Grajek and R. Marecik. 2002. Prospects and limitations of phytoremediation for the removal of persistent pesticides in the environment. Environ. Sci. Pollut. Res. 9: 4–17.

Chen, Z., Y. Chen, J. Vymazal, L. Kule and M. Koželuh. 2017. Dynamics of chloroacetanilide herbicides in various types of mesocosm wetlands. Sci. Total Environ. 577: 386–394.

Christofilopoulos, S., E. Syranidou, G. Gkavrou, E. Manousaki and N. Kalogerakis. 2016. The role of halophyte Juncus acutus L. in the remediation of mixed contamination in a hydroponic greenhouse experiment. J. Chem. Technol. Biotechnol. 91: 1665–1674.

Clarke, R.M. and E. Cummins. 2015. Evaluation of "Classic" and emerging contaminants resulting from the application of biosolids to agricultural lands: A review. Hum. Ecol. Risk Assess. 21: 492–513.

Coleman, J.O.D., M.M.A. Blake-Kalff and T.G.E. Davies. 1997. Detoxification of xenobiotics by plants: Chemical modification and vacuolar compartmentation. Trends Plant Sci. 2: 144–151.

Collins, C., M. Fryer and A. Grosso. 2006. Plant uptake of non-ionic organic chemicals. Environ. Sci. Technol. 40: 45–52.

Compant, S., K. Saikkonen, B. Mitter, A. Campisano and J. Mercado-Blanco. 2016. Editorial special issue: soil, plants and endophytes. Plant Soil 405: 1–11.

Das, K. and A. Roychoudhury. 2014. Reactive oxygen species (ROS) and response of antioxidants as ROS-scavengers during environmental stress in plants. Front. Environ. Sci. 2: 53.

Davies, L.C., C.C. Carias, J.M. Novais and S. Martins-Dias. 2005. Phytoremediation of textile effluents containing azo dye by using *Phragmites australis* in a vertical flow intermittent feeding constructed wetland. Ecol. Eng. 25: 594–605.

de Carvalho, R.F., R.H. Bromilow and R. Greenwood. 2007. Uptake and translocation of non-ionised pesticides in the emergent aquatic plant parrot feather Myriophyllum aquaticum. Pest Manag. Sci. 63: 798–802.

De Gara, L., M.C. de Pinto and F. Tommasi. 2003. The antioxidant systems vis-a-vis reactive oxygen species during plant-pathogen interaction. Plant Physiol. Biochem. 41: 863–870.

Dettenmaier, E.M., W.J. Doucette and B. Bugbee. 2009. Chemical hydrophobicity and uptake by plant roots. Environ. Sci. Technol. 43: 324–329.

Dietz, A.C. and J.L. Schnoor. 2001. Advances in phytoremediation. Environ. Health Perspect. 109: 163–168.

Dordio, A. and A.J.P. Carvalho. 2015. Removal of Pharmaceuticals in Conventional Wastewater Treatment Plants. pp. 1–44. *In*: Barret, L.M. [ed.]. Wastewater Treatment: Processes, Management Strategies and Environmental/Health Impacts. Nova Science Publishers, Hauppauge, NY, USA.

Dordio, A. and A.J.P. Carvalho. 2011. Phytoremediation: An option for removal of organic xenobiotics from water. pp. 51–92. *In*: Golubev, I.A. [ed.]. Handbook of Phytoremediation. Nova Science Publishers, Hauppauge, NY, USA.

Dordio, A.V., M. Belo, D.M. Teixeira, A.J.P. Carvalho, C.M.B. Dias, Y. Picó et al. 2011a. Evaluation of carbamazepine uptake and metabolization by *Typha* spp., a plant with potential use in phytotreatment. Bioresour. Technol. 102: 7827–7834.

Dordio, A.V. and A.J.P. Carvalho. 2013. Organic xenobiotics removal in constructed wetlands, with emphasis on the importance of the support matrix. J. Hazard. Mater. 252-253: 272–292.

Dordio, A., R. Ferro, D. Teixeira, A.J.P. Carvalho, A.P. Pinto and C.M.B. Dias. 2011b. Study on the use of *Typha* spp. for the phytotreatment of water contaminated with ibuprofen. Intern. J. Environ. Anal. Chem. 91: 654–667.

Dordio, A.V., C. Duarte, M. Barreiros, A.J.P. Carvalho, A.P. Pinto and C.T. da Costa. 2009. Toxicity and removal efficiency of pharmaceutical metabolite clofibric acid by *Typha* spp.—Potential use for phytoremediation? Bioresour. Technol. 100: 1156–1161.

Doucette, W.J., J.K. Chard, B.J. Moore, W.J. Staudt and J.V. Headley. 2005. Uptake of sulfolane and diisopropanolamine (DIPA) by cattails (Typha latifolia). Microchem. J. 81: 41–49.

Du, L.F. and W.K. Liu. 2012. Occurrence, fate, and ecotoxicity of antibiotics in agro-ecosystems. A review. Agron. Sustain. Dev. 32: 309–327.

Eapen, S., S. Singh and S.F. D'Souza. 2007. Advances in development of transgenic plants for remediation of xenobiotic pollutants. Biotechnol. Adv. 25: 442–451.

Ebele, A.J., M. Abou-Elwafa Abdallah and S. Harrad. 2017. Pharmaceuticals and personal care products (PPCPs) in the freshwater aquatic environment. Emerging Contam. 3: 1–16.

El Shahawi, M.S., A. Hamza, A.S. Bashammakh and W.T. Al Saggaf. 2010. An overview on the accumulation, distribution, transformations, toxicity and analytical methods for the monitoring of persistent organic pollutants. Talanta 80: 1587–1597.

Faure, M., A. San Miguel, P. Ravanel and M. Raveton. 2012. Concentration responses to organochlorines in *Phragmites australis*. Environ. Pollut. 164: 188–194.

Feng, N.X., J. Yu, H.M. Zhao, Y.T. Cheng, C.H. Mo, Q.Y. Cai et al. 2017. Efficient phytoremediation of organic contaminants in soils using plant-endophyte partnerships. Sci. Total Environ. 583: 352–368.

Fent, K., A.A. Weston and D. Caminada. 2006. Ecotoxicology of human pharmaceuticals. Aquat. Toxicol. 76: 122–159.

Ferrat, L., C. Pergent-Martini and M. Romeo. 2003. Assessment of the use of biomarkers in aquatic plants for the evaluation of environmental quality: application to seagrasses. Aquat. Toxicol. 65: 187–204.

Forni, C., A. Cascone, M. Fiori and L. Migliore. 2002. Sulphadimethoxine and Azolla filiculoides Lam.: a model for drug remediation. Water Res. 36: 3398–3403.

Garanzini, D.S. and M.L. Menone. 2015. Azoxystrobin causes oxidative stress and DNA damage in the aquatic macrophyte Myriophyllum quitense. Bull. Environ. Contam. Toxicol. 94: 146–151.

Gavrilescu, M., K. Demnerová, J. Aamand, S. Agathos and F. Fava. 2015. Emerging pollutants in the environment: present and future challenges in biomonitoring, ecological risks and bioremediation. New Biotech. 32: 147–156.

Gill, S.S. and N. Tuteja. 2010. Reactive oxygen species and antioxidant machinery in abiotic stress tolerance in crop plants. Plant Physiol. Biochem. 48: 909–930.

Gothwal, R. and T. Shashidhar. 2015. Antibiotic pollution in the environment: A review. Clean Soil Air Water 43: 479–489.

Groom, C.A., A. Halasz, L. Paquet, L. Olivier, C. Dubois and J. Hawari. 2002. Accumulation of HMX (octahydro-1,3,5,7-tetranitri-1,3,5,7-tetrazocine) in indigenous and agricultural plants grown in HMX-contaminated anti-tank firing-range soil. Environ. Sci. Technol. 36: 112–118.

Haman, C., X. Dauchy, C. Rosin and J.F. Munoz. 2015. Occurrence, fate and behavior of parabens in aquatic environments: A review. Water Res. 68: 1–11.

Hamilton, C.E. and T.L. Bauerle. 2012. A new currency for mutualism? Fungal endophytes alter antioxidant activity in hosts responding to drought. Fungal Diversity 54: 39–49.

Hamilton, C.E., P.E. Gundel, M. Helander and K. Saikkonen. 2012. Endophytic mediation of reactive oxygen species and antioxidant activity in plants: a review. Fungal Diversity 54: 1–10.

Hardoim, P.R., L.S. van Overbeek, G. Berg, A.M. Pirttila, S. Compant, A. Campisano et al. 2015. The hidden world within plants: ecological and evolutionary considerations for defining functioning of microbial endophytes. Microbiol Mol. Biol. Rev. 79: 293–320.

Hijosa-Valsero, M., R. Sidrach-Cardona, A. Pedescoll, O. Sanchez and E. Becares. 2017. Role of bacterial diversity on PPCPs removal in constructed wetlands. *In*: Stefanakis, A. [ed.]. Constructed Wetlands for Industrial Wastewater Treatment. Wiley-Blackwell, Hoboken, NJ.

Hijosa-Valsero, M., C. Reyes-Contreras, C. Domínguez, E. Bécares and J.M. Bayona. 2016. Behaviour of pharmaceuticals and personal care products in constructed wetland compartments: Influent, effluent, pore water, substrate and plant roots. Chemosphere 145: 508–517.

Hsu, F.C., R.L. Marxmiller and A.Y.S. Yang. 1990. Study of root uptake and xylem translocation of cinmethylin and related-compounds in detopped soybean roots using a pressure chamber technique. Plant Physiol. 93: 1573–1578.

Hussein, A. and M. Scholz. 2017. Dye wastewater treatment by vertical-flow constructed wetlands. Ecol. Eng. 101: 28–38.

Ijaz, A., A. Imran, M. Anwar ul Haq, Q.M. Khan and M. Afzal. 2016. Phytoremediation: recent advances in plant-endophytic synergistic interactions. Plant Soil 405: 179–195.

Imfeld, G., M. Braeckevelt, P. Kuschk and H.H. Richnow. 2009. Monitoring and assessing processes of organic chemicals removal in constructed wetlands. Chemosphere 74: 349–362.

Iori, V., M. Zacchini and F. Pietrini. 2013. Growth, physiological response and phytoremoval capability of two willow clones exposed to ibuprofen under hydroponic culture. J. Hazard. Mater. 262: 796–804.

Jones, K.C. and P. de Voogt. 1999. Persistent organic pollutants (POPs): state of the science. Environ. Pollut. 100: 209–221.

Kadlec, R.H. and S.D. Wallace. 2009. Treatment Wetlands. CRC Press, Boca Raton, FL.

Kang, J.W., Z. Khan and S.L. Doty. 2012. Biodegradation of trichloroethylene by an endophyte of hybrid poplar. Appl. Environ. Microbiol 78: 3504–3507.

Komives, T. and G. Gullner. 2005. Phase I xenobiotic metabolic systems in plants. Z. Naturforsch. C 60: 179–185.

Korte, F., G. Kvesitadze, D. Ugrekhelidze, M. Gordeziani, G. Khatisashvili, O. Buadze et al. 2000. Organic toxicants and plants. Ecotox. Environ. Safe. 47: 1–26.

Kvesitadze, G., G. Khatisashvili, T. Sadunishvili and J.J. Ramsden. 2006. The fate of organic contaminants in the plant cell. pp. 103–165. *In*: Kvesitadze, G., G. Khatisashvili, T. Sadunishvili and J.J. Ramsden [eds.]. Biochemical Mechanisms of Detoxification in Higher Plants. Springer, Berlin.

Lapworth, D.J., N. Baran, M.E. Stuart and R.S. Ward. 2012. Emerging organic contaminants in groundwater: A review of sources, fate and occurrence. Environ. Pollut. 163: 287–303.

Li, H.Y., D.Q. Wei, M. Shen and Z.P. Zhou. 2012. Endophytes and their role in phytoremediation. Fungal Diversity 54: 11–18.

Li, W.C. 2014. Occurrence, sources, and fate of pharmaceuticals in aquatic environment and soil. Environ. Pollut. 187: 193–201.

Li, Y.F., G.B. Zhu, W.J. Ng and S.K. Tan. 2014a. A review on removing pharmaceutical contaminants from wastewater by constructed wetlands: Design, performance and mechanism. Sci. Total Environ. 468: 908–932.

Li, Z., H. Xiao, S. Cheng, L. Zhang, X. Xie and Z. Wu. 2014b. A comparison on the phytoremediation ability of triazophos by different macrophytes. J. Environ. Sci. 26: 315–322.

Limmer, M.A. and J.G. Burken. 2014. Plant Translocation of organic compounds: Molecular and physicochemical predictors. Environ. Sci. Technol. Lett. 1: 156–161.

Liu, L., Y.h. Liu, C.x. Liu, Z. Wang, J. Dong, G.f. Zhu et al. 2013. Potential effect and accumulation of veterinary antibiotics in *Phragmites australis* under hydroponic conditions. Ecol. Eng. 53: 138–143.

Luo, Y., W. Guo, H.H. Ngo, L.D. Nghiem, F.I. Hai, J. Zhang et al. 2014. A review on the occurrence of micropollutants in the aquatic environment and their fate and removal during wastewater treatment. Sci. Total Environ. 473-474: 619–641.

Lv, T., Y. Zhang, L. Zhang, P.N. Carvalho, C.A. Arias and H. Brix. 2016. Removal of the pesticides imazalil and tebuconazole in saturated constructed wetland mesocosms. Water Res. 91: 126–136.

Macek, T., M. Mackova and J. Kas. 2000. Exploitation of plants for the removal of organics in environmental remediation. Biotechnol. Adv. 18: 23–34.

Masella, R., R. Di Benedetto, R. Vari, C. Filesi and C. Giovannini. 2005. Novel mechanisms of natural antioxidant compounds in biological systems: Involvement of glutathione and glutathione-related enzymes. J. Nutr. Biochem. 16: 577–586.

Matamoros, V., L.X. Nguyen, C.A. Arias, V. Salvadó and H. Brix. 2012. Evaluation of aquatic plants for removing polar microcontaminants: A microcosm experiment. Chemosphere 88: 1257–1264.

Mishra, S., S. Srivastava, R.D. Tripathi, R. Kumar, C.S. Seth and D.K. Gupta. 2006. Lead detoxification by coontail (*Ceratophyllum demersum* L.) involves induction of phytochelatins and antioxidant system in response to its accumulation. Chemosphere 65: 1027–1039.

Mittler, R. 2002. Oxidative stress, antioxidants and stress tolerance. Trends Plant Sci. 7: 405–410.

Mittler, R., S. Vanderauwera, M. Gollery and F. Van Breusegem. 2004. Reactive oxygen gene network of plants. Trends Plant Sci. 9: 490–498.

Nunes, B., G. Pinto, L. Martins, F. Gonçalves and S.C. Antunes. 2014. Biochemical and standard toxic effects of acetaminophen on the macrophyte species Lemna minor and Lemna gibba. Environ. Sci. Pollut. Res. 21: 10815–10822.

Olette, R., M. Couderchet, S. Biagianti and P. Eullaffroy. 2008. Toxicity and removal of pesticides by selected aquatic plants. Chemosphere 70: 1414–1421.

Pal, A., K.Y.-H. Gin, A.Y.-C. Lin and M. Reinhard. 2010. Impacts of emerging organic contaminants on freshwater resources: Review of recent occurrences, sources, fate and effects. Sci. Total Environ. 408: 6062–6069.

Paraiba, L.C. 2007. Pesticide bioconcentration modelling for fruit trees. Chemosphere 66: 1468–1475.

Pascal-Lorber, S., S. Despoux, E. Rathahao, C. Canlet, L. Debrauwer and F. Laurent. 2008. Metabolic Fate of [14C] Chlorophenols in Radish (Raphanus sativus), Lettuce (Lactuca sativa), and Spinach (Spinacia oleracea). J. Agric. Food Chem. 56: 8461–8469.

Pi, N., J.Z. Ng and B.C. Kelly. 2017. Bioaccumulation of pharmaceutically active compounds and endocrine disrupting chemicals in aquatic macrophytes: Results of hydroponic experiments with Echinodorus horemanii and Eichhornia crassipes. Sci. Total Environ. 601: 812–820.

Pietrini, F., D. Di Baccio, J. Acena, S. Pérez, D. Barceló and M. Zacchini. 2015. Ibuprofen exposure in Lemna gibba L.: Evaluation of growth and phytotoxic indicators, detection of ibuprofen and identification of its metabolites in plant and in the medium. J. Hazard. Mater. 300: 189–193.

Pilon-Smits, E. 2005. Phytoremediation. Annu. Rev. Plant Biol. 56: 15–39.

Pomati, F., A.G. Netting, D. Calamari and B.A. Neilan. 2004. Effects of erythromycin, tetracycline and ibuprofen on the growth of *Synechocystis* sp. and Lemna minor. Aquat. Toxicol. 67: 387–396.

Prasad, S.M., A. Singh and P. Singh. 2015. Physiological, biochemical and growth responses of Azolla pinnata to chlorpyrifos and cypermethrin pesticides exposure: a comparative study. Chem. Ecol. 31: 285–298.

Raja, W., P. Rathaur, S.A. John and P.W. Ramteke. 2012. Photosynthetic, Biochemical and Enzymatic Investigation of *Azolla microphylla* in Response to an Insecticide-Hexachlorohexahydro-Methano-Benzodioxathiepine-Oxide. Our Nature 10: 145–155.

Reinhold-Hurek, B. and T. Hurek. 2011. Living inside plants: bacterial endophytes. Curr. Opin. Plant Biol. 14: 435–443.

Renner, R. 2002. The Kow controversy. Environ. Sci. Technol. 36: 410A–413A.

Reyes-Contreras, C., M. Hijosa-Valsero, R. Sidrach-Cardona, J.M. Bayona and E. Bécares. 2012. Temporal evolution in PPCP removal from urban wastewater by constructed wetlands of different configuration: A medium-term study. Chemosphere 88: 161–167.

Rmiki, N.-E., Y. Lemoine and B. Schoefs. 1999. Carotenoids and stress in higher plants and algae. pp. 465–482. *In*: Pessarakli, M. [ed.]. Handbook of Plant and Crop Stress. CRC Press, Boca Raton, FL.

Rubin, E. and A. Ramaswami. 2001. The potential for phytoremediation of MTBE. Water Res. 35: 1348–1353.

San Miguel, A., P. Ravanel and M. Raveton. 2013a. A comparative study on the uptake and translocation of organochlorines by *Phragmites australis*. J. Hazard. Mater. 244-245: 60–69.

San Miguel, A., P. Schröder, R. Harpaintner, T. Gaude, P. Ravanel and M. Raveton. 2013b. Response of phase II detoxification enzymes in *Phragmites australis* plants exposed to organochlorines. Environ. Sci. Pollut. Res. 20: 3464–3471.

Sandermann, H. 1992. Plant-metabolism of xenobiotics. Trends Biochem. Sci. 17: 82–84.

Sandermann, H., Jr. 1994. Higher plant metabolism of xenobiotics: the 'green liver' concept. Pharmacogenetics 4: 225–241.

Santoyo, G., G. Moreno-Hagelsieb, Ma. del Carmen Orozco-Mosqueda and B.R. Glick. 2016. Plant growth-promoting bacterial endophytes. Microbiol. Res. 183: 92–99.

Sauvêtre, A. and P. Schröder. 2015. Uptake of Carbamazepine by rhizomes and endophytic bacteria of *Phragmites australis*. Front. Plant Sci. 6: 83.

Schröder, P. and C. Collins. 2002. Conjugating enzymes involved in xenobiotic metabolism of organic xenobiotics in plants. Int. J. Phytorem. 4: 247–265.

Shehzadi, M., M. Afzal, M.U. Khan, E. Islam, A. Mobin, S. Anwar et al. 2014. Enhanced degradation of textile effluent in constructed wetland system using Typha domingensis and textile effluent-degrading endophytic bacteria. Water Res. 58: 152–159.

Shelef, O., A. Gross and S. Rachmilevitch. 2013. Role of plants in a constructed wetland: Current and new perspectives. Water 5: 405–419.

Sicbaldi, F., G.A. Sacchi, M. Trevisan and A.A.M. Delre. 1997. Root uptake and xylem translocation of pesticides from different chemical classes. Pestic. Sci. 50: 111–119.

Smirnoff, N. 2005. Antioxidants and reactive oxygen species in plants. Blackwell Publishing Ltd, Oxford, UK.

Susarla, S., V.F. Medina and S.C. McCutcheon. 2002. Phytoremediation: An ecological solution to organic chemical contamination. Ecol. Eng. 18: 647–658.

Syranidou, E., S. Christofilopoulos, G. Gkavrou, S. Thijs, N. Weyens, J. Vangronsveld et al. 2016. Exploitation of endophytic bacteria to enhance the phytoremediation potential of the wetland helophyte *Juncus acutus*. Front. Microbiol. 7: 1016.

Toro-Vélez, A.F., C.A. Madera-Parra, M.R. Peña-Varón, W.Y. Lee, J.C. Cruz, W.S. Walker et al. 2016. BPA and NP removal from municipal wastewater by tropical horizontal subsurface constructed wetlands. Sci. Total Environ. 542, Part A: 93–101.

Truu, J., M. Truu, M. Espenberg, H. Nõlvak and J. Juhanson. 2015. Phytoremediation and plant-assisted bioremediation in soil and treatment wetlands: A review. Open Biotechnol. J. 9: 85–92.

Truu, M., J. Juhanson and J. Truu. 2009. Microbial biomass, activity and community composition in constructed wetlands. Sci. Total Environ. 407: 3958–3971.

Vafaei, F., A.R. Khataee, A. Movafeghi, S.Y. Salehi Lisar and M. Zarei. 2012. Bioremoval of an azo dye by Azolla filiculoides: Study of growth, photosynthetic pigments and antioxidant enzymes status. Int. Biodeterior. Biodegrad. 75: 194–200.

Verlicchi, P. and E. Zambello. 2014. How efficient are constructed wetlands in removing pharmaceuticals from untreated and treated urban wastewaters? A review. Sci. Total Environ. 470-471: 1281–1306.

Vymazal, J. and T. Brezinová. 2015. The use of constructed wetlands for removal of pesticides from agricultural runoff and drainage: A review. Environ. Int. 75: 11–20.

Vymazal, J., H. Brix, P.F. Cooper, M.B. Green and R. Haberl. 1998. Constructed wetlands for wastewater treatment in Europe. Backhuys Publishers, Leiden, The Netherlands.

Wang, C., S. Zheng, P. Wang and J. Qian. 2014. Effects of vegetations on the removal of contaminants in aquatic environments: A review. J. Hydrodyn. B 26: 497–511.

Wang, C., Q. Zhou, L. Zhang, Y. Zhang, E. Xiao and Z. Wu. 2013. Variation characteristics of chlorpyrifos in nonsterile wetland plant hydroponic system. Int. J. Phytorem. 15: 550–560.

Wang, Q., C. Li, R. Zheng and X. Que. 2016. Phytoremediation of chlorpyrifos in aqueous system by riverine macrophyte, Acorus calamus: toxicity and removal rate. Environ. Sci. Pollut. Res. 23: 16241–16248.

Wani, Z.A., N. Ashraf, T. Mohiuddin and S. Riyaz-Ul-Hassan. 2015. Plant-endophyte symbiosis, an ecological perspective. Appl. Microbiol. Biotechnol. 99: 2955–2965.

Weyens, N., D. van der Lelie, S. Taghavi and J. Vangronsveld. 2009. Phytoremediation: plant-endophyte partnerships take the challenge. Curr. Opin. Biotechnol. 20: 248–254.

Wilson, P.C., T. Whitwell and S.J. Klaine. 2000. Metalaxyl and simazine toxicity to and uptake by typha latifolia. Arch. Environ. Contam. Toxicol. 39: 282–288.

Wojtaszek, P. 1997. Oxidative burst: An early plant response to pathogen infection. Biochem. J. 322: 681–692.

Yan, Q., G. Feng, X. Gao, C. Sun, J.s. Guo and Z. Zhu. 2016. Removal of pharmaceutically active compounds (PhACs) and toxicological response of Cyperus alternifolius exposed to PhACs in microcosm constructed wetlands. J. Hazard. Mater. 301: 566–575.

Zezulka, Š., M. Kummerová, P. Babula and L. Vánová. 2013. Lemna minor exposed to fluoranthene: Growth, biochemical, physiological and histochemical changes. Aquat. Toxicol. 140-141: 37–47.

Zhang, D.Q., R.M. Gersberg, W.J. Ng and S.K. Tan. 2014. Removal of pharmaceuticals and personal care products in aquatic plant-based systems: A review. Environ. Pollut. 184: 620–639.

Zhang, D.Q., T. Hua, R.M. Gersberg, J. Zhu, W.J. Ng and S.K. Tan. 2012. Fate of diclofenac in wetland mesocosms planted with Scirpus validus. Ecol. Eng. 49: 59–64.

Zhang, Y., T. Lv, P.N. Carvalho, C.A. Arias, Z. Chen and H. Brix. 2016. Removal of the pharmaceuticals ibuprofen and iohexol by four wetland plant species in hydroponic culture: plant uptake and microbial degradation. Environ. Sci. Pollut. Res. 23: 2890–2898.

Zhu, X., L. Jin, K. Sun, S. Li, W. Ling and X. Li. 2016. Potential of endophytic Bacterium Paenibacillus sp. PHE-3 Isolated from Plantago asiatica L. for Reduction of PAH contamination in plant tissues. Int. J. Environ. Res. Public Health 13: 633.

Zhu, X., X. Ni, J. Liu and Y. Gao. 2014. Application of endophytic bacteria to reduce persistent organic pollutants contamination in plants. Clean Soil Air Water 42: 306–310.

Glossary

4	Number of electrons involved in the reaction (equation 2, Chapter 14)
8	Number of electrons involved in the reaction (equation 1, Chapter 14)
$[CH_3COO^-]$	Acetate concentration (mol/L) (equation 1, Chapter 14)
$[H^+]$	Protons concentration (mol/L) (equations 1, 2, Chapter 14)
$[HCO_3^-]$	Bicarbonate concentration (mol/L) (equation 1, Chapter 14)
A	Wetland surface area (m^2) (Chapter 7)
a	From 0 m of wetland input (Chapter 19)
ACA	Catalan Water Agency in Catalonia (Chapter 11)
ACC-HUMAN	Food chain model to predict the levels of lipophilic organic contaminants in humans (Czub and McLachlan 2004) (Chapter 15)
Accretion	An increase by natural growth or by gradual external addition; growth in size or extent
ACN:MeOH	Solvent mixture acetonitrile:methanol (Chapter 3)
Ag-NPs	Silver nanoparticles (Chapter 13)
AL	Anaerobic lagoon (Chapter 19)
AMD	Acid mine drainage (Chapter 2)
AMSL	Above mean sea level (Chapter 9)
Anammox	Anaerobic ammonia oxidation
ANOVA	Analysis of variance (several chapters)
ANP	Protected Natural Area for its name in Spanish *Área Natural Protegida* (Chapter 1)
AP	Anaerobic pond (Chapter 6)
ARBs	Antibiotic resistant bacteria
Archaea	Plural of archaeon. Any of a group of microorganisms that resemble bacteria but are different from them in certain aspects of their chemical structure, such as the composition of their cell walls. Archaea usually live in extreme, often very hot or salty environments, such as hot mineral springs or deep-sea hydrothermal vents, but some are also found in animal digestive systems. The archaea are considered a separate kingdom in some classifications, but a division of the prokaryotes (Monera) in others. Some scientists believe that archaea were the earliest forms of cellular life. Also called *archaebacterium*

ARGs	Antibiotic resistance genes
A_s	Surface area of the artificial wetland, m^2 (equation 5, Chapter 18)
ATZ	Atrazine
AW or CW	Artificial or constructed wetlands
AW-MFC	Artificial wetland-microbial fuel cell (Chapters 3, 14)
B	Biofilter (Chapter 19)
B	Boreal (Chapter 19)
B	Between 2–14 m of wetland input (Chapter 19)
b	Interception with y axis of a straight line (Table 2, Chapter 3)
b	Width of the artificial wetland cell, m (Chapter 18)
BAC	Bacterial artificial chromosomes (Chapter 8)
BCFs	Bioconcentration factors (Chapter 15)
BDL	Below Detection Limit
BFT	Bio Floc Technology (Chapter 12)
BOD	Biochemical oxygen demand (Several chapters)
BOD_5	Biochemical oxygen demand measured during five days (most chapters)
BPA and NP	Bisphenol A and nonylphenols (Chapter 13)
BS	Buffer Strip (Chapter 11)
bssA	Anaerobic enzyme benzyl succinate synthase (Chapter 2)
BTEX	Benzene-toluene-ethylbenzene-xylene mixtures (Chapter 2)
B_V	Bioconcentration factor in the plants (equation 2, Chapter 15)
C	Organic compound concentration in the water phase (Equation 1, Chapter 20)
c_0	Concentration added before adsorption test, mg L^{-1} (equation 2, Chapter 5)
CA	Cyanuric acid (Chapter 3)
CAD/1S	Consortium designed for organic matter, ammonium, and phosphate removal, for the case of domestic wastewater (Chapter 8)
CAM/G+	Consortium designed for lead, chromium, and mercury removal, for the case of domestic wastewater contaminated with metals (Chapter 8)
CBOD	Carbonaceous biochemical oxygen demand
C_e	Experimental concentration in shoot or root (mg/kg fresh mass) (Chapter 15)
C_e	Mean effluent concentration, mg L^{-1} (Chapter 7)
Ce	Concentration of the pollutant in the effluent, mg L^{-1} (equation 2, Chapter 18)
c_e	Heavy metals concentration quantified in the aqueous solution called the equilibrium concentration after three days (equation 2, Chapter 5)
CECs	Contaminants of emerging concern (Chapters 13, 15)
CENHICA	Cuban National Center of Hydrology and Water Quality in Spanish (Chapter 8)
Cer	*Ceratophyllum demersum* (Chapter 19)

CF	Bioconcentration factor (equation 1, Chapter 15)
C_f	Concentration in the wetland outlet (Chapter 10)
CFU	Colonies forming units per 100 milliliters (Chapter 6)
CFU	Colonies forming units accounting per milliliter (Chapter 3)
C_i	Inlet concentration of the response variable (cypermethrin, COD, TSS) (Chapter 10)
$C_{i,f}$	Cypermethrin, P-PO$_4^{3-}$, N-NH$_4^+$, and BOD$_5$ concentrations in mg L^{-1} at the inlet (i) and outlet (*final* in Spanish) of wetlands, respectively (equation 3, Chapter 10)
ci	Concentration of heavy metal studied (Chapter 5)
CIDS	Conventional Irrigation and Drainage System (Chapter 7)
CITMA	Cuban Ministry of Science, Technology, and Environment in Spanish (Chapter 8)
CALTOX	CALifornia TOXic Substances control (Chapter 15)
CLEA	Contaminated Land Exposure Assessment (Chapter 15)
C_m	Concentration in the model in shoot or root (mg/kg fresh mass) (Chapter 15)
C_0	Mean influent concentration, mg L^{-1} (Chapter 7)
Co	Concentration of the pollutant in the influent, mg L^{-1} (equation 2, Chapter 18)
COD	Chemical oxygen demand (several chapters)
C_{pt}	Calculated mass of chemical per unit mass of plant (mg kg^{-1} FM plant) (Chapter 15)
C_r	Ratio between the predicted concentration in the model (C_m) in shoot or root (mg/kg fresh mass) and the experimental (C_e) concentration in shoot or root (mg/kg fresh mass) (Chapter 15)
CRT/SRT	Cell/Solids retention time
c_s	Adsorbed concentration, mg L^{-1} (equation 2, Chapter 5)
C_{som}	Soil organic matter normalised chemical concentration in soil (mg kg^{-1} DM soil) (Chapter 16)
CSOIL	Exposure model for human risk assessment of soil contamination (Chapter 15)
Csp	*Canna* sp. (Chapter 19)
CSTR	Continuously stirred tank reactor (Chapter 2)
CT	Artificial or constructed wetlands *tezontle* uninoculated and sterile control, without vegetation (Chapter 5)
CV	Coefficient of variation (Chapters 3, 15)
CW or AW	Constructed or artificial wetlands (all chapters)
CWC	AW lab reactors used as controls neither planted nor inoculated (Chapter 5) and the other three were inoculated but not planted (CWI)
CWI	AW lab reactors used as controls inoculated but not planted (Chapter 5)
CWP	AW lab reactors planted with *Phragmites australis* (Chapter 5)
CWP_I	AW lab reactors planted with *Phragmites australis* (Chapter 5)

CWS	Constructed wetland systems
Cyp-B	Intermittent mode with *C. alternifolius* (Fig. 2, Chapter 10)
Cyp-C	Continuous flow with *C. alternifolius* (Fig. 2, Chapter 10)
DBPs	Disinfection-by-products (trihalomethanes [THMs] and haloketones [HKs]) (Chapter 13)
DBPs, nitrogenous	Nitrogenous disinfection-by-products (haloacetonitriles [HANs] and trihalonitromethanes [TNMs]) (Chapter 13)
DDT	Dichlorodiphenyltrichloroethane (Chapters 3, 15)
DEA	Deethylatrazine (Chapter 3)
DIA	Deisopropylatrazine (Chapter 3)
Diel, diurnal	The distinction is that "**diurnal**" and "nocturnal" describe when something with a daily cycle preferentially happens during the day or night hours, respectively. **Diel** describes something that has a 24-hour cycle, but is ambiguous as to when the "peak" of the cycle occurs (Chapters 2 and 10)
DIDEATZ	Deisopropyl-deethylatrazine (Chapter 3)
Digestion	This term according to Merriam Webster is the action, process, or power of digesting: such as (a) the process of making food absorbable by mechanically and enzymatically breaking it down into simpler chemical compounds in the digestive tract, and (b) the process in sewage treatment by which organic matter in sludge is decomposed by anaerobic bacteria -archaea- with the release of a burnable mixture of gases. Therefore, the use of the adjective anaerobic to digestion is unnecessary (a grammatical form called pleonasm), and in this book it will not be used (note of first editor)
DM	Dry mass (equation 1, models, Chapter 15)
DNA	Deoxyribonucleic acid
DNRA	Dissimilatory nitrate reduction to ammonia
DO	Dissolved oxygen (most chapters)
DOM	Dissolved organic matter
D_{ow}	Measurement of hydrophobicity for ionic substances. It takes into account the molecular fraction and the Kow value of the substance (Chapter 15)
E1, E2, E3 and EE2	Estrone, 17β-estradiol, estriol, and 17α-ethinylestradiol (Chapter 13)
EAAW	Electrochemically assisted artificial wetlands (Chapter 14)
E_{an}	Anode potential (Volts, V) (equation 1, Chapter 14)
E_{an}^0	Anode standard potential (298 K, 1 M) (Volts, V) (equation 1, Chapter 14)
EC	Electrical conductivity
ECD	Electron Capture Detector (Chapter 9)
E_{cat}	Cathode potential (Volts, V) (equation 3, Chapter 14)
E_{cat,O_2}	Cathode potential (Volts, V) (equation 2, Chapter 14)
E_{cat}^0	Cathode standard potential (298 K and atmospheric pressure) (Volts, V) (equation 2, Chapter 14)
E_{cell}	Microbial fuel cell voltage (Volts, V) (equation 3, Chapter 14)

Ectoparasite	A parasite, such as a flea, that lives on the outside of its host (Chapter 10)
EDCs	Endocrine disrupting compounds (Chapter 13)
Eh	Redox potential (Chapter 16)
EI, PE	Equivalent inhabitant, person or population equivalent
El	*Elodea canadensis* (Chapter 19)
EPS	Extracellular polymeric substances also known as biofilm
ET	Evapotranspiration is the sum of evaporation and plant transpiration from the Earth's land and ocean surface to the atmosphere
EU	European Union (Chapter 11)
EUSES	European Union System for the Evaluation of New and Existing Substances (Chapter 15)
F	Faraday constant ($9.65*10^4$ C/mol) (equation 1, Chapter 14)
FC	Fecal coliform organisms, fecal coliforms
Fermentation	Term coined by Louis Pasteur to describe the anaerobic conversion of glucose by *Saccharomyces cerevisiae* to produce ethyl alcohol (ethanol) and carbon dioxide. Some authors have tended to generalize this term to other biochemical reactions but considering the importance that Pasteur gave to this specific bioconversion it should be only used for this reaction (Note of the first editor)
FGAC	Ascending gross filter (Chapter 10)
FGDi	Descending dynamic coarse filter while the second unit was an ascending gross filter (Chapter 10)
FID	Flame ionization detector (Chapter 9)
$F_{i,f}$	Flow at the inlet (i) and outlet (*final* in Spanish) of wetlands, respectively (equation 3, Chapter 10)
FiP	Fish pond (Chapter 6)
FM	Fresh mass (equation 1, Chapter 15)
f_{OC}	Fraction of organic carbon in the soil (dimensionless) (Chapter 15)
FP	Facultative pond (Chapter 6)
f_{pm}	Mass fraction of water in the plant (g g^{-1}) (Chapter 15)
f_{pom}	Total mass fraction of the organic matter in the plant (g g^{-1}) (Chapter 15)
FS	Fecal streptococci
FT	Feed Tank (Chapter 19)
FTWs	Floating treatment wetlands (Chapter 7)
FWS	Free water surface system (Chapters 7, 17)
FWS CW	Free Water Surface Constructed Wetlands (several chapters)
GC	Gas chromatography (Chapter 3)
GC-FID	Gas chromatography and a flame ionization detector (Chapter 3)
GEF International Waters	Global Environmental Facilities International Waters (United Nations Development Program) (Chapter 8)
GEF/UNDP	Global Environment Facility for Cuba (Suárez et al. 2012) (Chapter 8)

GHG	Global Climate Change Gases (greenhouse gases): carbon dioxide (CO_2), methane (CH_4) and nitrous oxide (N_2O), among others (Mander et al. 2008; IPCC Glossary: https://www.ipcc.ch/ipccreports/tar/wg1/518.htm) (Chapters 7, 9 and 19)
GIAHS	Globally Important Agricultural Heritage Systems as part of a Global Partnership Initiative on conservation and adaptive management launched by the Food and Agriculture Organization (FAO) of the United Nations (Chapter 1)
Gr	Gravel (Chapter 4)
H	Height of the artificial wetland cell, m (Chapter 18)
HA	Hydroxiatrazine (Chapter 3)
HDPE	High density polyethylene used for artificial wetlands lining
HF	Horizontal flow (Chapter 2)
HFCW	Horizontal Flow Constructed Wetland (Chapter 2), Subsurface Horizontal Flow Constructed Wetland (Chapter 11)
HFR	High-loaded wetlands
HLR	Hydraulic loading rate (in most chapters)
HPLC	High performance liquid chromatography (Chapter 3)
HPV	High production volume chemicals (Chapter 13)
HRT	Hydraulic residence time (in most chapters)
HSSF	Horizontal sub-surface flow (Chapter 2)
HSSF	Horizontal sub-surface flow wetlands (several chapters)
HYDRUS, STELLA, PHWAT or *RetrascoCodeBright*	Specialized software packages used by mechanistic models (Chapter 2)
IA	Infectivity Assay
I.C.	Interval of confidence (Table 2, Chapter 3)
ICTs	Information and Communication Technologies (Chapter 11)
Illumina MiSeq high-throughput sequencing	The *MiSeq* benchtop sequencer enables targeted and microbial genome applications, with *high*-quality *sequencing*, simple data analysis, and cloud storage (https://www.illumina.com/content/dam/illumina-marketing/documents/products/illumina_sequencing_introduction.pdf) (Chapter 11)
INRH	Cuban "*Instituto Nacional de Recursos Hidráulicos*" in Spanish (National Institute of Hydraulic Resources) (Chapter 8)
InT	Inlet Tank (Chapter 11)
Integrons	Genetic mechanisms that allow bacteria to adapt and evolve rapidly through the acquisition, stockpiling, and differential expression of new genes. These genes are embedded in a specific genetic structure called gene cassette (a term that is lately changing to integron cassette) that generally carries one promoterless ORF (*open reading frame*) together with a recombination site (*attC*). Integron cassettes are incorporated to the *attI* site of the integron platform by site-specific recombination reactions mediated by the integrase (Chapter 15)

Iohexol	Iodinated X-ray contrast media (Chapter 13)
Ips	*Iris pseudacorus* (Chapter 19)
IT	Artificial or constructed wetlands inoculated *tezontle*, without vegetation (Chapter 5)
Je	*Juncus effesus* (Chapter 19)
k	First-order decay rate constant, d^{-1} (Chapter 7)
k'	Degradation constant (Chapter 3)
K_{20}	K_T depends on the pollutant to be eliminated. In case of not having values from the literature the following equation can be applied, considering its value at 20°C (equation 6, Table 20, Chapter 18)
k-C*	First-order model, kinetic constant k and concentration C (Kadlec and Knight 1996) (various chapters)
K_{AW}	Air-water partition constant (Chapter 15)
K_{lip}	Lipid-water partition coefficient (Chapter 15)
K_{OA}	Octanol-air partition constant (Chapter 15)
K_{pom}	Chemical partition coefficient between plant organic matter and water (dimensionless) (Chapter 15)
k_s	Hydraulic conductivity (Table 2, Chapter 18)
K_{som}	Chemical partition coefficient between soil organic matter and water (dimensionless)
K_T	First order reaction constant, dependent on temperature, d^{-1} (equation 2, Chapter 18)
K_{OC}	Organic carbon-water partition coefficient for the contaminant (cm^3/DM) (Chapter 15)
K_{ow}	Octanol-water partition coefficient (Chapters 15, 16)
KRRBCW	Kaoping River Rail Bridge Constructed Wetland (Chapter 7)
L	Length of the artificial wetland cell, m (equation 3, Chapter 18)
LCA	Life cycle assessment (Chapter 19)
LD	Limit of detection (Table 2, Chapter 3)
LECA	Light expanded clay aggregates (Chapter 13)
LFR	Low-loaded wetlands
LMEWP	Lake Manzala Engineered Wetland Project (Egypt) (Chapter 6)
$Load_e$ and $Load_s$	Inlet (*entrada* in Spanish) and outlet *salida* (outlet in Spanish) loads of the response variables (cypermethrin, $P\text{-}PO_4^{3-}$, $N\text{-}NH_4^+$, BOD_5) in each constructed wetland expressed in mg m^{-2} d^{-1} (Chapter 10)
LQ	Limit of quantification (Table 2, Chapter 3)
LxWxd	AW or CW dimensions (length, width, depth) (various chapters)
M	Mediterranean (Chapter 19)
M	Slope of a straight line (Table 2, Chapter 3)
M	*Tezontle* or river stone dry matter inside reactor, mg (equation 2, Chapter 5)

Mass *versus* weight	Differences between mass and weight:

Mass Characteristics	Weight Characteristics
1. It is the matter quantity of any body.	1. It is the force that causes bodies to fall.
2. It is a scalar magnitude.	2. It is a vector magnitude.
3. It is measured in a scale.	3. It is measured in a dynamometer.
4. Its value is constant, independent of altitude and latitude.	4. It varies according to its position, that is, it depends on altitude and latitude.
5. Its measuring units in the International System are the gram (g) and the kilogram (kg).	5. Its measuring units in the International System are the dyne and the Newton.
6. Mass suffers accelerations.	6. Weight produces accelerations.

(http://cienciasprimeroeso.blogspot.mx/2015/04/masa-versus-peso.html)

MCB	Monochlorobenzene (Chapter 2)
MCPA herbicide	A widely used systemic herbicide which belongs to the group of related synthetic herbicides, the chlorophenoxy herbicides (4-chloro-2-methyphenoxyacetic acid) (Dordio et al. 2007, Chapter 13)
MDL	Method detection limit calculated as three standard mean deviations of ten blank solutions (Chapter 10)
Mexico	Place where the god *Metl* is. From *Metl*, maguey (agave), *xictli*, navel, and *co*, location or place, that is, the location of the god with navel in the form of maguey (agave). This god is also known as *Huitzilopochtli*, left handed hummingbird, from *huitzillin*, hummingbird and *opochtli*, left handside and had a blue-green hummingbird helmet. His hummingbird helmet was the one item that consistently defined him as Huitzilopochtli, the sun god, in artistic renderings (Read, K.A. and J.J. Gonzalez. 2002. Mesoamerican Mythology: A Guide to the Gods, Heroes, Rituals, and Beliefs of Mexico and Central America. Oxford University Press. Oxford. P. 195) (Chapters 1 and 3)
MFC-LH	Microbial fuel cells with living hydrophytes (Chapter 14)
MM	Molecular mass (Chapter 15)
MM	Multiparameter Water Quality Meter (Chapter 11)
MMI	Total mass of metal in influent, mg (Chapter 5)
MP	Maturation pond (Chapter 6)
M-P-M	Marsh-Pond-Marsh (Chapter 19)
MS	Mass spectrometry (Chapter 3)
MSBSE	Magnetic stir bar sorption extraction (Chapter 3)
MSPD	Matrix solid phase dispersion (Chapter 3)
MTBE	Methyl tert butyl ether

N	Porosity–available space for water flow through the support media of the artificial wetland cell left by vegetation roots and other solids-, generally expressed between 0 and 1.0 (equation 3, Chapter 18)
Neonicotinoids	A new group of pesticides (Chapter 13), of a class of broad-spectrum insecticides having a chemical structure similar to that of nicotine and acting on the central nervous system of insects by selectively binding to nicotinic acetylcholine receptors: • Neonicotinoids are neurotoxins that, as the name implies, chemically resemble nicotine. They are considered safer for humans than many other classes of pesticides, because they interfere with neural pathways that are more common in insects than in mammals ... (Elizabeth Kolbert, *New Yorker*, 6 Aug. 2007) • French and German beekeepers blame their losses on insecticides called neonicotinoids—but France banned them 10 years ago and its bees are still dying (Debora MacKenzie, *New Scientist*, 14 Feb. 2009) (https://www.merriam-webster.com/dictionary/neonicotinoid)
NER	Net Energy Recovery (Chapter 3)
n.d.	No data (Chapter 6)
NPs	Nanoparticles (Chapter 13)
Nereda	Nereda is a wastewater treatment technology invented by Mark van Loosdrecht of the Delft University of Technology in the Netherlands. The technology is based on aerobic granulation and is a modification of the activated sludge process (Chapter 18).
NES	Cuban National Environmental Strategy, adopted in 1997 (Chapter 8)
NES 2007–2010	Organizations of the State Central Administration (OSCA) and the Territorial Strategies of Cuba (Chapter 8)
NH_4^+	Ammonium ion
NH_4-N	Ammonia nitrogen
nirS and nosz	Microbial genes as biological markers of denitrification (Chapter 11)
NMR	Nuclear magnetic resonance (Chapter 3)
NO_3-N	Nitrates nitrogen
NP, NPs	Nanoparticle(s) (Chapter 13)
NPS	Non point source pollutants from agricultural drainage (Chapter 7)
n.r.	Not reported (Chapter 12)
n.s.	Not specified (various chapters)
NTU	Nephelometric Turbidity Unit signifies that the instrument is measuring scattered light from the sample at a 90-degree angle from the incident light (Chapter 11).
Ny	*Nymphaea* sp. (Chapter 19)
O	Oceanic (Chapter 19)
Oasis HLB cartridges	Hydrophilic lipophilic balance solid sorbent, a universal polymeric reversed-phase sorbent [poly(divinylbenzene-co-N-vinylpyrrolidone)] (Chapter 3)

OMPs	Organic contaminants at trace level concentration ($\leq$ pp)
OPEC	Organization of the Petroleum Exporting Countries
ORP	Oxidation-reduction potentials (Chapter 7)
OutT	Outlet Tank (Chapter 11)
O&M costs	Costs associated with operation and maintenance (Chapter 7)
Pa	*Phragmites australis* (Chapter 19)
PAE, PAEs	Phthalic acid ester(s) (Chapter 13)
PAHO, *OPS*	Panamerican Health Organization, *Organización Panamericana de la Salud* (Chapter 8)
PAHs	Polycyclic aromatic hydrocarbons (Chapter 15)
PCBs	Polychlorinated biphenyls (Chapter 15)
PCDDs	Polychlorinated Dibenzo Dioxins (Chapter 15)
PCDFs	Polychlorinated Dibenzo Furans (Chapter 15)
PCE	Perchloroethylene (Chapter 2)
PCPs	Personal care products (Chapter 13)
PE, pe, EI	Person or population equivalent, equivalent inhabitant (several chapters)
PEDWS	Paddy Eco-ditch and Wetland System (Chapter 7)
PFBR	Planted fixed bed reactor (Chapter 2)
Phm	*Phragmites mauritianus* (Chapter 19)
Phr	*Phragmites australis* (Chapter 19)
PkC*	P is the hydraulic flow efficiency, k is the pollutant specific rates of biological degradation, and C* is the desired outflow concentration (Co) in relation to inflow pollutant levels, and system production of pollutant from degrading biological material in the system (Chapter 11)
Pleonasm	The use of more words than those necessary to denote mere sense (as in the man he said): redundancy
PLR	Pollutant loading rate, g m^2 d^{-1} (Chapter 7)
pO_2	Oxygen partial pressure (Pa) (equation 1, Chapter 14)
POM	Particulate organic matter, usually defined as material which is large enough to be retained on 0.45 mm filter (Kadlec and Wallace 2009) (Chapter 17)
POP	Persistent Organic Pollutants (Chapter 16)
PPCPs	Pharmaceuticals and personal care products (Chapter 15)
PVC	Polyvynil chloride used for artificial wetlands lining
Pz	Piezometer (Chapter 11)
Q	Average flow rate through the reactor, m^3 s^{-1} (Chapter 7)
Q	Flow rate, m^3 s^{-1} (equation 1, Chapter 18), average flow rate through the reactor, m^3 s^{-1} (equation 4, Chapter 18)
QBR	Riparian Forest Quality Index (Chapter 11)
QDs	Cadmium sulphide/zinc sulphide (CdS/ZnS) quantum dots (Chapter 13)

Q_e	Flow rate at the exit of the reactor, $m^3\ s^{-1}$ (equation 4, Chapter 18)
Q_o	Flow rate at the entrance of the reactor, $m^3\ s^{-1}$ (equation 4, Chapter 18)
R	Universal constant of gases (8.314 J/mol K) (equation 1, Chapter 14)
r	Pearson correlation coefficient (Chapters 3, 10)
R^2	Coefficient of determination
R%	Removal percentage
RAIDAR	Risk Assessment IDentification And Ranking (Chapter 15)
RAS	Recirculating aquaculture systems (Chapter 12)
RCF	Root concentration factor (Chapter 20)
RDX	Hexahydro-1,3,5-trinitro-1,3,5-triazine
RE	Removal efficiency (%), $(C_0 - C_e)/C_0 \times 100$ (Tables 1, 2, Chapter 7, Chapter 9)
REACH	Registration, Evaluation, Authorisation, and Restriction of Chemicals (Chapter 15)
Redox	Oxidation-reduction or reduction-oxidation potential
Residence time *versus* retention time	Residence time is usually applied to fluids whereas retention time is applied to solids, being biological or inorganic, passing through a vessel
RFC	Calculated soil-to-plant root concentration factor (µg/g FM plant over µg/g DM soil) (Chapter 15)
ROL	Radial Oxygen Loss: It is one of the most important functional characteristics of submerged plants, which may relate to their adaptability to the wetland support media and to their ability to remove nutrients in anoxic conditions. Radial oxygen loss (ROL) has been suggested to be a major process to protect plants exposed to the anaerobic by-products of soil anaerobiosis. (http://www.pjoes.com/pdf/24.4/Pol.J.Environ.Stud.Vol.24.No.4.1795-1802.pdf) (Chapter 9)
RSC	Artificial or constructed wetlands sterile no-inoculated river stone, pebbles, control, without vegetation (Chapter 5)
RSD	Relative standard deviations (Chapter 10)
RSI	Artificial or constructed wetlands inoculated river stone, pebbles, without vegetation (Chapter 5)
RT-PCR	Real-Time Reverse-Transcription Polymerase Chain Reaction
RYA/YCM	Dioxin-like activity (Fig. 2, Chapter 13)
SAG	Scientific Advisory Group (Chapter 1)
SAV	Submerged aquatic vegetation, presence of photosynthetic algae (Chapter 17)
SCF	Calculated soil-to-plant stem concentration factor (µg/g FM plant over µg/g DM soil) (Chapter 15)
Sch	*Schoenoplectus* sp. (Chapter 19)
Sch	*Schoenoplectus californicus* (Chapter 19)
Sci	*Scirpus* sp. (Chapter 19)

SF	Surface flow wetlands. Flow of water or wastewater over the surface of the ground (several chapters) (Brix, H., R.H. Kadlec, R.L. Knight, J. Vymazal, P. Cooper and R. Haberl. **Constructed wetlands for pollution control: Processes, Performance, Design and Operation**. 2000 IWA Pub.) (several chapters)
SFCW	Surface flow constructed wetlands (Chapter 13)
SF CWs	Surface flow constructed wetlands (Chapter 13)
SOM	Soil organic matter (Chapter 15)
Spar	*Sparganium americanum* (Chapter 19)
SPE	Solid Phase Extraction (Chapter 3)
SPE-GC-FID	Solid phase extraction technique followed by gas chromatography using a flame ionization detector (Chapter 3)
SRT, CRT	Solids/Cell Retention Time
Ss	*Scirpus silvaricus* (Chapter 19)
SSF	Sub-surface flow wetlands (Chapter 2)
SSFS	Subsurface flow system (Chapter 7)
ST	Sedimentation tank (Chapters 11 and 19)
STN, STP, ORP, WC, SOD, TOM	Sediments total nitrogen, sediments total phosphorus, oxidation-reduction potentials, water content, sediments oxygen demand, total organic matter (Table 3, Chapter 7)
Sts	*Stenotaphrum secundatum* (Chapter 19)
T	Temperature (K) (equation 1, Chapter 14)
T	Rate of plant transpiration (equation 1, Chapter 20)
T	Temperate (Chapter 19)
t	Hydraulic residence time, d (Chapter 7 and equation 2, Chapter 18)
t_1	Initial operation day, d (Chapter 5)
$t_{1/2}$	Half life time (Chapter 3)
Ta	*Typha angustifolia* (Chapter 19)
t_f	Total days of feeding, d (Chapter 5)
TCD	Thermal conductivity detector (Chapter 3)
TCE	Trichloroethylene
TGD	Technical Guidance Document (Chapter 15)
Thy-B	Intermittent mode with *T. latifolia* (Fig. 2, Chapter 10)
Thy-C	Continuous flow with *T. latifolia* (Fig. 2, Chapter 10)
TKN, TKNL^{-1}	Total Kjeldahl Nitrogen, Total Kjeldahl Nitrogen per liter
Tl	*Typha latifolia* (Chapter 19)
TMP	Trimethoprim (Chapter 15)
TNT	2,4,6-trinitrotoluene
TP	Total phosphorus
TPFBB	Three Phase Fluidized Bed Bioreactor (Chapter 19)
TPL^{-1}	Total phosphorus per liter
TS	Septic tank (Chapter 19)

TSCF	Transpiration stream concentration factor (dimensionless ratio between the concentration of the organic compound in the liquid of the transpiration stream (xylem sap) and the bulk concentration in the root zone solution) (Dietz and Schnoor 2001, Doucette et al. 2005; Kvesitadze et al. 2006) (Equation 1, Chapter 20)
TSCF	Transpiration stream concentration factor (equation 3, Chapter 15)
TSS	Total suspended solids (in most chapters)
TWW	Treated wastewater (Chapter 13)
TVA-technology	Tennessee Valley Authority
Tzc	Coarse volcanic gravel (Chapter 4)
Tzf	Fine volcanic gravel (Chapter 4)
U	Rate of organic compound assimilation (Equation 1, Chapter 20)
$U = d_{60}/d_{10}$	Uniformity coefficient, with filter sand composition of d_{10} 0.25–1.2 mm and d_{60} 1–4 mm of diameter (Brix and Arias 2005) (Chapter 2)
UASB	Upflow Anaerobic Sludge Blanket reactors
UN	United Nations (Chapter 2)
UNDP	United Nations Development Program
UNEP	United Nations Environment Program
UNICEF/WHO	United Nations International Children's Emergency Fund/World Health Organization
UP	Unplanted (Chapter 19)
US$	United States dollars
UV	Ultraviolet (Chapter 3)
V	Constructed wetland volume, L (Chapter 5)
V	Reactor volume, m³ (equation 1, Chapter 18)
v	Solution volume added, L (equation 2, Chapter 5)
VBFWs	Vertical-baffled flow wetlands (Chapter 7)
VF	Vertical flow (several chapters): Intermittently dosed reed bed system in which the flow is predominantly down-flow. The system will be under-drained and because of its aerobic nature will be better for nitrification (Brix, H., R.H. Kadlec, R.L. Knight, J. Vymazal, P. Cooper and R. Haberl. **Constructed Wetlands for Pollution Control: Processes, Performance, Design and Operation**. 2000 IWA Pub.)
VFCW	Vertical Flow Constructed Wetland (Chapter 2), Subsurface Vertical Flow Constructed Wetland (Chapter 11)
VSSF	Vertical sub-surface flow wetlands
W	Warm (Chapter 19)
W	Width of the artificial wetland cell, m (equation 3, Chapter 18)
w.e.	Without effect (Chapter 19)
Wetland Area	Area in cm² (equation 3, Chapter 10)
WHO	World Health Organization
WSP	Waste Stabilization Pond
WSSV	White Spot Syndrome Virus (Chapter 12)

WWTP, WWTPs	Wastewater treatment plant(s) (several chapters)
Xochimilco	*Xochimilco* is a name that comes from the *Nahuatl* or Aztec language, *xochitl*=flower, *milli*=fertile land or sowing and *co*=place or site (locative termination), according to Cabrera (2002) and, commonly, it means place where flowers grow ("*sementera de flores*" in Spanish) (Chapters 1 and 3)
y	Depth of the artificial wetland cell, m (equation 3, Chapter 18)
Zl	*Zizania latifolia* (Chapter 19)

Greek symbols

α_{pt}	Ratio of the contaminant concentrations in plant water and soil interstitial water (≤ 1) (Chiou et al. 2001) (Chapter 15)
θ	Temperature coefficients for the first-order kinetic model (Kadlec and Knight 1996)
$\varnothing$	Diameter

Index